OXFORD MONOGRAPHS ON GEOLOGY AND GEOPHYSICS NO. 20

OXFORD MONOGRAPHS ON GEOLOGY AND GEOPHYSICS

Radiation and Cloud Processes in the Atmosphere

Theory, Observation, and Modeling

K. N. Liou

University of Utah

New York   Oxford
OXFORD UNIVERSITY PRESS
1992

Oxford University Press

Oxford   New York   Toronto
Delhi   Bombay   Calcutta   Madras   Karachi
Kuala Lumpur   Singapore   Hong Kong   Tokyo
Nairobi   Dar es Salaam   Cape Town
Melbourne   Auckland

and associated companies in
Berlin   Ibadan

Library of Congress Cataloging-in-Publication Data
Liou, Kuo-Nan.
Radiation and cloud processes in the atmosphere:
theory, observation, and modeling / K. N. Liou.
p.   cm. — (Oxford monograph on geology and geophysics; no. 20)
Includes bibliographical references and index.
ISBN 0-19-504910-1
1. Atmospheric radiation.   2. Cloud physics.
3. Atmospheric physics.   4. Radiative transfer.
I. Title.   II. Series.
QC912.3.L57   1992   551.57'6—dc20   91-24300

2 4 6 8 9 7 5 3 1

Printed in the United States of America
on acid-free paper

PREFACE

The prime regulator of the radiation field in the atmosphere is clouds. The uncertainty surrounding cloud–climate feedback is very great because of the uncertainty in our understanding of clouds and radiation processes in the atmosphere. An approximate and yet reliable treatment of radiation and cloud physics appears to be essential if we are to make a quantum advance on the present weather prediction models for medium- and long-range forecasting. Remote sensing of cloud types, compositions, and locations in the atmosphere using radiometric information gathered from space requires both a fundamental understanding of the theory of radiative transfer and the ability to make appropriate simplifications of the theory. A correct treatment of clouds is critical to achieving accurate retrieval of temperature and humidity profiles from satellite measurements. Innovative ideas and the development of new measurement techniques for monitoring global biogeochemical change must rely on some aspects of radiation theory and on our knowledge of clouds. It is not possible to separate clouds from the discussion of radiative transfer in planetary atmospheres and in atmospheric processes.

There has been a significant advance in the field of atmospheric radiation and remote sensing in the past 20 years. This advance is a result of the basic principles developed for the treatment of radiative processes, the availability of radiometric data obtained from the ground, the air, and space, and the availability of supercomputers. However, the problem of clouds in radiative transfer, and the intricate interactions involving cloud and radiation processes in the atmosphere and atmospheric models, are still subjects of contemporary research.

A number of books have been written that address the question of "atmospheric radiation" on a fundamental level: *Atmospheric Radiation, Theoretical Basis*, by Goody and Yung (1989), *Radiation in the Atmosphere*, by Kondratyev (1969), and *An Introduction to Atmospheric Radiation*, by Liou (1980). Clouds play a very minor role in these texts. There is a further need to integrate radiative transfer and cloud physics in a coherent manner and to bridge the gaps between cloud–radiation and dynamic processes. At any rate, that has been my justification for writing a new monograph. In this book, I have attempted to focus the discussion on the physical principles and approximations that are required to develop a specific subject in radiative transfer, cloud physics, and atmospheric models. Graduate students and research scientists with adequate background in physics and mathematics should be able to follow the physical deductions and to fill the gaps in the mathematical analyses.

I am indebted to S. C. Ou, Y. Takano, and Q. Fu for providing some of the numerical data presented and for assisting me in proofreading the equations developed in the text. I thank S. Asano, T. Charlock, M. D. Chou, J. E. Geisler, A. Heymsfield, S. Krueger, A. Lacis, E. Smith, and H. Sundqvist for reading various chapters and for offering many constructive suggestions for improvements. I also thank R. M. Goody for his encouragement and for his wisdom on the unpublished materials. During the preparation of this monograph, my research programs have benefitted by the continuous support of the National Science Foundation, NASA, and the Air Force Office of Scientific Research.

Salt Lake City   K. N. Liou
September 1991

CONTENTS

Radiation and Cloud Processes in the Atmosphere

1

INTRODUCTION

Almost all the energy that drives the earth's atmosphere and ocean currents originates from the sun. The land and ocean surfaces absorb $\sim$44% of the incoming solar fluxes, while the atmosphere absorbs $\sim$26% of these fluxes. Over a sufficiently long period of time, the temperature of the earth–atmosphere system is determined in such a manner that outgoing thermal infrared fluxes are emitted and balanced by absorbed solar fluxes. Radiative processes are the only means by which the earth and atmosphere exchange energy with space; therefore, an understanding of weather and climate processes must begin with a detailed investigation of radiative processes and the radiation balance of the earth–atmosphere system.

1.1 Radiative equilibrium of the earth–atmosphere system

Earth is one of nine planets in the Solar System. Steady rotation eastward around the axis of its poles takes 24 hours with respect to the sun. At the same time, orbital motion in an ellipse with the sun at one focus takes approximately 365 days to complete: The distance between the earth and the sun varies. These two motions are the most important factors that determine the amount of solar radiant energy reaching the earth and, subsequently, the climate and climatic changes of the earth–atmosphere system. The total radiant energy from the sun varies only slightly. Hence, a term referred to as the *solar constant* is used for the total radiant energy of all wavelengths received per unit time and unit area at the top of the earth's atmosphere, corrected to the mean sun–earth distance r_0. Let the radius of the visible solar disk be a_s and the temperature of the solar photosphere be $T_\odot$. On the basis of the energy conservation principle, and assuming that the Stefan–Boltzmann law for a black body is applicable, the solar constant is given by

$$S = \sigma T_\odot^4 (a_s/r_0)^2, \tag{1.1}$$

where σ is the Stefan–Boltzmann constant. The solar constant is a function of the emitting temperature of the solar photosphere; it has been inferred from satellite measurements over a period of more than a decade. The best typical value is about

1365 W m^{-2} (Mecherikunnel et al., 1988). The long-term solar constant trend may be linked to solar activities, which are primarily associated with the sunspot cycle. The small variability of the solar constant, and its effect on the climate and climatic changes of the earth–atmosphere system, are of fundamental concern in the field of climate modeling and sensitivity analysis.

The amount of solar flux available to the earth–atmosphere system as a function of time is determined by the orbital characteristics of the earth about the sun. There are three ways in which the earth's orbit about the sun varies. These are *eccentricity*, defined as the ratio of the distance between the two foci to the major axis of the ellipse; *obliquity*, produced by the tilt of the axis of the earth's rotation with respect to its plane of rotation about the sun; and *periodic precession*, due to the wobbling motion of the earth in which its closest position about the sun advances each year. The variations of solar fluxes caused by these three parameters are fundamental to the development of the Milankovitch (1941) theory of climate change.

The equilibrium temperature of the earth–atmosphere system may be estimated if the solar constant is known. Again, on the basis of the conservation of absorbed solar and emitted thermal infrared energy, and assuming a global albedo of $\bar{r}$, the equilibrium temperature may be expressed by

$$T_e = [S(1 - \bar{r})/4\sigma]^{1/4}, \tag{1.2}$$

where the factor of 4 accounts for the difference between the absorption and emission areas. On a global scale, and considering the earth–atmosphere as a unit, the equilibrium temperature depends on the solar constant, which represents the energy available at the top of the atmosphere, and on the global albedo, which is the internal reflecting power of the earth–atmosphere system. The latter parameter is obviously related to the composition of the atmosphere and the earth's surface characteristics. A value of ~0.3 for the global albedo was derived from data produced by a satellite radiation budget experiment (Jacobowitz et al., 1984). Figure 1.1 shows all the significant components within the earth's atmosphere that are pertinent to the reflecting power of the system. These include scattering by molecules (primarily N_2 and O_2); aerosols (including those from volcanic sources); various types of clouds; and reflection by different surfaces, including the oceans, which cover about 70% of the earth's surface, and various land surfaces such as ice, snow, desert, vegetation, forests, and mountains. Interactions between multiple scattering and absorption processes in the atmosphere and surface reflection determine the amount of solar flux reflected back to space.

1.2 Atmospheric composition and absorbing gases

The earth's atmosphere is composed of two groups of gases, one with nearly permanent concentrations and another with variable concentrations. The atmosphere

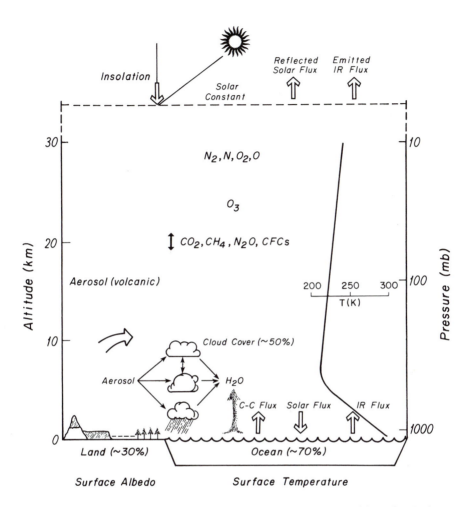

FIG. 1.1 Systematic diagram describing the composition of the atmosphere and the surface. Various fluxes at the surface and the standard atmospheric profile are also shown.

also contains various kinds of aerosols, clouds and precipitation, which are highly variable in space and time. Table 1.1 lists the chemical formula and volume ratio for the concentrations of permanent and variable gases. Nitrogen, oxygen, and argon account for more than 99.96% of the atmosphere by volume. The permanent gases have virtually constant volume ratios up to an altitude of about 60 km.

Although listed as a permanent constituent, the carbon dioxide concentration has been observed to increase by about 0.4% per year as a result of the combustion of fossil fuels, absorption and release by the oceans, and photosynthesis. Figure 1.2 shows the mean monthly values of CO_2 at Mauna Loa, Hawaii, for the period from 1958 to 1990. The seasonal variation is due to the summer growing season of the Northern Hemisphere during which the CO_2 concentration decreases.

Table 1.1 The composition of the atmosphere[a]

Permanent constituents		Variable constituents	
Constituent	% by volume	Constituent	% by volume
Nitrogen (N_2)	78.084	Water vapor (H_2O)	0–0.04
Oxygen (O_2)	20.948	Ozone (O_3)	$0-12 \times 10^{-4}$
Argon (Ar)	0.934	Sulfur dioxide (SO_2)[b]	0.001×10^{-4}
Carbon dioxide (CO_2)	0.034	Nitrogen dioxide (NO_2)[b]	0.001×10^{-4}
Neon (Ne)	18.18×10^{-4}	Ammonia (NH_3)[b]	0.004×10^{-4}
Helium (He)	5.24×10^{-4}	Nitric oxide (NO)[b]	0.0005×10^{-4}
Krypton (Ke)	1.14×10^{-4}	Hydrogen sulfide (H_2S)[b]	0.00005×10^{-4}
Xenon (Xe)	0.089×10^{-4}	Nitric acid vapor (HNO_3)	Trace
Hydrogen (H_2)	0.5×10^{-4}	Chlorofluorocarbons	Trace
Methane (CH_4)	1.7×10^{-4}	$CFCl_3$, CF_2Cl_2,	
Nitrous oxide (N_2O)[b]	0.3×10^{-4}	CH_3CCl_3, CCl_4, etc.	
Carbon monoxide (CO)[b]	0.08×10^{-4}		

[a] After the U.S. Standard Atmosphere (1976) with modifications.

[b] Concentration near the earth's surface.

A number of measurement series indicate that the atmospheric methane concentration, with a present value of $\sim$1.7 ppmv, has increased by 1–2% per year (Blake and Rowland, 1988), and that it may have been increasing for a long period of time. The most likely cause of the increase in the CH_4 concentration is greater biogenic emissions associated with a rising human population. The production of rice paddies seems to be another prime source of CH_4. There is no direct evidence of an increase in carbon monoxide concentration. However, deforestation, biomass burning, and modification of CH_4 sources could lead to changes in the atmospheric CO concentration. There is also some evidence for an increase in nitrous oxide.

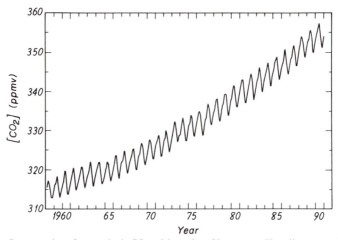

FIG. 1.2 Concentration of atmospheric CO_2 at Mauna Loa Observatory, Hawaii, expressed as a mole fraction in parts per million of dry air for the period 1958–1990 (courtesy of P. Tans, Geophysical Monitoring for Climate Change, Environmental Research Laboratory, National Oceanic and Atmospheric Administration).

A possible global increase of ∼0.2% per year in N_2O has been suggested (Weiss, 1981). This increase is attributed to the combustion of fossil fuels and, in part, to fertilizer denitrification.

The amounts of variable gases listed in Table 1.1 are small, but they are extremely important in the radiation budget of the atmosphere. Water vapor is the major radiative and dynamic element in the earth's atmosphere. The H_2O concentration varies significantly with both space and time. The spatial distribution of tropospheric H_2O is determined by the local hydrological cycle via evaporation, condensation, and precipitation, and by large-scale transport processes. Figure 1.3 shows the zonally averaged values of the mean specific humidity in units of g kg^{-1} as functions of pressure (Peixóto and Oort, 1983). Specific humidity decreases rapidly with pressure, almost following an exponential function. Specific humidity also decreases with latitude. More than 50% of water vapor is concentrated below ∼850 mb, while more than 90% is confined to the layers below ∼500 mb. The variability of the vertical H_2O concentration not illustrated in the figure shows a bimodal distribution with a maximum in the subtropics of both hemispheres below ∼700 mb. The variability is very small in the equatorial region and poleward of ∼60°. The stratospheric H_2O concentration is relatively small, with a value of ∼3–4 ppmv in the lower stratosphere. It has been suggested that H_2O in the lower stratosphere is controlled by the temperature of the tropical tropopause, and by the formation and dissipation of cirrus anvils due to outflow from cumulonimbus (Doherty et al., 1984).

The ozone concentration also varies significantly with space and time, and occurs principally at altitudes from ∼15 to 30 km, an area referred to as the *ozone layer*. The maximum ozone concentration occurs at ∼20–25 km, depending on latitude and season. Atmospheric ozone is continually created and destroyed by photochemical processes associated with solar ultraviolet radiation. The absorption of deadly solar ultraviolet radiation by the ozone layer is essential to life on earth. Many photochemical reactions associated with O_3 involve H_2O, CH_4, and CO. Figure 1.4 shows the zonally averaged, seasonal cross section of total ozone in units of 10^{-3} cm of the gas reduced to the standard temperature and pressure condition. Significant variations in total ozone in terms of latitude and season are evident. Maximum total ozone occurs in the polar night.

Nitrogen oxides ($NO_x = NO, NO_2$), although not significant absorption gases in the infrared, appear to be important in the determination of both tropospheric and stratospheric O_3 concentrations. Atmospheric NO_x are emitted by transportation and combustion processes at the surface and by high-flying aircraft in the upper troposphere and lower stratosphere. In the stratosphere, the major source of NO_x is the dissociation of N_2O by excited oxygen atoms. In the lower troposphere, the major source of NO_x appears to be anthropogenic combustion of fossil fuels and biomass burning. Chlorofluorocarbons are also recognized as presenting a potential threat to the ozone layer. Large amounts of these chemicals are produced by industry

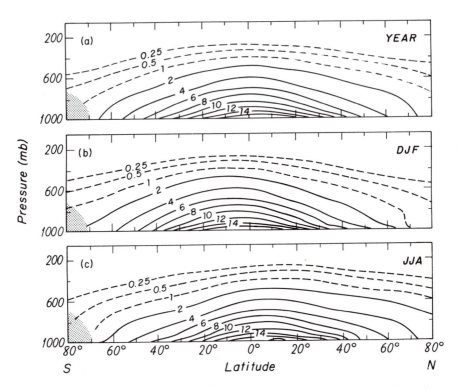

Fig. 1.3 Vertical cross sections of the specific humidity in units of g/kg. (a) annual; (b) December, January, and February (DJF); (c) June, July, and August (JJA) (after Peixóto and Oort, 1983).

and are used in solvents, refrigerants, and spray-can propellants. Chlorofluorocarbons include $CFCl_3$ (methylchloroform), CF_2Cl_2 (CFC-113), CF_2Cl (CFC-114), CF_2ClCF_3 (CFC-115), CHF_2Cl (CFC-22), and CH_3Cl (methylchloride).

Sulfur dioxide in the stratosphere is largely produced by volcanic eruptions. SO_2 and other sulfur-based gases are believed to be the primary precursors of stratospheric aerosols. Emissions of SO_2 from the surface may be important in the formation of tropospheric aerosols and, hence, are related to the production of acid rain through cloud and precipitation processes.

The atmosphere continuously contains aerosol particles ranging in size from $\sim 10^{-3}$ to $\sim 20\ \mu$m. These aerosols are known to be produced by natural processes as well as by human activity. Natural aerosols include volcanic dust, smoke from forest fires, particles from sea spray, wind-blown dust, and small particles produced by the chemical reactions of natural gases. Primary man-made aerosols include particles directly emitted during combustion processes and particles formed from gases emitted during combustion. The atmospheric aerosol concentration varies with locality; the largest concentration generally occurs in urban and desert areas. In normal conditions, the background aerosol concentration would have a visibility

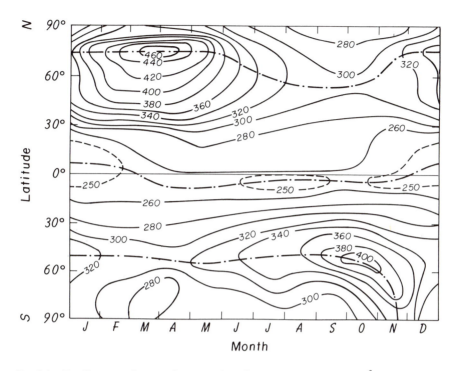

FIG. 1.4 Zonally averaged, seasonal cross section of total ozone in units of 10^{-3} cm of the gas reduced to the standard temperature and pressure condition (after Brasseur and Solomon, 1984; original diagram from London, J., 1980: Radiative energy sources and sinks in the stratosphere and mesophere. *Proceeding of the NATO Advanced Study Institute on Atmospheric Ozone: Its Variation and Human Influences*, M. Nicolet and A. C. Aikin, Eds., U.S. Department of Transportation, Washington, D. C., pp. 703–721).

of ~20–50 km. Aerosol concentrations generally decrease rapidly with height in the troposphere. Some aerosols are effective condensation and ice nuclei upon which cloud particles may form. For the hygroscopic type, the sizes of aerosols depend on relative humidity. Thin layers of aerosols are observed to persist for a long period of time in some altitudes of the stratosphere.

Clouds are global in nature and regularly cover ~50% of the earth. There are various types of clouds. Some clouds, such as cirrus in the tropics and stratus in the Arctic and near the coastal areas, are climatologically persistent. The microphysical composition of clouds in terms of particle size distribution and cloud thickness varies significantly with cloud type. Some clouds generate precipitation. A detailed description of cloud climatology, cloud compositions, and associated cloud dynamics is given in Chapter 4. In addition, the radiative properties of clouds are presented in Chapter 5.

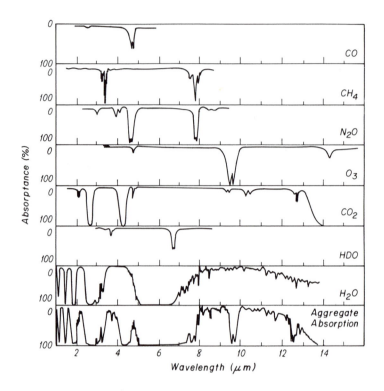

FIG. **1.5**   Absorption of infrared radiation by atmospheric gases (from Handbook of Geophysics and Space Environments, 1965).

1.3   Global energy budget

Figure 1.5 shows the infrared absorptance of several of the aforementioned gases. In the thermal infrared region, absorption of H_2O essentially covers the entire spectrum. CO_2 and O_3 exhibit significant absorption in the 15 and 9.6 μm regions. H_2O, CO_2, and O_3 are radiatively active gases of primary importance to the earth's atmosphere. In addition, CH_4 and N_2O show strong absorption bands in the 7–8 μm region. Other gases with absorption bands in the thermal infrared region are SO_2, NH_3, and CFCs. The above gases are referred to as *greenhouse gases* because of their absorption properties in the thermal infrared. Discussions of the thermal infrared and solar radiative processes involving the optically active gases are presented in Chapters 2 and 3.

Aerosols absorb and scatter sunlight. The significance of absorption relative to scattering is determined by chemical composition and particle size distribution. Aerosols are usually considered to be important for their influence on solar radiation. Water droplets and ice crystals are relatively transparent in visible light, but absorb near-infrared radiation in the solar spectrum. Clouds have a profound influence on both solar and terrestrial radiation, as illustrated in Chapters 5 and 6.

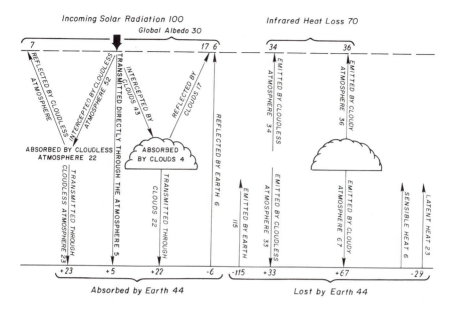

FIG. 1.6   The energy balance of the earth and the atmosphere system. The incoming solar energy is taken to be 100 units. Under equilibrium conditions, the incoming solar energy at the top of the atmosphere is balanced by the reflected solar energy and thermal ir heat loss. At the surface, the energy balance involves sensible and latent heat components.

Moreover, as pointed out previously, there are numerous types of land surfaces, which vary significantly in their reflecting properties with respect to incoming solar fluxes. Variations in one or all of the above radiative components may lead to significant and irreversible changes in weather and climate.

In view of the aforementioned atmospheric and surface components, an accurate definition of the global albedo, in connection with the global equilibrium temperature, is quite complicated. If we assume that the global albedo is ~0.3 and employ the standard solar constant, we obtain an equilibrium temperature value of about 255 K for the earth–atmosphere system. However, the standard temperature profile depicted in Fig. 1.1 shows that the earth's surface has a higher temperature, with a global mean of about 288 K. The global lapse rate in the troposphere, according to the standard atmospheric profile, is about $6.5 \, \text{K km}^{-1}$. Two basic questions may be asked: First, why is the surface temperature higher than the global equilibrium temperature? Second, what are the mechanisms for the formation and structure of the tropospheric and stratospheric temperature fields depicted in Fig. 1.1? In order to answer these questions, we must understand the internal global radiation budget and convective processes.

An approximate annual heat budget of the earth–atmosphere system is presented in Fig. 1.6. The solar flux available to the earth–atmosphere system, averaged for the entire year, is represented by 100 units. Using a solar constant of

1365 W m^{-2}, the average solar flux at the top of the atmosphere is $\sim$341 W m^{-2}. Basically, Fig. 1.6 consists of three sections: one dealing with solar flux and the manner in which it is apportioned in the atmosphere, the second concerning the infrared flux and its distribution, and the third dealing with nonradiative processes. Of the 100 units of incoming solar flux, $\sim$26 are absorbed within the atmosphere, $\sim$22 by cloud-free air, and $\sim$4 by clouds. A total of $\sim$30 units are reflected back to space, including $\sim$7 from cloudless atmospheres, $\sim$17 from cloudy atmospheres, and $\sim$6 directly from the earth's surface. The remaining 44 units are absorbed by the earth's surface.

The earth–atmosphere system emits thermal infrared radiation according to the temperature and composition distributions within the system. The upward flux from the warmer surface accounts for $\sim$115 units. The colder troposphere emits both upward and downward fluxes, with $\sim$70 and $\sim$100 units at the top and surface, respectively. The net upward flux at the surface, which is the difference between the flux emitted by the surface and the downward flux from the atmosphere reaching the surface, is $\sim$15 units. As a result of thermal emission, the atmosphere loses $\sim$ 55 units. With absorption of only $\sim$26 units of incoming solar flux, the net radiative loss from the atmosphere amounts to $\sim$29 units. This deficit is balanced by upward fluxes of latent and sensible heat. The average annual ratio of sensible to latent heat loss at the surface, the so-called *Bowen ratio*, has a global value of $\sim$0.27. Thus the latent and sensible heat components should be $\sim$23 and $\sim$6 units, respectively, in order to produce an overall balance at the surface. The atmosphere experiences a net radiative cooling that must be balanced by the latent heat of condensation released in precipitation processes and by the convection and conduction of sensible heat from the underlying surface. If there were no latent and sensible heat transfer, the earth's surface would have a temperature much higher than the observed value of 288 K. Radiative equilibrium and radiation budget are discussed in Chapter 6.

1.4   Vertical temperature structure and the greenhouse effect

It is evident from Fig. 1.6 that incoming solar flux at the top of the atmosphere must be balanced by reflected solar and emitted infrared fluxes over a sufficiently long period of time, and that the only energy exchanges with space occur by means of radiative processes. However, at the surface, equilibrium must be achieved by the balance between net radiative fluxes and convective fluxes of sensible and latent heat.

Vertical fluxes of sensible and latent heat, graphically illustrated in Fig. 1.1, are governed by the motions of various scales. Within about 1 mm adjacent to the surface, an area referred to as the *molecular boundary layer*, the principal mechanisms for the transport of sensible and latent heat fluxes are molecular conduction and diffusion. In the layer immediately above and within about a few tens of meters of the surface—the so-called *surface layer*—sensible and latent heat fluxes are transferred upward by means of eddies. Based on the theory of turbulence, the surface layer is characterized by strong vertical wind shear, with wind speed proportional to the logarithm of height. Above and within about 1 km of this layer is the *mixed layer*, where convectively driven thermals assume the prime role of transporting the vertical sensible and latent heat fluxes. In the free atmosphere, that is, from the mixed layer to the tropopause, the upward transport of sensible and latent heat is governed by deep cumulus convection, as well as synoptic and planetary-scale circulations. In this region essentially all conversion of latent to sensible heat takes place via irreversible condensation processes. The preceding scales of vertical motion associated with the transport of sensible and latent heat are generally classified as *convection*.

The determination of a globally averaged temperature profile is an involved process and will be covered comprehensively in the section that deals with thermal (radiative plus convective) equilibrium in Chapter 6. In the following we shall briefly describe the procedure by which radiative and thermal equilibrium temperature profiles may be obtained. Basically, net fluxes of solar and thermal infrared radiation can be computed by prescribing various gaseous and cloud fields. From the net flux divergence, the heating and cooling rates due to solar and infrared flux exchanges may be evaluated. Subsequently, the radiative equilibrium temperature profile may be computed by an iterative and time-marching method until the temperatures converge within a preset range of accuracy. Likewise, if we include the heating rate profile produced by convective processes associated with latent and sensible heat transports from the surface to the atmosphere, the thermal equilibrium temperature may be obtained.

Figure 1.7 shows the climatological temperature profile, which is produced by the equilibrium between radiative and convective processes. Also shown is a temperature profile under the condition of radiative equilibrium. The radiative equilibrium temperature is too warm near the surface and too cold in the tropopause. The stratospheric temperature profile is largely determined by radiative equilibrium, basically through the absorption of solar fluxes by ozone and through the emission of infrared fluxes by carbon dioxide. In the troposphere and on a one-dimensional global scale, the only mechanism that may bring the system into thermal equilibrium is evidently the vertical transport of latent and sensible heat fluxes by eddies, since mean vertical motions are generally negligible. The convective nature of the earth–atmosphere system is fundamental in weather and climate systems, as well as in the numerical modeling of such systems.

Compared to the climatological surface temperature of 288 K, the global equilibrium temperature is colder by about 33 K. The earth's atmosphere is relatively transparent with respect to incoming solar fluxes. About half of these fluxes are absorbed by the surface, as shown in Fig. 1.6. On the other hand, the emitted thermal infrared fluxes from the surface are largely trapped by CO_2, H_2O, and O_3, as well

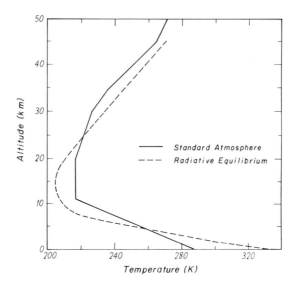

FIG. 1.7   Temperature profile computed from radiative equilibrium. This profile is compared with the climatological temperature profile.

as by clouds. The trapping of thermal infrared radiation emitted from the surface is typical of the earth's atmosphere and is referred to as the *greenhouse effect*. We may express the surface temperature in terms of the equilibrium temperature in the form

$$T_s = T_e + \gamma H, \tag{1.3}$$

where γ is the lapse rate, and H is defined as the effective emission height for the greenhouse effect. Assuming a conventional lapse rate of $6.5\,\mathrm{K\,km^{-1}}$, we find $H \approx 5\,\mathrm{km}$. Convective activities that take place primarily in the boundary layer reduce the greenhouse effect. Using a radiative equilibrium temperature for the surface of 335 K, this reduction in terms of the effective emission height is as much as a factor of 2.4. Increases in CO_2 as well as CH_4, N_2O, and CFCs concentrations are known to have a significant impact on climate through the greenhouse effect.

1.5   Radiation balance, general circulation, and thermal structure of the atmosphere

Figure 1.8 shows the radiation balance at the top of the atmosphere in terms of the zonally averaged patterns for the absorbed solar and emitted infrared fluxes observed from satellites. As illustrated in this diagram, there is a gain of radiative energy between about 40° N and 40° S, whereas there are losses in polar regions. This pattern is largely caused by the sharp decrease in insolation during the winter season and the high surface albedo in polar regions. In addition, the outgoing infrared flux is only slightly latitudinally dependent, owing to the larger amount of

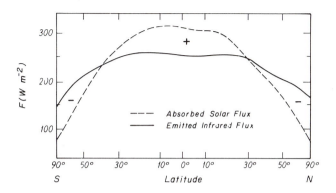

Fig. 1.8 Zonally averaged components of the absorbed solar flux and emitted thermal infrared flux at the top of the atmosphere. + and − denote energy gain and loss, respectively (after Vonder Haar and Suomi, 1971, with modifications).

atmospheric water vapor and the higher and colder clouds in the tropics. As a result of radiative energy excess and deficit, the equator-to-pole temperature gradient is generated, and, subsequently, a growing store of zonal mean available potential energy is produced. In the equatorial region, warm air expands upward and creates a poleward pressure gradient force at the upper altitudes, where air flows poleward from the equator, while in the upper levels air cools and sinks in the subtropical high-pressure belts (∼30°) and returns to the equator at the surface. Kinetic energy is generated as a result of the work done by the horizontal pressure gradient force.

This thermally driven circulation between the equator and subtropics is referred to as the *Hadley cell*. Because of the earth's rotation, air flowing toward the equator at the surface is deflected to the west and creates easterly trade winds. In the upper level of the Hadley cell, the Coriolis deflection of poleward-moving air generates westerly winds.

In the polar regions, similar thermally driven circulation is found. Cold air shrinks downward, producing a poleward-directed pressure gradient force and motion in the lower altitudes. The sinking motion over the poles results in airflow in the upper level towards the equator and into the low-pressure belts (∼60°). Thus a Hadley circulation develops between the poles and the subpolar low-pressure regions. Here, the effect of the Coriolis force is the same; that is, easterly winds are produced at the surface, and westerly winds are produced aloft. In the Hadley circulation, the atmosphere may be regarded as an engine that absorbs net heat from a high-temperature reservoir and releases it to a low-temperature reservoir. The temperature differences generate the available potential energy, which in turn is partly converted to kinetic energy to overcome the effect of friction.

Poleward zonal thermal winds at the upper altitudes become unstable in middle latitudes—an effect referred to as *baroclinic instability*—and generate a reverse circulation. Here, warm air sinks in the subtropical highs, and cold air rises in the subpolar lows in which westerly winds prevail at all levels. The meridional circulation in this region cannot be explained by the direct heating and cooling effects, as in the Hadley circulation, and cannot generate kinetic energy. The maintenance of westerlies in middle latitudes is explained by the continuous transfer of angular momentum from the tropics, influenced by large-scale wave disturbances. The baroclinic waves transport heat poleward and intensify until heat transport is balanced by the radiation deficit in polar regions. The relevance and importance of radiative heating in general circulation models are discussed in Section 7.4.

From the preceding discussion, it is evident that in the mean there is a radiation excess in tropical regions and a radiation deficit at middle and high latitudes. It is also evident that there are thermally driven circulations caused either directly or indirectly by horizontal heating and cooling gradients in a stable stratified atmosphere. Thus, in addition to the upward transport of sensible and latent heat by convective activities, there should be a substantial transport of sensible and latent heat as well as momentum at middle latitudes by large-scale circulations in the atmosphere.

The importance of large-scale horizontal transports on the temperature distribution is illustrated in Fig. 1.9. Figure 1.9(a) illustrates temperatures obtained from pure radiative balance computations (Ou and Liou, 1984). Under radiative equilibrium, the tropical surface temperature is well above 330 K, whereas temperatures as low as 200 K are observed in the Arctic. The vertical eddy transport is then incorporated into the computation of thermal equilibrium temperatures, the results of which are shown in Fig. 1.9(b). The vertical eddy flux reduces the surface temperature somewhat so that the lapse rate becomes smaller. In this case, the meridional temperature gradient is greatly reduced. However, in comparison with the zonally averaged climatological temperature displayed in Fig. 1.9(c) (Oort and Rasmusson, 1971), the tropical surface, with a temperature of $\sim$310 K, is still too warm and the Arctic is still too cold ($\sim$210 K). The horizontal transport must be accounted for in order to produce thermal equilibrium temperatures close to observed data.

The transport of latent heat by atmospheric circulation is a direct result of the hydrological cycle, which determines the transport of water substances in the earth–atmosphere system. On the basis of the water mass conservation principle, the total annual precipitation must be balanced by the total annual evaporation over the globe. However, there are large imbalances in localized regions. In the subtropical belts of both hemispheres there is an excess of evaporation over precipitation due to the domination of oceanic anticyclones. In the equatorial belt and at the latitudes above about 40°, on the other hand, precipitation exceeds evaporation. This is because the equatorial regions correspond to the location of the Intertropical Convergence Zone (ITCZ) and the maximum monsoon rainfall and because higher latitudes correspond to regions associated with baroclinic storm activities. Thus the subtropical belts serve as source regions for water vapor, whereas equatorial

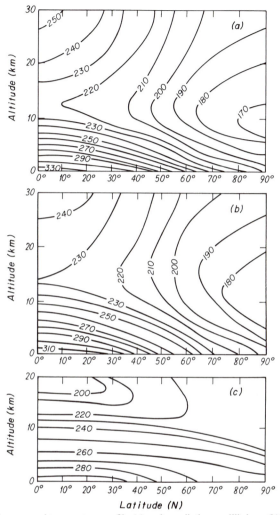

FIG. 1.9   Zonally averaged temperature profile (a) under radiative equilibrium, (b) under radiative plus convective (vertical equilibrium); and (c) from climatological data for annual mean (Oort and Rasmussen, 1971).

regions and high-latitude areas are moisture sinks in the atmosphere. The general circulation of the atmosphere removes water vapor from source regions into sink regions, resulting in the transport of latent heat fluxes. On the earth's surface, river runoff, leakage through subsurface aquifers, and ocean currents all serve to transport liquid water from regions of excess precipitation to regions of excess evaporation.

Transport of sensible heat in terms of potential energy has double maxima at about 10° and 50° in both hemispheres. Latent heat is transferred both equatorward and poleward from about 20° N and S where the evaporation maxima are located. Of the total transport in both hemispheres, the sensible and latent heat fluxes account for about 50–60% and 20–25%, respectively. The remaining horizontal transport is due to ocean currents. Oceanic transport of energy that occurs principally at low latitudes accounts for about 20–25%.

Over a climatic time scale, these poleward energy transports are required to balance the radiative energy component and to achieve an overall equilibrium within the earth–atmosphere system. Chapter 6 presents the atmosphere in radiative and thermal equilibrium. Finally, the role of radiation and cloud processes in atmospheric models, including one-dimensional, two-dimensional, and general circulation models, is discussed in Chapter 7.

REFERENCES

Blake, D. R., and F. S. Rowland, 1988: Continuing worldwide increase in tropospheric methane, 1978–1987. *Science*, **239**, 1129–1131.

Brasseur, G., and S. Solomon, 1984: *Aeronomy of the Middle Atmosphere: Chemistry and Physics of the Stratosphere and Mesophere*. Reidel, Boston, 441 pp.

Doherty, G. M., R. E. Newell, and E. F. Danielson, 1984: Radiative heating rates near the stratospheric formation. *J. Geophys. Res.*, **89**, 1380–1384.

Handbook of Geophysics and Space Environment, 1965: Bedford, Mass., U.S. Air Force Cambridge Laboratories, Office of Aerospace Research.

Jacobowitz, H., R. J. Tighe, and Nimbus 7 ERB Experiment Team, 1984: The earth radiation budget derived from the Nimbus 7 ERB experiment. *J. Geophys. Res.*, **89**, 4997–5010.

Mecherikunnel, A. T., R. B. Lee III, H. L. Kyle, and E. R. Major, 1988: Intercomparison of solar total irradiance data from recent spacecraft measurements. *J. Geophys. Res.*, **93**, 9503–9509.

Milankovitch, M., 1941: *Canon of Insolation and the Ice-Age Problem*. Königlich Serbische Akademie, Belgrade, 484 pp. (English translation by the Israel Program for Scientific Translation, published by the U.S. Department of Commerce and the National Science Foundation).

Oort, A. H., and E. M. Rasmusson, 1971: Atmospheric circulation statistics. *NOAA Prof. Paper*, **5**, Washington, D.C., U.S. Govt. Printing Office, 323 pp.

Ou, S. C., and K. N. Liou, 1984: A two-dimensional radiative–turbulent climate model. I: Sensitivity to cirrus radiative properties. *J. Atmos. Sci.*, **41**, 2289–2309.

Peixóto, J. P., and A. H. Oort, 1983: The atmospheric branch of the hydrological cycle and climate. In *Variations in the Global Water Budget*, A. Street-Perrott et al., Eds., Reidel, Boston, pp.5–65.

U.S. National Oceanic and Atmospheric Administration, 1976: *U.S. Standard Atmosphere, 1976*. NOAA-S/T76-1562, Washington D.C., U.S. Govt. Printing Office, 227 pp.

Vonder Haar, T. H., and V. E. Suomi, 1971: Measurements of the earth's radiation budget from satellites during a five-year period. *J. Atmos. Sci.*, **28**, 305–314.

Weiss, R. W., 1981: The temporal and spatial distribution of tropospheric nitrous oxide. *J. Geophys. Res.*, **86**, 7185–7195.

2

THEORY AND PARAMETERIZATION OF
THERMAL INFRARED RADIATIVE TRANSFER

The equilibrium temperature of the earth–atmosphere system is about 255 K, as discussed in Chapter 1, and the general range of temperatures in the earth's atmosphere can be seen in Fig. 1.7. According to Planck's law, the intensity (also referred to as radiance) emitted from a blackbody may be expressed in the wavenumber domain in the form

$$B_\nu(T) = \frac{2h\nu^3 c^2}{\exp(hc\nu/KT) - 1}, \qquad (2.0.1)$$

where the intensity is the flux of energy in a given direction per unit time per unit wave-number range per unit solid angle, ν the wave number in cm^{-1}, T the absolute temperature, h Planck's constant, K the Boltzmann constant, and c the velocity of light.

We shall confine our discussion to the condition of local thermodynamic equilibrium (LTE) under which Kirchhoff's law is applicable. In the localized portion, the atmosphere absorbs radiation of a particular wave number (or wavelength), and at the same time also emits radiation of the same wavenumber. Because there is equilibrium, the amount of radiation absorbed must be equal to that emitted. Otherwise, equilibrium would not be possible as to define a uniform temperature, and this would violate the second law of thermodynamics. A blackbody absorbs the maximum possible radiation, and it has to emit that same amount of radiation. The rate at which emission takes place is a function of temperature and wave number in accord with Eq. (2.0.1). A gray body is characterized by incomplete absorption and emission, and is said to have emissivity less than unity. It is in the context of LTE that we begin our discussion on the thermal infrared (also referred to as ir or long-wave) radiative transfer.

Figure 2.1 shows the Planck intensity as a function of wave number ν or wavelength λ $(= 1/\nu)$, for various temperatures of the earth's atmosphere. The wave number corresponding to the maximum of the Planck intensity can be obtained by differentiating Eq. (2.0.1) with the result $(1/\nu)_m = \alpha/T$, where α is the

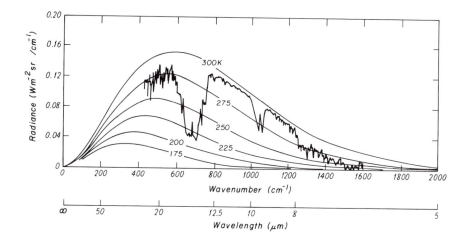

FIG. 2.1 Observed thermal ir absorption spectrum (after Kunde et al., 1974) and theoretical Planck intensity curves for a number of the earth's atmospheric temperatures.

Wien displacement constant. This maximum shifts toward small wave numbers as the temperature decreases (Wien's displacement law). Integrating Eq. (2.0.1) over all wave numbers, we find the Planck flux

$$\pi B(T) = \pi \int_0^\infty B_\nu(T)\, d\nu = \sigma T^4 \ . \tag{2.0.2}$$

This is the Stefan–Boltzmann law, and σ is the Stefan–Boltzmann constant.

Also illustrated in Fig. 2.1 is an atmospheric emission spectrum measured from the Infrared Interferometer Spectrometer (IRIS) instrument on board the Nimbus IV satellite (Kunde et al., 1974). This spectrum suffices to illustrate that certain portions of the thermal ir radiation emitted by the earth and the atmosphere are trapped by a number of minor gases described in Chapter 1. In this chapter, we shall first introduce the fundamental equation that governs the transfer of radiation in planetary atmospheres.

2.1 Fundamentals of thermal infrared radiative transfer

Consider an absorbing and emitting medium. A pencil of radiation traversing this medium will be weakened by the interaction with matter through absorption. At the same time, this radiation may be strengthened by thermal emission from the medium. This pencil of radiation is usually represented by the intensity (or radiance), I_ν, in the field of radiative transfer. The general equation for radiative transfer in an absorbing and emitting medium can be written in terms of the differential change in the intensity in the form

$$-\frac{1}{k_\nu \rho_a}\frac{dI_\nu}{ds} = I_\nu - J_\nu, \qquad (2.1.1)$$

where k_ν denotes the absorption coefficient, ρ_a is the density of absorbing gases, s is the slant path, and J_ν is the source function. For applications to the radiation budget of the planet, it suffices to consider the intensity as being independent of time, so that Eq. (2.1.1) may be expressed by

$$-\frac{1}{k_\nu \rho_a}(\mathbf{\Omega} \cdot \nabla)I_\nu(\mathbf{s}, \mathbf{\Omega}) = I_\nu(\mathbf{s}, \mathbf{\Omega}) - J_\nu(\mathbf{s}, \mathbf{\Omega}). \qquad (2.1.2)$$

Thus the intensity is a function of position $\mathbf{s}$ in three-dimensional space and the directional vector denoted by $\mathbf{\Omega}$. The differential operator, $\mathbf{\Omega} \cdot \nabla$, can be expressed in Cartesian, spherical, and cylindrical coordinates. For example, in Cartesian coordinates, we write

$$\mathbf{\Omega} \cdot \nabla = \Omega_x \frac{\partial}{\partial x} + \Omega_y \frac{\partial}{\partial y} + \Omega_z \frac{\partial}{\partial z}, \qquad (2.1.3)$$

with the directional cosines given by

$$\Omega_x = (1 - \mu^2)^{1/2}\cos\phi, \qquad \Omega_y = (1 - \mu^2)^{1/2}\sin\phi, \qquad \Omega_z = \mu, \quad (2.1.4)$$

where $\mu = \cos\theta$, and θ and ϕ are the zenith and azimuthal angles in polar coordinates, respectively.

In the discussion of the transfer of thermal ir radiation in terrestrial atmospheres, it is commonly assumed that, in localized portions, the atmosphere is in thermodynamic equilibrium, as well as being plane-parallel. The first assumption allows us to use the Planck intensity for the source function by virtue of the Kirchhoff law. The so-called plane-parallel assumption implies that variations in the intensity and atmospheric parameters (temperature and gaseous profiles) are permitted only in the vertical direction (e.g., height or pressure). Under this assumption, absorption and emission processes would be symmetrical with respect to the azimuthal angle. It follows that the intensity is a function of the vertical position and zenith angle. Under these conditions, the basic equation that governs thermal ir radiation in the height coordinate may be written in the form

$$-\mu \frac{dI_\nu(z, \mu)}{k_\nu \rho_a\, dz} = I_\nu(z, \mu) - B_\nu(z), \qquad (2.1.5)$$

where the Planck intensity, $B_\nu(z) = B_\nu(T(z))$.

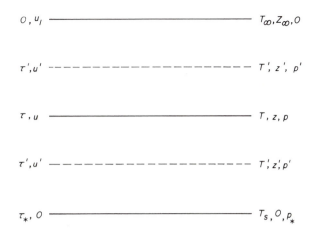

FIG. 2.2   Coordinate systems in τ, u, T, p, and z for ir radiative transfer in plane-parallel atmospheres. u is the path length for absorbing gases. The total path length is denoted by u_1. T_∞ and z_∞ are temperature and height, respectively, at TOA. The surface temperature, $T_s = T(\tau_*)$. The surface pressure is denoted by p_*.

Because of the height dependence of both the gaseous density and absorption coefficient, it is convenient to define a physical parameter referred to as the *normal optical depth*, or simply *optical depth*, in the form

$$\tau = \int_z^{z_\infty} k_\nu(z')\rho_a(z')\,dz' = \int_0^p k_\nu(p')q(p')\,\frac{dp'}{g} , \qquad (2.1.6)$$

where z_∞ denotes the height at the top of the atmosphere (TOA) and we have introduced the pressure coordinate p, using the hydrostatic equation; $q = \rho_a/\rho$, the gaseous mixing ratio, and ρ is the air density. Figure 2.2 shows the coordinate systems in optical depth, height, and pressure. The coordinate systems in temperature and path length will be discussed in Section 2.7 concerning the broadband emissivity approach for flux calculations. The differential optical depth can be readily obtained from Eq. (2.1.6) with the form $d\tau = -k_\nu(z)\rho_a(z)\,dz = k_\nu(p)q(p)dp/g$. In terms of the τ coordinate, Eq. (2.1.5) may be rewritten as follows:

$$\mu\frac{dI_\nu(\tau,\mu)}{d\tau} = I_\nu(\tau,\mu) - B_\nu(\tau). \qquad (2.1.7)$$

For the upward intensity, the zenith angle $0 \le \theta \le \pi/2$; that is, $0 \le \mu \le 1$. However, for the downward intensity, $\pi/2 \le \theta \le \pi$. In this case, we may set $\mu = -\mu$ for convenience in radiative transfer analyses.

Equation (2.1.7) represents a first-order differential equation. In order to solve both upward and downward components for an atmosphere with a total optical depth of τ_*, two boundary conditions are required. Under the plane-parallel assumption, these conditions are isotropic emissions from both the surface and TOA.

In general, the earth's surface may be considered a blackbody in the infrared, so that $I_\nu(\tau_*, \mu) = B_\nu(T(\tau_*))$. In addition, we may allow for a possible source of downward emission at TOA, and write $I_\nu(0, -\mu) = B_\nu(\text{TOA})$. Normally, $B_\nu(\text{TOA}) \approx 0$. Subject to the preceding boundary conditions, the formal solutions for upward and downward intensities are given by

$$I_\nu^+(\tau, \mu) = B_\nu(\tau_*)e^{-(\tau_* - \tau)/\mu}$$
$$+ \int_\tau^{\tau_*} B_\nu(\tau')e^{-(\tau' - \tau)/\mu}\frac{d\tau'}{\mu}, \quad (2.1.8a)$$
$$I_\nu^-(\tau, -\mu) = B_\nu(\text{TOA})e^{-\tau/\mu}$$
$$+ \int_0^\tau B_\nu(\tau')e^{-(\tau - \tau')/\mu}\frac{d\tau'}{\mu}. \quad (2.1.8b)$$

We shall now define the monochromatic transmittance (also referred to as transmission function) so that the exponential attenuation may be expressed by

$$T_\nu(\tau/\mu) = e^{-\tau/\mu}. \quad (2.1.9a)$$

The differential form is

$$\frac{dT_\nu(\tau/\mu)}{d\tau} = \frac{-1}{\mu}e^{-\tau/\mu}. \quad (2.1.9b)$$

The formal solutions for the intensities can then be expressed by

$$I_\nu^+(\tau, \mu) = B_\nu(\tau_*)T_\nu\left[(\tau_* - \tau)/\mu\right]$$
$$- \int_\tau^{\tau_*} B_\nu(\tau')\frac{d}{d\tau'}T_\nu\left[(\tau' - \tau)/\mu\right]d\tau', \quad (2.1.10a)$$
$$I_\nu^-(\tau, -\mu) = B_\nu(\text{TOA})T_\nu(\tau/\mu)$$
$$+ \int_0^\tau B_\nu(\tau')\frac{d}{d\tau'}T_\nu\left[(\tau - \tau')/\mu\right]d\tau'. \quad (2.1.10b)$$

Equation (2.1.10a) can be applied to remote sensing from space if we set $\tau(\text{TOA}) = 0$.

For atmospheric heating rate calculations, the required quantities are upward and downward fluxes, which are the sum of directional intensities from the upper and lower hemispheres, respectively. In accordance with the plane-parallel assumption, we have

$$F_\nu^\pm(\tau) = 2\pi \int_0^1 I_\nu^\pm(\tau, \pm\mu)\mu\, d\mu. \quad (2.1.11)$$

On noting the angular integration and in order to obtain the fluxes, we may define a physical parameter, referred to as *slab* or *diffuse transmittance*, in the form

$$T_\nu^f(\tau) = 2 \int_0^1 T_\nu(\tau/\mu)\mu\, d\mu. \quad (2.1.12)$$

With the help of this definition, the expressions for fluxes may now be written in the form

$$F_\nu^+(\tau) = \pi B_\nu(\tau_*)T_\nu^f(\tau_* - \tau)$$
$$- \int_\tau^{\tau_*} \pi B_\nu(\tau')\frac{d}{d\tau'}T_\nu^f(\tau' - \tau)\,d\tau', \qquad (2.1.13a)$$

$$F_\nu^-(\tau) = \pi B_\nu(\text{TOA})T_\nu^f(\tau)$$
$$+ \int_0^\tau \pi B_\nu(\tau')\frac{d}{d\tau'}T_\nu^f(\tau - \tau')\,d\tau'. \qquad (2.1.13b)$$

The upward flux at a given level is a result of two sources: the surface emission that is attenuated to that level and the emission contributions from the atmospheric layers characterized by Planck fluxes multiplied by the weighting function, $dT_\nu^f/d\tau$. Likewise, the downward flux at a given level is produced by the contributions from the emission at TOA and from the atmospheric layers. The emission at TOA can be neglected, as pointed out previously.

Finally, to account for the contributions from all wave numbers in the thermal ir spectrum, an integration of the monochromatic flux with respect to wave number must be performed. Since τ is a function of wave number, we use the height coordinate and write

$$F^\pm(z) = \int_0^\infty F_\nu^\pm(z)\,d\nu. \qquad (2.1.14)$$

At this point, the transfer of thermal ir radiation in plane-parallel atmospheres (without clouds) is formally solved. Conditional to the definition of diffuse transmittance, the computation of atmospheric fluxes involves solving the integrations over wave number and along the optical depth.

The net flux at a given level is a result of the difference between upward and downward fluxes:

$$F_\nu(\tau) = F_\nu^+(\tau) - F_\nu^-(\tau). \qquad (2.1.15)$$

The divergence (or convergence) of net fluxes must be proportional to the differential heating/cooling rates. Thus the local rate of temperature change due to monochromatic ir radiative transfer may be expressed by

$$\left(\frac{\partial T}{\partial t}\right)_\nu = -\frac{1}{\rho C_p}\frac{dF_\nu(z)}{dz} = \frac{k_\nu q}{C_p}\frac{dF_\nu(\tau)}{d\tau}, \qquad (2.1.16)$$

where C_p denotes the specific heat at constant pressure. The total heating rate, covering the entire thermal ir spectrum, is then

$$\left(\frac{\partial T}{\partial t}\right)_{ir} = \int_0^\infty \left(\frac{\partial T}{\partial t}\right)_\nu d\nu = -\frac{1}{\rho C_p}\frac{dF(z)}{dz}. \qquad (2.1.17)$$

where the total net flux F is the difference between the total upward and downward fluxes denoted in Eq. (2.1.14). The exchange of thermal ir fluxes due to absorption and emission of atmospheric gases leads to differential heating and cooling, which is intimately connected with the energetics and dynamics of the atmosphere. Having outlined the fundamentals associated with thermal ir radiation, we shall now discuss the absorption characteristics of atmospheric gases in the earth's atmosphere.

2.2   Absorption characteristics of atmospheric gases

In this section, the properties of absorbing gases in the thermal infrared region are discussed. We first present the absorption coefficient in terms of line formation, line shape and width, and line intensity. The absorption spectra of various radiatively active gases are then identified.

2.2.1   Absorption coefficient

The efficiency of absorption by a particular molecular species is expressed by its spectral absorption coefficient, k_ν. In reference to the discussion on the general radiative transfer equation, the attenuation of monochromatic intensity along a distance s through a medium with density ρ_a, due to absorption alone, may be expressed by

$$I_\nu(s) = I_\nu(0)T_\nu(\tau) = I_\nu(0)e^{-\tau}, \qquad (2.2.1)$$

where T_ν is the transmittance defined in Eq. (2.1.9a) and the optical depth

$$\tau = \int k_\nu \, du = \int k_\nu \, d\ell, \qquad (2.2.2)$$

defines the absorption coefficient in units of $cm^2 \, g^{-1}$ or $cm^{-1} \, atm^{-1}$, where u and ℓ denote path lengths in units of $g \, cm^{-2}$ and cm atm, respectively, and $1 \, g \, cm^{-2} \cong 2.24 \times 10^4 / M$ cm atm, where M is the molecular weight for individual gas. The absorption coefficient of a spectral line at position ν_0 is governed by the line intensity (or line strength) S and line-shape factor f and may be written in the form

$$k_\nu = Sf(\nu - \nu_0). \qquad (2.2.3a)$$

The shape factor is normalized to unity so that

$$S = \int_{-\infty}^{\infty} k_\nu \, d\nu. \qquad (2.2.3b)$$

2.2.2   Absorption line formation

A molecule has three forms of internal energy: rotational, E_r, vibrational, E_v, and electronic, E_e. The energy in each form is quantized and has only discrete values. These forms of energy are specified by one or more quantum numbers, and are additional to the kinetic energy of molecular translation, E_t. The translation energy is not quantized but exists in a continuous range of values. If we neglect the interaction among forms of energy, the total energy of an isolated molecule at any instant may be written

$$E = E_r + E_v + E_e + E_t. \tag{2.2.4a}$$

Absorption and emission of radiation takes place when the molecule undergoes transitions from one energy state to another. In absorption, the molecule captures a photon and undergoes a transition to a higher level of internal energy. As a result, spectral absorption lines are produced. In emission, the molecule releases a photon and makes a transition to a lower level, thereby producing spectral emission lines. When a transition occurs, the frequency, $\tilde{\nu}$ of the absorbed or emitted quantum is given by Planck's relation:

$$\Delta E = h\tilde{\nu} \tag{2.2.4b}$$

The most general transition involves simultaneous changes in the three forms of energy.

In radiative transitions, the molecule must couple with an electromagnetic field so that energy exchanges can take place. This coupling is generally provided by the electric dipole moment of the molecule. If the effective centers of the positive and negative charges of the molecule have nonzero separation, then the dipole moment exists. Radiatively active gases in the infrared, such as H_2O and O_3, have permanent electric dipole moments due to their asymmetrical charge distributions. N_2 and O_2, however, are inactive in the infrared because of their symmetrical charge distributions. However, they have weak magnetic dipole moments that allow radiative activities to take place in the ultraviolet and, to a lesser extent, in the visible region.

Rotational energy changes are relatively small, with a minimum on the order of $1 \, cm^{-1}$. For this reason, pure rotational lines occur in the microwave and far-infrared spectra. Many of the rotational energy levels above the lowest level are populated at terrestrial temperatures. Changes in vibrational energy are generally greater than $600 \, cm^{-1}$, which is much larger than the minimum changes in rotational energy. Thus vibrational transitions never occur alone but are coupled with simultaneous rotational transitions. This coupling gives rise to a group of lines known as the vibrational–rotational band in the intermediate infrared spectrum. An electronic transition typically involves a few electron volts of energy. Because a high-energy photon is required for the transition, absorption and emission usually occur in the ultraviolet or visible spectrum. Atoms can produce line spectra associated with electronic energy. Molecules, however, can have two additional types of energy, leading to complex band systems.

Table 2.1 Range of temperature-dependent index n for the line half-width

Molecule	n	
	min.	max.
H_2O	0.64	
CO_2	0.75	0.79
O_3	0.76	
N_2O	0.64	0.82
CH_4	0.63	1.00

2.2.3   Absorption line shape

2.2.3.1   Lorentz profile

In the lower atmosphere, radiative transitions are sufficiently disturbed by molecular collisions to cause a broadening of spectral lines, an effect referred to as *pressure broadening*. Such molecular collisions destroy the coherency of the wave train absorbed or emitted during the transition. Using Fourier analyses of the wave trains and the effective lifetime of collisions that are directly proportional to the mean free path of the molecules, one can derive the absorption coefficient, which is given in the form

$$k_\nu = \frac{S}{\pi} \frac{\alpha}{(\nu - \nu_0)^2 + \alpha^2} = S f(\nu - \nu_0), \qquad (2.2.5)$$

where f is the Lorentz line shape, and the half-width α, is inversely proportional to the mean free time between collisions. The half-width is a function of pressure and temperature and can be expressed by

$$\alpha(p, T) = \alpha_0 \left(\frac{p}{p_0} \right) \left(\frac{T_0}{T} \right)^n, \qquad (2.2.6)$$

where α_0 denotes the reference half-width for a reference temperature T_0 (273 K) and pressure p_0 (1013 mb). The air-broadened half-width is the product of density, velocity, and the optical cross section. The density is inversely proportional to temperature, while the velocity is proportional to $T^{1/2}$. The optical cross section is a function of T. Thus the index n in Eq. (2.2.6) is given by $n = \frac{1}{2} - m$. When $m = 0$, $n = \frac{1}{2}$, which is known as the classical value. Table 2.1 lists the range of n for various optically active gases (Rothman et al., 1987). For most other gases, $n = \frac{1}{2}$. Under the reference condition, α_0 ranges from about 0.01 to 0.1 cm^{-1} for most radiatively active gases in the earth's atmosphere. The term α_0 generally depends on the spectral line, but for CO_2 it is fairly constant, with a value of about 0.07 cm^{-1}.

The Lorentz line profile is fundamental in atmospheric radiative transfer. This line profile is directly proportional to pressure. Since the pressure from the surface to 40 km varies by an order of 3, the importance of the pressure dependence of absorption coefficients is evident. The dependence on temperature is less significant in view of the possible temperature variation in the earth's atmosphere. The mean free path derived from kinetic theory assumes that a molecule is a hard, elastic sphere. However, a molecule is surrounded by a force field so that its diameter is not a single absolute value. The observed half-widths of collision-broadened lines are often two or three times greater than those calculated from kinetic theory. Also, careful laboratory measurements reveal that in the far wings the line shapes do not always obey the Lorentz profile. Line widths are also affected by the distribution of the total molecular population among various species. Collisions between like molecules (self-broadening) produce greater linewidths than do collisions between unlike molecules (foreign broadening). The second case predominates in atmospheric thermal infrared radiative transfer because all the radiatively active gases have relatively low concentrations.

The above discussion points out the uncertainty in the line shape and width. For this reason, a modified Lorentz line shape has been suggested, in the form

$$f_m(\nu - \nu_0) = f(\nu - \nu_0)\chi(\nu - \nu_0), \qquad (2.2.7a)$$

where χ is defined as a cutoff function. For CO_2, χ is found to be significantly less than 1 for $|\nu - \nu_0| > 3 \, cm^{-1}$. For H_2O, χ could be greater than 1 (Burch et al., 1969).

In the vicinity of strong Q branches in the CO_2 15 μm band (Subsection 2.2.5.1), significant discrepancies have been detected between observed spectra and simulated spectra using the available line parameters. These discrepancies are attributed to the distortion of the line shape due to *line coupling*, also referred to as *line mixing*. Due to rotationally inelastic collisions that change the magnitude and direction of the rotational angular momentum of radiating molecules, line intensity is transferred from one line to another, as though they are "coupled" or "mixed." Line coupling is important for molecules with small rotational constants. A review of line mixing of atmospheric absorbers has been provided by Spänkuch (1989). In order to represent the line shape due to coupling properly, a modification of the Lorentz line shape may be made in the form

$$f(\nu - \nu_0) = \frac{1}{\pi} \frac{\beta(\nu - \nu_0) + \alpha}{(\nu - \nu_0)^2 + \alpha^2}, \qquad (2.2.7b)$$

where β is the coupling coefficient, which is temperature dependent. In principle, this coefficient can be determined through laboratory measurements. When $\beta = 0$, f reduces to the Lorentz shape.

2.2.3.2 Doppler profile

In the upper atmosphere, above about 40 km, the air density is low but the temperature increases, and the molecular velocities become significant. The molecular velocity components along any direction of observation produce Doppler broadening of spectral lines. Under the condition that the transitional states of molecules are in thermodynamic equilibrium, the probability of the velocity components is given by the Maxwell–Boltzmann distribution. The absorption coefficient at the shifted frequency must be proportional to this probability, with the result

$$f_D(\nu - \nu_0) = \frac{1}{\alpha_D \sqrt{\pi}} \exp\left(-\frac{(\nu - \nu_0)^2}{\alpha_D^2}\right). \tag{2.2.8}$$

The Doppler width

$$\alpha_D = \frac{\nu_0}{c}(2RT)^{1/2}, \tag{2.2.9}$$

where R is the individual gas constant and c the velocity of light. The half-width is directly proportional to the square root of the temperature, and at the half-maximum it is given by $\alpha_D(\ell n\, 2)^{1/2}$. The Doppler width can be evaluated once the position of the line associated with a specific molecule is known. The absorption coefficient of a Doppler line is governed by the exponential function and is more intense at the line center and much weaker in the wings than the Lorentz line shape.

2.2.3.3   Voigt profile

In the altitude region extending from about 20 to 50 km, effective line shapes are determined by both collision- and Doppler-broadening processes. We must add the Doppler shift component to the pressure broadened lines at wave numbers $\nu' - \nu_0$, in order to combine the two effects. The Doppler line redistributes the Lorentz line at wavenumber ν' to ν. The line shapes for pressure and Doppler broadening may then be expressed by $f(\nu' - \nu_0)$ and $f_D(\nu - \nu')$, respectively. To account for all the possible thermal velocities, a convolution of the Lorentz and Doppler line shapes can be performed to obtain

$$\begin{aligned}
f_v(\nu - \nu_0) &= \int_{-\infty}^{\infty} f(\nu' - \nu_0) f_D(\nu - \nu') d\nu' \\
&= \frac{1}{\pi^{3/2}} \frac{\alpha}{\alpha_D} \int_{-\infty}^{\infty} \frac{1}{(\nu' - \nu_0)^2 + \alpha^2} \\
&\quad \times \exp\left(-\frac{(\nu - \nu')^2}{\alpha_D^2}\right) d\nu'.
\end{aligned} \tag{2.2.10}$$

This line shape is referred to as the *Voigt profile*.

To simplify the representation of the Voigt profile, we let $t = (\ell n\, 2)^{1/2}(\nu - \nu')/\alpha_D'$; $y = (\ell n\, 2)^{1/2}\alpha/\alpha_D'$; $x = (\ell n\, 2)^{1/2}(\nu - \nu_0)/\alpha_D'$; and the Doppler half-width, $\alpha_D' = \alpha_D(\ell n\, 2)^{1/2}$. Thus, we have

$$f_v(\nu - \nu_0) = \frac{1}{\alpha_D'}\left(\frac{\ell n\, 2}{\pi}\right)^{1/2} K(x, y), \tag{2.2.11}$$

where the Voigt function is defined by

$$K(x, y) = \frac{y}{\pi} \int_{-\infty}^{\infty} \frac{1}{y^2 + (x - t)^2} e^{-t^2} dt. \tag{2.2.12}$$

The Voigt profile satisfies the requirement of normalization such that

$$\int_{-\infty}^{\infty} f_v(\nu - \nu_0) d(\nu - \nu_0) = 1 \tag{2.2.13}$$

When $\alpha \to 0$ (i.e., $y \to 0$), using the definition of δ-function we find

$$K(x, 0) = e^{-x^2}. \tag{2.2.14}$$

In this case the line shape is pure Doppler. On the other hand, when $\alpha_D \to 0$ (i.e., $x \to \infty$, $y \to \infty$), we have

$$f_v(\nu - \nu_0)\big|_{x/y=(\nu-\nu_o)/\alpha} = \frac{1}{\pi} \frac{1}{(\nu - \nu_0)^2 + \alpha^2}. \tag{2.2.15}$$

The line shape in this case is pure Lorentz. It is clear from the foregoing analysis that in the two limiting cases the Voigt profile reduces to Lorentz and Doppler shapes.

Many attempts have been made to simplify the computation of the Voigt function (Armstrong, 1967). A simple closed-form approximation that is valid over a variety of parameters has been given by Fomichev and Shved (1985) in the form

$$f_v(\nu - \nu_0) = \left(\frac{\ell n\, 2}{\pi}\right)^{1/2} \frac{1}{\alpha_v} (1 - \xi) \exp\left(-\ell n\, 2\eta^2\right)$$
$$+ \frac{1}{\pi\alpha_v} \xi \frac{1}{1 + \eta^2} - \frac{1}{\pi\alpha_v} \xi(1 - \xi) \left(\frac{1.5}{\ell n\, 2} + 1 + \xi\right)$$
$$\times \left(0.066 \exp(-0.4\eta^2) - \frac{1}{40 - 5.5\eta^2 + \eta^4}\right), \tag{2.2.16}$$

where $\xi = \alpha/\alpha_v$, $\eta = (\nu - \nu_0)/\alpha_v$, and the Voigt half-width is given by

$$\alpha_v = 0.5 \left[\alpha + (\alpha^2 + 4\ell n\, 2\alpha_D^2)^{1/2}\right]$$
$$+ 0.05\alpha \left(1 - \frac{2\alpha}{\alpha + (\alpha^2 + 4\ell n 2\alpha_D^2)^{1/2}}\right). \tag{2.2.17}$$

This approximation yields an accuracy generally within about 3%.

The relative shapes of Lorentz, Doppler, and Voigt profiles are shown in Fig. 2.3. These shapes correspond to a pressure and temperature at 30 km for H_2O,

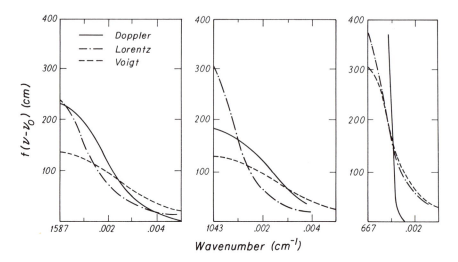

FIG. 2.3   Lorentz, Doppler, and Voigt profiles corresponding to a pressure and temperature at 30 km for infrared lines centered at 1587 cm^{-1} (H_2O), 1043 cm^{-1} (O_3), and 667 cm^{-1} (CO_2) (after Kuhn and London, 1969).

O_3 and CO_2 lines centered at 1587, 1043, and 667cm^{-1}, respectively (Kuhn and London, 1969). The Lorentz half-widths used are 0.1, 0.08, and 0.064 cm^{-1}, while the Doppler widths used are 2.27×10^{-3}, 2.89×10^{-3}, and 6.08×10^{-4} cm^{-1}. For these three gases, the Lorentz profile should be adequate in representing the line shape up to about 20 km, while the Doppler profile is valid above about 50 km. Between 20 to 50 km, the Voigt profile must be used in radiative transfer calculations. With respect to the Lorentz profile, the effect of the Voigt profile is to decrease absorption at the line center, while increasing it in the wings.

2.2.4   Absorption line intensity

The line intensity is related to the probability of a single molecule in its lower energy state that could absorb a photon and undergo a transition to the upper energy state, and the relative population of lower and upper energy states. Based on quantum theory, the line strength for a specific molecule that undergoes a transition from the upper state j to the lower state i may be expressed in terms of the transition probability R_{ij} in the form (Rothman et al., 1987)

$$S(T) = \frac{8\pi^3}{3hc}\nu_0 \left[1 - \exp(-hc\nu_0/KT)\right] \frac{g_i I_a}{Q(T)}$$
$$\times \exp(-hcE_i/KT)|R_{ij}|^2 \times 10^{-36} \left[cm^{-1}/\left(mol\,cm^{-2}\right)\right], (2.2.18)$$

where the line-center wave number $\nu_0 = \nu_0(i,j)$, E_i is the energy of the lower state of transition, g_i is the nuclear spin degeneracy of the lower level, $Q(T)$ is the total internal partition sum, and I_a is the natural isotopic abundance. The transition probability is given by

$$R_{ij} = \int \psi_i^* \mathbf{M} \psi_j dV, \qquad (2.2.19)$$

where dV is the volume element, $\mathbf{M}$ the matrix form of the dipole moment, $\psi_{i,j}$ are wave functions of the lower and upper states, and ψ_i^* is the conjugate of ψ_i.

According to the Maxwell–Boltzmann distribution, the internal partition function may be expressed by

$$Q(T) = \sum_n g_n \exp\left(\frac{-E_n}{KT}\right), \qquad (2.2.20)$$

where the internal energy E_n is the sum of electronic, vibrational, and rotational energies, and g_n is the internal statistical weight. For atmospheric conditions, the internal partition function due to vibration $Q_v \approx 1$. Thus the internal partition function is primarily associated with the rotational energy and is proportional to the temperature in the forms (Herzberg, 1945)

$$Q(T) \cong \begin{cases} T, & \text{linear molecules} \\ T^{3/2}, & \text{nonlinear molecules} \end{cases} \qquad (2.2.21)$$

Thus the line intensity in Eq. (2.2.18) for a given temperature may be expressed in terms of that for a reference temperature T_r in the form

$$S(T) = S(T_r) \left(\frac{T_r}{T}\right)^n \frac{1 - \exp(-hc\nu_0/KT)}{1 - \exp(-hc\nu_0/KT_r)} \exp\left[-\frac{hcE_i}{K}\left(\frac{1}{T} - \frac{1}{T_r}\right)\right],$$
$$(2.2.22a)$$

where $n = 1$ for linear molecules such as CO_2, N_2O, CO, O_2, etc., and $n = \frac{3}{2}$ for nonlinear molecules such as H_2O, O_3, CH_4, etc. If the units for line intensity are in $cm^{-2} atm^{-1}$, then $n = 2$ and $\frac{5}{2}$ for linear and nonlinear molecules, respectively. Because the term involving the ratio of the two exponential functions is very close to 1, we may simplify the expression for the line intensity as follows:

$$S(T) \cong S(T_r) \left(\frac{T_r}{T}\right)^n \exp\left[-\frac{hcE_i}{K}\left(\frac{1}{T} - \frac{1}{T_r}\right)\right]. \qquad (2.2.22b)$$

2.2.5 *The absorption spectrum of atmospheric gases in the thermal infrared*

2.2.5.1 Carbon dioxide

Spectroscopic evidence indicates that the CO_2 molecule has a linear symmetrical configuration, with the carbon atom in the middle and oxygen atoms on each side. It has one rotational constant. Due to its linear symmetry, the CO_2 molecule has no permanent electric dipole moment and, hence, no pure rotational transitions.

Because of vibrational symmetries, the symmetrical stretch mode, ν_1, is radiatively inactive at its fundamental. A fundamental represents a transition from the ground state to the first excited state. The fundamental for ν_1 has been identified in the Raman spectrum near 7.5 μm.

The bending mode, ν_2, is degenerate and consists of ν_{2a} and ν_{2b} vibrations at the same frequency. The CO_2 15 μm band represents this particular vibration. Owing to perpendicular vibration, the ν_2 fundamental transition is coupled with rotational transitions corresponding to changes in the quantum number $\Delta J = -1, 0, +1$. The spectral lines produced by these changes are referred to as P, Q, and R branches, respectively. For P and R branches, the lines have wave numbers lower and greater, respectively, than those of the line center. In the case of the Q branch, the lines are clustered near the center of the 15 μm band.

The natural isotopes for the carbon atom are ^{12}C and ^{13}C with relative abundances of 98.892 and 1.108%, respectively. For the oxygen atom, the isotopes are ^{16}O (99.758%), ^{17}O (0.0373%), and ^{18}O (0.2039%). Thus several CO_2 isotopes are present in the atmosphere. The most significant of these for the radiation problem are $^{12}C^{16}O^{16}O$, $^{13}C^{16}O^{16}O$ and $^{12}C^{16}O^{18}O$. The fundamental ν_2 vibration wave numbers for these isotopes are centered at 667.4, 648.52, and 662.39 cm^{-1}, respectively.

Besides the ν_2 fundamental band, numerous combination bands have been detected in the 15 μm region. Simultaneous transitions in two of the vibration modes are possible, resulting in weak combination (or difference) frequencies. There are also numerous hot bands in the 15 μm CO_2 band. These bands are produced by transitions between excited levels and are significant in cooling rate calculations in the upper atmosphere (see Figure 2.19).

Due to the asymmetric stretching vibration, the ν_3 fundamental transition has a wave number centered at 4.3 μm. The isotopes $^{12}C^{16}O^{16}O$, $^{13}C^{16}O^{16}O$, and $^{12}C^{16}O^{18}O$ have ν_3 fundamental wave numbers centered at 2349.16, 2283.48, and 2333 cm^{-1}, respectively. In addition, the combination band $\nu_1 + \nu_3 - 2\nu_2$ is centered at 2429.37 cm^{-1} and occurs when the ν_3 transition originates at the vibrational level $v = 2$ or higher. Because of parallel vibration, the Q branch corresponding to $\Delta J = 0$ does not appear.

There are several overtone and combination bands of carbon dioxide besides the 15 and 4.3 μm bands. At atmospheric temperatures, most of the molecular population is in the vibrational level with quantum number $v = 0$. When transitions take place between nonadjacent levels ($\Delta v = 2, 3, 4$), weaker overtone frequencies are produced. Two moderately strong bands appear in the solar spectrum and are centered at 1063.8 and 961.0 cm^{-1}. Both are parallel bands and are responsible for the CO_2 laser emission at about 10.6 μm. The bands near 5 μm consist of the $3\nu_2$ band at 5.2 μm and several combination bands at 4.8 μm. Other CO_2 bands in the solar spectral region will be discussed in Chapter 3, which deals with solar radiative transfer.

Table 2.2 Vibrational and rotational transitions for CO_2, H_2O, O_3, CH_4, N_2O, and CFCs

Band	$\nu(cm^{-1})$	Transition	Band Interval (cm^{-1})
CO_2 15 μm	667	ν_2; P,Q,R, hot bands Combination	550–800
CO_2 10.6 μm	{ 961.0 1063.8	Overtone and combination	850–1100
CO_2 4.3 μm	2349	ν_3; P,R Overtone and combination	2100–2400
H_2O rotation (pure)		P,R	$\sim$0–1000
H_2O 6.3 μm	1594.8	ν_2; P,R	640–2800
H_2O continuum		Far wings of very strong lines, $(H_2O)_2$	$\sim$200–1200
O_3 9.6 μm	{ 1110 1043	ν_1; P,R ν_3; P,R	950–1200 600–800
O_3 14.2 μm	705	ν_2; P,R	600–800
O_3 4.8 μm	2105	Overtone and combination	
CH_4 7.6μm	1306.2	ν_4	950–1650
N_2O 7.9 μm	1285.6	ν_1	1200–1350
N_2O 17.0 μm	588.8	ν_2	520–660
N_2O 4.5 μm	2223.5	ν_3	2120–2270
CFCs			$\sim$700–1300

Table 2.2 summarizes the CO_2 bands that are important in thermal infrared radiation. The line position, line half-width, line strength, and lower energy state for the CO_2 molecule have been comprehensively documented in the HITRAN data base (Rothman et al., 1987) and the ATMOS data base (Brown et al., 1987). According to these studies, the accuracies of the line position vary from 0.0001 to 0.05 cm^{-1}. The accuracies of the line strengths are more difficult to assess, but it has been suggested that they vary from 2% for strong lines to 10% for weak or blended lines.

2.2.5.2   Water vapor

The water vapor molecule forms an isosceles triangle that is obtuse. This molecular shape, referred to as an asymmetric top (bent triatomic) configuration, has three rotational constants. Using these constants and the selection rules $\Delta J = \pm 1$, the pure rotational lines can be computed. The spacings of rotational lines for the three degrees of rotational freedom tend to be irregular and relatively wide.

The hydrogen atom has two isotopic forms, ^1H and ^2D, with relative abundances of 99.9851 and 0.0149%, respectively. The isotropic forms of water vapor that are important in infrared radiative transfer are HH^{16}O, HH^{18}O, HD^{16}O, and HD^{18}O. Each of these molecules has a different vapor pressure, and their abundances are a function of the hydrological cycle. The H_2O molecule has three fundamental vibration modes (Fig. 2.4). The bending vibration, ν_2, has the lowest

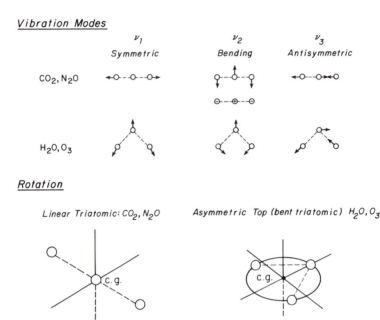

FIG. 2.4 Vibrational modes of triatomic atmospheric molecules and the axes of rotational freedom for linear and symmetrical top molecules.

wave number; both ν_1 and ν_3 have wave numbers about twice the wave number for ν_2.

The pure H_2O rotational band ranges from 0 to $1000\,cm^{-1}$. This band is important in the generation of tropospheric cooling. The ν_2 fundamental band at $1594.78\,cm^{-1}$ ($6.25\,\mu m$) is the most important vibrational–rotational band of water vapor. The two other fundamental bands, ν_1 and ν_3, are found to be close to one another and are centered at $2.7\,\mu m$. The H_2O isotopes $HH^{18}O$, $HH^{17}O$ and $HD^{16}O$ have been identified in both the rotational and ν_2 bands. There are numerous overtone and combination bands associated with water vapor that have been identified in the solar spectrum. These bands will be discussed in Chapter 3.

The region from 800 to $1200\,cm^{-1}$, the so-called *thermal infrared window*, contains the moderately strong $9.6\,\mu m$ band of ozone, which will be discussed in the following. Apart from the ozone band, absorption is continuous and is primarily due to water vapor species. The attenuation due to the water vapor continuum in the $10\,\mu m$ window remains a theoretical mystery. It has been suggested that the continuum results from the accumulated absorption of the distant wings of water vapor lines, principally in the far-infrared part of the spectrum (e.g., see Clough et al., 1980). This absorption is caused by the collision broadening between absorbing molecules (H_2O–H_2O) and between absorbing and nonabsorbing molecules (H_2O–N_2). There is evidence that contributions to continuous absorption may be caused by water dimer ($H_2O)_2$. However, this possibility has yet to be verified. Absorption by water dimer depends significantly on the water vapor pressure and temperature. While accurate and well-controlled measurements are required in order to account for the water vapor continuum in real atmospheric situations, limited experimental measurements have been used to develop empirical parameterizations.

The absorption coefficient for the water vapor continuum is the sum of the self- and foreign-broadening components and may be written

$$k_\nu = \sigma_s \left[p_w + \frac{\sigma_n}{\sigma_s}(p - p_w) \right], \tag{2.2.23}$$

where p_w and p denote the water vapor partial pressure and the ambient pressure (in atm), respectively, and σ_s and σ_n are the self- and foreign-broadening coefficients for water vapor, respectively.

The self-broadening coefficient has been derived empirically based on available laboratory measurements in the form (Roberts et al., 1976)

$$\sigma_s(\nu, T_r) = a + be^{-\beta\nu}, \tag{2.2.24}$$

where $T_r = 296\,K$, $a = 4.18$, $b = 5578$, and $\beta = 7.87 \times 10^{-3}$. The self-broadening coefficient, σ_s ($cm^2\,g^{-1}\,atm^{-1}$), is displayed in Fig. 2.5(a). The temperature dependence of this coefficient has been found to vary as

$$\sigma_s(\nu, T) = \sigma_s(\nu, T_r) \exp[c(T_r/T - 1)], \tag{2.2.25}$$

with $c = 6.08$. The attenuation coefficient increases as the temperature decreases. This implies that more absorption occurs at colder temperatures for a fixed amount of water vapor. Likewise, less absorption occurs at warmer temperatures. In real atmospheres, colder conditions are generally associated with less water vapor. Thus the temperature dependence on the attenuation in the 8 to 12 μm region has two competing effects via the total amount of water vapor path length and the absorption coefficient.

The second term in Eq. (2.2.23) represents the ratio of the self- to foreign-broadening coefficients. At $T_r = 296\,K$, a value of 0.002 has been suggested based on laboratory measurements. Subject to more accurate and comprehensive experiments, it may be assumed that this value does not vary with temperature.

Based on limited laboratory measurements, the temperature dependence of the absorption coefficient in the 3–5 μm region may be expressed by (Kneizys et al., 1980)

$$\sigma_s(\nu, T) = \sigma_s(\nu, T_r) \exp[d(T_r/T - 1)], \tag{2.2.26}$$

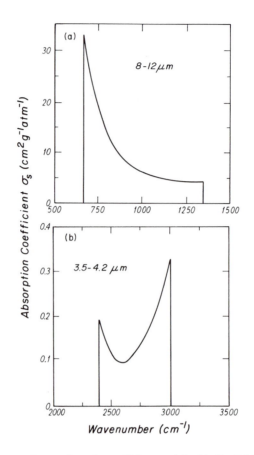

FIG. 2.5   Water vapor continuum absorption coefficient σ_s defined in Eq. (2.2.23) at a temperature of 296 K for (a) 8–12 μm and (b) 3.5–4.2 μm (after Kneizys et al., 1980).

where the empirical coefficient $d = 4.56$. σ_s at $T_r = 296$ K is displayed in Fig. 2.5(b). A value of 0.12 cm^2 g^{-1} atm^{-1} has been suggested for the nitrogen-broadening coefficient σ_n at a temperature of 428 K. Although the aforementioned laboratory data were not formally published, they nevertheless represent the best information available at this time. A recent review on the subject of the water vapor absorption coefficient in the 8–12 μm window region listed numerous references on the measurements and theory of continuum absorption (Grant, 1990).

The line-by-line data for H_2O molecules have been compiled in the HITRAN and ATMOS data bases. The accuracies of the positions and strengths of the H_2O lines are within ± 0.005 cm^{-1} and 5–20%, respectively.

2.2.5.3   Ozone

The ozone molecule has an asymmetric top configuration similar to the water vapor molecule, but with a different apical angle. Thus it has a relatively strong rotational spectrum. The three isotopes, $^{16}O_3$, $^{16}O^{18}O^{16}O$, and $^{16}O^{16}O^{18}O$ have fundamental bands at the 9.6 and 14.27 μm bands. The ν_1 and ν_3 fundamental vibration modes are centered at 1110 and 1043 cm^{-1} and constitute the well-known 9.6 μm ozone band. The ν_2 fundamental band, centered at 705 cm^{-1} (14.27 μm), is well-masked by the strong CO_2 15 μm band and appears to be less significant in atmospheric radiative transfer. There is also a strong band of ozone at 4.75 μm that is produced by overtone and combination transitions. The electronic bands of ozone will be discussed in Chapter 3.

The accuracies of ozone line positions vary from band to band, ranging from $\sim$0.3 to $\sim$5 cm^{-1}. Ozone line strength accuracies are difficult to define, and errors of 10–20% in relative strengths should be expected (Brown et al., 1987).

2.2.5.4 Methane

The methane molecule has a spherical top configuration. It has no permanent electric dipole moment and, hence, no pure rotational spectrum. There are four fundamental vibration modes. Of these, only ν_3 and ν_4, centered at 3020.3 and 1306.2 cm^{-1}, are active in the infrared spectrum. The ν_4 fundamental band of CH_4 is important in the climatic greenhouse effect. The inactive ν_1 and ν_2 fundamental bands are centered at 2914.2 and 1526 cm^{-1}. Methane also possesses a rich spectrum of overtone and combination bands that have been identified in the solar spectrum. The line parameters of methane are available for three isotopes: $^{12}CH_4$, $^{13}CH_4$, and $^{12}CH_3D$. Line strength accuracies range from $\pm$3% for strong lines to $\pm$30% for blended and weak lines.

2.2.5.5 Nitrous oxide

The nitrous oxide molecule has a linear and asymmetric structure, with the configuration NNO. Like carbon dioxide, it has a single rotational constant and a detectable rotational spectrum. Numerous bands produced by the fundamental, overtone, and combination frequencies exist in the infrared. The three fundamental frequencies are centered at 1285.6 cm^{-1} (ν_1), 588.8 cm^{-1} (ν_2), and 2223.5 cm^{-1} (ν_3). The ν_1 fundamental band of nitrous oxide overlaps the ν_4 fundamental band of methane. The line positions and strengths for the majority of the transitions in the 900–2800 cm^{-1} region have accuracies of 0.0001 cm^{-1} for line positions and 2–5% for line strengths.

2.2.5.6 Chlorofluorocarbons

The methylchloride (CH_3Cl) molecule has two bands of interest in atmospheric infrared radiative transfer: the ν_3 band at 732 cm^{-1} and the ν_2 band at 1350 cm^{-1}. In the ν_8 region, which is centered at 1161 cm^{-1} and has a band at 1095 cm^{-1}, Q branch features have been found for dichlorodifluoromethane (CF_2Cl_2). For the trichlorofluoromethane ($CFCl_3$) molecule, the ν_1 and ν_4 fundamental transitions are active and centered at 848 and 1085 cm^{-1}, respectively. The methylchloroform

(CH_3CCl_3) molecule has a narrow Q branch associated with the ν_2 fundamental band at 1384.5 cm^{-1}. The carbon tetrachloride (CCl_4) molecule has an active band in the ν_3 region near 796 cm^{-1}. Absorption of these anthropogenic trace gases is primarily located in the window region. Thus their potential increases can make the atmospheric window "dirty" and may lead to significant greenhouse effects.

Table 2.2 summarizes the preceding absorption bands in the thermal infrared region in terms of the spectral band, central wave number, primary transitions, and band interval.

2.3   Computation of transmittance and application

2.3.1   *Effects of line shape*

The monochromatic transmittance for an nonhomogeneous path length is defined by

$$T_\nu(u) = e^{-\tau} = \exp\left(-\int_u k_\nu(u)\,du\right), \qquad (2.3.1)$$

where the path length for an absorber with density ρ_a is given by

$$u = \int_0^z \rho_a(z')\,dz'. \qquad (2.3.2)$$

The absorption coefficient is a function of pressure and temperature. To illustrate the effects of the line shape on transmittance, we shall assume a homogeneous path length so that

$$T_\nu(u) = e^{-k_\nu u}. \qquad (2.3.3)$$

Based on the definitions of several variables given in Subsection 2.2.3.3 and letting $z = S/\sqrt{\pi}\alpha_D$, expressions of the absorption coefficients for Lorentz, Doppler, and Voigt profiles may be simplified as follows:

$$k_\nu(\text{Lorentz}) = \frac{z}{\sqrt{\pi}} \frac{y}{x^2 + y^2}, \qquad (2.3.4a)$$

$$k_\nu(\text{Doppler}) = ze^{-x^2}, \qquad (2.3.4b)$$

$$k_\nu(\text{Voigt}) = zK(x, y), \qquad (2.3.4c)$$

where $y = \alpha/\alpha_D$ (the ratio of the Lorentz half-width to the Doppler width) and $x = (\nu - \nu_0)/\alpha_D$. The transmittance as a function of x is illustrated in Fig. 2.6, where we let $zu = 10$. Also, y is set at 0.1 and 1, representing the relative magnitudes of the Lorentz to Doppler width. For the case where the Doppler broadening dominates ($y = 0.1$), the transmittances of Voigt and Doppler lines are quite similar. When the pressure broadening becomes more significant ($y > 1$), the transmittance of a Voigt line is close to that of a Lorentz line.

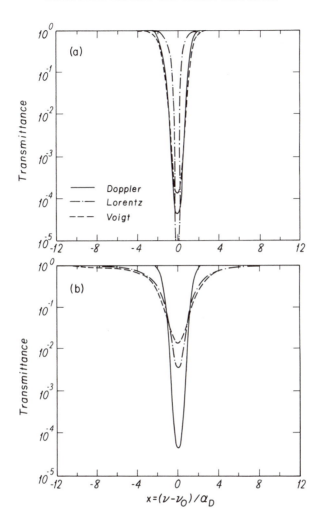

FIG. 2.6   Comparison of transmittance for three line shapes (a) with $y = 0.1$ and $zu = 10$, and (b) $y = 1$ and $zu = 10$.

2.3.2   Diffusivity factor in infrared radiative transfer

For flux calculations, we require the diffuse (slab) transmittance defined in Eq. (2.1.12) as follows:

$$T_\nu^f(\tau) = 2 \int_0^1 T_\nu(\tau/\mu)\mu \, d\mu = 2 \int_0^1 e^{-\tau/\mu}\mu \, d\mu. \qquad (2.3.5)$$

Generally, a four point Gauss quadrature will give an accurate result for an integration over the cosine of the zenith angle. For many atmospheric applications, however, the diffuse transmittance curve, as a function of path length, is similar to the transmittance curve. We may postulate that

$$T_\nu^f(\tau) = T_\nu(\tau/\bar{\mu}), \tag{2.3.6}$$

where $1/\bar{\mu}$, the inverse of the cosine of the mean emergent angle, is referred to as the *diffusivity factor*. To obtain this value with an analytic form, we shall consider an isothermal atmosphere so that the flux is proportional to

$$F \sim 2 \int_0^{\tau_1} \int_0^1 \frac{d}{d\tau'} T_\nu(\tau'/\mu)\mu \, d\mu \, d\tau'. \tag{2.3.7}$$

On using the mean value theorem and substituting the expression for the transmittance, we find

$$F \sim \int_0^{\tau_1} \frac{d}{d\tau'} e^{-\tau'/\bar{\mu}} \, d\tau' = e^{-\tau_1/\bar{\mu}} - 1. \tag{2.3.8}$$

But if we perform the integration with respect to τ' first, we have

$$F \sim 2 \int_0^1 (e^{-\tau_1/\mu} - 1)\mu \, d\mu. \tag{2.3.9}$$

By equating Eqs. (2.3.8) and (2.3.9), we obtain

$$e^{-\tau_1/\bar{\mu}} - 2E_3(\tau_1) = 0, \tag{2.3.10}$$

or

$$1/\bar{\mu} = -\ln[2E_3(\tau_1)]/\tau_1, \tag{2.3.11}$$

where E_3 denotes an exponential integral of the third kind:

$$E_3(\tau_1) \equiv \int_1^\infty e^{-\tau_1 x}/x^3 dx.$$

Figure 2.7 shows the diffusivity factor as a function of the optical depth. The diffusivity factor decreases with increasing optical depth. For very large optical depths ($\tau_1 > 20$), it is approximately equal to 1. For very small optical depths ($\tau_1 \to 0$), it has a value close to 2. In Fig. 2.8, we show the diffuse transmittance as a function of the optical depth for three cases. The first is an exact calculation using Eq. (2.1.12), while the other two cases use $\bar{\mu}$ of 0.6 and 0.5; that is, $1/\bar{\mu} = 1.66$ and 2. For a diffusivity factor of 1.66, the transmittance curve is very close to the diffuse transmittance curve for $\tau_1 < 1$. This diffusivity factor was originally proposed by

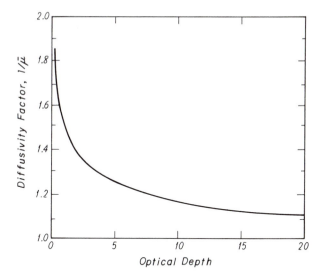

FIG. 2.7 The diffusivity factor as a function of the optical depth.

Elsasser (1942). It has been found to be a useful and reliable approximation in the computation of radiative fluxes and heating rates.

2.3.3 Line-by-line integration

Absorption line parameters have been computed for the major gases in the earth's atmosphere based on the available fundamental data discussed in Subsection 2.2.5.

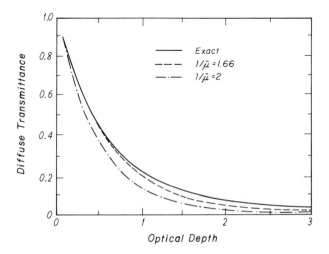

FIG. 2.8 The diffuse transmittance as a function of the optical depth for three cases.

Laboratory data are also available for a limited number of spectral intervals. As discussed in Subsection 2.2.5, line parameters have been compiled over the range 0 to $17,900\,\text{cm}^{-1}$ by scientists at the Air Force Geophysics Laboratory (Rothman et al., 1987). Data for more than 100,000 lines are presented. The absorption line parameters are listed in data form in terms of line position in cm^{-1}, line intensity in $\text{cm}^{-1}/(\text{mol cm}^{-2})$ at 296 K, air-broadened half-width in $\text{cm}^{-1}/\text{atm}$ at 296 K, and lower-state energy in cm^{-1}. The molecular species include H_2O, CO_2, O_3, N_2O, CO, CH_4, and O_2, etc.

For a given wave number and species, contributions to transmittance arise from absorption coefficients for N lines. The optical depth is then

$$\tau = \sum_{j=1}^{N} \tau_j = \int_u \sum_{j=1}^{N} k_{\nu,j}(u)\, du. \qquad (2.3.12)$$

Thus the absorption coefficient can be expressed in terms of line strength and line shape in the form

$$k_\nu(p, T) = \sum_{j=1}^{N} S_j(T) f_{\nu,j}(p, T). \qquad (2.3.13)$$

In the following, we shall discuss problems associated with the spectral resolution and line cutoff points.

In order to resolve individual lines, the absorption coefficient should be computed at wave number intervals that are smaller than the line half-width. In the upper stratosphere, absorption and emission processes are dominated by CO_2 and O_3. Broadening of the absorption lines is primarily due to the Doppler effect. The Doppler half-width in the 15 μm CO_2 and 9.6 μm O_3 bands is ~ 0.0005–$0.001\,\text{cm}^{-1}$. The spectral interval in these two bands covers about $400\,\text{cm}^{-1}$. Thus the absorption must be calculated at more than half a million points if individual lines are to be resolved.

In the troposphere, absorption due to H_2O predominates. H_2O lines cover essentially the entire infrared spectrum with a spectral region of about $15,000\,\text{cm}^{-1}$. These lines are broadened by collisions, and their half-widths are $> 0.01\,\text{cm}^{-1}$. Computations must be performed at about one million points to resolve H_2O lines. For each point, there are numerous lines and atmospheric conditions that must be considered for applications to atmospheric radiative transfer. The computer time required for line-by-line calculations, even with the availability of a supercomputer, is formidable. This is especially evident for flux calculations in which an integration over all absorption bands is necessary. Based on numerical experimentation, the spectral resolutions of 0.01, 0.002, and $0.005\,\text{cm}^{-1}$ are adequate for the computation of infrared fluxes for H_2O, O_3, and CO_2, respectively (Chou and Kouvaris, 1986).

In the case of Lorentz line shapes, it is important to cut off the contribution of significant lines at computational points in a line-by-line program. The far wings of a pressure-broadened absorption line in the infrared have been suggested to be sub-Lorentzian. Two approaches have been used to compute the far wing contribution. The first approach is to multiply the Lorentz profile by an empirical function χ, such that $\chi = 1$ at the line center and $\chi = 0$ at some distance from the center (Burch et al., 1969). The second is to use the Lorentz profile for all wave numbers but to cut the lines off at some distance from the line center. The lines can be cut off at a constant distance from the center (Fels and Schwarzkopf, 1981), or they can be cut off at a distance varying with the half-width, that is , the cutoff wave number, $\nu_c = \beta\alpha$, with β a constant (Chou and Kouvaris, 1986). The latter method demonstrates that computations of absorption are affected only slightly by cutting a line off at a wave number 190 times the Lorentz half-width from the center.

From Eqs. (2.3.1) and (2.3.12), we find that the monochromatic transmittance at a wave number ν is a product of the transmittance for each absorption line and may be written

$$T_\nu = T_{\nu,1} T_{\nu,2} \dots T_{\nu,N}, \tag{2.3.14}$$

where

$$T_{\nu,j} = \exp\left(-\int_u k_{\nu,j}(u)\,du\right), \qquad j = 1, 2, \dots, N, \tag{2.3.15}$$

and N is the total number of lines that contribute to the transmittance. The multiplication principle is valid only for monochromatic calculations.

Over a spectral interval, the calculation of the transmittance involves the following integral:

$$
\begin{aligned}
T_{\bar\nu}(u) &= \int_{\Delta\nu} \exp\left(-\int_u k_\nu(u)\,du\right) \frac{d\nu}{\Delta\nu} \\
&= \int_{\Delta\nu} \exp\left(-\int_u \sum_j k_{\nu,j}(u)\,du\right) \frac{d\nu}{\Delta\nu} \\
&= \sum_i \exp\left(-\int_u \sum_j k_{i,j}(u)\,du\right) \frac{\Delta\nu_i}{\Delta\nu} \\
&= \sum_i \exp\left(-\sum_n \sum_j k_{i,j}(p_n, T_n)\,\Delta u_n\right) \frac{\Delta\nu_i}{\Delta\nu}. \tag{2.3.16}
\end{aligned}
$$

In these equations, ν_i denotes the discrete wave number in the spectral interval; j $(j = 1, \dots, N)$ is the index of the absorption coefficient for the jth line, which makes a contribution at ν_i; and p_n, T_n, and u_n denote the pressure, temperature, and path length at level n. Thus, in order to calculate the spectral transmittance, $T_{\bar\nu}(u)$, exactly, adequate and reliable summations must be performed to cover the absorption lines, spectral interval, and the nonhomogeneous path. Moreover, the absorption coefficient is a function of the line strength, which is a function of temperature via Eq. (2.2.22b), and the half-width, which is a function of pressure and temperature via Eqs. (2.2.6) and (2.2.9). Exact line-by-line calculations for the spectral transmittance are very tedious and require a significant amount of computer time. All the infrared radiative transfer theories that have been developed are essentially intended to simplify and economize the computation of the spectral transmittance by circumventing the integration over the spectral interval and nonhomogeneous path length.

Radiance observations in the atmosphere by aircraft and at TOA by satellites provide a means of crosschecking the "exact" line-by-line method. The radiance spectra have been measured by IRIS on board the Nimbus III and IV satellites mentioned at the beginning of this chapter. This instrument measured spectra with $5\,cm^{-1}$ resolution of the vertically outgoing spectral radiances from 400 to $2000\,cm^{-1}$ with a reported accuracy and precision of about 1%. Each spectral radiance arises from a cone with a well-defined circular area on the earth of about 150 km in diameter. In order to perform theoretical computations, it is necessary to obtain collocated temperature and gaseous profiles; the humidity profile is particularly important. For a very high spectral resolution such as that of IRIS, variations in the Planck intensity can be neglected. The upwelling spectral radiance at TOA ($\mu = 1$) from Eq. (2.1.10a) may be written in the form

$$I_{\bar{\nu}} = B_{\bar{\nu}}(\tau_*)T_{\bar{\nu}}(\tau_*) + \int_{\tau_*}^{0} B_{\bar{\nu}}(\tau') \frac{\partial}{\partial \tau'} T_{\bar{\nu}}(\tau')\,d\tau', \qquad (2.3.17)$$

where τ_* is the total optical depth of the atmosphere. Given the spectral transmittance from the line-by-line method, the upwelling radiance can be calculated.

Comparisons between observed and computed radiances have been made by Conrath et al. (1970) and Kunde et al. (1974). The agreement between theory and observation in these studies is generally within about 10%. Uncertainties in the theoretical computation include the line-by-line spectroscopic data and local variations in temperature and humidity profiles. In addition, radiosonde data used may not coincide with the area where radiances were measured. Smith et al. (1987) have designed a high-resolution interferometer sounder, referred to as HIS, to measure the infrared spectrum from 3.5 to 17 μm with very high spectral resolution and radiometric accuracy. Spectral measurements using this instrument have been obtained from the NASA ER-2 aircraft flying at $\sim$20 km. Figure 2.9 illustrates a typical comparison between observed and theoretical spectra near Boston, Massachusetts, on July 16, 1986. The theoretical spectra were obtained from line-by-line calculations using the Portland, Maine, radiosonde. The spectral resolution for the 600–1100, 1100–2000, and 2000–2500 cm^{-1} regions are 0.5, 0.8, and 1.6 cm^{-1}, respectively. The computed spectrum closely matches the

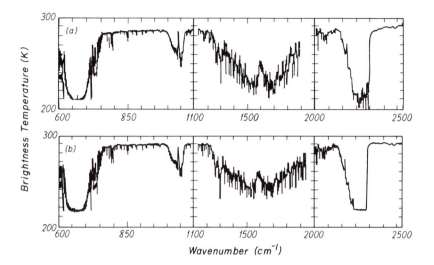

FIG. 2.9 Comparison of brightness temperature spectra (a) observed at Boston, Massachusetts, on July 16, 1986, and (b) calculated using radiosonde observations from Portland, Maine (after Smith et al., 1987).

observed spectra, which clearly show the features of the 15 μm CO_2, 9.6 μm O_3, and 4.3 μm CO_2 bands. Theoretical calculations at the 4.3 μm CO_2 band do not produce line structure at the band center.

2.3.4 *Principle of infrared remote sounding from space*

Equation (2.3.17) represents the fundamental forward radiative transfer equation that may be used to develop methodologies for determining temperature and gaseous profiles from radiance observations at TOA.

King (1956) first proposed that atmospheric temperature profiles might be inferred from satellite ir radiance measurements. King also pointed out that the angular radiance distribution is the Laplace transform of the Planck intensity distribution as a function of optical depth. Kaplan (1959) demonstrated for the first time that vertical resolution of the temperature field may be inferred from the spectral distribution of atmospheric emissions. Observations in the band center see only the very top layer of the atmosphere, whereas observations in the wings see deeper into the atmosphere. By selecting a set of sounding wave numbers that will sense various portions of the atmosphere, a continuous temperature profile may be constructed. In view of the variables involved in Eq. (2.3.17), in order to obtain interpretable radiances emitted from the earth and the atmosphere, the source of emission must be a relatively abundant gas of known and uniform distribution. The use of such a source gas would eliminate the uncertainty in the gaseous concentration profile that is used to determine temperature from radiance measurements. Carbon dioxide and molecular oxygen are the two gases in the earth's atmosphere that have uniform abundance in terms of their mixing ratios below about 100 km, and that, at the same time, exhibit significant emission spectra that can be utilized for radiance observations. As discussed in Subsection 2.2.5.1, CO_2 molecules produce important rotational and vibrational spectra in the 15 and 4.3 μm regions, which have been clearly identified in Fig. 2.9. Molecular oxygen, a major constituent in the atmosphere with a relative volume abundance of ~ 0.21, also shows a microwave spin-rotational band in the 60 GHz region. With microwave technology, this band has been used to infer atmospheric temperature profiles, especially for cloudy conditions. A comprehensive review on the subject of satellite remote sounding of atmospheric parameters, including temperature and humidity profiles, has been provided by Isaacs et al. (1986).

The forward radiative transfer equation presented previously [Eq. (2.3.17)] may be written in the pressure coordinate (see Fig. 2.2) as follows:

$$I_{\bar{\nu}} = B_{\bar{\nu}}(p_*)T_{\bar{\nu}}(p_*) + \int_{p_*}^{0} B_{\bar{\nu}}(p)\frac{\partial T_{\bar{\nu}}(p)}{\partial p}dp. \qquad (2.3.18)$$

Consider the pressure integration from p_* to $p \to \infty$ and write

$$\Delta I_{\bar{\nu}} = \int_{\infty}^{p_*} B_{\bar{\nu}}(p)\frac{\partial T_{\bar{\nu}}(p)}{\partial p}dp. \qquad (2.3.19)$$

The layer below the surface may be taken as an infinite isothermal emitter with a temperature T_s. Because $T_{\bar{\nu}}(\infty) = 0$, the term $\Delta I_{\bar{\nu}}$ represents the surface contribution in Eq. (2.3.18). In the band center, $\Delta I_{\bar{\nu}} \to 0$, while in the wing regions this term can be an important source of upwelling radiance. The weighting (kernel) function, signifying the weight of the Planck intensity contribution to upwelling radiance, may be defined in the logarithm of pressure such that

$$K_{\bar{\nu}}(p) = -\frac{\partial T_{\bar{\nu}}(p)}{\partial \ln p}. \qquad (2.3.20)$$

For each sounding channel $\bar{\nu}$, there must be a peak weighting function located at $p = \bar{p}$ that gives the maximum contribution of Planck intensity to upwelling radiance. We may set $K_{\bar{\nu}}(p) \equiv K(p, \bar{p})$. Moreover, over a small spectral interval, the Planck function does not vary significantly with wave number, and, therefore, its dependence on wave number may be omitted. Using the foregoing information, the basic equation to be applied in remote sounding from space may be written in the form

$$I(\bar{p}) = \int_{0}^{\infty} B(p)K(p, \bar{p})\, dp/p, \qquad (2.3.21)$$

where the peak of the weighting function, $\bar{p}$, replaces the channel index $\bar{\nu}$.

Equation (2.3.21) is the well-known Fredholm equation of the first kind, which is fundamental in the retrieval of temperature profiles from space using infrared technology. The universal inversion problem is to determine a profile of the Planck function, $B(p)$, given the weighting functions $K(p, \bar{p})$ and observed radiances $I(\bar{p})$. In practice, since only finite $\bar{p}$ values may be chosen, the solution for $B(p)$ is mathematically ill conditioned, even if $K(p, \bar{p})$ and $I(\bar{p})$ are given without any errors. Constraints of one kind or another, involving numerous inversion algorithms, are required to obtain physically meaningful temperature profiles.

In order to determine temperature profiles using infrared sounding channels successfully, it is necessary to construct a set of weighting functions that can suitably cover the vertical domain of the atmosphere. In principle, an ideal weighting function would be a delta function. Weighting functions can be constructed using known line-by-line absorption data and a program for transmittance calculations. As an example, Fig. 2.10 displays the weighting functions for the High Resolution Infrared Radiation Sounder 2 (HIRS2) temperature channels aboard operational satellites. There are five channels in both the 15 and 4.3 μm CO_2 bands. The peak in the weighting function in each curve, $\bar{p}$, represents the maximum contribution to upwelling radiance. In the stratosphere, the weighting function peaks are much broader than those in the troposphere; this condition makes it difficult to retrieve stratospheric temperature, and makes it even more difficult to retrieve the temperature of the tropopause. In order to resolve the vertical structure of the temperature profile comprehensively, Chahine et al. (1984) have proposed the Advanced Moisture and Temperature Sounder (AMTS) channels, which would yield a spectral resolution substantially higher than that of HIRS2. The weighting functions for AMTS are compared with those of HIRS2 in Fig. 2.10. Clearly, AMTS has appreciably higher vertical resolution throughout the atmosphere than HIRS2, and its weighting functions are particularly sharp in the mid-lower troposphere. It is the goal of future atmospheric infrared sounder observations to provide an overall accuracy of $\sim$1 K with a vertical resolution of 1 km in the temperature retrieval. The search for efficient, physically based retrieval schemes requires continuous research efforts. Observed radiances frequently contain cloud contributions. The classification of radiance data and the removal of cloud contamination in radiance data, allowing temperature retrieval to be carried out successfully, are practical and difficult research areas that call for innovative approaches.

The information content of gaseous concentrations is hinted at in the transmittance term via Eq. (2.3.16). Once the temperature profile has been determined, it is possible, in principle, to infer a specific gaseous concentration profile from a set of radiance observations in the absorption band of that gas. This inference may be accomplished using the basic forward radiative transfer equation, Eq. (2.3.21). For example, the humidity profile may be inferred using radiances observed in the H_2O 6.3 μm vibrational–rotational band (see Subsection 2.2.5.2). However, the retrieval method used to derive gaseous profiles depends heavily on numerical iteration procedures, and the accuracy of retrieval is subject to further improvement.

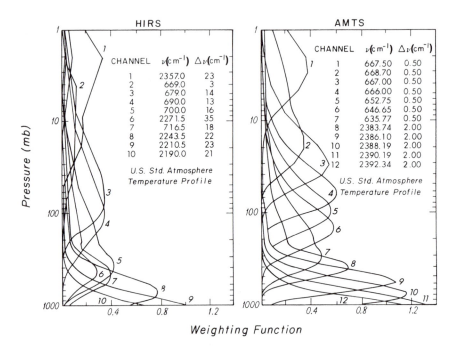

Weighting Function

FIG. 2.10   Weighting functions for HIRS2 aboard operational satellites and for proposed AMTS (after Phillips et al., 1988).

2.4   Band models

Band models are approaches that simplify the computation of the spectral trans-mittance. It is assumed that the atmosphere is homogeneous so that analytical expressions may be developed. We first define spectral absorptance in the form

$$A_{\bar{\nu}}(u) = 1 - T_{\bar{\nu}} = \frac{1}{\Delta \nu} \int_{\Delta \nu} \left(1 - e^{-k_\nu u}\right) d\nu. \qquad (2.4.1)$$

The quantity $A_{\bar{\nu}} \Delta \nu$ is referred to as the equivalent width $W(u)$. It is the width of an infinitely strong line of rectangular shape, which would have been the same as the actual absorption of a single line (Fig. 2.11). The concept of equivalent width plays an important role in the development of band models.

Using a single Lorentz line, introducing two new variables, $x = Su/2\pi\alpha$ and $\tan y/2 = (\nu - \nu_0)/\alpha$, extending the wave-number integration from $-\infty$ to ∞, and performing integration by part, we can derive an analytic expression for the equivalent width in the form

$$W = A_{\bar{\nu}} \Delta \nu = 2\pi\alpha L(x) = 2\pi\alpha x e^{-x} \left[I_0(x) + I_1(x)\right], \qquad (2.4.2)$$

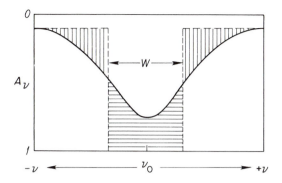

FIG. 2.11   The concept of equivalent width, $W = A_{\bar{\nu}} \Delta \nu$, where $A_{\bar{\nu}}$ represents the spectral absorptance for a spectral interval $\Delta \nu$.

where $L(x)$ is known as the Ladenburg and Reiche function, and the modified Bessel function of the first kind of order n is defined as

$$I_n(x) = i^{-n} J_n(ix), \qquad (2.4.2a)$$

where $i = \sqrt{-1}$ and the integral representation of the Bessel function of order n is given by

$$J_n(x) = \frac{i^{-n}}{\pi} \int_0^{\pi} e^{ix \cos \theta} \cos n\theta \, d\theta. \qquad (2.4.2b)$$

In the case of weak line absorption, either k_ν or u is small, so that $k_\nu u \ll 1$. Spectral absorptance is approximately given by

$$A_{\bar{\nu}} \cong \frac{1}{\Delta \nu} \int_{-\infty}^{\infty} k_\nu u \, d\nu = \frac{Su}{\Delta \nu}, \qquad (2.4.3)$$

by virtue of the definition of line intensity, regardless of the line shape. Absorptance, under the limits of weak-line approximations, is directly proportional to the path length and is called the region of *linear absorption*. On the other hand, if $Su/\pi \alpha \gg 1$, absorptance approaches 1 in the line center region. We may omit the half-width α in the denominator of the Lorentz profile. It follows that

$$A_{\bar{\nu}}(u) = \frac{1}{\Delta \nu} \int_{-\infty}^{\infty} \left[1 - \exp\left(\frac{-S\alpha u}{\pi (\nu - \nu_0)^2} \right) \right] d\nu. \qquad (2.4.4)$$

Letting $y = S\alpha u / \pi (\nu - \nu_0)^2$, we find

$$A_{\bar{\nu}}(u) = \frac{1}{\Delta \nu} \sqrt{S\alpha u / \pi} \int_0^{\infty} (1 - e^{-y}) y^{-3/2} \, dy$$

$$= 2\sqrt{S\alpha u} / \Delta \nu, \qquad (2.4.5)$$

where the integral is equal to $2\Gamma(\frac{1}{2}) = 2\sqrt{\pi}$, with Γ denoting the Γ function. Absorptance is therefore proportional to the square root of the path length and is in the region of the so-called *square root absorption*. Approximations for the weak and strong line limits can also be derived directly from Eq. (2.4.2).

In summary, the equivalent width for a spectral line is

$$W = A_{\bar{\nu}}(u)\,\Delta\nu = \begin{cases} Su, & \text{weak-line approximation,} & (2.4.6) \\ 2\sqrt{S\alpha u}, & \text{strong-line approximation.} & (2.4.7) \end{cases}$$

The weak- and strong-line limits are useful in the development of approximations for infrared radiation transfer.

2.4.1 *Statistical band model*

On inspection of the water vapor rotational band, the only common feature over a $25\,\mathrm{cm}^{-1}$ range is the apparent random line positions (Goody, 1952, 1964a). Hence, the absorption of a band with certain random properties should be considered.

We let $\Delta\nu$ be a spectral interval consisting of n lines of mean distance δ, so that $\Delta\nu = n\delta$. Let $p(S_i)$ be the probability that the ith line has an intensity S_i, and let p be normalized such that

$$\int_0^\infty p(S_i)\,dS_i = 1, \qquad i = 1, \ldots, n. \tag{2.4.8}$$

It is assumed that any line has an equal probability of being anywhere in the interval $\Delta\nu$. The mean transmittance is found by averaging over all line positions and intensities. Hence,

$$\begin{aligned} T_{\bar{\nu}} &= \frac{1}{(\Delta\nu)^n} \int_{\Delta\nu} d\nu_1 \ldots \int_{\Delta\nu} d\nu_n \\ &\times \int_0^\infty p(S_1)e^{-k_1 u}\,dS_1 \ldots \int_0^\infty p(S_n)e^{-k_n u}\,dS_n, \end{aligned} \tag{2.4.9}$$

where k_n denotes the absorption coefficient for the nth line. Since all the integrals are alike, we have

$$\begin{aligned} T_{\bar{\nu}} &= \left[\frac{1}{\Delta\nu} \int d\nu \int_0^\infty p(S)e^{-ku}\,dS \right]^n \\ &= \left[1 - \frac{1}{\Delta\nu} \int d\nu \int_0^\infty p(S)(1 - e^{-ku})\,dS \right]^n. \end{aligned} \tag{2.4.10}$$

The averaged equivalent width for n absorption lines may be defined by

$$\bar{W} = \int_0^\infty p(s) \int_{\Delta\nu} (1 - e^{-ku})\,d\nu\,dS. \tag{2.4.11}$$

Noting that $\Delta\nu = n\delta$, the spectral transmittance may then be written in terms of the averaged equivalent width in the form

$$T_{\bar{\nu}}(u) = \left[1 - \frac{1}{n}\left(\frac{\bar{W}}{\delta}\right)\right]^n. \tag{2.4.12}$$

Since $\lim_{n\to\infty}(1 - x/n)^n \to e^{-x}$, we have

$$T_{\bar{\nu}}(u) = e^{-\bar{W}/\delta}. \tag{2.4.13}$$

Let the lines be of different intensities, and consider a Poisson distribution for the probability of their intensities in the form

$$p(S) = \frac{1}{\bar{S}}e^{-S/\bar{S}}, \tag{2.4.14a}$$

with

$$\bar{S} = \int_0^\infty Sp(S)\,dS, \tag{2.4.14b}$$

where $\bar{S}$ is defined as the mean line intensity. We note that $p(S)$ is normalized to 1, as is required. Inserting this probability function into Eq. (2.4.11) for the averaged equivalent width, and writing $k_\nu = Sf_\nu$, with f_ν the line-shape factor, we find

$$\bar{W} = \int_{\Delta\nu} \frac{\bar{S}f_\nu u}{1 + \bar{S}f_\nu u}\,d\nu. \tag{2.4.15}$$

By using the Lorentz line shape for f_ν and performing the wave-number integration in the domain $(-\infty, \infty)$, without introducing significant errors for the integral confined in the $\Delta\nu$ interval, the spectral transmittance for randomly distributed Lorentz lines is given by

$$T_{\bar{\nu}}(u) = \exp\left[-\frac{\bar{S}}{\delta}u\left(1 + \frac{\bar{S}}{\alpha\pi}u\right)^{-1/2}\right]. \tag{2.4.16}$$

Thus the spectral transmittance for the random model originally proposed by Goody (1952) can be expressed as a function of two parameters, $\bar{S}/\delta$ and $\bar{S}/\alpha\pi$, apart from the path length u. For a given spectral interval $\Delta\nu$, these two parameters may be derived by fitting the random model with line-by-line data. For computations of fluxes in the atmosphere, the device of the spectral interval must ensure that the Planck flux does not vary significantly.

On the basis of our preceding discussion, the averaged equivalent width for the random model is

$$\bar{W} = \bar{S}u\left(1 + \frac{\bar{S}u}{\alpha\pi}\right)^{-1/2}. \tag{2.4.17a}$$

The averaged equivalent width for n individual lines with equivalent widths of W_i is simply

$$\bar{W} = \sum_{i=1}^{n} \frac{W_i}{n}. \qquad (2.4.17b)$$

The spectral transmittance given in Eq. (2.4.16) is general and should be valid under the limits of weak- and strong-line approximations. Thus for weak-line approximations where $\bar{S}u/\alpha\pi \ll 1$, we have

$$\frac{\bar{W}}{\delta} \approx \frac{\bar{S}}{\delta} u. \qquad (2.4.18)$$

However, by using the equivalent width for a spectral line denoted by the subscript i on the limits of weak-line approximations, we have

$$\bar{W} = \frac{1}{n}\sum_{i=1}^{n} W_i(\text{weak}) \approx \frac{1}{n}\sum_{i=1}^{n} S_i u. \qquad (2.4.19)$$

From Eqs. (2.4.18) and (2.4.19), we obtain

$$\frac{\bar{S}}{\delta} = \sum_{i=1}^{n} \frac{S_i}{\Delta\nu} = a_{\bar{\nu}}, \qquad (2.4.20)$$

where $a_{\bar{\nu}}$ is so defined. Following the same procedure under the limits of strong-line approximations, we find

$$\frac{\sqrt{\pi\alpha\bar{S}}}{\delta} = 2\sum_{i=1}^{n} \frac{\sqrt{S_i\alpha_i}}{\Delta\nu} = b_{\bar{\nu}}^*. \qquad (2.4.21)$$

If we define $b_{\bar{\nu}} = \pi\alpha/\delta$, then $b_{\bar{\nu}}^{*2} = a_{\bar{\nu}}b_{\bar{\nu}}$. It follows that

$$b_{\bar{\nu}} = \left(2\sum_{i=1}^{n} \sqrt{S_i\alpha_i}\right)^2 \bigg/ \left(\Delta\nu\sum_{i=1}^{n} S_i\right). \qquad (2.4.22)$$

Thus the spectral transmittance may be written in the form

$$T_{\bar{\nu}}(u) = \exp\left[-a_{\bar{\nu}}u(1 + ua_{\bar{\nu}}/b_{\bar{\nu}})^{-1/2}\right]. \qquad (2.4.23)$$

The band parameters $a_{\bar{\nu}}$ and $b_{\bar{\nu}}$ are functions of two variables: $\sum_i S_i$ and $\sum_i \sqrt{S_i\alpha_i}$. These variables can be computed from the AFGL line-by-line data described in Rothman et al. (1987) for the CO_2, O_3, H_2O, CH_4, and N_2O bands. Half-width and line strength are both functions of temperature, as shown in Eqs. (2.2.6) and (2.2.22b). Thus, temperature variations must be included in the calculations.

We may use a narrow band of 5 cm^{-1} in producing these two variables for applications to band models. Wider-band variables may be obtained by linearly averaging the narrow band values.

The spectral transmittance for the random model depends on two parameters, $a_{\bar{\nu}}$ and $a_{\bar{\nu}}/b_{\bar{\nu}}$. Examination of these parameters reveals that a further simplification may be obtained by defining a parameter referred to as the *generalized absorption coefficient*, $\ell_{\bar{\nu}}$, such that $a_{\bar{\nu}} = a\ell_{\bar{\nu}}$ and $a_{\bar{\nu}}/b_{\bar{\nu}} = b\ell_{\bar{\nu}}$, where the coefficients a and b are certain constants. Thus a one-parameter representation of the spectral transmittance for the random model may be expressed by

$$T_{\bar{\nu}}(u) = \exp\left[-a\ell_{\bar{\nu}}u(1 + b\ell_{\bar{\nu}}u)^{-1/2}\right]. \qquad (2.4.24)$$

The generalized absorption coefficient was originally presented by Elsasser (1942). Under the limits of strong- and weak-line approximations, we have

$$T_{\bar{\nu}}(u) \cong \begin{cases} \exp(-a\ell_{\bar{\nu}}u), & \text{weak line} \\ \exp(-c\ell_{\bar{\nu}}^*u^*), & \text{strong line,} \end{cases} \qquad (2.4.25)$$

where $c = a/\sqrt{b}$, $\ell_{\bar{\nu}}^* = \sqrt{\ell_{\bar{\nu}}}$, and $u^* = \sqrt{u}$. Transmittance is now expressed by a simple exponential function in terms of the generalized absorption coefficient, which can be obtained from the band parameters using statistical fitting procedures.

2.4.2 *Modified statistical band model*

The random model described above utilizes the probability distribution function given in Eq. (2.4.14a). In many cases, it has been found that this exponential intensity distribution substantially underestimates the number of low intensity lines (Malkmus, 1967). The line intensity is governed by the Boltzmann factor in the form $S \sim \exp(-hcE/KT)$, viz. Eq. (2.2.18), where E is the lower energy level. Thus, dE/dS is proportional to S^{-1}. The probability distribution function $p(S)$ must be proportional to dn/dS, where n is the number density of energy levels. However, dn/dE is approximately constant since, in many cases, the energy levels are approximately equally spaced. It follows that $p(S) \sim (dE/dS)(dn/dE) \sim dE/dS \sim S^{-1}$. Thus the S^{-1} dependence remains a dominating influence on the probability distribution function. For this reason, it is necessary to use a normalized probability distribution function for line intensity in the form

$$p(S) = \frac{1}{\ell n r}\frac{1}{S}\left(e^{-S/S_m} - e^{-Sr/S_m}\right), \qquad (2.4.26)$$

where S_m is the maximum value of S and r is the ratio of the maximum and minimum values of S. In Fig. 2.12, a histogram of the line intensities for the 15 μm CO_2 band (450–900 cm^{-1}) is shown. Line intensities within $\sim$20% of the mean line intensity are grouped together. Each bar in the figure represents the number of

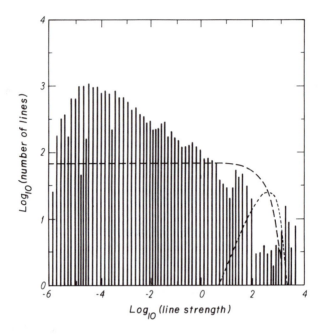

FIG. 2.12   A histogram (vertical bars) of all CO_2 lines at wave numbers between 450 and 900 cm^{-1}. Each bar represents the number of lines in a given line group. Lines with strengths within 20% of the mean strength of a given group are gathered into that group. Analytic line-strength distributions obtained with the Goody (dotted line) and Malkmus (long-dashed line,) models are also shown (after Crisp et al., 1986, with modifications).

lines in each group (Crisp et al., 1986). The Poisson distribution function for line intensity is denoted by dots, while the Malkmus distribution function is denoted by the dashed line. It is apparent that the Poisson representation would significantly underestimate lines with lower intensities.

The probability distribution function defined in Eq. (2.4.26) is normalized to 1 by noting that

$$\int_0^\infty \frac{1}{S} \left(e^{-aS} - e^{-bS} \right) dS = \ell n(b/a). \tag{2.4.27}$$

The mean line intensity is

$$\bar{S} = \int_0^\infty S p(S) \, dS = \frac{r-1}{r \, \ell n r} S_m. \tag{2.4.28}$$

After inserting Eq. (2.4.26) into Eq. (2.4.11), the averaged equivalent width defined in Eq. (2.4.11) is

$$\bar{W} = \frac{1}{\ell n r} \int_{\Delta \nu} \ell n \left(\frac{1 + f_\nu u S_m}{1 + f_\nu u S_m / r} \right) d\nu. \tag{2.4.29}$$

Using the Lorentz line shape and noting that

$$\int \ell n(\nu^2 + a^2)\, d\nu = \nu \ell n(\nu^2 + a^2) - 2\nu + 2a \tan^{-1} \frac{\nu}{a}, \qquad (2.4.30)$$

an analytical solution for Eq. (2.4.29) may be derived and is given by

$$\bar{W} = \frac{2\pi\alpha}{\ell nr}\left[\left(1 + \frac{u\bar{S}}{\pi\alpha}\frac{r\ell nr}{r-1}\right)^{1/2} - \left(1 + \frac{u\bar{S}}{\pi\alpha}\frac{\ell nr}{r-1}\right)^{1/2}\right]. \qquad (2.4.31)$$

In the limits of weak-line approximations, $u\bar{S}$ is small, so that by expanding the radicals, we get $\bar{W} = u\bar{S}$. This is the same as the term in Eq. (2.4.6). In the limits of strong-line approximations, $u\bar{S}$ is large, and we have

$$\bar{W}(\text{strong}) = 2\sqrt{\pi\alpha u\bar{S}}\,\chi(r), \qquad (2.4.32)$$

where

$$\chi(r) = \frac{1}{\sqrt{\ell nr}}\frac{\sqrt{r}-1}{\sqrt{r-1}}. \qquad (2.4.33)$$

The strong line approximation, however, differs from that in Eq. (2.4.7) by a factor $\sqrt{\pi}\chi(r)$. We may define an equivalent S_e and δ_e in the form

$$S_e = \bar{S}/\pi\chi^2(r), \qquad \delta_e = \delta/\pi\chi^2(r), \qquad (2.4.34)$$

so that the optical depth

$$\frac{\bar{W}}{\delta}(\text{weak}) \cong \frac{\bar{S}}{\delta}u = \frac{S_e}{\delta_e}u, \qquad (2.4.35)$$

and

$$\frac{\bar{W}}{\delta}(\text{strong}) \cong \frac{2}{\delta}\sqrt{\pi\alpha u\bar{S}}\,\chi(r) = \frac{2}{\delta_e}\sqrt{\alpha u S_e}. \qquad (2.4.36)$$

The definitions of S_e and δ_e allow us to express the equivalent width given in Eq. (2.4.31) as follows:

$$\frac{\bar{W}}{\delta} = \frac{2\alpha}{\delta_e}\left(\frac{\sqrt{r}+1}{\sqrt{r}-1}\right)\left[\left(1 + \frac{S_e u}{\alpha(1+1/\sqrt{r})^2}\right)^{1/2} - \left(1 + \frac{S_e u}{\alpha(1+\sqrt{r})^2}\right)^{1/2}\right].$$

$$(2.4.37)$$

The averaged equivalent width, $\bar{W}$, should be insensitive to large r. In the limit $r \to \infty$, we obtain

$$\frac{\bar{W}}{\delta} = \frac{2\alpha}{\delta_e}\left[\left(1 + \frac{S_e u}{\alpha}\right)^{1/2} - 1\right]. \qquad (2.4.38)$$

Again, spectral transmittance depends on two parameters, $c_{\bar{\nu}} = 2\alpha/\delta_e$ and $d_{\bar{\nu}} = S_e/\alpha$. In order to obtain these parameters from basic line parameters, we follow the same procedures as those developed for the Goody random model. Utilizing the limits of weak- and strong-line approximations, we find

$$\frac{S_e}{\delta_e} = \sum_i \frac{S_i}{\Delta \nu},$$

$$\frac{\sqrt{\alpha S_e}}{\delta_e} = \sum_i \frac{\sqrt{\alpha_i S_i}}{\Delta \nu}. \tag{2.4.39}$$

Thus,

$$c_{\bar{\nu}} = \frac{2\alpha}{\delta_e} = 2 \left(\sum_i \sqrt{\alpha_i S_i} \right)^2 \Big/ \left(\Delta \nu \sum_i S_i \right),$$

$$d_{\bar{\nu}} = \frac{S_e}{\alpha} = \left(\sum_i S_i \right)^2 \Big/ \left(\sum_i \sqrt{\alpha_i S_i} \right)^2. \tag{2.4.40}$$

Spectral transmittance may then be expressed by

$$T_{\bar{\nu}}(u) = \exp \left\{ -c_{\bar{\nu}}[(1 + d_{\bar{\nu}} u)^{1/2} - 1] \right\}. \tag{2.4.41}$$

This is referred to as the *Malkmus model*. The model has also been expressed in terms of the mean line intensity $\bar{S}$ and mean line spacing δ. By inserting S_e and δ_e, defined in Eqs. (2.4.34) and (2.4.33), into Eq. (2.4.38) for the equivalent width, we obtain

$$T_{\bar{\nu}}(u) = \exp \left\{ -\frac{2\pi\alpha}{\delta} \chi^2 \left[\left(1 + \frac{\bar{S} u}{\pi\alpha} \frac{1}{\chi^2} \right)^{1/2} - 1 \right] \right\}. \tag{2.4.42}$$

As defined in Eq. (2.4.33), χ depends on the ratio of the maximum and minimum values of S. If we match the results in the limits of strong- and weak-line approximations with those from Goody's statistical model, we find $\chi = \frac{1}{2}$. In this case, $\pi\alpha/2\delta = c_{\bar{\nu}}$ and $4\bar{S}/\pi\alpha = d_{\bar{\nu}}$, as defined in Eq. (2.4.40).

2.4.3   Statistical band model for simplified Voigt lines

As pointed out in Subsection 2.2.3.3, there is rather extensive literature on the approximate methods for evaluating the Voigt function. However, in order to incorporate the Voigt line shape in band models, the shape factor must be simplified so that analytical expressions may be derived. The shape of a Voigt line may be parameterized as a rectangular Doppler core with Lorentzian wings, as follows (Fels, 1979):

$$f_v(\nu) = \begin{cases} C, & |\nu| \leq \nu_0 \\ \alpha/\pi\nu^2, & |\nu| > \nu_0, \end{cases} \tag{2.4.43}$$

where ν is the distance from the line center, and the width may be assumed to be

$$\nu_0 = \frac{2}{\pi}(1 + \zeta)\alpha + \beta\alpha_D. \tag{2.4.44}$$

The constants ζ and β can be chosen to give the proper equivalent widths in the Doppler and Lorentz limits. For the 15 μm CO_2 and 9.6 μm O_3 bands, $\beta = 1.25$ and $\zeta = 1.0$ give reasonable results. When $\alpha = 0$, the Doppler line is rectangular with a half-width $\beta\alpha_D$. Normalization of the shape factor gives the height of the rectangular line core as follows:

$$C = \frac{1}{2\nu_0} - \frac{\alpha}{\pi}\frac{1}{\nu_0^2}. \tag{2.4.45}$$

For Goody's statistical model, the optical depth is [Eq. (2.4.15)]

$$\frac{\bar{W}}{\delta} = \frac{1}{\delta}\int_{-\infty}^{\infty} \frac{u\bar{S}f_v(\nu)}{1 + u\bar{S}f_v(\nu)}\, d\nu. \tag{2.4.46}$$

Inserting the simplified Voigt profile, we obtain

$$\begin{aligned} \frac{\bar{W}}{\delta} &= \frac{2}{\pi}(ua_{\bar{\nu}}b_{\bar{\nu}})^{1/2}\tan^{-1}\left(\frac{\delta}{\pi\nu_0}(ua_{\bar{\nu}}b_{\bar{\nu}})^{1/2}\right) \\ &+ \frac{2\nu_0 C ua_{\bar{\nu}}}{1 + \nu_0 C ua_{\bar{\nu}}(\delta/\nu_0)}, \end{aligned} \tag{2.4.47}$$

where the band parameters $a_{\bar{\nu}}$ and $b_{\bar{\nu}}$ have been defined in Eqs. (2.4.20) and (2.4.22). These parameters and the mean line spacing δ can be computed from line data.

For the Malkmus statistical model, it suffices to use a probability distribution function for line intensity in the form

$$p(S) \sim \frac{1}{S}e^{-S/\bar{S}} \tag{2.4.48}$$

to derive an analytical expression for transmittance. With this function, the equivalent optical depth may be written

$$\frac{\bar{W}}{\delta} = \frac{1}{\delta}\int_{-\infty}^{\infty} \ell n\left[1 + u\bar{S}f_v(\nu)\right]\, d\nu. \tag{2.4.49}$$

Substituting the Voigt profile given in Eq. (2.4.43) leads to

$$\frac{\bar{W}}{\delta} = \frac{1}{2}\left(\frac{\nu_0}{\delta}\right)\ell n\left[\frac{1 + 4ua_{\bar{\nu}}\nu_0 C(\delta/\nu_0)}{1 + (4/\pi^2)ua_{\bar{\nu}}b_{\bar{\nu}}(\delta/\nu_0)^2}\right]$$
$$+ \frac{2}{\pi}(ua_{\bar{\nu}}b_{\bar{\nu}})^{1/2}\tan^{-1}\left(\frac{2\delta}{\pi\nu_0}(ua_{\bar{\nu}}b_{\bar{\nu}})^{1/2}\right). \qquad (2.4.50)$$

Again, $a_{\bar{\nu}} = \bar{S}/\delta$ and $b_{\bar{\nu}} = \pi\alpha/\delta$, have been defined in Eqs. (2.4.20) and (2.4.22). Both the Goody and Malkmus models, along with the simplified Voigt profile, produce heating rate errors within $\sim$0.03 K day^{-1} for the O_3 9.6 μm band when the two-parameter approximation (Subsection 2.5.2) is used.

2.5   Physical adjustment of spectral transmittance for nonhomogeneous atmospheres

In the analysis of theoretical band models, we have assumed that the absorption coefficient is independent of temperature and pressure. This allows us to develop analytic forms for transmittance in terms of simple exponential functions. However, as is evident from Section 2.2, the absorption coefficient is a strong function of pressure and temperature. These effects must be accounted for when applications to realistic atmospheres are required. Before we present numerous ways in which pressure and temperature effects may be incorporated in band models, let us review the definition of the optical depth, given by

$$\tau = \int_u k_\nu(p, T)\,du. \qquad (2.5.1)$$

The primary objective of approximate solutions for nonhomogeneous path lengths is to reduce the transfer problem to that of a homogeneous path with some sort of averaged values, $\tilde{u}$, $\tilde{T}$, and $\tilde{p}$, so that the optical depth may be accurately computed.

2.5.1   Scaling approximation

It has been common practice to assume that $\tilde{T}$ and $\tilde{p}$ can be assigned fixed standard values and that the transfer problem consists solely of determining the optimum representation for $\tilde{u}$. Consequently, transmittance may be treated as a function of a single variable (i.e., $\tilde{u}$), for a nonhomogeneous path. The variable $\tilde{u}$ represents the scaled amount of the absorber, and the approximation for the transmittance calculation is referred to as a *one-parameter scaling approximation*.

In the troposphere and lower stratosphere, the Lorentz line shape may be used for absorption lines. In reference to Section 2.3, the absorption coefficient at wave number ν, as a function of pressure and temperature, may be written in the form

$$k_\nu(p, T) = \sum_i \frac{S_i(T)}{\pi}\frac{\alpha_i(p, T)}{(\nu - \nu_{0i})^2 + \alpha_i^2(p, T)}, \qquad (2.5.2)$$

where the summation is over all absorption lines. The half-width is defined in Eq. (2.2.6). At the line center, $\nu = \nu_0$, $k_\nu \sim 1/\alpha_i(p, T) \sim 1/p$. However, in the far wing, $|\nu - \nu_{0i}| \gg \alpha$, so that $k_\nu \sim \alpha_i(p, T) \sim p$. Under normal atmospheric conditions, the half-widths of absorption lines are small compared to the mean line spacing. Thus absorption in the wings predominates. Moreover, at atmospheric levels where water vapor cooling is important, absorption near the centers of most water vapor lines becomes saturated. For this reason, errors made in the absorption coefficient in the vicinity of line centers will not significantly affect atmospheric transmittance. On the basis of these arguments, it is appropriate to approximate the pressure and temperature dependence of the absorption coefficient by using its behavior in the line wings. We wish to separate the absorption coefficient $k_\nu(p_r, T_r)$ for a reference pressure p_r and temperature T_r, from the pressure and temperature dependence. Based on Eq. (2.5.2), we write

$$k_\nu(p_r, T_r) = \sum_i \frac{S_i(T_r)}{\pi} \frac{\alpha_i(p_r, T_r)}{(\nu - \nu_{0i})^2 + \alpha_i^2(p_r, T_r)}. \qquad (2.5.3)$$

Combining Eqs. (2.5.2) and (2.5.3) and utilizing the expression for the half-width, we find

$$k_\nu(p, T) = k_\nu(p_r, T_r) \left(\frac{p}{p_r}\right) R_\nu(T, T_r, p_r), \qquad (2.5.4)$$

where

$$R_\nu(T, T_r, p_r) = \left(\frac{T_r}{T}\right)^{1/2} \sum_i \frac{S_i(T)\alpha_i(p_r, T_r)}{(\nu - \nu_{0i})^2} \bigg/ \sum_i \frac{S_i(T_r)\alpha_i(p_r, T_r)}{(\nu - \nu_{0i})^2}. \qquad (2.5.5)$$

In an inspection of R_ν for various temperatures in the rotational band of water vapor, Chou and Arking (1980) have noted that R_ν is less sensitive to temperature in the band center than in the wing region, and that its variation with wave number is much smaller when compared with the variation in the absorption coefficient (Fig. 2.13). As a good approximation, we may use a mean value, $R_{\bar\nu}$, over a spectral interval to replace R_ν in the form

$$R_{\bar\nu}(T, T_r, p_r) = \frac{1}{\Delta\nu} \int_{\Delta\nu} R_\nu(T, T_r, p_r)\, d\nu. \qquad (2.5.6)$$

A separation of the pressure and temperature dependence on the absorption coefficient is now possible, leading to

$$k_\nu(p, T) \approx k_\nu(p_r, T_r) \left(\frac{p}{p_r}\right) R_{\bar\nu}(T, T_r, p_r). \qquad (2.5.7)$$

The aforementioned analyses offer a theoretical foundation and numerical approximation for the scaling approximation. The scaled path length from Eq. (2.5.1) is then

$$\tilde{u} = \int_u \left(\frac{p}{p_r}\right) R_{\bar\nu}(T, T_r, p_r)\, du, \qquad (2.5.8)$$

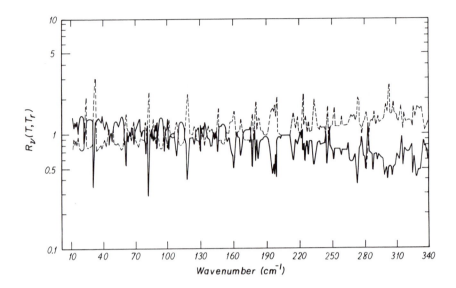

FIG. 2.13   Effect of temperature on the absorption coefficient $R_\nu(T, T_r)$ in the band-center region with $T_r = 225$ K. The dashed curve is for $T = T_r + 40$ K, and the solid curve is for $T = T_r - 40$ K (after Chou and Arking, 1980).

so that

$$\tau = k_\nu(p_r, T_r)\tilde{u}. \qquad (2.5.9)$$

The only approximation required in the scaling is the averaging procedure over the wave number to obtain $R_{\bar{\nu}}$, which depends on the reference pressure and temperature. From Eq. (2.5.5), $R_{\bar{\nu}} = 1$, when $p = p_r$ and $T = T_r$. Although p_r and T_r could be fixed, the best approximation would be a set of p_r and T_r for the spectral regions that make a significant contribution to cooling. $R_{\bar{\nu}}$ is close to 1 for the band center and ranges from about $\frac{1}{2}$ to 2 for the band wings.

For many atmospheric applications involving the computation of cooling rates due to water vapor, it has been suggested that

$$R_{\bar{\nu}}(T, T_r) \cong \left(\frac{T_r}{T}\right)^{1/2}. \qquad (2.5.10)$$

In this case, the scaled path length is

$$\tilde{u} = \int_u \left(\frac{p}{p_r}\right) \left(\frac{T_r}{T}\right)^{1/2} du. \qquad (2.5.11)$$

This scaling approximation for flux and cooling-rate computations was originally proposed by Elsasser and Culbertson (1960). A more general form for the scaled path length may be written

$$\tilde{u} = \int_u \left(\frac{p}{p_r}\right)^n \left(\frac{T_r}{T}\right)^m du, \qquad (2.5.11a)$$

where the indices n and m depend on the absorbing gases and spectral regions.

2.5.2 Two-parameter approximation

In view of the optical depth defined in Eq. (2.5.1), we wish to find an adjusted absorption coefficient such that

$$\tau = \int_u k_\nu(p, T) \, du = k_\nu(\tilde{p}, \tilde{T}) u. \qquad (2.5.12)$$

Since the absorption coefficient is determined by line intensity and line shape, and if the Lorentz line profile is used, we may write

$$k_\nu(\tilde{p}, \tilde{T}) = \sum_i \tilde{S}_i \tilde{f}_{\nu i} = \sum_i \frac{\tilde{S}_i}{\pi} \frac{\tilde{\alpha}_i}{(\nu - \nu_{0i})^2 + \tilde{\alpha}_i^2}. \qquad (2.5.13)$$

Two adjusted parameters, $\tilde{S}$ and $\tilde{\alpha}$, are needed to satisfy Eq. (2.5.12) for the optical depth. This is referred to as the *two-parameter approximation*.

In reference to Eq. (2.4.1), the average equivalent width for a nonhomogeneous path is

$$\bar{W} = A_{\bar{\nu}}(u) \, \Delta\nu = \int_{\Delta\nu} \left[1 - \exp\left(-\int_u \sum_i S_i f_{\nu i} \, du \right) \right] d\nu. \qquad (2.5.14)$$

In the weak-line limit, we have

$$\bar{W} \cong \int_{\Delta\nu} \int_u \sum_i S_i f_{\nu i} \, du \, d\nu = \int_u \sum_i S_i \, du, \qquad (2.5.15)$$

regardless of the line shape. Moreover, using the Lorentz profile, the average equivalent width given in Eq. (2.5.14) may be written in the form

$$\bar{W} = \int_{\Delta\nu} \left[1 - \exp\left(-\int_u \sum_i \frac{S_i}{\pi} \frac{\alpha_i}{(\nu - \nu_{0i})^2 + \alpha_i^2} \, du \right) \right] d\nu. \qquad (2.5.16)$$

In the strong-line limit, the absorption near the line-center region is nearly saturated. The average equivalent width in this case is not very sensitive to the absorption coefficient. Hence, we may select any half-width, say α', to replace α in the denominator of Eq. (2.5.16). To find the α' that also satisfies the weak-line limit, we must have

$$\bar{W} \cong \int_{\Delta\nu} \int_u \sum_i \frac{S_i}{\pi} \frac{\alpha_i}{(\nu - \nu_{0i})^2 + \tilde{\alpha}_i^2} \, du \, d\nu$$

$$\cong \int_u \sum_i \frac{S_i}{\pi} \int_{-\infty}^{\infty} \frac{\alpha_i}{(\nu - \nu_{0i})^2 + \tilde{\alpha}_i^2} \, d\nu \, du$$

$$= \int_u \sum_i \frac{\alpha_i}{\tilde{\alpha}_i} S_i \, du, \tag{2.5.17}$$

where $\tilde{\alpha}$ is the specific α' that fulfills the requirements of both strong- and weak-line limits. Since $\alpha_i/\tilde{\alpha}_i \cong p/\tilde{p}$, combining Eqs. (2.5.17) and (2.5.15) gives

$$\tilde{p} = \int_u p \sum_i S_i \, du \bigg/ \int_u \sum_i S_i \, du$$

$$= \int_u p\bar{S} \, du \bigg/ \int_u \bar{S} \, du, \tag{2.5.18}$$

where $\bar{S}$ is the mean line intensity. The effect of pressure on absorption is through the line half-width. The preceding portion of scaling is to adjust the pressure variation along the path.

Furthermore, from the equivalent width given in Eqs. (2.5.14) and (2.5.15), we may select an adjusted $\tilde{S}_i$ such that

$$\int_u \sum_i S_i \, du = u \sum_i \tilde{S}_i. \tag{2.5.19}$$

This defines the scaled mean line intensity, regardless of the line shape, in the form

$$\tilde{\bar{S}} = \int \bar{S} \, du/u. \tag{2.5.20a}$$

The effect of temperature on absorption is through the line strength. This part of scaling is to account for the nonisothermal path. Alternatively, we may select a reference $\bar{S}$, say $\bar{S}_r$, and adjust the path length u such that

$$\tilde{u} = \int_u \frac{\bar{S}}{\bar{S}_r} \, du. \tag{2.5.20b}$$

Equations (2.5.18) and (2.5.20) constitute the so-called *Curtis–Godson (CG) approximation* for nonhomogeneous atmospheres independently proposed by Curtis (1952) and Godson (1953). Extensive cooling-rate computations were performed by Walshaw and Rodgers (1963) to check this approximation. Errors introduced by the CG approximation are less than a few percent for the H_2O rotational and $15\,\mu m$ CO_2 bands. This accuracy is determined by comparison with the results computed from a line-by-line integration. The CG approximation, however, is less satisfactory for the $9.6\,\mu m$ O_3 band.

The CG relations can be deduced from a Taylor series expansion of the line shape. Let us use pressure as the vertical coordinate. Based on hydrostatic equilibrium, the path length and pressure are related via $du = -q(p)\,dp/g$, where q is the mixing ratio and g the gravitational acceleration. Thus the optical depth may be expressed in the p coordinate as follows:

$$\tau = \int_p \sum_i S_i(p) f_{\nu i}(p) q(p) \frac{dp}{g}, \tag{2.5.21}$$

where the temperature dependence is suppressed for simplicity of presentation. We expand the line shape in a Taylor series with respect to a scaled pressure $\tilde{p}$, which is within the integration limits for pressure in the form

$$f_{\nu i}(p) = f_{\nu i}(\tilde{p}) + \left(\frac{df_{\nu i}}{dp}\right)_{\tilde{p}} (p - \tilde{p}) + \frac{1}{2}\left(\frac{d^2 f_{\nu i}}{dp^2}\right)_{\tilde{p}} (p - \tilde{p})^2 + \dots . \tag{2.5.22}$$

By substituting this expression in Eq. (2.5.21), we obtain

$$\begin{aligned}
\tau = &\sum_i f_{\nu i}(\tilde{p}) \int_p S_i(p) q(p) \frac{dp}{g} \\
&+ \sum_i \left(\frac{df_{\nu i}}{dp}\right)_{\tilde{p}} \int_p S_i(p)(p - \tilde{p}) q(p) \frac{dp}{g} \\
&+ \frac{1}{2}\sum_i \left(\frac{d^2 f_{\nu i}}{dp^2}\right)_{\tilde{p}} \int_p S_i(p)(p - \tilde{p})^2 q(p) \frac{dp}{g} + \dots .
\end{aligned} \tag{2.5.23}$$

The second and higher derivative terms should be small. This is especially evident for the far wings, where $|\nu - \nu_0| \gg \alpha$, so that the line shape is linearly dependent on the half-width and, hence, on pressure. Thus, in order to minimize the errors in computations of the optical depth, we may select a $\tilde{p}$ so that the first-derivative term vanishes:

$$\int_p S_i(p)(p - \tilde{p}) q(p) \frac{dp}{g} = 0 \tag{2.5.24}$$

for each line. Summations can be performed over all lines to obtain the scaled pressure in terms of the mean line intensity in the form

$$\tilde{p} = \int_p \bar{S}(p) p q(p)\,dp \bigg/ \int_p \bar{S}(p) q(p)\,dp . \tag{2.5.25}$$

Moreover, we may select a scaled mean line intensity such that

$$\tilde{S} = \int_p \bar{S}(p)q(p)\,dp \bigg/ \int_p q(p)\,dp. \tag{2.5.26}$$

Using the definition of the path length, Eqs. (2.5.25) and (2.5.26) are exactly the same as Eqs. (2.5.18) and (2.5.20).

Subject to the condition that the first-derivative term is set to zero, the optical depth may be written

$$\tau = \tau_{CG} + \frac{1}{2}\sum_i \left(\frac{d^2 f_{\nu i}}{dp^2}\right)_{\tilde{p}} \int_u S_i(p)(p - \tilde{p})^2\,du + \dots, \tag{2.5.27}$$

where

$$\tau_{CG} = \left(\sum_i \tilde{S}_i f_{\nu i}(\tilde{p})\right)\Delta u = \left(\sum_i k_{\nu i}(\tilde{p}, \tilde{S})\right)\Delta u. \tag{2.5.28}$$

The CG approximation is a first-order adjustment in which the first term in the Taylor series is minimized regardless of the line shape. This accounts for the fact that it works well for both Lorentz and Voigt profiles. The interpretation of the CG approximation by means of the Taylor series expansion has been discussed by Armstrong (1968).

As pointed out by Goody (1964b), the Curtis–Godson technique for a non-homogeneous atmosphere was first discussed by van de Hulst (1945), who also demonstrated that the Voigt profile may be readily incorporated in this technique. Consider the cosine transformation of the optical depth in the form

$$\tilde{g}(t) = \int_{-\infty}^{\infty} \tau(\nu)\cos\nu t\,d\nu = \int_u g(t)\,du, \tag{2.5.29}$$

where

$$\tau(\nu) = \frac{1}{\pi}\int_0^{\infty} \tilde{g}(t)\cos\nu t\,dt, \tag{2.5.30a}$$

$$g(t) = \int_{-\infty}^{\infty} k_\nu \cos\nu t\,d\nu. \tag{2.5.30b}$$

On substituting the expressions of the absorption coefficient for Lorentz, Doppler, and Voigt line shapes, we find

$$g(t) = \begin{cases} S\exp(-\alpha t), & \text{Lorentz} \\ S\exp(-\alpha_D^2 t^2/4), & \text{Doppler} \\ S\exp[-(\alpha t + \alpha_D^2 t^2/4)], & \text{Voigt.} \end{cases} \tag{2.5.31}$$

In order to derive the expression for the Voigt profile, we have set $\cos \nu t = \cos[(\nu - \nu') + \nu']t$. By using the Voigt profile and defining the scaled path length $\tilde{u}$ and the scaled line widths, $\tilde{\alpha}$ and $\tilde{\alpha}_D$, we obtain

$$
\begin{aligned}
\tilde{g}(t) &= \int_u S \exp\left[-\left(\alpha t + \frac{\alpha_D^2 t^2}{4}\right)\right] du \\
&\equiv S_r \tilde{u} \exp\left[-\left(\frac{\tilde{\alpha} t + \tilde{\alpha}_D^2 t^2}{4}\right)\right].
\end{aligned}
\tag{2.5.32}
$$

If we take the first two terms in the expansion, we have

$$
\begin{aligned}
&\int_u S\left[1 - \left(\alpha t + \frac{\alpha_D^2 t^2}{4}\right) + \dots\right] du \\
&\equiv S_r \tilde{u}\left[1 - \left(\frac{\tilde{\alpha} t + \tilde{\alpha}_D^2 t^2}{4}\right) + \dots\right].
\end{aligned}
\tag{2.5.33}
$$

Matching the powers of t in the expansion yields

$$
\tilde{u} = \int_u \left(\frac{S}{S_r}\right) du,
\tag{2.5.34}
$$

$$
\tilde{\alpha} = \frac{1}{\tilde{u}} \int_u \alpha \left(\frac{S}{S_r}\right) du,
\tag{2.5.35}
$$

$$
\tilde{\alpha}_D^2 = \frac{1}{\tilde{u}} \int_u \alpha_D^2 \left(\frac{S}{S_r}\right) du.
\tag{2.5.36}
$$

The Doppler width α_D is a function of temperature only. Thus, if the temperature does not vary greatly along the nonhomogeneous path, we may set $\tilde{\alpha}_D = \alpha_D$. Equations (2.5.34) and (2.5.35) represent the Curtis–Godson relations. In principle, the order of the approximation may be increased by considering high order expansions [see also Eq. (2.5.27)]. However, it is yet to be demonstrated that applications of the higher-order approximations to nonhomogeneous atmospheres can be successfully accounted for in band models and line-by-line calculations.

2.5.3   Three-parameter approximation

The equivalent width for a nonhomogeneous path may be written in the form

$$
W = A_{\bar{\nu}}(u) \Delta \nu = \int_{-\infty}^{\infty} \left[1 - \exp\left(-\int_u k_\nu \, du\right)\right] d\nu.
\tag{2.5.37}
$$

For hydrostatic equilibrium, $du = -q dp/g$. As a good approximation, the half-width for a Lorentz line, $\alpha \cong \alpha_2 p/p_2$, where α_2 is the half-width at the pressure p_2. For a single Lorentz line, we have

$$
W = \int_{-\infty}^{\infty} \left[1 - \exp\left(-\int_{\alpha_1}^{\alpha_2} \frac{Sqp_2}{\pi g \alpha_2} \frac{\alpha}{\alpha^2 + \nu^2} d\alpha\right)\right] d\nu,
\tag{2.5.38}
$$

where $\alpha_2(p_2) > \alpha_1(p_1)$. We define a dimensionless parameter as follows:

$$\lambda = \frac{Sqp_2}{2\pi g\alpha_2}. \tag{2.5.39}$$

Under the condition that λ is independent of α, that is, that an isothermal atmosphere with an absorber of constant mixing ratio exists, an integration over α can be performed so that

$$W = 2\int_0^\infty \left[1 - \left(\frac{\nu^2 + \alpha_1^2}{\nu^2 + \alpha_2^2}\right)^\lambda\right] d\nu. \tag{2.5.40}$$

By introducing the variable transformations $\nu = \alpha_2 \tan(\phi/2)$ and $\cos\phi = 2x - 1$, an integration over the wave number can be carried out analytically with the result (Yamamoto and Aida, 1970):

$$W = 2\pi\lambda\alpha_2 \left[F(-\lambda, \frac{1}{2}, 1, z) - (p_1/p_2)^2 F(1 - \lambda, \frac{1}{2}, 1, z)\right], \tag{2.5.41}$$

where $z = 1 - (p_1/p_2)^2$ and the hypergeometric function is defined by

$$F(\alpha, \beta, \gamma, z) = \frac{\Gamma(\gamma)}{\Gamma(\beta)\Gamma(\gamma - \beta)} \int_0^1 x^{\beta-1}(1 - x)^{\gamma-\beta-1}$$
$$\times (1 - zx)^{-\alpha} dx. \tag{2.5.42}$$

The equivalent width of a single Lorentz line along a nonhomogeneous path may be expressed in terms of a known mathematical function. This expression is useful for the development of a three-parameter approximation for nonhomogeneous atmospheres.

As described in Subsection 2.5.2, the CG two-parameter approximation for a nonhomogeneous path length gives the exact solution in the strong- and weak-line limits of absorption. For the intermediate range of path lengths, this approximation may overestimate absorption due to a larger scaled half-width. The absorption of a single line for a nonhomogeneous path length would better match that for a homogeneous path length if a parameter were introduced to reduce the scaled half-width. It is appropriate, then, to redefine the scaled half-width and line strength in the forms

$$\widetilde{\alpha}^\epsilon = \int_u S\alpha^\epsilon \, du \bigg/ \int_u S \, du, \tag{2.5.43}$$

$$\widetilde{S} = \int_u S \, du \bigg/ \int_u du = \widetilde{S}_{\text{CG}}. \tag{2.5.44}$$

The introduction of a third parameter, $\epsilon \leq 1$, allows us to readjust the scaled half-width. In the limits of weak-line approximations ($\epsilon = 0$), absorption is independent of the line width. In the limits of strong-line approximations ($\epsilon = 1$), the scaled half-width reduces to the CG approximation as follows:

$$\tilde{\alpha}_{CG} = \int_u S\alpha \, du \bigg/ \int_u S \, du. \tag{2.5.45}$$

In order to obtain ϵ, we consider an isothermal atmosphere consisting of an absorber with a constant mixing ratio. In this case, the equivalent width, W_{exact}, of a single Lorentz line for a nonhomogeneous path length can be exactly expressed in terms of the hypergeometric function given in Eq. (2.5.42). The equivalent width of a single Lorentz line for a homogeneous path length is given by Eq. (2.4.2). We may define scaled $\tilde{S}$ and $\tilde{\alpha}$ in this equation so that

$$W_{exact} = 2\pi\tilde{\alpha}xe^{-x}\left[I_0(x) + I_1(x)\right], \tag{2.5.46}$$

where

$$x = \tilde{S}u/2\pi\tilde{\alpha}. \tag{2.5.47}$$

It is convenient to introduce a similar variable in terms of the CG parameters in the form

$$x_{CG} = \tilde{S}_{CG}u/2\pi\tilde{\alpha}_{CG}$$
$$= x\tilde{\alpha}/\tilde{\alpha}_{CG}. \tag{2.5.48}$$

Thus $\tilde{\alpha}$ can be determined from Eq. (2.5.46) in terms of x_{CG}, which is directly proportional to the dimensionless parameter λ. Subsequently, the third parameter, ϵ, can be evaluated from Eq. (2.5.43) as a function of x_{CG}. In the computation, $p_2 = p_0$ (surface pressure) and $p_1 = 0$ may be used. The exact relation between ϵ and x_{CG} is unknown. However, we may derive their relation based on curve fitting from numerical computations. A general form may be assumed:

$$\epsilon = \left(\frac{x_{CG}}{1 + x_{CG}}\right)^n, \tag{2.5.49}$$

where x_{CG} can be computed from the CG two-parameter relations and n is a constant that must be determined from the distribution of the absorbing gas. For H_2O and CO_2, $n = 1.6$, and for O_3, $n = 3.2$. Equations (2.5.43), (2.5.44), and (2.5.49) constitute the three-parameter approximation for nonhomogeneous atmospheres developed by Yamamoto et al. (1972). Although the preceding method is for a single absorption line, the methodology is applicable to statistical band models using the concept of equivalent width for randomly distributed lines in a spectral interval.

2.5.4 Applications of two-parameter approximation to band models

From Eqs. (2.4.20) and (2.4.22), the statistical band parameters for a given temperature and pressure may be written in the forms

$$\frac{\bar{S}(T)}{\delta} = \frac{\sum_i S_i(T)}{\Delta\nu}, \tag{2.5.50}$$

$$\frac{\pi\alpha(T,p)}{\delta} = \frac{\left(2\sum_i \sqrt{S_i(T)\alpha_i(T,p)}\right)^2}{\Delta\nu \sum_i S_i(T)}. \tag{2.5.51}$$

Thus the scaled path length defined in Eq. (2.5.20b) in terms of the line parameters is

$$\tilde{u} = \int_u \frac{\bar{S}(T)}{\bar{S}(T_r)} \, du = \int_u \left[\frac{\sum_i S_i(T)}{\sum_i S_i(T_r)}\right] du. \tag{2.5.52}$$

The CG approximation for the scaled half-width may be written

$$\tilde{\alpha} = \int_u \alpha\bar{S} \, du \Big/ \int_u \bar{S} \, du. \tag{2.5.53}$$

Using Eqs. (2.5.53) and (2.5.51), we have

$$\frac{\pi\tilde{\alpha}}{\delta} = \int_u \left(2\sum_i \sqrt{S_i\alpha_i}\right)^2 du \Big/ \left(\Delta\nu \int_u \sum_i S_i \, du\right). \tag{2.5.54}$$

We shall find a reference pressure and temperature so that

$$\left(\frac{\pi\tilde{\alpha}}{\delta}\right) \Big/ \left(\frac{\pi\alpha}{\delta}\right)_r = \int_u \left[\frac{\left(\sum_i \sqrt{S_i(T)\alpha_i(T,p)}\right)^2}{\left(\sum_i \sqrt{S_i(T_r)\alpha_i(T_r,p_r)}\right)^2}\right] \frac{du}{\tilde{u}}. \tag{2.5.55}$$

To simplify the computational procedure, we may separate the pressure and temperature dependence in the half-width and write $\alpha_i(T,p) = \alpha_i(T)p/p_0$, where $\alpha_i(T) = \alpha_i(T_0, p_0)(T_0/T)^{1/2}$, and p_0 and T_0 denote the pressure and temperature for the reference condition. In this manner, we obtain

$$\left(\frac{\pi\tilde{\alpha}}{\delta}\right) = \left(\frac{\pi\alpha}{\delta}\right)_r \frac{\tilde{p}}{p_r}, \tag{2.5.56}$$

where

$$\tilde{p} = \int_u p \frac{\left(\sum_i \sqrt{S_i(T)\alpha_i(T)}\right)^2}{\left(\sum_i \sqrt{S_i(T_r)\alpha_i(T_r)}\right)^2} \frac{du}{\tilde{u}}. \tag{2.5.57}$$

The temperature-dependent terms in Eqs. (2.5.52) and (2.5.57) may be parameterized in terms of the polynomial in the forms

$$\ell n \left[\frac{\sum_i S_i(T)}{\sum_i S_i(T_r)} \right] \cong \sum_{n=0}^{N} c_n (T - T_r)^n , \qquad (2.5.58a)$$

$$\ell n \left[\frac{\left(\sum_i \sqrt{S_i(T)\alpha_i(T)} \right)^2}{\left(\sum_i \sqrt{S_i(T_r)\alpha_i(T_r)} \right)^2} \right] \cong \sum_{n=0}^{N} c'_n (T - T_r)^n , \qquad (2.5.58b)$$

where c_n and c'_n denote empirical constants. The number of terms in the polynomial fitting is subject to the required accuracy. Usually $N = 2$ is sufficient for most applications.

Using the preceding parameters, the spectral transmittance for a nonhomogeneous atmosphere using the statistical band model presented in Eq. (2.4.16) may be expressed by

$$T_{\tilde{\nu}}(u) = \exp \left[- \left(\frac{\bar{S}}{\delta} \right)_r \tilde{u} \left(\frac{1 + (\bar{S}/\delta)_r \tilde{u}}{(\pi\alpha/\delta)_r \tilde{p} / p_r} \right)^{-1/2} \right] . \qquad (2.5.59)$$

In this manner, the nonhomogeneous atmosphere is approximately accounted for by scaling the path length $\tilde{u}$ and pressure, $\tilde{p}$. The line-by-line calculations are carried out for two parameters, $(\bar{S}/\delta)$ and $(\pi\alpha/\delta)$, at a reference pressure and temperature. Similar applications may be made to the modified statistical models discussed in Subsections 2.4.2 and 2.4.3.

The two-parameter approximation for pressure and temperature corrections in connection with random band models has been widely used in infrared radiative transfer calculations. We shall present the manner in which this approximation is used for flux and heating-rate calculations. For a small spectral interval in which the Planck flux is about constant, the spectral upward and downward fluxes from Eqs. (2.1.13a) and (2.1.13b) using a diffusivity factor $1/\bar{\mu}$ (see Subsection 2.3.2) may be expressed in the z coordinate in the form

$$F_{\tilde{\nu}}^+(z) = \pi B_{\tilde{\nu}}(0) T_{\tilde{\nu}}(z/\bar{\mu}) + \int_0^z \pi B_{\tilde{\nu}}(z') \frac{d}{dz'} T_{\tilde{\nu}} \left(\frac{z - z'}{\bar{\mu}} \right) dz' , \qquad (2.5.60a)$$

$$F_{\tilde{\nu}}^-(z) = \int_{z_\infty}^z \pi B_{\tilde{\nu}}(z') \frac{d}{dz'} T_{\tilde{\nu}} \left(\frac{z' - z}{\bar{\mu}} \right) dz' . \qquad (2.5.60b)$$

To cover an appropriate spectral interval, $\Delta\nu$, for heating-rate calculations for a specific gas, we may divide this interval into N subintervals, $\Delta\nu_i (i = 1, \ldots, N)$. Thus the total upward and downward fluxes may be written

$$F^\pm(z) = \sum_{i=1}^{N} F_{\tilde{\nu},i}^\pm(z) \Delta\nu_i . \qquad (2.5.61)$$

Measurements of infrared fluxes and cooling rates in the atmosphere are difficult to make, and the observed results available in the literature are limited. Kuhn (1963) has measured total upward infrared fluxes using the radiometersonde designed by Suomi and Kuhn (1958) and compared them with calculated fluxes from the same soundings. The computations used the broadband emissivity approach (see Subsection 2.7.2) and a simple one-parameter pressure correction (see Subsection 2.5.1). The agreement between the computed and observed upward fluxes is reasonable below about 400 mb. However, downward fluxes were not presented. Ellingson and Gille (1978) have compared the vertical profiles of the upward, downward, and net fluxes and cooling rates measured by radiometersondes during the Barbados Oceanographic and Meteorological Experiment (BOMEX) with those from calculations using simultaneously observed temperature and water vapor profiles. Figure 2.14 shows a comparison of the upward and net fluxes. The computational program used the Goody random model with spectral intervals of $40\,\text{cm}^{-1}$. The two-parameter approximation (see Subsection 2.5.2) was used to account for pressure and temperature variations, along with a diffusivity factor of 1.66 for the flux calculations (see Subsection 2.3.2). A comparison of the computed and observed upward fluxes shows that the computations experience the same general changes with altitude as do the observations. There are a number of regions where the observed and calculated fluxes agree to within the estimated standard deviation of the observations due to random error. Differences between the observed and theoretically computed upward fluxes could arise from the uncertainties in meteorological and spectroscopic data. Comparisons of the computed and observed downward fluxes are not reliable because of the unknown stratospheric water vapor, systematically low observed humidity, and the presence of aerosols. However, the calculated and observed net fluxes agree to within the random error of the observations over large layers.

2.6 k-distribution method

2.6.1 The fundamentals

The k-distribution method for the computation of infrared radiative transfer is based on the grouping of gaseous spectral transmittances according to the absorption coefficient k_ν. In a homogeneous atmosphere, spectral transmittance is independent of the ordering of k for a given spectral interval. Hence, the wave number integration may be replaced by an integration over the k space. If the normalized probability distribution function for k_ν in the interval $\Delta\nu$ is given by $f(k)$ and its minimum and maximum values are $k_{\min}$ and $k_{\max}$, respectively, then the spectral transmittance may be expressed by

$$T_{\bar{\nu}}(u) = \int_{\Delta\nu} e^{-k_\nu u}\,\frac{d\nu}{\Delta\nu} = \int_0^\infty e^{-ku} f(k)\,dk, \qquad (2.6.1)$$

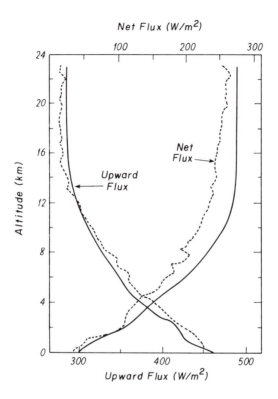

Fig. 2.14 Model-calculated (solid lines) and radiometersonde observed (dots) fluxes for the ascent from the NOAA research vessel Discoverer on June 3, 1969 (after Ellingson and Gille, 1978).

where we have set $k_{min} \to 0$ and $k_{max} \to \infty$, for mathematical convenience, and

$$\int_0^\infty f(k)\,dk = 1. \qquad (2.6.2)$$

Moreover, a cumulative probability function may be defined in the form

$$g(k) = \int_0^k f(k)\,dk, \qquad (2.6.3)$$

where $g(0) = 0$, $g(k \to \infty) = 1$, and $dg(k) = f(k)\,dk$. By definition, $g(k)$ is a monotonically increasing and smooth function in k space. By using the g function, the spectral transmittance can be written

$$T_{\tilde\nu}(u) = \int_0^1 e^{-k(g)u}\,dg \cong \sum_{j=1}^{M} e^{-k(g_j)u}\,\Delta g_j. \qquad (2.6.4)$$

From Eq. (2.6.3), since $g(k)$ is a smooth function in k space, the inverse will also be true here; that is, $k(g)$ is a smooth function in g space. Consequently, the integration in g space, which replaces the tedious wave-number integration, can be evaluated by a finite sum of exponential terms, as shown in Eq. (2.6.4).

Figure 2.15(a) shows k_ν as a function of ν in a portion of the 9.6 μm O_3 band at a pressure of 30 mb and a temperature of 220 K. Figure 2.15(b) shows the probability distribution $f(k)$ as a function of k for this band [see the following for an evaluation of $f(k)$]. In Fig. 2.15(c), the cumulative probability function $g(k)$ is shown as a function of k. We may then compute $k(g)$ as a function of g from Eq. (2.6.3). This curve is illustrated in Fig. 2.15(d). Since g is a smooth monotonic function, a few quadrature points will suffice to achieve a high degree of accuracy in the transmittance computations. The physical foundation for the k distribution is quite simple, but it offers an advantage in the computation of infrared flux transfer. The k distribution method has been used by Arking and Grossman (1972) in a discussion of the line shape effect on the equilibrium temperature of planetary atmospheres. It has also been discussed by Domoto (1974) on some aspects of the theoretical foundation and the Laplace transforms for a number of band models. The idea of scrambling and ranking absorption lines was described in the work of Ambartsumian (1936) on stellar atmospheres.

2.6.2   Evaluation of the probability function

There are three approaches that can be used to evaluate the probability function $f(k)$ and the cumulative probability function $g(k)$.

1. From Eq. (2.6.1), $f(k)$ is simply the inverse of the Laplace transform of spectral transmittance,

$$f(k) = L^{-1}[T_{\bar{\nu}}(u)], \qquad (2.6.5)$$

where the notation L^{-1} denotes the inverse of the Laplace transform. If the spectral transmittance can be expressed in terms of the analytic band models discussed previously, and if the inverse Laplace transform can be performed, it is possible to obtain analytic expressions for the probability distribution function. In particular, the inverse Laplace transform can be derived for the Malkmus random model given in Eq. (2.4.41). After the transformation we have

$$f(k) = \frac{1}{2\sqrt{\pi}} c_{\bar{\nu}} d_{\bar{\nu}}^{1/2} k^{-3/2} \exp\left(c_{\bar{\nu}} - k/d_{\bar{\nu}} - c_{\bar{\nu}}^2 d_{\bar{\nu}}/4k\right), \qquad (2.6.6)$$

where $c_{\bar{\nu}}$ and $d_{\bar{\nu}}$ are band parameters defined in Eq. (2.4.40). Using this form the cumulative probability function can be derived and is given by

$$g(k) = \frac{1}{2} e^{2c_{\bar{\nu}}} \operatorname{erfc}\left[\left(\frac{c_{\bar{\nu}}}{2}\right)^{1/2} \left(\frac{1}{y} + y\right)\right]$$
$$+ \frac{1}{2} \operatorname{erfc}\left[\left(\frac{c_{\bar{\nu}}}{2}\right)^{1/2} \left(\frac{1}{y} - y\right)\right], \qquad (2.6.7)$$

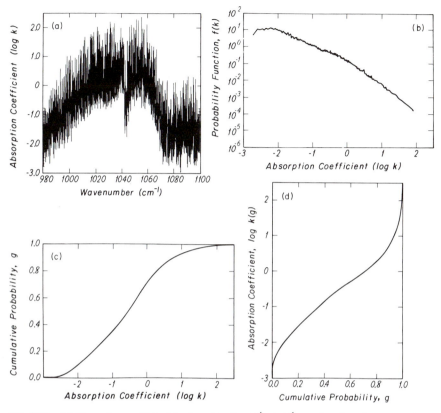

FIG. 2.15 (a) Absorption coefficient k_ν in units of cm^{-1} atm^{-1} as a function of wave number with a resolution of 0.05 cm^{-1} in the 9.6 μm O$_3$ band. $p = 30$ mb; $T = 220$ K. (b) The probability function $f(k)$ of the absorption coefficient. (c) The cumulative probability function for $f(k)$ shown in (b), plotted as a function of k. (d) Same as (c), except that values of the absorption coefficient are expressed as a function of g.

where $y = [2k/(c_{\bar\nu}d_{\bar\nu})]^{1/2}$. The notation erfc is the complementary error function given by erfc$(x) = 1 - $ erf(x), and the error function is defined by

$$\text{erf}(x) = \frac{2}{\sqrt{\pi}} \int_0^x e^{-x^2}\, dx. \tag{2.6.8}$$

From Eq. (2.6.7), k can then be numerically computed once g is given.

The inverse Laplace transform cannot be performed on the Goody random model. Consequently, $f(k)$ cannot be derived analytically. However, under the limits of weak-line approximations, we have

$$f(k) = L^{-1}(e^{-a_{\bar\nu}u}) = \delta(k - a_{\bar\nu}), \tag{2.6.9a}$$

where δ denotes the delta function. In the case of strong-line limits, we find

$$f(k) = L^{-1} \left\{ \exp \left[-(a_{\tilde{\nu}} b_{\tilde{\nu}} u)^{1/2} \right] \right\}$$
$$= \frac{1}{2} (a_{\tilde{\nu}} b_{\tilde{\nu}} / \pi)^{1/2} k^{-3/2} \exp(-a_{\tilde{\nu}} b_{\tilde{\nu}} / 4k). \qquad (2.6.9b)$$

2. From line-by-line data, $f(k)$ and $g(k)$ can be computed directly. We may divide the spectral interval $\Delta\nu$ at all minima and maxima of k_ν to obtain subintervals over which k_ν is monotonic. Let $\Delta\nu_i$ be the subintervals, where $i = 1, 2, \ldots, N$. Within each subinterval, invert the function of k_ν and denote it by $\nu_i(k)$, where k is in the range of $k^i_{\min}$ and $k^i_{\max}$. The probability distribution function can then be computed from

$$f(k) = \frac{1}{\Delta\nu} \sum_{i=1}^{N} \left| \frac{d\nu_i}{dk} \right| \left[h(k - k^i_{\min}) - h(k - k^i_{\max}) \right], \qquad (2.6.10a)$$

where h is the step function such that $h(x) = 1$ if $x > 0$ and $h(x) = 0$ if $x < 0$. In practice, $f(k)$ can be obtained by a simple statistical procedure in which k_ν is calculated at a number of points in $\Delta\nu$, with a wavenumber resolution less than the line half-width. Alternatively, we may evaluate $g(k)$ based on Eq. (2.6.3) in the form

$$g(k) = \frac{1}{\Delta\nu} \sum_{i=1}^{N} \left\{ \left| \nu_i(k^i_{\max}) - \nu_i(k^i_{\min}) \right| h(k - k^i_{\max}) \right.$$
$$\left. + \left| \nu_i(k) - \nu_i(k^i_{\min}) \right| \left[h(k - k^i_{\min}) - h(k - k^i_{\max}) \right] \right\}. \qquad (2.6.10b)$$

Again, a statistical procedure can be devised to obtain $g(k)$, a monotonically increasing function.

3. From Eq. (2.6.4), spectral transmittance can be expressed in terms of a set of exponential sums in the form

$$T_{\tilde{\nu}}(u) = e^{-k_1 u} \Delta g_1 + e^{-k_2 u} \Delta g_2 \ldots + e^{-k_M u} \Delta g_M. \qquad (2.6.11)$$

If spectral transmittance is given as a function of the path length, k_j and Δg_j ($j = 1, 2, \ldots, M$) may be computed from an appropriate numerical fitting program. After the fitting, the wave-number dependent information is lost, and the arrangement of k_j ($j = 1, 2, \ldots, M$) in the spectral interval is arbitrary. However, the largest k_j must be in the band center, while the smallest k_j must be in the wing. Using this principle, k_j may be rearranged in empirical order in the spectral band. The exponential sum fitting can be carried out for different pressures and temperatures, and, by using the empirical arrangements for k_j, the transfer of spectral radiation through a nonhomogeneous atmosphere may be computed.

The numerical fitting technique to find the equivalent absorption coefficients and weights has been used in the past in connection with the incorporation of absorption processes in a scattering atmosphere. Kondratyev (1969), Lacis and Hansen (1974), Liou and Sasamori (1975), and Ackerman et al. (1976) have employed this technique to account for absorption of water vapor in the near-infrared solar spectrum in the calculation of solar flux and heating-rate profiles for aerosol and Rayleigh atmospheres. Roewe and Liou (1978) have used this method to combine absorption due to gases, and scattering and absorption due to cloud particles in broadband radiative transfer calculations. Numerical techniques to obtain the accurate equivalent absorption coefficient k_j and weight Δg_j have been reported by Arvett and Hummer (1965) and Wiscombe and Evans (1977).

2.6.3 Correlated k-distribution method

The basic theory for the k-distribution method has been developed for a homogeneous path length as in the case of band models. We must now consider how this method can be applied to nonhomogeneous atmospheres. The optical depth for a nonhomogeneous path length is

$$\tau = \int_0^u k_\nu \, du = \sum_i k_{\nu,i} \, \Delta u_i. \tag{2.6.12}$$

We wish to express spectral transmittance in the form

$$T_{\bar\nu}(u) = \int_{\Delta\nu} e^{-\tau} \, d\nu = \int_0^\infty e^{-k^* u} f(k^*) \, dk^*, \tag{2.6.13}$$

where

$$k^* = \frac{\tau}{u} = \sum_i k_i a_i, \tag{2.6.14}$$

where $a_i = \Delta u_i / u$, and $f(k^*)$ is the probability density function for k^*. The cumulative probability function associated with k^* is given by

$$g^*(k^*) = \int_0^{k^*} f(k^{*'}) dk^{*'}. \tag{2.6.15}$$

For a given temperature and pressure, we may express the cumulative probability function in the form

$$g(k_i) = \int_0^{k_i} f(k') \, dk'. \tag{2.6.16}$$

Hence, Eq. (2.6.14) can be expressed in the form

$$T_{\bar\nu}(u) = \int_0^1 \exp\left(-u \sum_i k_i(g) a_i\right) dg^*. \tag{2.6.17}$$

Here, the cumulative probability function g^*, defined in Eq. (2.6.16) for a nonhomogeneous path, depends on the temperature and pressure along the nonhomogeneous path. One possibility for obtaining g^* is to find $f(k^*)$ through an averaging method such that $f(k^*) \cong \sum_i f(k_i)a_i$. However, this method is quite involved and may not appear to be a practical approach.

We may approximate Eq. (2.6.17) by assuming that the cumulative probability function is independent of temperature and pressure so that

$$T_{\bar{\nu}}(u) \cong \int_0^1 \exp\left(-u\sum_i k_i(g)a_i\right) dg. \qquad (2.6.18)$$

This is referred to as the *correlated k-distribution method* for nonhomogeneous atmospheres proposed by Lacis and Oinas (1991). In this approach, the g function is correlated throughout the atmosphere and is independent of temperature and pressure. We must now determine the deviation of the expression in Eq. (2.6.18) from that in Eq. (2.6.17) or Eq. (2.6.13).

For a single line of any line shape in a spectral interval $\Delta\nu$, the cumulative probability function is given by

$$g(k) = \frac{2}{\Delta\nu} \int_0^k \left|\frac{\partial\nu}{\partial k}\right|_{k=k'} dk' = \frac{2}{\Delta\nu}\left|\nu(k) - \nu(0)\right|, \qquad (2.6.19)$$

where the spectral interval is set from $-\Delta\nu/2$ to $\Delta\nu/2$, and the line center is set at $\nu(k_{\max}) = 0$. Thus, $|\nu(k_{\min})| = \Delta\nu/2$ and $dg(k) = 2\,d\nu(k)/\Delta\nu$. The correlated k distribution is exact as in the case of a single line that repeats itself periodically. For absorption lines that are randomly distributed in a spectral interval $\Delta\nu$, and for which an averaged equivalent width corresponding to a single line may be realized, the correlated k distribution may also be viewed as an exact method. The general condition under which this method is applicable to nonhomogeneous atmospheres for a group of absorption lines is that the strong and weak lines must occur at the same wave numbers at all altitudes. Under this condition, $g(k_i \to \infty) = g^*(k \to \infty) = 1$ and $g(k_i \to 0) = g^*(k \to 0) = 0$. These approximations are analogous to the CG relations, which are derived under the limits of strong- and weak-line approximations in the context of band models.

The correlated k-distribution method for the computation of transmittances along nonhomogeneous paths appears to offer an approach that is superior to the CG approximation, because it is exact in the case of a single line. In addition, the effects of gaseous absorption can be directly incorporated in multiple-scattering computations using the equivalent absorption coefficients, k_i, derived for a given spectral interval. The correlated k-distribution method requires a set of k_i for a number of pressure and temperature values that adequately covers the nonhomogeneous atmosphere. This requirement can entail more computational effort than

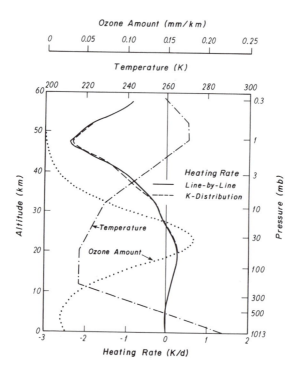

FIG. 2.16   Atmospheric heating/cooling by the 9.6 μm ozone band for the indicated temperature distribution and ozone concentration. The solid line is from line-by-line integration, and the dashed line from the correlated k-distribution method (after Hansen et al., 1983).

is required by band models. Goody et al. (1989) have discussed the accuracy of the correlated k-distribution method and illustrated that a 1% accuracy can be achieved by this method in comparison with line-by-line results.

An example of the applicability of the correlated k distribution is the transfer of thermal infrared radiation in the 9.6 μm O_3 band. Because the ozone profile is characterized by the combination of large concentration at low pressure and small concentration at high pressure, the two-parameter CG and other scaling approximations do not yield accurate flux and cooling-rate profiles for ozone. Figure 2.16 shows the ozone heating rate as a function of altitude computed by line-by-line and correlated k-distribution methods. The k-distribution method requires only six quadrature points to perform the integration over nonhomogeneous atmospheres, as compared to the millions of points required in the line-by-line integration to resolve the rapid variations of absorption lines in the atmosphere. In general, heating rates computed from the correlated k-distribution method deviate only slightly from those computed from the line-by-line program. Using the temperature and ozone profiles depicted in the figure, ozone generates a maximum cooling rate of $2.5 \, K \, d^{-1}$ at ~ 50 km.

2.6.4   *Overlap consideration*

If two absorbing gases overlap in a spectral interval, such as H_2O rotational and $15 \, \mu m$ CO_2 lines, an appropriate treatment must be undertaken to compute the spectral transmittance. Let u_w and u_c denote the path lengths for H_2O and CO_2, and let $k_{\nu 1}$ and $k_{\nu 2}$ be their respective absorption coefficients. Then the monochromatic transmittance for these two gases by definition can be written in the form

$$T_\nu(u_w, u_c) = \exp\left[-(k_{\nu 1} u_w + k_{\nu 2} u_c)\right] = T_\nu(u_w) T_\nu(u_c). \qquad (2.6.20)$$

Over a spectral interval of $\Delta \nu$ we have

$$T_{\bar\nu}(u_w, u_c) = \int_{\Delta\nu} T_\nu(u_w) T_\nu(u_c) \, \frac{d\nu}{\Delta\nu}. \qquad (2.6.21)$$

We shall now consider a small spectral interval (e.g., $\Delta\nu \le 5 \, \text{cm}^{-1}$) in which H_2O and CO_2 absorption lines are statistically independent such that

$$\int_{\Delta\nu} [T_\nu(u_w) - T_{\bar\nu}(u_w)] \, [T_\nu(u_c) - T_{\bar\nu}(u_c)] \, \frac{d\nu}{\Delta\nu} = 0. \qquad (2.6.22)$$

It follows that

$$\int_{\Delta\nu} T_\nu(u_w) T_\nu(u_c) \, \frac{d\nu}{\Delta\nu} = T_{\bar\nu}(u_w) T_{\bar\nu}(u_c), \qquad (2.6.23a)$$

that is,

$$\int_{\Delta\nu} \exp\left[-(k_{\nu 1} u_w + k_{\nu 2} u_c)\right] \frac{d\nu}{\Delta\nu} = \int_{\Delta\nu} e^{-k_{\nu 1} u_w} \, \frac{d\nu}{\Delta\nu}$$
$$\times \int_{\Delta\nu} e^{-k_{\nu 2} u_c} \, \frac{d\nu}{\Delta\nu}. \qquad (2.6.23b)$$

Using the k-distribution method for the computation of $T_{\bar\nu}(u_w)$ and $T_{\bar\nu}(u_c)$ via Eq. (2.6.4), we have

$$T_{\bar\nu}(u_w, u_c) = \int_0^1 \int_0^1 \exp\left[-(k_1 u_w + k_2 u_c)\right] dg_1 \, dg_2$$
$$\cong \sum_{m=1}^{M} \sum_{n=1}^{N} e^{-\tau_{mn}} \Delta g_{1m} \Delta g_{2n}, \qquad (2.6.24)$$

where M and N denote the number of k values for H_2O and CO_2 in the band interval $\Delta\nu$ and the optical depth is given by

$$\tau_{mn} = k_{1m} u_w + k_{2n} u_c. \qquad (2.6.25)$$

Subject to the condition given in Eq. (2.6.23b), the spectral transmittance for overlapping lines can be computed from Eqs. (2.6.24) and (2.6.25). There are $N \times M$ computations required to obtain the optical depth. Applications to non-homogeneous atmospheres may be carried out using the correlated k-distribution method in a manner described in Section 2.6.3.

2.7 Broadband approaches to flux and cooling-rate computations

2.7.1 Broadband emissivity based on the radiation chart concept

The essence of the broadband emissivity approach for the calculation of infrared fluxes and heating rates is to use the temperature directly in terms of the Stefan–Boltzmann law instead of the Planck function. The earlier development of broadband emissivity methods was based on the concept of the radiation chart from which the flux integration may be carried out by using graphs and tables (Elsasser, 1942; Möller, 1943; Yamamoto, 1952). In reference to Eq. (2.1.13) for upward and downward fluxes, an integration by parts may be performed to obtain

$$F_\nu^+(\tau) = \pi B_\nu(\tau) + \int_\tau^{\tau_*} T_\nu^f(\tau' - \tau) \frac{d\pi B_\nu(\tau')}{d\tau'} \, d\tau', \quad (2.7.1a)$$

$$F_\nu^-(\tau) = \pi B_\nu(\tau) - \pi B_\nu(0) T_\nu^f(\tau)$$
$$- \int_0^\tau T_\nu^f(\tau - \tau') \frac{d\pi B_\nu(\tau')}{d\tau'} \, d\tau'. \quad (2.7.1b)$$

In Eq. (2.7.1), the surface temperature and the air temperature immediately above the surface are assumed to be the same. In terms of the path-length coordinate illustrated in Fig. 2.2, and after rearrangements, we have

$$F_\nu^+(u) = \pi B_\nu(T_s) + \int_0^u \left[1 - T_\nu^f(u - u')\right] \frac{d\pi B_\nu(u')}{du'} \, du', \quad (2.7.2a)$$

$$F_\nu^-(u) = \pi B_\nu(u_1) \left[1 - T_\nu^f(u_1 - u)\right]$$
$$+ \int_{u_1}^u \left[1 - T_\nu^f(u' - u)\right] \frac{d\pi B_\nu(u')}{du'} du'. \quad (2.7.2b)$$

Note that transmittance notation has a subscript ν. In view of the integral expressions given in these equations, we may define a parameter in the form

$$R(u, T) = \int_0^\infty \left[1 - T_\nu^f(u)\right] \frac{d\pi B_\nu(T)}{dT} \, d\nu. \quad (2.7.3)$$

By performing an integration over the wave number on the flux equations and changing path lengths to their corresponding temperatures, we obtain

$$F^+(u) = \int_0^\infty F_\nu^+(u) \, d\nu$$

$$= \sigma T_s^4 + \int_{T_s}^{T} R\big([u(T) - u(T')], T'\big)\, dT', \tag{2.7.4}$$

$$F^-(u) = \int_0^\infty F_\nu^-(u)\, d\nu$$

$$= \int_0^\infty \pi B_\nu(T_\infty) \left\{ 1 - T_\nu^f[u(T_\infty) - u(T)] \right\}\, d\nu$$

$$+ \int_{T_\infty}^{T} R([u(T') - u(T)], T')\, dT'. \tag{2.7.5}$$

If the diffuse transmittance is independent of temperature so that $dT_\nu^f/dT \approx 0$, then

$$\int_0^\infty \pi B_\nu(T_\infty) \left\{ 1 - T_\nu^f[u(T_\infty) - u(T)] \right\}\, d\nu$$

$$= \int_0^{T_\infty} R\big([u(T') - u(T)], T'\big)\, dT', \tag{2.7.6}$$

where we note that $B_\nu(T = 0) = 0$. Using Eq. (2.7.6), which is subject to the assumption that $dT_\nu^f/dT = 0$, the downward flux may subsequently be expressed by

$$F^-(u) \cong \int_0^{T} R\big([u(T') - u(T)], T'\big)\, dT'. \tag{2.7.7}$$

Moreover, a dimensionless averaged emissivity may be defined in the form

$$\bar{\epsilon}^f(u, T) = R(u, T) \Big/ \int_0^\infty \frac{d\pi B_\nu(T)}{dT}\, d\nu = \frac{R(u, T)}{4\sigma T^3}$$

$$\cong \sum_i [1 - T_{\nu,i}^f(u)] \frac{d\pi B_{\bar\nu,i}(T)}{dT} \frac{\Delta\nu_i}{4\sigma T^3}. \tag{2.7.8}$$

Using the averaged emissivity, the upward and downward fluxes may now be written

$$F^+(u) = \sigma T_s^4 + \int_{T_s}^{T} \bar{\epsilon}^f\big([u(T) - u(T')], T'\big)\, 4\sigma T'^3\, dT', \tag{2.7.9a}$$

$$F^-(u) \cong \int_0^{T} \bar{\epsilon}^f\big([u(T') - u(T)], T'\big)\, 4\sigma T'^3\, dT'. \tag{2.7.9b}$$

If the spectral transmittances are known via band models or line-by-line calculations, the averaged emissivities for a number of temperatures may then be computed. Earlier efforts to design a numerical infrared radiation scheme for applications to general circulation experiments used the emissivity equations given in Eq. (2.7.9). Using the radiation tables listed by Elsasser and Culbertson (1960) to obtain the averaged emissivity, Sasamori (1968) has noted inaccuracy in the computation of downward fluxes when Eq. (2.7.9b) is employed. The temperature dependence of the diffuse transmittance must be accounted for, and the correct form of Eq. (2.7.5) must be used in the computation of downward fluxes. Sasamori developed empirical formulas for mean flux emissivity for the first term on the right-hand side of Eq. (2.7.5); use of these formulas gave correct downward fluxes. In the following we discuss the approach to the calculation of broadband fluxes using isothermal broadband emissivity.

2.7.2 Isothermal broadband emissivity

Based on Eq. (2.1.13), the total upward and downward fluxes in the u coordinate may be written

$$F^+(u) = \int_0^\infty \pi B_\nu(T_s) T_\nu^f(u)\, d\nu$$
$$+ \int_0^\infty \int_0^u \pi B_\nu(u') \frac{dT_\nu^f(u-u')}{du'}\, du'\, d\nu, \qquad (2.7.10a)$$

$$F^-(u) = \int_0^\infty \int_{u_1}^u \pi B_\nu(u') \frac{dT_\nu^f(u'-u)}{du'}\, du'\, d\nu. \qquad (2.7.10b)$$

In order to express the upward and downward fluxes in terms of broadband emissivity, we may define isothermal broadband flux emissivity in the form

$$\epsilon^f(u, T) = \int_0^\infty \pi B_\nu(T)[1 - T_\nu^f(u)] \frac{d\nu}{\sigma T^4}. \qquad (2.7.11)$$

We shall consider a plane-parallel atmosphere and divide this atmosphere into a number of layers such that each of them may be thought of as an isothermal layer. In this manner, we may express the following differential operator in finite-difference form:

$$\int_0^\infty \pi B_\nu(T) \frac{dT_\nu^f(u)}{du}\, d\nu \cong \int_0^\infty \pi B_\nu(T)$$
$$\times \frac{T_\nu^f(u + \Delta u/2) - T_\nu^f(u - \Delta u/2)}{\Delta u} d\nu. \qquad (2.7.12)$$

Using Eqs. (2.7.11) and (2.7.12), the total upward and downward fluxes may be expressed by

$$F^+(u) \cong \sigma T_s^4[1 - \epsilon^f(u, T_s)]$$
$$- \int_0^u \sigma T^4(u')\{[\epsilon^f((u-u'-\Delta u'/2), T(u'))$$
$$- \epsilon^f((u-u'+\Delta u'/2), T(u'))]/\Delta u'\}\, du', \qquad (2.7.13a)$$

$$F^-(u) \cong \int_{u_1}^{u} \sigma T^4(u') \{ [\epsilon^f((u' - u + \Delta u'/2), T(u'))$$
$$- \epsilon^f((u' - u - \Delta u'/2), T(u'))]/\Delta u' \} \, du'. \qquad (2.7.13b)$$

In practice, Eq. (2.7.13) can be evaluated by numerical means. This can be done by using appropriate summations over u'. These two finite-difference equations may be written in the forms

$$F^+(u) \cong \sigma T_s^4 [1 - \epsilon^f(u, T_s)] - \int_0^u \sigma T^4(u')$$
$$\times \frac{d\epsilon^f(u - u', T(u'))}{du'} \, du', \qquad (2.7.14a)$$

$$F^-(u) \cong \int_u^{u_1} \sigma T^4(u') \frac{d\epsilon^f(u' - u, T(u'))}{du'} \, du'. \qquad (2.7.14b)$$

Computations of fluxes and cooling rates that use the broadband emissivity as defined in Eq. (2.7.11) have been performed by numerous researchers. We shall now discuss the manner in which isothermal broadband emissivities are constructed.

In numerical calculations, Eq. (2.7.11) for broadband flux emissivity is expressed in terms of a finite sum over spectral bands $\Delta \nu_i$ for three principal absorbers. Let u_1 ($= u_w$), u_2 ($= u_c$), and u_3 ($= u_o$) denote the path lengths for H_2O, CO_2, and O_3, respectively. Then we write

$$\epsilon^f(u_j, T) = \sum_i \pi B_{\bar\nu,i}(T)[1 - T_{\bar\nu,i}^f(u_j)] \frac{\Delta \nu_i}{\sigma T^4}, \qquad j = 1, 2, 3. \qquad (2.7.15)$$

As a good approximation, we may replace diffuse spectral transmittance with spectral transmittance, $T_{\bar\nu}(u/\bar\mu)$, with the diffusivity factor, $1/\bar\mu = 1.66$, discussed previously. The spectral transmittance can be generated from data via detailed line-by-line computations or band models for a homogeneous path.

The total emission of an isothermal atmosphere is the sum of individual emissions due to various gases. However, overlaps of the H_2O rotational and 15 μm CO_2 absorption lines are significant. Thus it is necessary to make a proper correction to circumvent the overestimate of H_2O and CO_2 emissions. The emissivity for the overlap region can be expressed exactly by

$$\epsilon^f(u_w, u_c, T) = \int_0^\infty \pi B_\nu(T)[1 - T_\nu(\bar u_w, \bar u_c)] \frac{d\nu}{\sigma T^4}, \qquad (2.7.16)$$

where $\bar u_w = u_w/\bar\mu$ and $\bar u_c = u_c/\bar\mu$. From the definition of monochromatic transmittance, its value for two absorbing gases is the product of the individual value for each absorbing gas. It follows that

$$T_\nu(\bar u_w, \bar u_c) = T_\nu(\bar u_w) T_\nu(\bar u_c). \qquad (2.7.17)$$

With this relation, which is valid only for monochromatic radiation, we may express the emissivity for the overlap region in terms of the individual emissivity in the form

$$\epsilon^f(u_w, u_c, T) = \epsilon^f(u_w, T) + \epsilon^f(u_c, T) - \Delta\epsilon^f(u_w, u_c, T), \qquad (2.7.18)$$

where the correction term is

$$\Delta\epsilon^f(u_w, u_c, T) = \int_0^\infty \pi B_\nu(T)[1 - T_\nu(\bar{u}_w)][1 - T_\nu(\bar{u}_c)]\frac{d\nu}{\sigma T^4}$$

$$\cong \sum_i \pi B_{\bar\nu,i}(T) \int_{\Delta\nu_i} [1 - T_\nu(\bar{u}_w)][1 - T_\nu(\bar{u}_c)]\frac{d\nu}{\sigma T^4}. \quad (2.7.19)$$

If the variation in either $T_\nu(\bar{u}_w)$ or $T_\nu(\bar{u}_c)$ is smaller than their products, we may carry out a wave number integration over either one to obtain

$$\Delta\epsilon^f(u_w, u_c, T) \cong \sum_i \pi B_{\bar\nu,i}(T)[1 - T_{\bar\nu,i}(\bar{u}_w)][1 - T_{\bar\nu,i}(\bar{u}_c)]\frac{\Delta\nu_i}{\sigma T^4}. \quad (2.7.20)$$

This appears to be a good approximation and has been used to evaluate the H_2O–CO_2 overlap in the context of the broadband emissivity for flux computations. The computation of broadband flux emissivities for H_2O, CO_2, and O_3 has been correctly carried out by Staley and Jurica (1970, 1972) using the generalized absorption coefficients presented by Elsasser and Culbertson (1960).

The isothermal broadband emissivity is constructed on the basis of the spectral transmittance values that are computed for a reference temperature. In the broadband flux expressions [Eqs. (2.7.10a) and (2.7.10b)], the monochromatic transmittance terms are associated with the path length integration in which temperature varies along the integration path. Consequently, use of the isothermal broadband emissivity in flux calculations requires improvements. A nonisothermal broadband emissivity approach has been discussed by Ramanathan and Downey (1986) in which the narrow band transmittance is treated more exactly by incorporating a temperature correction. As discussed in the following parameterization subsection, we may account for the temperature as well as pressure variations by properly adjusting the path length in the flux calculations. A review of various methods that can be used to construct the broadband emissivity for flux and heating rate calculations has been presented by Chou et al. (1991).

2.7.3 Parameterization of broadband flux emissivity

In this section, we shall present a parameterization approach for the calculation of fluxes and heating rates based on broadband emissivity. In this approach, the infrared spectrum is divided into five bands: the (1) H_2O rotational band, (2) H_2O continuum band, (3) H_2O vibrational–rotational band, (4) CO_2 rotational–vibrational band, and (5) O_3 rotational–vibrational band. The overlap between bands (1) and (4) is denoted as band (6). Thus the broadband flux emissivity may be written in the form

$$\epsilon^f(u, T) = \sum_{i=1}^{5} \epsilon_i^f(\tilde{u}_i, T) + \epsilon_6^f(\tilde{u}_w, \tilde{u}_c, T), \qquad (2.7.21)$$

where the individual broadband flux emissivity has been defined in Eq. (2.7.15). In Eq. (2.7.21), $u_1 = u_2 = u_3 = u_w$ (the H_2O path length), $u_5 = u_o$ (the O_3 path length), $u_4 = u_c$ (the CO_2 path length), and $\tilde{u}_i$ denotes the pressure- and temperature-corrected path length for the respective absorbing gaseous component. In principle, the overlap correction may be constructed from Eq. (2.7.20). However, it is a function of three parameters, u_w, u_c, and T, and the computations would be quite involved. In practice, the overlap band may be taken as $\epsilon_6^f \cong -\epsilon_1^f(\tilde{u}_w)\epsilon_4^f(\tilde{u}_c)$, where the broadband emissivity values for the water vapor rotational and vibrational–rotational bands, as functions of temperature and path length, can be built from the Malkmus model assuming the Lorentz line shape. For water vapor continuum absorption in the window region, the empirical formula presented in Subsection 2.2.5.2 is used. For the three water vapor bands, the pressure–temperature scaling method proposed by Liou and Ou (1981) appears to be adequate. The calculations for these bands are as follows:

$$\tilde{u}_1 \cong \int_0^z \frac{p(z')}{p_0} \exp\left\{ A'[T(z') - T_a] + B'[T(z') - T_a]^2 \right\} \rho_w(z') \, dz', \quad (2.7.22)$$

$$\tilde{u}_2 \cong \int_0^z \frac{p_w(z')}{p_{0w}} \exp\left(-\frac{1800}{T_b T(z')} [T(z') - T_b] \right) \rho_w(z') \, dz', \qquad (2.7.23)$$

$$\tilde{u}_3 \cong \int_0^z \frac{p(z')}{p_0} \left(\frac{T_c}{T(z')} \right)^{1/2} \rho_w(z') \, dz', \qquad (2.7.24)$$

where $A' = 10.6278 \times 10^{-3} \, \text{K}^{-1}$; $B' = -44.6152 \times 10^{-6} \, \text{K}^{-2}$; $T_a = 260 \, \text{K}$; $T_b = 296 \, \text{K}$; $T_c = 300 \, \text{K}$; $p_0 = 1013 \, \text{mb}$; p_w and p_{0w} are vapor pressures at temperatures T and T_b, respectively; and ρ_w is the water vapor density. For ozone, no adjustment has been made because of the uncertainty in applying pressure and temperature corrections.

The broadband emissivities for the preceding four bands may be parameterized in terms of polynomial functions in the form

$$\epsilon_i^f(\tilde{u}_i, T) = \exp\left[\sum_{n=0}^{3} c_{ni} \bar{u}_i^n \right], \qquad (2.7.25)$$

where

$$\bar{u}_i = (2 \log_{10} \tilde{u}_i - \bar{a}_i)/\bar{b}_i,$$
$$\bar{a}_i = \log_{10}(\tilde{u}_{i,\max} \tilde{u}_{i,\min}),$$
$$\bar{b}_i = \log_{10}(\tilde{u}_{i,\max}/\tilde{u}_{i,\min}).$$

Table 2.3 Coefficients of broadband emissivity values for water vapor and ozone bands

$T(K)$	i	c_{0i}	c_{1i}	c_{2i}	c_{3i}
220	1	−2.04744	3.16892	−2.46364	0.90375
	2	−8.61009	9.59458	−0.92840	−1.49967
	3	−5.40949	4.40819	−2.92984	0.53672
	5	−6.68758	4.30955	−0.74780	−0.55775
260	1	−2.24944	3.30262	−2.45860	0.84656
	2	−9.24965	9.32921	−0.60115	−0.81971
	3	−4.44616	4.41186	−3.01151	0.47593
	5	−6.28714	4.32859	−0.72052	−0.54514
300	1	−2.41200	3.39922	−2.46510	0.79550
	2	−9.08980	9.07803	−0.80627	−0.74675
	3	−3.86718	4.42569	−3.06243	0.43230
	5	−6.08221	4.34229	−0.70102	−0.53622

The coefficients c_{ni} were obtained by the least-squares method. The quantity $\bar{u}_i$ is expressed as a linear function of $\log_{10} \tilde{u}_i$, so that $-1 < \bar{u}_i < 1$. Errors are smaller for the intermediate $\bar{u}_i$ values than for $\bar{u}_i \sim \pm 1$, due to the characteristics of the least-squares method. The prescription of $\tilde{u}_{i,\max}$ and $\tilde{u}_{i,\min}$ was done according to the range for realistic atmospheres. For water vapor, $\tilde{u}_{i,\max} = 10\,\mathrm{g\,cm}^{-2}$ and $\tilde{u}_{i,\min} = 10^{-7}\,\mathrm{g\,cm}^{-2}$. Table 2.3 lists the coefficients of the broadband emissivity equation, c_{ni}, for the water vapor ($i = 1, 2, 3$) and ozone bands ($i = 5$). These coefficients were computed from the line-by-line data for three temperatures. For other temperatures, proper interpolations may be performed to obtain needed coefficients for radiative transfer calculations.

Since the CO_2 15 μm band covers a small spectral interval of about $250\,\mathrm{cm}^{-1}$, the preceding approach for deriving the broadband emissivity is not adequate. Other wide-band approaches are required. Based on line-by-line transmittance data presented by Fels and Schwarzkopf (1981), Ou and Liou (1983) have shown that broadband flux emissivity for the CO_2 15 μm band can be parameterized in terms of two exponential forms:

$$\epsilon_4^f(\tilde{u}_4, T) = \begin{cases} \exp\left(\sum_{n=0}^{3} a_n \tilde{u}_4'^n\right), & \tilde{u}_4 \geq 10^{-4}\,\mathrm{g\,cm}^{-2} \\ \exp\left(b_0 + b_1 \tilde{u}_4'\right), & \tilde{u}_4 < 10^{-4}\,\mathrm{g\,cm}^{-2}, \end{cases} \qquad (2.7.26)$$

where $\tilde{u}_4' = (2 \log_{10} \tilde{u} + 7.69897)/6.30103$. The coefficient values are $a_0 = -4.00893 + f(T)$, $a_1 = 4.39828$, $a_2 = -3.07709$, $a_3 = 0.94529$, $b_0 = -4.00360 + f(T)$, and $b_1 = 5.13453$, where the temperature-dependent term is

$$f(T) = \ell n\{h(T)g(T)/[h(T_0)g(T_0)]\}.$$

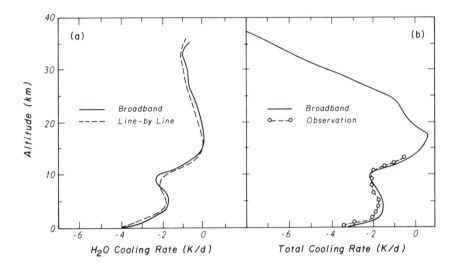

Fig. 2.17 (a) Comparison of cooling-rate profiles for water vapor computed from broadband emissivity and line-by-line methods. A tropical atmospheric profile is used in these calculations. (b) Comparison of total cooling-rate profiles for a clear tropical atmosphere computed from the broadband emissivity method with observations in the tropics reported by Cox (1969).

The quantities $h(T)$ and $g(T)$ are given by

$$h(T) = \frac{1 - T_{\bar{\nu}}(\tilde{u}_4, T)}{1 - T_{\bar{\nu}}(\tilde{u}_4, T_0)} = 1 + A\Delta T(1 - B\Delta T), \qquad (2.7.27)$$

$$g(T) = \int_{\nu_1}^{\nu_2} \pi B_\nu(T) \frac{d\nu}{\sigma T^4}, \qquad (2.7.28)$$

where $T_{\bar{\nu}}(\tilde{u}_4, T)$ is the transmittance; $\Delta T = T - T_0$; $A = 1.833 \times 10^{-4}$; $B = 1.364 \times 10^{-2}$; $T_0 = 250\,\text{K}$; $\nu_1 = 500\,\text{cm}^{-1}$; and $\nu_2 = 850\,\text{cm}^{-1}$. Only the coefficients a_0 and b_0 are temperature dependent, since, by using the definition of $T_{\bar{\nu}}(\tilde{u}_4, T)$ presented by Fels and Schwarzkopf (1981), ϵ^f can be expressed by

$$\epsilon^f(\tilde{u}_4, T) = h(T)g(T)[1 - T_{\bar{\nu}}(\tilde{u}_4, T_0)]. \qquad (2.7.29)$$

In this manner, the dependence of temperature and path length on emissivity has been separated.

By matching the results from band models, the pressure- and temperature-corrected path length that is appropriate for CO_2 may be written in the form

$$\tilde{u}_4 = 2c / \left\{ [1 + 4(c^2/u^2 + c/\bar{u})]^{1/2} - 1 \right\}, \qquad (2.7.30a)$$

where the constant $c = 3.7551 \times 10^{-4}\,\text{g cm}^{-2}$ and

$$\bar{u}_4 = \int_0^z \frac{p(z')}{p_0} \left(\frac{T_0}{T(z')} \right)^{1/2} \rho_c(z')\,dz', \qquad (2.7.30b)$$

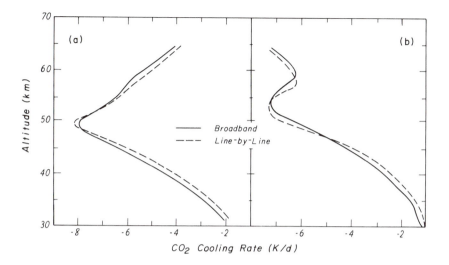

FIG. 2.18 Comparison of the infrared cooling rates due to the 15 μm CO_2 band computed from the broadband parameterization scheme (Ou and Liou, 1983) with those from the line-by-line method of Fels and Schwazkopf (1981). (a) Standard temperature profile; and (b) polar temperature profile.

where ρ_c is the carbon dioxide density and $T_0 = 273$ K. In Eq. (2.7.30) a combination of u and $\bar{u}$ is used, because a single pressure-scaled path length that could provide accurate flux emissivity values is not possible.

In Fig. 2.17(a), ir cooling rates due to water vapor computed from the broadband emissivity approach are presented and compared with those computed from a line-by-line program. Tropical water vapor and temperature profiles are employed in the calculation. The broadband emissivity approach reproduces the cooling rate below 20 km with accuracies generally within about 0.3 K d^{-1}. The results from this approach are reasonably good, as compared to those from line-by-line calculations. Atmospheric cooling is largely produced by the rotational band of water vapor. Near the tropical surface, large cooling on the order of 3–4 K d^{-1} is primarily due to water vapor continuum.

Total cooling rates due to H_2O, CO_2, and O_3 computed from the broadband parameterization program are illustrated in Fig. 2.17(b). Also presented in this figure for comparison is the observed cooling rate profile in the tropics as presented by Cox (1969). Observed cooling rates are derived from the finite differences of net fluxes, which are the sum of measured upward and downward fluxes from aircraft. Errors in these measurements and in the required analyses can propagate into the determination of the cooling rate. Moreover, the atmospheric water vapor and temperature profiles corresponding to the observed cooling rate may not be exactly the same as those for the tropical atmosphere used in theoretical calculations. Nevertheless, the comparison serves as a useful and independent check on the reliability of theoretical results. Theoretical cooling rates are in general agreement with the observed values that are available for the troposphere. In a clear tropical atmosphere, strong cooling in the upper atmosphere due to CO_2 and O_3 is shown. The small positive maximum at ~ 17 km is produced by O_3 coupled with the temperature inversion at that region. Cooling in the troposphere is due primarily to H_2O, as discussed previously.

Figures 2.18(a) and (b) show infrared cooling rates due to the CO_2 15 μm band computed from the foregoing parameterization program using standard and polar temperature profiles. Included in these figures are results calculated from a line-by-line program and presented by Fels and Schwarzkopf (1981). From these comparisons, the broadband results differ from the line-by-line results by no more than 5%. Cooling rates produced by CO_2 absorption and emission depend significantly on the temperature profile, as shown in these two diagrams.

2.7.4 Newtonian cooling approximation

The general circulation of the middle atmosphere from ~ 15 to 90 km is driven by differential vertical and horizontal radiative heating. Absorption of solar insolation in this region is primarily generated by O_3 and O_2 (see Subsection 3.8). This absorption is largely balanced by radiative cooling throughout much of the stratosphere and parts of the mesosphere produced principally by emission and absorption of thermal ir radiation due to CO_2 and O_3. As discussed in the preceding subsection, the computations of the ir cooling rate due to CO_2 and O_3 are extremely involved, even with the parameterized emissivity approach. For applications to the middle atmosphere, the shape and overlap of absorption lines, the variation in line intensity with temperature, and departures from LTE at high altitudes must be taken into account in the computations. The breakdown of LTE, which has not been addressed, is associated with the departure of the absorption coefficient and source function from their equilibrium values by the action of incident radiation (induced emission).

Many attempts have been made to develop accurate yet efficient methods to compute ir cooling rates in the middle atmosphere (Drayson, 1967; Dickinson, 1973; Wehrbein and Leovy, 1982) in connection with dynamic models. The accuracy of computed cooling rates is subject to the accuracy of the input absorption line data. The other limitation to accurate computation of radiative heating and cooling rates in the middle atmosphere has been the lack of accurate data on the atmospheric temperature and absorbing gaseous profiles (Gille and Lyjak, 1986). The most important aspect of radiative cooling from the viewpoint of dynamic circulation is its variation with variations in temperature. The method that is based on the cooling-to-space approximation has been widely used by modelers and is introduced in the following.

Consider the upward and downward fluxes given in Eqs. (2.1.13a) and

(2.1.13b) for a spectral interval in which the variation in Planck fluxes can be neglected. In the height coordinate, we may write

$$F_{\bar{\nu}}^{+}(z) = \pi B_{\bar{\nu}}(0) T_{\bar{\nu}}^{f}(z) + \int_{0}^{z} \pi B_{\bar{\nu}}(z') \frac{d}{dz'} T_{\bar{\nu}}^{f}(z - z') \, dz', \quad (2.7.31)$$

$$F_{\bar{\nu}}^{-}(z) = \int_{z_{\infty}}^{z} \pi B_{\bar{\nu}}(z') \frac{d}{dz'} T_{\bar{\nu}}^{f}(z' - z) \, dz', \quad (2.7.32)$$

where z_{∞} denotes the height at TOA and $z = 0$ denotes the surface. The net flux at a given level is then

$$F_{\bar{\nu}}(z) = F_{\bar{\nu}}^{+}(z) - F_{\bar{\nu}}^{-}(z) = \pi B_{\bar{\nu}}(0) T_{\bar{\nu}}^{f}(z)$$
$$+ \int_{0}^{z_{\infty}} \pi B_{\bar{\nu}}(z') \frac{d}{dz'} T_{\bar{\nu}}^{f}(|z - z'|) \, dz'. \quad (2.7.33)$$

The cooling rate defined in Eq. (2.1.17) for a spectral interval is given by

$$\left(\frac{\partial T}{\partial t} \right)_{\bar{\nu}} = -\frac{1}{\rho C_p} \frac{dF_{\bar{\nu}}(z)}{dz} = -\frac{1}{\rho C_p} \left(\pi B_{\bar{\nu}}(0) \frac{d}{dz} T_{\bar{\nu}}^{f}(z) \right.$$
$$\left. + \int_{0}^{z_{\infty}} \pi B_{\bar{\nu}}(z') \frac{d}{dz'} \frac{d}{dz} T_{\bar{\nu}}^{f}(|z - z'|) \, dz' \right). \quad (2.7.34)$$

Consider an atmosphere with an isothermal temperature profile such that local cooling rates are produced solely from the emission of a local layer; then we have

$$\left(\frac{\partial T}{\partial t} \right)_{\text{space}} = -\frac{1}{\rho C_p} \pi B_{\bar{\nu}}(z) \frac{d}{dz} T_{\bar{\nu}}^{f}(z_{\infty} - z). \quad (2.7.35)$$

This is referred to as the *cooling-to-space approximation* in which the cooling rates are dependent on the local temperature and independent of the temperatures of other levels. Although it is imperfect, this simple approximation gives good results under a number of conditions (Rodgers and Walshaw, 1966). The cooling rate may be expressed in terms of a cooling-to-space term that depends only on the temperature at that level and a term representing the exchange of radiation at that level with all other levels. This latter term is a function of the entire temperature profile. Thus we may write

$$Q = \left(\frac{\partial T}{\partial t} \right)_{\text{ir}} = \left(\frac{\partial T}{\partial t} \right)_{\text{space}} + \left(\frac{\partial T}{\partial t} \right)_{\text{layer exchange}} \quad (2.7.36)$$

In the following we develop a parameterization scheme based on the cooling-to-space approximation in which perturbations in radiative cooling may be expressed in terms of temperature perturbations. Let T_o denote the standard temperature and ΔT a small temperature perturbation. Then by differentiating the cooling rate with respect to temperature and expressing it in a finite-difference form, we have

$$\frac{Q(T_o + \Delta T) - Q(T_o - \Delta T)}{2\Delta T} = a_o(z), \tag{2.7.37}$$

where the term a_o is called the *Newtonian cooling coefficient*, which when it is multiplied by ΔT gives the cooling rate deviation from a standard value Q_o. The preceding discussion provides the foundation for the Newtonian cooling approximation for the calculation of cooling rates in upper atmospheres, where the cooling-to-space approximation is most suitable for cooling calculations.

To correct for variation in temperature in order to obtain a Newtonian cooling coefficient that is valid over a wide range of temperatures for cooling rates in 30–70 km, Dickinson (1973) has developed a general parameterized form based on a second-order Taylor series expansion on the temperature difference as follows:

$$a(z) = a_o(z)\{1 + b[T(z) - T_o(z)]\}, \tag{2.7.38}$$

where b is an empirical coefficient given by

$$b = \begin{cases} f(T_o), & p > 0.2\,\text{mb} \\ f(T_o) + 0.04[1 - 5p(\text{mb})], & p < 0.2\,\text{mb} \end{cases} \tag{2.7.39}$$

with $f(T_o) = 0.033/[T_o(z) - 135]$, and $\Delta T = 0.1$ K in Eq. (2.7.37). In the development of the Newtonian cooling approximation for the efficient calculation of infrared cooling rates between 30 and 70 km, Dickinson (1973) has carried out line-by-line calculations for the 15 μm CO_2 and 9.6 μm O_3 bands using the line parameters available at that time. Figure 2.19 shows the cooling-rate profiles produced by different types of CO_2. These include the fundamental and all the first and second hot bands for $^{12}C^{16}O_2$ and other isotopic bands. The hot bands are important contributors to cooling between 50 to 70 km. The Newtonian cooling coefficients derived from the standard atmosphere at various levels are given in Table 2.4. Intermediate values may be obtained by linear interpolation. Using the Newtonian cooling approximation, the cooling rate deviation for a given level may be written as follows:

$$\Delta Q(z) = a(z)[T(z) - T_o(z)]. \tag{2.7.40}$$

This provides an efficient means of calculating cooling rates and can be effectively incorporated into dynamic–chemical models. The accuracy of the preceding parameterization is within about 0.5 Kd^{-1}, and the scheme is applicable to the region from $\sim$30 to 70 km.

2.7.5 *Theory of remote sounding of flux and cooling rates*

Infrared flux exchanges in the atmosphere and atmospheric cooling produced by these exchanges are significant energy sources in the dynamic and thermodynamic

Table 2.4 Newtonian cooling coefficient and cooling rate for the standard atmospheric temperature profile[a]

$\log_{10}[1/p(mb)]$	z (km)	T(K)	$a_0(d^{-1})$	$Q_0(Kd^{-1})$
4.5	78.5	200	0.016	−1.7
3.125	70.5	216	0.062	0.7
2.2	64.0	234	0.125	4.2
1.8	61.5	241	0.172	7.6
1.0	56.0	258	0.200	9.5
0.5	52.0	269	0.220	12.1
0.0	48.0	271	0.212	11.4
−0.8	41.8	255	0.135	6.7
−2.1	32.6	230	0.080	2.8
−3.0	26.5	223	0.060	2.0

[a] After Dickinson (1973).

processes of the atmosphere (see Subsections 6.5.4 and 7.4.2). In the troposphere under clear conditions, ir cooling is associated with absorption and emission of water vapor molecules, as illustrated in the preceding subsection. In particular,

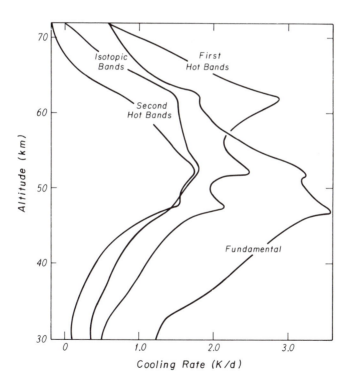

FIG. 2.19 Infrared cooling rate given by different types of CO_2 bands for $^{12}C^{16}O_2$, including the fundamental, all the first hot bands, and all the fundamental bands of other isotopes (isotopic bands) (after Dickinson, 1973). The computations were based on the line data available at that time.

this cooling is primarily produced by the H_2O rotational band and H_2O continuum (near the surface). In the middle atmosphere, cooling is largely produced by the CO_2 15 μm band and, to a lesser degree, by the O_3 9.6 μm band, as pointed out in the preceding subsections.

We have presented in various sections of this chapter the fundamentals of thermal ir radiation transfer and the physical means upon which parameterizations for the entire ir spectrum may be undertaken to economize and optimize the computational efforts. The input data for ir calculations in clear atmospheres are basically temperature and water vapor profiles. (For middle-atmosphere applications, other absorbing gaseous profiles are also needed). If these profiles may be reliably determined from the radiance data derived from satellite measurements, ir fluxes and cooling rates can be computed using a radiative transfer methodology. Radiation algorithms for cooling-rate calculations usually require very large amounts of computer time, and the discrepancies between different algorithms, even with the same input data, are notorious. Direct measurement of ir fluxes and cooling rates from space could have important advantages. An exploration of remote sounding of cooling-rate profiles from space has been made by Liou and Xue (1988) using the H_2O rotational band and simplified expressions for transmittance. In the following, we present the general principles for the determination of the cooling-rate profile and surface ir flux.

Consider the ir cooling rate defined in Eq. (2.1.17) and rewrite this equation in the form

$$\frac{dF(z)}{dz} = -\rho C_p \overset{\bullet}{T}(z), \tag{2.7.41}$$

where $\overset{\bullet}{T} = (\partial T/\partial t)_{ir}$, denoting the cooling rate for a spectral band. There is no obvious relationship between atmospheric cooling rates and radiances at TOA. However, it seems appropriate to assume a direct link between the two, at least to the extent that both are related to temperature and specific gaseous profiles. As demonstrated in Subsection 2.3.4, the retrieval of the profile of any physical parameter requires a set of weighting functions that adequately cover the atmosphere. Consider the following weighting function in the z coordinate:

$$K_j(z) = \rho C_p T_j(u(z)), \tag{2.7.42}$$

where the transmittance for a spectral sub-band (or channel), $\Delta \nu_j$, corresponding to an absorber path length $u(z)$ above the level z, is defined by (see Subsection 2.6.1)

$$T_j(u) = \int_{\Delta \nu_j} e^{-k_\nu u}\, \frac{d\nu}{\Delta \nu} = \int_0^1 e^{-k(g)u}\, dg. \tag{2.7.43}$$

We may introduce a proper pressure- and temperature-correction scheme to adjust the path length.

A convolution of the cooling-rate profile and weighting function defined in Eq. (2.7.42) leads to

$$\int_0^\infty \dot{T}(z) K_j(z) \, dz = -\int_0^\infty T_j(u(z)) \frac{dF(z)}{dz} \, dz. \tag{2.7.44}$$

The expression on the right-hand side of Eq. (2.7.44) can be transformed to the path-length coordinate. To relate this expression to radiances at TOA, several mathematical manipulations are required. First, based on the equations for upward and downward fluxes, the divergence of net flux can be expressed in terms of spectral transmittance $T_{\tilde{\nu}}(u)$, which may be defined in a form similar to channel transmittance, given in Eq. (2.7.43). Second, we note that the integrals associated with spectral and channel transmittances are linear operators. Third, we utilize the mean value theorem and carry out rearrangements of the integrations of path length, u and u', and various terms. The above operations lead to the final result in the form

$$\int_0^{u_1} T_j(u) \frac{dF(u)}{du} \, du = I(\bar{\mu})\alpha + I_j(\mu)\beta, \tag{2.7.45}$$

where u_1 is the total path length, $I(\bar{\mu})$ is the radiance for the spectral band at a mean angle $\cos^{-1} \bar{\mu}$, $I_j(\mu)$ is the channel radiance at an emergent angle, $\cos^{-1} \mu$, and α and β are coefficients associated with the absorption coefficients that may be determined exactly from numerical means. The value of $\bar{\mu}$ varies from ~ 0.5 to 0.6.

Combining Eqs. (2.7.44) and (2.7.45), we obtain the basic Fredholm equation of the first kind for the sounding of cooling rate profiles:

$$I(\bar{\mu})\alpha + I_j(\mu)\beta = -\int_0^\infty \dot{T}(z) K_j(z) \, dz. \tag{2.7.46}$$

Two sets of measurements are needed to derive the spectral cooling rate profiles: spectral radiance and channel radiances, which can both be measured at an emergent angle, $\cos^{-1} \bar{\mu}$. In addition, we must examine the required weighting functions. A number of weighting functions, based on Eq. (2.7.42), have been calculated in the H_2O rotational band for a $20 \, cm^{-1}$ interval using a line-by-line program. Illustrated in Fig. 2.20 are seven weighting functions whose peaks are spaced within $\sim 2 \, km$ in height. The four uppermost and three lowermost channels could be used to retrieve cooling rates for spectral intervals of 20–$500 \, cm^{-1}$ and 500–$800 \, cm^{-1}$, respectively. The preceding theory would be ideal for the retrieval of the cooling rate produced by CO_2 in the middle atmosphere, since there would be no cloud problem. However, appropriate weighting functions must be found in the stratosphere and lower mesosphere for practical applications.

The determination of surface radiative fluxes from available satellite radiance data has been a subject of considerable interest in recent years in view of the importance of surface–atmosphere interactions in climate. A direct observational study

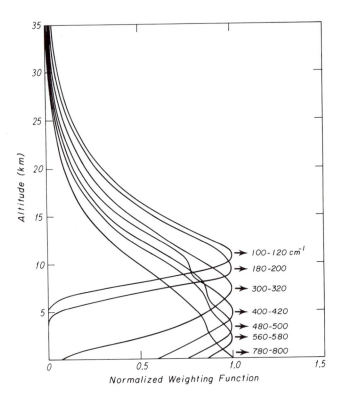

FIG. 2.20 Normalized weighting functions for atmospheric cooling-rate retrieval in the H_2O rotational band (20–800 cm^{-1}). The weighting functions presented here are products of air density and channel transmittances for a 20 cm^{-1} interval using standard atmospheric temperature and water vapor profiles.

Table 2.5 Components of ir radiation budget (W m^{-2})

H_2O rot	H_2O continuum	H_2O vib–rot	O_3	CO_2	H_2O–CO_2 overlap	Total cooling	$F(z_\infty)$	$F(0)$
−128	−4	−9	3	−26	23	−141	250	109

of surface ir fluxes could be made using the following conceptual approach. Based on the definition of cooling rate in Eq. (2.7.41), we may perform an integration of this equation from the surface ($z = 0$) to TOA ($z = z_\infty$) to get

$$F(0) = F(z_\infty) + \int_0^{z_\infty} \rho C_p \dot{T}(z) \, dz. \qquad (2.7.47)$$

The broadband ir flux at TOA, $F(z_\infty)$, has been routinely derived from satellite measurements (see Section 6.4). Thus, if the cooling rate profile $\dot{T}(z)$ is retrievable from radiance observations, it would be possible to evaluate the surface radiative flux $F(0)$ based solely on measurements from space. Table 2.5 shows the contribution of individual bands to the net ir flux and cooling rates, using the standard atmospheric profile. If we omit the contributions due to O_3, CO_2, and overlap, the total cooling would be accurate within $\sim 4\%$ (or $\sim 5\,\mathrm{W\,m^{-2}}$), depending on the temperature profile. Thus, as a first approximation, the total atmospheric cooling may be obtained by measuring the cooling profile produced by water vapor.

The history of infrared soundings from space, as briefly discussed in Subsection 2.3.4, is one of slow evolution of retrieval methodologies and of the selection of sounding channels. To be successful, the ideas presented in this subsection would require refinements, and many additional aspects of the practical issues, such as cloud problems and infrared technology, will need to be examined further.

REFERENCES

Ackerman, T. P., K. N. Liou, and C. B. Leovy, 1976: Infrared radiative transfer in polluted atmospheres. *J. App. Meteor.*, **15**, 28–35.

Ambartzumian, V. A., 1936: The effect of absorption lines on the radiative equilibrium of the outer layers of the stars. *Pub. Observ. Astronom.*, **6**, Univ. Leningrad, Leningrad, 7.

Arking, A., and K. Grossman, 1972: The influence of line shape and band structure on temperatures in planetary atmospheres. *J. Atmos. Sci.*, **29**, 937–949.

Armstrong, B. H., 1967: Spectrum line profiles: The Voigt function. *J. Quant. Spectrosc. Radiat. Transfer*, **7**, 61–88.

Armstrong, B. H., 1968: Analysis of the Curtis–Godson approximation and radiation transmission through inhomogeneous atmospheres. *J. Atmos. Sci.*, **25**, 312–322.

Avrett, E. H., and D. G. Hummer, 1965: Non-coherent scattering. II. Line formation with a frequency independent of source function. *Mon. Not. Roy. Astron. Soc.*, **130**, 295–331.

Brown, L. R., C. B. Farmer, C. P. Rinsland, and R. A. Toth, 1987: Molecular line parameters for the atmospheric trace molecule spectroscopy experiment. *Appl. Opt.*, **26**, 5154–5182.

Burch, D. E., D. A. Gryvnak, R. R. Patty, and C. E. Bartky, 1969: Absorption of infrared radiant energy by CO_2 and H_2O. IV. Shapes of collision-broadened CO_2 lines. *J. Opt. Soc. Amer.*, **59**, 267–280.

Chahine, M. T., N. L. Evans, V. Gilbert, and R. D. Haskins, 1984: Requirements for a passive IR advanced moisture and temperature sounder. *Appl. Opt.*, **23**, 979–989.

Chou, M. D., and A. Arking, 1980: Computation of infrared cooling rates in the water vapor bands. *J. Atmos. Sci.*, **37**, 855–867.

Chou, M. D., and L. Kouvaris, 1986: Monochromatic calculations of atmospheric radiative transfer due to molecular line absorption. *J. Geophys. Res.*, **91**, 4047–4055.

Chou, M. D., D. P. Kratz, and W. Ridgway, 1991: IR radiation parameterizations in numerical climate models. *J. Climate*, **4**, 424–437.

Clough, S .A., F. X. Kneizys, R. Davis, R. Gamache, and R. Tipping, 1980: Theoretical line shape for H_2O vapor: Application to the continuum. In *Atmospheric Water Vapor*, A. Deepak, T. D. Wilkerson, and L. H. Ruhnke, Eds., Academic Press, New York, pp. 25–46.

Conrath, B. J., R. A. Hanel, V. G. Kunde, and C. Prabhakara, 1970: The infrared interferometer experiment on Nimbus 3. *J. Geophys. Res.*, **75**, 5831–5857.

Cox, S. K., 1969: Observational evidence of anomalous infrared cooling in a clear tropical atmosphere. *J. Atmos. Sci.*, **26**, 1347–1349.

Crisp, D., S. B. Fels, and M. D. Schwarzkopf, 1986: Approximate methods for finding CO_2 15 μm band transmission in planetary atmospheres. *J. Geophys. Res.*, **91**, 11,851–11,866.

Curtis, A .R., 1952: Contribution to a discussion of "A statistical model for water vapor absorption", by R. M. Goody. *Quart. J. Roy. Meteor. Soc.*, **78**, 638–640.

Dickinson, R. E., 1973: Method of parameterization for infrared cooling between altitudes of 30 and 70 kilometers. *J. Geophys. Res.*, **78**, 4451–4457.

Domoto, G. A., 1974: Frequency integration for radiative transfer problems involving homogeneous non-gray gases: The inverse transmission function. *J. Quant. Spectrosc. Radiat. Transfer*, **14**, 935–942.

Drayson, S. R., 1967: Atmospheric transmission in the CO_2 bands between 12 μ and 18 μ. *Appl. Opt.*, **5**, 385–391.

Ellingson, R .G., and J. C. Gille, 1978: An infrared radiative transfer model. I. Model description and comparison of observations with calculations. *J. Atmos. Sci.*, **35**, 523–545.

Elsasser, W. M., 1942: *Heat Transfer by Infrared Radiation in the Atmosphere.* Harvard Meteor. Studies, **6**, Harvard University Press, Cambridge, Mass., 107 pp.

Elsasser, W. M., and M. F. Culbertson, 1960: Atmospheric radiation tables. *Meteor. Monogr.*, **4**, 1–43.

Fels, S. B., 1979: Simple strategies for inclusion of Voigt effects in infrared cooling rate calculations. *Appl. Opt.*, **18**, 2634–2637.

Fels, S. B., and M. D. Schwarzkopf, 1981: An efficient, accurate algorithm for calculating CO_2 15 μm band cooling rates. *J. Geophys. Res.*, **86**, 1205–1232.

Fomichev, V. I., and G. M. Shved, 1985: Parameterization of the radiative flux divergence in the 9.6 μm O_3 band. *J. Atmos. Terr. Phys.*, **47**, 1037–1049.

Gille, J. C., and L. V. Lyjak, 1986: Radiative heating and cooling rates in the middle atmosphere. *J. Atmos. Sci.*, **43**, 2215–2229.

Godson, W. L., 1953: The evaluation of infrared radiative fluxes due to atmospheric water vapor. *Quart. J. Roy. Meteor. Soc.*, **79**, 367–379.

Goody, R .M., 1952: A statistical model for water vapor absorption. *Quart. J. Roy. Meteor. Soc.*, **78**, 165–169.

Goody, R. M., 1964a: *Atmospheric Radiation. I: Theoretical Basis*. Oxford University Press, London, 436 pp.

Goody, R. M., 1964b: The transmission of radiation through an inhomogeneous atmosphere. *J. Atmos. Sci.*, **21**, 575–581.

Goody, R. M., R. West, L. Chen, and D. Crisp, 1989: The correlated-k method for radiation calculations in nonhomogeneous atmospheres. *J. Quant. Spectrosc. Radiat. Transfer*, **42**, 539–550.

Grant, W. B, 1990: Water vapor absorption coefficients in the 8–13 μm spectral region: A critical review. *Appl. Opt.*, **29**, 451–462.

Hansen, J. E., G. Russel, D. Rind, P. Stone, A. Lacis, S. Lebedeff, R. Ruedy, and L. Travis, 1983: Efficient three-dimensional global models for climate studies: Models I and II. *Mon. Wea. Rev.*, **111**, 609–662.

Herzberg, G., 1945: *Molecular Spectra and Molecular Structure*. Van Nostrand Reinhold, Princeton, New Jersey, 632 pp.

Isaacs, R. G., R. N. Hoffman, and L. D. Kaplan, 1986: Satellite remote sensing of meteorological parameters for global numerical weather prediction. *Rev. Geophys.*, **24**, 701–743.

Kaplan, L .D., 1959: Inference of atmospheric structure from remote radiation measurements. *J. Opt. Soc. Amer.*, **49**, 1004–1007.

King, J. I. F., 1956: The radiative heat transfer of planet earth. In *Scientific Uses of Earth Satellites*. University of Michigan Press, Ann Arbor, pp. 133–136.

Kneizys, F., J. Chetwynd, R. Fenn, E. Shettle, L. Abreu, R. McClatchey, W. Gallery, J. Selby, 1980: *Atmospheric Transmittance/Radiance Computer Code LOW-TRAN 5*. Scientific Report, AFGL-TR-80-0067, Air Force Geophysics Laboratory, 233 pp.

Kondratyev, K. Ya., 1969: *Radiation in the Atmosphere*. Academic Press, New York, 912 pp.

Kuhn, P. M., 1963: Soundings of observed and computed infrared flux. *J. Geophys. Res.*, **68**, 1415–1420.

Kuhn, W. R., and J. London, 1969: Infrared radiative cooling in the middle atmosphere (30–110 km). *J. Atmos. Sci.*, **26**, 189–204.

Kunde, V. G., B. J. Conrath, R. A. Hanel, W .C. Maguire, C. Prabhakara, and V. V. Salomonson, 1974: The Nimbus IV infrared spectroscopy experiment. 2. Comparison of observed and theoretical radiances from 425–1450 cm^{-1}. *J. Geophys. Res.*, **79**, 777–784.

Lacis, A. A., and J. E. Hansen, 1974: A parameterization for the absorption of solar radiation in the earth's atmosphere. *J. Atmos. Sci.*, **31**, 118–133.

Lacis, A. A., and V. Oinas, 1991: A description of the correlated k-distribution method for modeling non-grey gaseous absorption, thermal emission and multiple scattering in vertically inhomogeneous atmospheres. *J. Geophys. Res.*, **96**, 9027–9063.

Liou, K. N., and T. Sasamori, 1975: On the transfer of solar radiation in aerosol atmospheres. *J. Atmos. Sci.*, **32**, 2166–2177.

Liou, K. N., and S. C. Ou, 1981: Parameterization of infrared radiative transfer in cloudy atmospheres. *J. Atmos. Sci.*, **38**, 2707–2716.

Liou, K. N., and Y. Xue, 1988: Exploration of the remote sounding of infrared cooling rates due to water vapor. *Meteor. Atmos. Phys.*, **38**, 131–139.

Malkmus, W., 1967: Random Lorentz band model with exponential-tailed S^{-1} line intensity distribution function. *J. Opt. Soc. Amer.*, **57**, 323–329.

Möller, F., 1943: *Das Strahlungsdiagramm*. Reichsamt für Wetterdienst (Luftwaffe), Berlin.

Ou, S. C., and K. N. Liou, 1983: Parameterization of carbon dioxide 15 μm band absorption and emission. *J. Geophys. Res.*, **88**, 5203–5207.

Phillips, N., J. Susskind, and L. McMillin, 1988: Results of a joint NOAA/NASA sounder simulation study. *J. Atmos. Oceanic Tech.*, **5**, 44–56.

Ramanathan, V., and P. Downey, 1986: A nonisothermal emissivity and absorptivity formulation for water vapor. *J. Geophys. Res.*, **91**, 8649–8666.

Roberts, E., J. Selby, and L. Biberman, 1976: Infrared continuum absorption by atmospheric water vapor in the 8–12 μm window. *Appl. Opt.*, **15**, 2085–2090.

Rodgers, C. D., and C. D. Walshaw, 1966: The computation of infra-red cooling rate in planetary atmospheres. *Quart. J. Roy. Meteor. Soc.*, **92**, 67–92.

Roewe, D., and K .N. Liou, 1978: Influence of cirrus clouds on the infrared cooling rate in the troposphere and lower stratosphere. *J. Appl. Meteor.*, **17**, 92–106.

Rothman, L. S., R. R. Gamache, A. Goldman, L. R. Brown, R. A. Toth, H. M. Pickett, R. L. Poynter, J.-M. Flaud, C. Camy-Peyret, A. Barbe, N. Husson, C. P. Rinsland, and M. A. H. Smith, 1987: The HITRAN database: 1986 edition. *Appl. Opt.*, **26**, 4058–4097.

Sasamori, T., 1968: The radiative cooling calculation for application to general circulation experiments. *J. Appl. Meteor.*, **7**, 721–729.

Smith, W. L., H. E. Revercomb, H. B. Howell, H. M. Woolf, and D. D. LaPorte, 1987: The high resolution interferometer sounder (HIS). In *Atmospheric Radiation Progress and Prospects*, K. N. Liou and X. Zhou, Eds., Science Press, Beijing, pp. 271–281.

Spänkuch, D., 1989: Effects of line shapes and line coupling on the atmospheric transmittance. *Atmos. Res.*, **23**, 323–344.

Staley, D. O., and G. M. Jurica, 1970: Flux emissivity tables for water vapor, carbon dioxide and ozone. *J. Appl. Meteor.*, **9**, 365–372.

Staley, D. O., and G. M. Jurica, 1972: Effective atmospheric emissivity under clear skies. *J. Appl. Meteor.*, **11**, 349–356.

Suomi, V. E., and P. M. Kuhn, 1958: An economical net radiometer. *Tellus*, **10**, 160–163.

van de Hulst, H. C., 1945: Theory of absorption lines in the atmosphere of the earth. *Ann. Astrophys.*, **8**, 1–10.

Walshaw, C. D., and C. D., Rodgers, 1963: The effect of the Curtis–Godson approximation on the accuracy of radiative heating-rate calculations. *Quart. J. Roy. Meteor. Soc.*, **89**, 122–130.

Wehrbein, W. M., and C. B. Leovy, 1982: An accurate radiative heating and cooling algorithm for use in a dynamical model of the middle atmosphere. *J. Atmos. Sci.*, **39**, 1532–1544.

Wiscombe, W. J., and J .W. Evans, 1977: Exponential-sum fitting of radiative transmission functions. *J. Comp. Phys.*, **24**, 416–444.

Yamamoto, G., 1952: On a radiation chart. *Sci. Rep. Tohoku Univ. Ser. 5 Geophys.*, **4**, 9–23.

Yamamoto, G., and M. Aida, 1970: Transmission in a non-homogeneous atmosphere with an absorbing gas of constant mixing ratio. *J. Quart. Spectrosc. Radiat. Transfer*, **10**, 593–608.

Yamamoto, G., M. Aida, and S. Yamamoto, 1972: Improved Curtis–Godson approximation in a non-homogeneous atmosphere. *J. Atmos. Sci.*, **29**, 1150–1155.

3

THEORY AND PARAMETERIZATION OF SOLAR RADIATIVE TRANSFER

Electromagnetic radiation emitted from the sun's photosphere, having an equivalent blackbody temperature of about 6000 K, has its peak energy located at about 0.47 μm, according to Wien's displacement law, which has been discussed at the beginning of Chapter 2. The total solar flux that is available to a planet is commonly represented by the solar constant. The solar constant for the earth has been discussed in Section 1.1. The distribution of solar fluxes averaged over a certain period of time (e.g., one solar day) is referred to as *solar insolation*. Solar insolation is a function of latitude and the characteristics of the earth's orbit around the sun (see Section 6.1).

The solar wavelengths that are significant for the transfer of solar flux range from ~0.2 to 4 μm. Shown in Fig. 3.1 is an observed solar irradiance at the top of the atmosphere (TOA) with a spectral resolution of 20 cm^{-1}. Fluctuations in the ultraviolet (uv) and visible regions are due to absorption of various elements in the solar atmosphere. A temperature of about 6000 K fits the observed curve closely in the visible and near-infrared (ir) regions. In the uv region ($< 0.4\,\mu$m), the solar irradiance spectrum deviates significantly from the 6000 K Planck curve. Variations in the solar irradiance in the uv are due primarily to sunspot variations. The spectral solar irradiance available at the earth's surface in a clear atmosphere without aerosols and clouds is also shown in Fig. 3.1. The depletion of solar irradiance is due to the scattering of molecules and the absorption of various molecules and atoms, including atomic and molecular nitrogen and oxygen in the uv, ozone and molecular oxygen in the visible, and water vapor (and to a lesser degree carbon dioxide) in the near-ir. A detailed discussion of gaseous absorption in the solar spectrum will be given in Section 3.8.

In view of the spectra depicted in Figs. 2.1 and 3.1, the solar and thermal ir spectra may be separated into two independent regions at about 4 μm. The overlap between these two spectra is relatively insignificant. This distinction makes it possible to treat the transfer of solar radiation independent of the transfer of thermal ir radiation in terms of the source function (and vice versa). For the most part, the

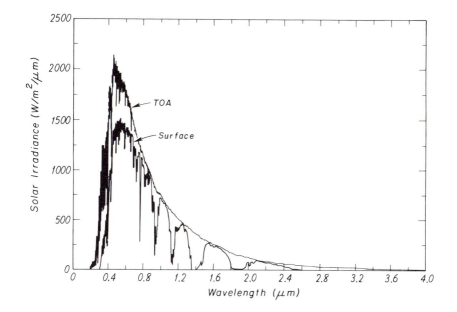

FIG. 3.1 Solar irradiance for a 20 cm^{-1} spectral interval at the top of the atmosphere and at the surface (solar zenith angle 60°) in a clear atmosphere based on the LOWTRAN 7 program (Kneizys et al., 1988).

emission contribution from the earth and the atmosphere can be neglected in the discussion of solar radiative transfer. The one exception to this rule involves the 3.7 μm window, discussed in Subsection 2.2.5.2. If this wavelength is to be used for remote sensing purposes, contributions from solar and thermal ir radiation sources must both be accounted for during daytime.

In the earth's atmosphere, the particulates responsible for scattering range from molecules ($\sim 10^{-4}$ μm), aerosols (~ 1 μm), water droplets (~ 10 μm), and ice crystals (~ 100 μm) to raindrops and hailstones (~ 1 cm). In view of the ubiquitous nature of aerosols and clouds, scattering plays the dominant role in the transfer of solar radiation. The principles and methodologies for radiative transfer presented in this chapter are primarily developed for plane-parallel atmospheres and an isotropic medium. Subjects relating to radiative transfer in clouds will be comprehensively addressed in Chapter 5.

3.1 Basic equations for solar radiative transfer

In the discussion of the transfer of solar radiation in planetary atmospheres, the plane-parallel assumption, described in Section 2.1, can be followed. The position of the sun, which may be considered as a point light source, must be accounted for in the formulation of the basic radiative transfer equation. The transfer problem may be divided into two parts consisting of direct and diffuse components. The

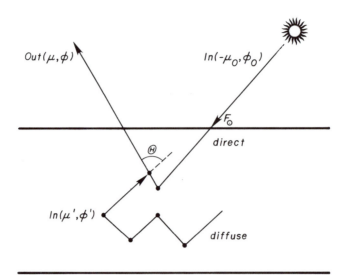

FIG. 3.2 The definitions of direct and diffuse radiation, and the scattering angle, Θ, with respect to solar radiative transfer in plane-parallel atmospheres. The monochromatic solar irradiance at TOA is denoted by $F_{\odot}$.

direct component is associated with the exponential attenuation of unscattered solar flux in the atmosphere. The diffuse component arises from light beams that undergo multiple-scattering events. In the polar coordinate, the directions defining the incoming and outgoing light beams may be expressed by (μ', ϕ') and (μ, ϕ), respectively, where $\mu = \cos\theta$, θ is the zenith angle and ϕ is the azimuthal angle. Let μ and $-\mu$ denote the upward and downward directions associated with the light beams. Thus the position of the sun may be denoted by $(-\mu_0, \phi_0)$, where $\mu_0 = \cos\theta_0$, and θ_0 and ϕ_0 denote the solar zenith and azimuthal angles, as seen in Fig. 3.2. For simplicity of presentation in this chapter, we shall omit the wavelength subscript, λ, in the radiative parameters.

Based on Eq. (2.1.2) and under the plane-parallel assumption, the basic equation governing the transfer of diffuse solar intensity may be written in the form

$$\mu \frac{dI(\tau, \mu, \phi)}{d\tau} = I(\tau, \mu, \phi) - J(\tau, \mu, \phi). \tag{3.1.1}$$

Three factors contribute to the source function: emission, multiple scattering of diffuse intensity, and single scattering of direct solar irradiance (flux density) at TOA, $F_{\odot}$, which is attenuated to the level τ. Energy emitted from the earth and the atmosphere with an equilibrium temperature of $\sim$255 K is practically negligible in comparison with that emitted from the sun. Thus, in the solar region, the source function is given by

$$
J(\tau, \mu, \phi) = \frac{\tilde{\omega}}{4\pi} \int_0^{2\pi} \int_{-1}^1 I(\tau, \mu', \phi') P(\mu, \phi; \mu', \phi') \, d\mu' \, d\phi'
$$
$$
+ \frac{\tilde{\omega}}{4\pi} F_\odot P(\mu, \phi; -\mu_0, \phi_0) e^{-\tau/u_0}, \tag{3.1.2}
$$

where P is the scattering phase function (or, simply, the phase function), which represents the angular distribution of scattered energy as a function of direction. The phase function redirects the incoming intensity in the direction (μ', ϕ') to the direction (μ, ϕ), and the integrals account for all possible scattering events within the 4π solid angle. The single-scattering albedo, $\tilde{\omega}$, is defined as the ratio of the scattering cross section σ_s to the extinction (scattering plus absorption) cross section, σ_e; that is,

$$
\tilde{\omega} = \frac{\sigma_s}{\sigma_e}. \tag{3.1.3}
$$

The single-scattering albedo, phase function, and extinction cross section are fundamental parameters in radiative transfer. These parameters are functions of the incident wavelength, particle size and shape, and refractive index with respect to wavelength. The first and second terms on the right-hand side of Eq. (3.1.2) represent the diffuse (multiple scattering) and direct (single scattering of the direct solar flux) contributions, respectively.

The phase function depends on the incoming and outgoing directions. For spherical particles or nonspherical particles randomly oriented in space, the phase function can be expressed in terms of the scattering angle: the angle defining the incoming and outgoing directions shown in Fig. 3.2. We may express the phase function in terms of a known mathematical function for the purpose of solving Eq. (3.1.2), the first-order differential–integral equation. The Legendre polynomials, by virtue of their unique mathematical properties, have been used extensively in the analysis of radiative transfer problems. In terms of Legendre polynomials, the phase function may be written in the form

$$
P(\cos \Theta) = \sum_{\ell=0}^N \tilde{\omega}_\ell P_\ell(\cos \Theta). \tag{3.1.4}
$$

The Legendre polynomials have the following orthogonal and recurrence properties:

$$
\int_{-1}^1 P_\ell(\mu) P_k(\mu) \, d\mu = \begin{cases} 0, & \ell \neq k \\ \frac{2}{2\ell+1}, & \ell = k \end{cases} \tag{3.1.5}
$$
$$
\mu P_\ell(\mu) = \frac{\ell+1}{2\ell+1} P_{\ell+1} + \frac{\ell}{2\ell+1} P_{\ell-1}. \tag{3.1.6}
$$

From the orthogonal equation, the expansion coefficient is given by

$$\tilde{\omega}_\ell = \frac{2\ell+1}{2} \int_{-1}^{1} P(\cos\Theta) P_\ell(\cos\Theta)\, d\cos\Theta, \qquad \ell = 0, 1, \ldots, N. \quad (3.1.7)$$

In the present notation, the phase function is normalized to unity, viz.,

$$\frac{1}{4\pi} \int_0^{2\pi} \int_{-1}^{1} P(\cos\Theta)\, d\cos\Theta\, d\phi = 1. \qquad (3.1.8)$$

Thus we have $\tilde{\omega}_0 = 1$. From spherical geometry, the cosine of the scattering angle can be expressed in terms of the incoming and outgoing directions in the form

$$\cos\Theta = \mu\mu' + (1-\mu^2)^{1/2}(1-\mu'^2)^{1/2}\cos(\phi'-\phi). \qquad (3.1.9)$$

Using Eq. (3.1.9), the phase function defined in Eq. (3.1.4) may be written

$$P(\mu, \phi; \mu', \phi') = \sum_{\ell=0}^{N} \tilde{\omega}_\ell P_\ell \left(\mu\mu' + (1-\mu^2)^{1/2}(1-\mu'^2)^{1/2}\cos(\phi'-\phi) \right).$$
$$(3.1.10)$$

Moreover, from the addition theorem for Legendre polynomials (see, e.g., Liou, 1980), we find

$$P(\mu, \phi; \mu', \phi') = \sum_{m=0}^{N} \sum_{\ell=m}^{N} \tilde{\omega}_\ell^m P_\ell^m(\mu) P_\ell^m(\mu') \cos m(\phi'-\phi), \qquad (3.1.11a)$$

where P_ℓ^m denotes the associated Legendre polynomials, and

$$\tilde{\omega}_\ell^m = (2-\delta_{0,m})\tilde{\omega}_\ell \frac{(\ell-m)!}{(\ell+m)!}, \qquad \ell = m, \ldots, N, 0 \leq m \leq N, \quad (3.1.11b)$$

$$\delta_{0,m} = \begin{cases} 1, & m=0 \\ 0, & \text{otherwise.} \end{cases}$$

In view of the expansion of the phase function, the diffuse intensity may also be expanded in the cosine series in the form

$$I(\tau, \mu, \phi) = \sum_{m=0}^{N} I^m(\tau, \mu) \cos m(\phi_0-\phi). \qquad (3.1.12)$$

Substituting Eqs. (3.1.11) and (3.1.12) into Eq. (3.1.2), and using the orthogonality of the associated Legendre polynomials, the equation of transfer splits into $(N+1)$ independent equations, and may be written

$$\mu \frac{dI^m(\tau, \mu)}{d\tau} = I^m(\tau, \mu) - (1+\delta_{0,m})\frac{\tilde{\omega}}{4} \sum_{\ell=m}^{N} \tilde{\omega}_\ell^m P_\ell^m(\mu)$$

$$\times \int_{-1}^{1} P_\ell^m(\mu') I^m(\tau, \mu') \, d\mu'$$

$$-\frac{\tilde{\omega}}{4\pi} \sum_{\ell=m}^{N} \tilde{\omega}_\ell^m P_\ell^m(\mu) P_\ell^m(-\mu_0) F_\odot e^{-\tau/\mu_0},$$

$$m = 0, 1, \ldots N. \tag{3.1.13}$$

Omitting the superscript 0 for simplicity of presentation, for $m = 0$ we have

$$\mu \frac{dI(\tau, \mu)}{d\tau} = I(\tau, \mu) - \frac{\tilde{\omega}}{2} \sum_{\ell=0}^{N} \tilde{\omega}_\ell P_\ell(\mu) \int_{-1}^{1} P_\ell(\mu') I(\tau, \mu') \, d\mu'$$

$$-\frac{\tilde{\omega}}{4\pi} \sum_{\ell=0}^{N} \tilde{\omega}_\ell P_\ell(\mu) P_\ell(-\mu_0) F_\odot e^{-\tau/\mu_0}. \tag{3.1.14}$$

This is the equation of transfer that is independent of the azimuthal angle. From Eq. (3.1.11a), the azimuth-independent phase function may be obtained from

$$P(\mu, \mu') = \frac{1}{2\pi} \int_0^{2\pi} P(\mu, \phi; \mu', \phi') \, d\phi'$$

$$= \begin{cases} \sum_{\ell=0}^{N} \tilde{\omega}_\ell P_\ell(\mu) P_\ell(\mu'), & m = 0, \\ 0, & m \neq 0. \end{cases} \tag{3.1.15}$$

Equation (3.1.14) may then be expressed in terms of the azimuth-independent phase function in the form

$$\mu \frac{dI(\tau, \mu)}{d\tau} = I(\tau, \mu) - \frac{\tilde{\omega}}{2} \int_{-1}^{1} I(\tau, \mu') P(\mu, \mu') \, d\mu'$$

$$-\frac{\tilde{\omega}}{4\pi} P(\mu, -\mu_0) F_\odot e^{-\tau/\mu_0}. \tag{3.1.16}$$

The azimuth-independent phase function defined in Eq. (3.1.15) has the following properties:

$$\frac{1}{2} \int_{-1}^{1} P(\mu, \mu') \, d\mu' = \tilde{\omega}_0 = 1, \tag{3.1.17}$$

$$\frac{1}{2} \int_{-1}^{1} P(\mu, \mu') \mu' \, d\mu' = \tilde{\omega}_1 \mu/3. \tag{3.1.18}$$

The monochromatic upward and downward diffuse fluxes at a given optical depth level, τ, are defined by

$$F_{\text{dif}}^{\pm}(\tau) = \int_0^{2\pi} \int_0^{\pm 1} I(\tau, \mu, \phi) \mu \, d\mu \, d\phi, \tag{3.1.19a}$$

where the superscripts $+$ and $-$ denote the upward and downward fluxes, respectively. Using Eq. (3.1.12), we find

$$F_{\text{dif}}^{\pm}(\tau) = 2\pi \int_0^{\pm 1} I(\tau, \mu)\mu \, d\mu. \tag{3.1.19b}$$

For solar flux computations, it suffices to consider the azimuthally averaged equation for the transfer of diffuse radiation. The direct downward solar flux at level τ is given by the exponential attenuation of the effective solar flux at TOA, $\mu_0 F_\odot$. Thus,

$$F_{\text{dir}}^-(\tau) = \mu_0 F_\odot e^{-\tau/\mu_0}. \tag{3.1.20}$$

The total upward and downward fluxes covering the entire solar spectrum, using the height coordinate, may be written

$$F^+(z) = \int_0^\infty F_{\text{dif}}^+(\tau) \, d\lambda, \tag{3.1.21}$$

$$F^-(z) = \int_0^\infty (F_{\text{dif}}^- + F_{\text{dir}}^-) \, d\lambda. \tag{3.1.22}$$

Thus the net flux is

$$F_s(z) = F^-(z) - F^+(z). \tag{3.1.23}$$

The heating rate due to the absorption of solar flux in the atmosphere is produced by the divergence of the net solar flux, and is given by

$$\left(\frac{\partial T}{\partial t}\right)_s = \frac{1}{\rho C_p} \frac{dF_s(z)}{dz}, \tag{3.1.24}$$

where ρ is the air density and C_p is the specific heat at constant pressure.

3.2   Exact methods for radiative transfer

3.2.1   *Discrete-ordinates method*

The discrete-ordinates method for radiative transfer has been elegantly developed by Chandrasekhar (1950) for applications to the transfer of radiation in planetary atmospheres. Liou (1973a) has demonstrated that the discrete-ordinates method is a useful and powerful method for the computation of radiation fields in aerosol and cloudy atmospheres. In essence, the method involves the discretization of the basic radiative transfer equation and the solution of a set of first-order differential equations. With the advance in numerical techniques for solving differential equations, the discrete-ordinates method has been found to be both efficient and accurate for calculations of scattered intensities and fluxes (Stamnes and Swanson, 1981; Stamnes and Dale, 1981).

In presenting the discrete-ordinates method, we begin with the general equation of transfer in terms of a set of independent components that have been given in Eq. (3.1.13), and rewrite these equations as follows:

$$\mu \frac{dI^m(\tau, \mu)}{d\tau} = I^m(\tau, \mu) - J^m(\tau, \mu). \tag{3.2.1}$$

The source function is given by

$$J^m(\tau, \mu) = (1 + \delta_{0,m})\frac{\tilde{\omega}}{4} \sum_{\ell=m}^{N} \tilde{\omega}_\ell^m(\mu) P_\ell^m(\mu) \int_{-1}^{1} P_\ell^m(\mu') I^m(\tau, \mu') \, d\mu'$$

$$+ \frac{\tilde{\omega}}{4\pi} \sum_{\ell=m}^{N} \tilde{\omega}_\ell^m P_\ell^m(\mu) P_\ell^m(-\mu_0) F_\odot e^{-\tau/\mu_0}. \tag{3.2.2}$$

To proceed with the solution of Eq. (3.2.1), we first discretize the equation by replacing μ with μ_i ($i = -n, \ldots, n$, with $n = 1, 2, \ldots$) and the integral,

$$\int_{-1}^{1} f(\mu) \, d\mu = \sum_{j=-n}^{n} f(\mu_j) a_j,$$

with the weight a_j. The homogeneous solution for the set of first-order differential equations may be written

$$I^m(\tau, \mu_i) = \sum_{j=-n}^{n} L_j^m \phi_j^m(\mu_i) e^{-k_j^m \tau}, \tag{3.2.3}$$

where ϕ_j^m and k_j^m denote the eigenvectors and eigenvalues, respectively, and L_j^m are coefficients to be determined from appropriate boundary conditions. On substituting Eq. (3.2.3) into the homogeneous part of Eq. (3.2.1), the eigenvectors may be expressed by

$$\phi_j^m(\mu_i) = \frac{(1 + \delta_{0,m})\tilde{\omega}}{4(1 + \mu_j k_j^m)} \sum_{\ell=m}^{N} \tilde{\omega}_\ell^m P_\ell^m(\mu_i) \sum_{\alpha=-n}^{n} a_\alpha P_\ell^m(\mu_\alpha) \phi_j^m(\mu_\alpha). \tag{3.2.4}$$

The particular solution may be written in the form

$$I_p^m(\tau, \mu_i) = Z^m(\mu_i) e^{-\tau/\mu_0}. \tag{3.2.5}$$

From Eq. (3.2.1) we have

$$Z^m(\mu_i) = \frac{\tilde{\omega}}{4(1 + \mu_i/\mu_0)} \sum_{\ell=m}^{N} \tilde{\omega}_\ell^m P_\ell^m(\mu_i)$$

$$\times \left(\sum_{\alpha=-n}^{n} a_\alpha P_\ell^m(\mu_\alpha) Z^m(\mu_\alpha) + P_\ell^m(-\mu_0)\frac{F_\odot}{\pi} \right). \tag{3.2.6}$$

Equations (3.2.4) and (3.2.6) are linear equations in ϕ_j^m and Z^m and may be solved numerically. The complete solution for Eq. (3.2.1) is the sum of the general solution for the associated homogeneous system of the differential equations and the particular solution. Thus,

$$I^m(\tau, \mu_i) = \sum_{j=-n}^{n} L_j^m \phi_j^m(\mu_i) e^{-k_j^m \tau} + Z^m(\mu_i) e^{-\tau/\mu_o}, \qquad i = -n, \ldots, n.$$

$$(3.2.7)$$

In order to determine the unknown coefficients, L_j^m, boundary conditions must be imposed. Assuming that there are no external radiation sources from above or below a layer with an optical depth of τ_*, we have

$$\left. \begin{array}{l} I^m(0, -\mu_i) = 0 \\ I^m(\tau_*, +\mu_i) = 0 \end{array} \right\} \text{ for } i = 1, \ldots, n \text{ and } m = 0, \ldots, N. \qquad (3.2.8)$$

A mathematical procedure from which the eigenvalues k_j^m may be calculated from a recurrence characteristic equation has been developed by Chandrasekhar (1950). The eigenvectors $\phi_j^m(\mu_i)$ may be expressed in terms of known functions, which contain the eigenvalues, and the particular solution is related to a known mathematical function, the so-called H function. The characteristic equation for the eigenvalues derived by Chandrasekhar is mathematically, as well as numerically, ambiguous. The method is unstable for highly peaked phase functions, as pointed out by Liou (1973a), who discovered that the solution of the characteristic equation may be formulated as an algebraic eigenvalue problem. Further, Asano (1975) has shown that the degree of the characteristic equation for the eigenvalues can be reduced by a factor of two because the solution for the eigenvalues may be obtained by solving a characteristic polynomial of degree n for k^2. Both of these authors have expanded the matrix in polynomial form to solve the characteristic equation for the eigenvalues corresponding to the associated homogeneous system of the differential equations. However, the expansion in polynomial form is not a stable numerical scheme for obtaining eigenvalues. To solve the algebraic eigenvalue problem, a well-developed numerical subroutine found in the IMSL User's Manual (1987) may be used to compute the eigenvalues and eigenvectors of a real general matrix in connection with the discrete-ordinates method. Stamnes and Dale (1981) have shown that azimuthally dependent scattered intensities may be computed accurately and efficiently using numerical methods.

In the discrete-ordinates method for radiative transfer, analytical solutions for the diffuse intensity are explicitly given for any optical depth. Thus the internal radiation field can be evaluated without additional computational effort. Moreover, useful approximations can be developed from this method for flux calculations.

We now confine our discussion to the transfer of solar fluxes, and consider the azimuth-independent component in the diffuse intensity component. On replacing the integral with a summation, Eq. (3.1.16) may be written in the form

$$\mu_i \frac{dI(\tau, \mu_i)}{d\tau} = I(\tau, \mu_i) - \frac{\tilde{\omega}}{2} \sum_{j=-n}^{n} I(\tau, \mu_j)P(\mu_i, \mu_j)a_j$$

$$- \frac{\tilde{\omega}}{4\pi} F_\odot P(\mu_i, -\mu_0)e^{-\tau/\mu_0}, \qquad i = -n, \dots, n, \quad (3.2.9)$$

where we may select the quadrature weights and points that satisfy $a_{-j} = a_j (\sum_j a_j = 2)$ and $\mu_{-j} = -\mu_j$. To simplify this equation, we may define

$$c_{i,j} = \frac{\tilde{\omega}}{2} a_j P(\mu_i, \mu_j) = \frac{\tilde{\omega}}{2} a_j \sum_{\ell=0}^{N} \tilde{\omega}_\ell P_\ell(\mu_i)P_\ell(\mu_j),$$

$$j = -n, \dots - 0, \dots, n, \qquad\qquad\qquad (3.2.10)$$

and

$$I(\tau, -\mu_0) = e^{-\tau/\mu_0} F_\odot/2\pi, \qquad\qquad (3.2.11)$$

where we set $a_{-0} = 1$ and the notation -0 is used to be consistent with the definition, $\mu_{-0} = -\mu_0$. By virtue of the definiton of Legendre polynomials, we have

$$c_{i,-j} = c_{-i,j}, \qquad c_{-i,-j} = c_{i,j}, \qquad i \neq -0. \qquad (3.2.12)$$

Moreover, we may define

$$b_{i,j} = \begin{cases} c_{i,j}/\mu_i, & i \neq j \\ (c_{i,j} - 1)/\mu_i, & i = j. \end{cases} \qquad (3.2.13)$$

It follows that $b_{i,j} = -b_{-i,-j}$, and $b_{i,-j} = -b_{-i,-j}$. Using the preceding definitions, Eq. (3.2.9) becomes

$$\frac{dI(\tau, \mu_i)}{d\tau} = \sum_j b_{i,j} I(\tau, \mu_j). \qquad (3.2.14)$$

We may separate the upward and downward intensities in the forms

$$\frac{dI(\tau, \mu_i)}{d\tau} = \sum_{j=1}^{n} b_{i,j} I(\tau, \mu_i) + \sum_{j=0}^{n} b_{i,-j} I(\tau, -\mu_j), \qquad (3.2.15a)$$

$$\frac{dI(\tau, -\mu_i)}{d\tau} = \sum_{j=1}^{n} b_{-i,j} I(\tau, \mu_j) + \sum_{j=0}^{n} b_{-i,-j} I(\tau, -\mu_j). \qquad (3.2.15b)$$

In terms of a matrix representation for the homogeneous part, we write

$$\frac{d}{d\tau} \begin{bmatrix} \mathbf{I}^+ \\ \mathbf{I}^- \end{bmatrix} = \begin{bmatrix} \mathbf{b}^+ & \mathbf{b}^- \\ -\mathbf{b}^- & -\mathbf{b}^+ \end{bmatrix} \begin{bmatrix} \mathbf{I}^+ \\ \mathbf{I}^- \end{bmatrix}, \tag{3.2.16}$$

where

$$\mathbf{I}^{\pm} = \begin{bmatrix} I(\tau, \pm\mu_1) \\ I(\tau, \pm\mu_2) \\ \cdot \\ \cdot \\ \cdot \\ I(\tau, \pm\mu_n) \end{bmatrix}, \tag{3.2.17}$$

and $\mathbf{b}^{\pm}$ denotes the elements associated with $b_{i,j}$ and $b_{i,-j}$. Since Eq. (3.2.16) is a first-order differential equation, we may seek a solution in the form

$$\mathbf{I}^{\pm} = \phi^{\pm} e^{-k\tau}. \tag{3.2.18}$$

Substituting Eq. (3.2.18) into Eq. (3.2.16) leads to

$$\begin{bmatrix} \mathbf{b}^+ & \mathbf{b}^- \\ -\mathbf{b}^- & -\mathbf{b}^+ \end{bmatrix} \begin{bmatrix} \phi^+ \\ \phi^- \end{bmatrix} = -k \begin{bmatrix} \phi^+ \\ \phi^- \end{bmatrix}. \tag{3.2.19}$$

Equation (3.2.19) may be solved as a standard eigenvalue problem. In the discrete-ordinates method for radiative transfer, the eigenvalues associated with the differential equations are all real and occur in pairs ($\pm k$) because of the symmetry of the $\mathbf{b}$ matrix. Thus the rank of the matrix may be reduced by a factor of 2. To accomplish this reduction, we rewrite Eq. (3.2.19) in the forms

$$\mathbf{b}^+ \phi^+ + \mathbf{b}^- \phi^- = -k\phi^+, \tag{3.2.20a}$$

$$\mathbf{b}^- \phi^+ + \mathbf{b}^+ \phi^- = k\phi^-, \tag{3.2.20b}$$

Adding and subtracting these two equations yield

$$(\mathbf{b}^+ - \mathbf{b}^-)(\mathbf{b}^+ + \mathbf{b}^-)(\phi^+ + \phi^-) = k^2(\phi^+ + \phi^-). \tag{3.2.21}$$

Hence, the eigenvectors of the original system, $\phi^{\pm}$, can now be obtained from the reduced system, $(\phi^+ + \phi^-)$, in terms of the eigenvalue k^2. As discussed by Chandrasekhar (1950), the Gaussian quadrature formula for the complete angular range, $-1 < \mu < 1$, is efficient and accurate for the discretization of the basic radiative transfer equation. However, the Gaussian quadrature can also be applied separately to the half-ranges, $-1 < \mu < 0$ and $0 < \mu < 1$, which are referred to as the *double-Gauss quadrature* and appear to offer numerical advantages when upward and downward radiation streams are treated separately.

Radiation from above

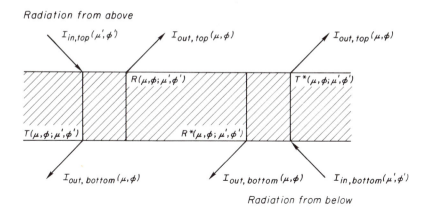

FIG. 3.3   Configurations for radiation incident from above and below, and the definitions of the reflection and transmission functions.

3.2.2   Adding method

The adding method has been demonstrated to be a powerful tool for multiple-scattering calculations. The principle for the method was stated by Stokes (1862) in a problem dealing with reflection and transmission by glass plates. Peebles and Plesset (1951) have developed the adding method theory for application to gamma-ray transfer. van de Hulst (1980) has presented a set of adding equations for multiple scattering that is now commonly used.

To introduce the adding method for radiative transfer, the reflection function R and the transmission function T must first be defined. Consider a light beam incident from above, as represented in Fig. 3.3. The reflected and transmitted intensities of this beam are expressed in terms of the incident intensity in the form

$$I_{\text{out,top}}(\mu, \phi) = \frac{1}{\pi} \int_0^{2\pi} \int_0^1 R(\mu, \phi; \mu', \phi') I_{\text{in,top}}(\mu', \phi') \mu' \, d\mu' \, d\phi', \quad (3.2.22)$$

$$I_{\text{out,bottom}}(\mu, \phi) = \frac{1}{\pi} \int_0^{2\pi} \int_0^1 T(\mu, \phi; \mu', \phi') I_{\text{in,top}}(\mu', \phi') \mu' \, d\mu' \, d\phi', \quad (3.2.23)$$

Likewise, if the light beam comes from below, as is also shown in Fig. 3.3, we write

$$I_{\text{out,bottom}}(\mu, \phi) = \frac{1}{\pi} \int_0^{2\pi} \int_0^1 R^*(\mu, \phi; \mu', \phi') I_{\text{in,bottom}}(\mu', \phi') \mu' \, d\mu' \, d\phi', \quad (3.2.24)$$

$$I_{\text{out,top}}(\mu, \phi) = \frac{1}{\pi} \int_0^{2\pi} \int_0^1 T^*(\mu, \phi; \mu', \phi') I_{\text{in,bottom}}(\mu', \phi') \mu' \, d\mu' \, d\phi', \quad (3.2.25)$$

where R^* and T^* are so defined, and the superscript $*$ signifies that radiation comes from below.

Consider the transfer of monochromatic solar radition. The incident solar intensity, in the present notation, may be written in the form

$$I_{in,top}(-\mu_0, \phi_0) = \delta(\mu' - \mu_0)\delta(\phi' - \phi_0)F_\odot, \qquad (3.2.26)$$

where δ is the Dirac delta function. Using Eq. (3.2.26), the reflection and transmission functions defined in Eqs. (3.2.22) and (3.2.23) are given by

$$R(\mu, \phi; \mu_0, \phi_0) = \pi I_{out,top}(\mu, \phi)/\mu_0 F_\odot, \qquad (3.2.27)$$

$$T(\mu, \phi; \mu_0, \phi_0) = \pi I_{out,bottom}(\mu, \phi)/\mu_0 F_\odot. \qquad (3.2.28)$$

Under the single-scattering approximation, the source function defined in Eq. (3.1.2) may be written in the form

$$J(\tau, \mu, \phi) = \frac{\tilde{\omega}}{4\pi} F_\odot P(\mu, \phi; -\mu_0, \phi_0)e^{-\tau/\mu_0}. \qquad (3.2.29)$$

Assuming that there are no diffuse intensities from the top and bottom of the layer with an optical depth $\Delta\tau$, then the radiation boundary conditions are as follows:

$$I_{in,top}(\mu, \phi) = 0$$

$$I_{in,bottom}(\mu, \phi) = 0. \qquad (3.2.30)$$

With these boundary conditions, the reflected and transmitted diffuse intensities due to single scattering can be derived directly from the basic radiative transfer equation. Thus the solutions for the reflection and transmission functions for an optical depth $\Delta\tau$ are given by

$$R(\mu, \phi; \mu_0, \phi_0) = \frac{\tilde{\omega}}{4(\mu + \mu_0)} P(\mu, \phi; -\mu_0, \phi_0)\left\{1 - \exp\left[-\Delta\tau\left(\frac{1}{\mu} + \frac{1}{\mu_0}\right)\right]\right\}, \qquad (3.2.31)$$

$$T(\mu, \phi; \mu_0, \phi_0) = \begin{cases} \dfrac{\tilde{\omega}}{4(\mu - \mu_0)} P(-\mu, \phi; -\mu_0, \phi_0)(e^{-\Delta\tau/\mu} - e^{-\Delta\tau/\mu_0}), & \mu \neq \mu_0 \\[2mm] \dfrac{\tilde{\omega}\Delta\tau}{4\mu_0^2} P(-\mu, \phi; -\mu_0, \phi_0)e^{-\Delta\tau/\mu_0}, & \mu = \mu_0. \end{cases} \qquad (3.2.32)$$

If we consider a layer in which $\Delta\tau$ is very small (e.g., $\Delta\tau \approx 10^{-8}$), Eqs. (3.2.31) and (3.2.32) may further be simplified in the forms

$$R(\mu, \phi; \mu_0, \phi_0) = \frac{\tilde{\omega}\Delta\tau}{4\mu\mu_0} P(\mu, \phi; -\mu_0, \phi_0), \qquad (3.2.33)$$

$$T(\mu, \phi; \mu_0, \phi_0) = \frac{\tilde{\omega}\Delta\tau}{4\mu\mu_0} P(-\mu, \phi; -\mu_0, \phi_0). \qquad (3.2.34)$$

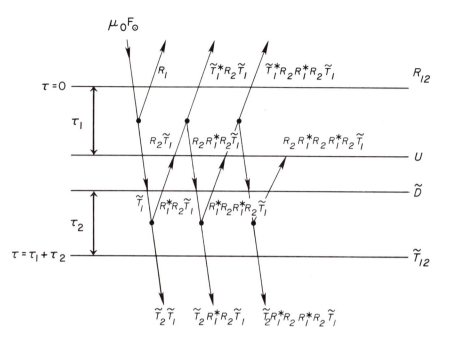

FIG. 3.4 Configuration of the adding method. The two layers of optical depths τ_1 and τ_2 are rendered, for convenient illustration, as though they were physically separated.

For a thin homogeneous layer, the reflection and transmission functions will be the same regardless of whether the light beam is incident from above or below. Thus, $R^* = R$ and $T^* = T$. However, when we proceed with the adding principle for radiative transfer, the reflection and transmission functions for combined layers will depend on the direction of the incoming light beam.

In reference to Fig. 3.4, consider two layers, one on top of the other. Let the reflection and total (direct and diffuse) transmission functions be denoted R_1 and $\tilde{T}_1$ for the first layer and as R_2 and $\tilde{T}_2$ for the second layer, respectively. We define $\tilde{D}$ and U for the combined total transmission and reflection functions between layers 1 and 2. In principle, the light beam may undergo an infinite number of scattering events. Upon tracing the light beam in the two layers, we find the combined reflection and total transmission functions as follows:

$$
\begin{aligned}
R_{12} &= R_1 + \tilde{T}_1^* R_2 \tilde{T}_1 + T_1^* R_2 R_1^* R_2 \tilde{T}_1 + T_1^* R_2 R_1^* R_2 R_1^* R_2 \tilde{T}_1 + \cdots \\
&= R_1 + \tilde{T}_1^* R_2 \left[1 + R_1^* R_2 + (R_1^* R_2)^2 + \cdots \right] \tilde{T}_1 \\
&= R_1 + \tilde{T}_1^* R_2 (1 - R_1^* R_2)^{-1} \tilde{T}_1, \qquad\qquad (3.2.35) \\
\tilde{T}_{12} &= \tilde{T}_2 \tilde{T}_1 + \tilde{T}_2 R_1^* R_2 \tilde{T}_1 + \tilde{T}_2 R_1^* R_2 R_1^* R_2 \tilde{T}_1 + \cdots \\
&= \tilde{T}_2 \left[1 + R_1^* R_2 + (R_1^* R_2)^2 + \cdots \right] \tilde{T}_1 \\
&= \tilde{T}_2 (1 - R_1^* R_2)^{-1} \tilde{T}_1. \qquad\qquad (3.2.36)
\end{aligned}
$$

Likewise, the expressions for U and $\tilde{D}$ are given by

$$
\begin{aligned}
U &= R_2\tilde{T}_1 + R_2 R_1^* R_2 \tilde{T}_1 + R_2 R_1^* R_2 R_1^* R_2 \tilde{T}_1 + \cdots \\
&= R_2 \left[1 + R_1^* R_2 + (R_1^* R_2)^2 + \cdots \right] \tilde{T}_1 \\
&= R_2 (1 - R_1^* R_2)^{-1} \tilde{T}_1, \tag{3.2.37} \\
\tilde{D} &= \tilde{T}_1 + R_1^* R_2 \tilde{T}_1 + R_1^* R_2 R_1^* R_2 \tilde{T}_1 + \cdots \\
&= \left[1 + R_1^* R_2 + (R_1^* R_2)^2 + \cdots \right] \tilde{T}_1 \\
&= (1 - R_1^* R_2)^{-1} \tilde{T}_1. \tag{3.2.38}
\end{aligned}
$$

In Eqs. (3.2.35)–(3.2.38), the infinite series is replaced by a single inverse function. We may define an operator in the form

$$
S = R_1^* R_2 (1 - R_1^* R_2)^{-1}. \tag{3.2.39}
$$

Thus, $(1 - R_1^* R_2)^{-1} = 1 + S$. From the preceding adding equations, we have

$$
\begin{aligned}
R_{12} &= R_1 + \tilde{T}_1^* U, &\text{(3.2.40a)} \\
\tilde{T}_{12} &= \tilde{T}_2 \tilde{D}, &\text{(3.2.40b)} \\
U &= R_2 \tilde{D}. &\text{(3.2.40c)}
\end{aligned}
$$

At this point, we wish to separate the diffuse and direct components of the total transmission function, which is defined by

$$
\tilde{T} = T + e^{-\tau/\mu'}, \tag{3.2.41}
$$

where $\mu' = \mu_0$ when transmission is associated with the incident solar beam, and $\mu' = \mu$ when it is associated with the emergent light beam in the direction μ. Using Eq. (3.2.41), we may separate the direct and diffuse components in Eqs. (3.2.38) and (3.2.40b) to obtain

$$
\begin{aligned}
\tilde{D} &= D + e^{-\tau_1/\mu_0} = (1 + S)(T_1 + e^{-\tau_1/\mu_0}) \\
&= (1 + S)T_1 + Se^{-\tau_1/\mu_0} + e^{-\tau_1/\mu_0}, \tag{3.2.42} \\
\tilde{T}_{12} &= (T_2 + e^{-\tau_2/\mu})(D + e^{-\tau_1/\mu_0}) \\
&= e^{-\tau_2/\mu} D + T_2 e^{-\tau_1/\mu_0} + T_2 D \\
&\quad + \exp\left[-\left(\frac{\tau_1}{\mu_0} + \frac{\tau_2}{\mu} \right) \right] \delta(\mu - \mu_0), \tag{3.2.43}
\end{aligned}
$$

where D, T_1, and T_2 denote the diffuse components only, and a delta function is added to the exponential term to signify that the direct transmission function is a function of μ_0 only. On the basis of the preceding analysis, a set of iterative equations for the computation of diffuse transmission and reflection for the two layers may be written in the forms

$$Q = R_1^* R_2, \tag{3.2.44a}$$

$$S = Q(1 - Q)^{-1}, \tag{3.2.44b}$$

$$D = T_1 + S T_1 + S e^{-\tau_1/\mu_0}, \tag{3.2.44c}$$

$$U = R_2 D + R_2 e^{-\tau_1/\mu_0}, \tag{3.2.44d}$$

$$R_{12} = R_1 + e^{-\tau_1/\mu} + T_1^* U, \tag{3.2.44e}$$

$$T_{12} = e^{-\tau_2/\mu} D + T_2 e^{-\tau_1/\mu_0} + T_2 D. \tag{3.2.44f}$$

The direct transmission function for the combined layer is given by $\exp[-(\tau_1 + \tau_2)/\mu_0]$. In these equations, the product of two functions implies an integration over the appropriate solid angle so that all possible multiple-scattering contributions are accounted for, as in the following example:

$$R_1^* R_2 = \frac{1}{\pi} \int_0^{2\pi} \int_0^1 R_1^*(\mu, \phi; \mu', \phi') R_2(\mu', \phi'; \mu_0, \phi_0) \mu' \, d\mu' \, d\phi'. \tag{3.2.45}$$

In the numerical computations, we may set $\tau_1 = \tau_2$. This is referred to as the *doubling method*. We start with an optical depth $\Delta\tau$ on the order of 10^{-8} and use Eqs. (3.2.33) and (3.2.34) to compute the reflection and transmission functions. Equations (3.2.44a–f) are subsequently employed to compute the reflection and transmission functions for an optical depth of $2\Delta\tau$. For the initial layers, $R_{1,2}^* = R_{1,2}$ and $T_{1,2}^* = T_{1,2}$. Using the adding equations, the computations may be repeated until a desirable optical depth is achieved.

For radiation emergent from below, R_{12}^* and T_{12}^* may be computed from a scheme analogous to Eq. (3.2.44). Let the incident direction be μ'; then the adding equations are as follows:

$$Q = R_2 R_1^*, \tag{3.2.46a}$$

$$S = Q(1 - Q)^{-1}, \tag{3.2.46b}$$

$$U = T_2^* + S T_2^* + S e^{-\tau_2/\mu'}, \tag{3.2.46c}$$

$$D = R_1^* U + R_1^* e^{-\tau_2/\mu'}, \tag{3.2.46d}$$

$$R_{12}^* = R_2^* + T_2 D + e^{-\tau_2/\mu} D, \tag{3.2.46e}$$

$$T_{12}^* = T_1^* U + e^{-\tau_1/\mu} U + T_1^* e^{-\tau_2/\mu'}. \tag{3.2.46f}$$

When polarization and azimuth dependence are neglected, the transmission function is the same regardless of whether radiation is from above or below; that is , $T^*(\mu, \mu') = T(\mu', \mu)$. This relation can be derived based on the Helmholtz principle of reciprocity in which the light beam may reverse its direction (Hovenier, 1969).

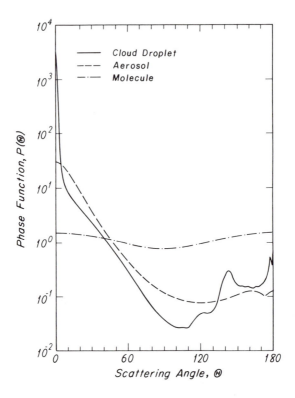

Fig. 3.5 Normalized phase functions for cloud droplets, aerosols, and molecules illuminated by a visible wavelength of 0.5 μm.

For practical applications, we begin with the computations of the reflection and transmission functions given in Eqs. (3.2.33) and (3.2.34). The phase function must be expressed as a function of the incoming and outgoing directions via Eq. (3.1.11a) in the form

$$P(\mu, \phi; \mu', \phi') = P^0(\mu, \mu') + 2 \sum_{\ell=1}^{N} P^m(\mu, \mu') \cos m(\phi' - \phi), \qquad (3.2.47)$$

where $P^m(\mu, \mu')$ $(m = 0, 1, \ldots, N)$ denotes the Fourier expansion coefficients. The number of terms required in the expansion depends on the sharpness of the forward diffraction peak in phase function (see Fig. 3.5).

The preceding adding equations for radiative transfer have been written in scalar forms involving diffuse intensity. However, these equations can be applied to the case that takes into account polarization in which the light beam is characterized by the Stokes parameters and the phase function is replaced by the phase matrix (see Subsection 5.1.2 for further discussion). The phase matrix must be expressed with respect to the local meridian plane in a manner defined in Eq. (5.5.6).

3.3 Two-stream and Eddington's approximations for radiative transfer

3.3.1 *Two-stream approximation*

On the basis of the analysis presented in Subsection 3.2.1 of the discrete-ordinates method for radiative transfer, Eq. (3.2.9) may be written in the form

$$\mu_i \frac{dI_i}{d\tau} = I_i - \sum_{j=-n}^{n} c_{i,j} I_j - c_{i,-0} I_\odot, \qquad i = -n, \ldots, n, \tag{3.3.1}$$

where $c_{i,j}$ and $I_\odot$ have been defined in Eqs. (3.2.10) and (3.2.11), respectively. The simplest way of solving Eq. (3.3.1) is to take $n = 1$ and let the expansion term in the Legendre polynomials be $N = 1$. After rearranging terms and denoting $I^+ = I_1 = I(\tau, \mu_1)$ and $I^- = I_{-1} = I(\tau, -\mu_1)$, two simultaneous equations may be written in the forms

$$\mu_1 \frac{dI^+}{d\tau} = I^+ - \tilde{\omega}(1-b)I^+ - \tilde{\omega}bI^- - s^- e^{-\tau/\mu_0}, \tag{3.3.2a}$$

$$-\mu_1 \frac{dI^-}{d\tau} = I^- - \tilde{\omega}(1-b)I^- - \tilde{\omega}bI^+ - s^+ e^{-\tau/\mu_0}, \tag{3.3.2b}$$

where $\mu_1 = 1/\sqrt{3}$, $a_1 = a_{-1} = 1$ from the Gauss formula, and

$$s^\pm = F_\odot \tilde{\omega}(1 \pm 3g\mu_1\mu_0)/4\pi, \tag{3.3.2c}$$

$$b = (1-g)/2. \tag{3.3.2d}$$

The asymmetry factor g is defined by

$$g = \frac{\tilde{\omega}_1}{3} = \frac{1}{2} \int_{-1}^{1} P(\cos\Theta) \cos\Theta \, d\cos\Theta. \tag{3.3.3}$$

The asymmetry factor is the first moment of the phase function. It is derived from Eq. (3.1.4) by using the orthogonal property of the Legendre polynomials. Note that the zero moment of the phase function is equal to $\tilde{\omega}_0(=1)$. For isotropic scattering, g is zero, as it is for Rayleigh scattering. The asymmetry factor increases as the diffraction peak of the phase function sharpens. Conceivably, the asymmetry factor may be negative if the phase function peaks in backward directions (90–180°). For Mie particles, whose phase function has a generally sharp peak at the 0° scattering angle, the asymmetry factor denotes the relative strength of forward scattering. Parameters b and $(1-b)$ can be interpreted as the integrated fraction of the energy that is backscattered and forward scattered, respectively. Thus it is apparent in Eq. (3.3.2) that the multiple-scattering contribution in the two-stream approximation is represented by upward and downward intensities weighted by appropriate fractions of the forward or backward phase function. The upward intensity is strengthened by its coupling with the forward fraction (0–90°) of the phase function plus the downward intensity, which appears in the backward fraction (90–180°) of the phase function. A similar argument is valid for the downward intensity.

The form of Eq. (3.3.2), without the direct solar source term, was first presented by Schuster (1905). Schuster's formulations have been used by Neiburger (1949) for solar reflection, absorption, and transmission measurements from California stratus clouds, and have been further discussed by Herman and Abraham (1960). An application of the two-stream approximation to planetary atmospheres has been presented by Sagan and Pollack (1967) for interpreting observed visual and near-ir reflectivity from the clouds of Venus.

The solutions of two first-order nonhomogeneous differential equations [Eqs. (3.3.2a) and (3.3.2b)] can be derived by straightforward analysis. Let $\tilde{\omega} \neq 1$; then we obtain

$$I^+ = I(\tau, \mu_1) = Kve^{k\tau} + Hue^{-k\tau} + \epsilon e^{-\tau/\mu_0}, \qquad (3.3.4a)$$

$$I^- = I(\tau, -\mu_1) = Kue^{k\tau} + Hve^{-k\tau} + \gamma e^{-\tau/\mu_0}, \qquad (3.3.4b)$$

where

$$k^2 = (1 - \tilde{\omega})(1 - \tilde{\omega}g)/\mu_1^2, \qquad (3.3.4c)$$

$$v = (1 + a)/2, \qquad u = (1 - a)/2, \qquad (3.3.4d)$$

$$a^2 = (1 - \tilde{\omega})/(1 - \tilde{\omega}g), \qquad (3.3.4e)$$

$$\epsilon = (\alpha + \beta)/2, \qquad \gamma = (\alpha - \beta)/2, \qquad (3.3.4f)$$

$$\alpha = Z_1\mu_0^2/(1 - \mu_0^2k^2), \qquad \beta = Z_2\mu_0^2/(1 - \mu_0^2k^2), \qquad (3.3.4g)$$

$$Z_1 = -\frac{(1 - \tilde{\omega}g)(s^- + s^+)}{\mu_1^2} + \frac{s^- - s^+}{\mu_1\mu_0}, \qquad (3.3.4h)$$

$$Z_2 = -\frac{(1 - \tilde{\omega})(s^- - s^+)}{\mu_1^2} + \frac{s^- + s^+}{\mu_1\mu_0}. \qquad (3.3.4i)$$

The terms $\pm k$, in Eq. (3.3.4c) are the eigenvalues for the solution of the differential equations, and u and v represent the eigenfunctions, which are defined by the similarity parameter a in Eq. (3.3.4e) (see Section 3.4 for discussion on the similarity principle in radiative transfer). For conservative scattering, $\tilde{\omega} = 1$. Simpler solutions can be derived from Eqs. (3.3.2a) and (3.3.2b) with one of the eigenvalues, $k = 0$. In practice, however, we may set $\tilde{\omega} = 0.99999$ in Eqs. (3.3.4a–i) and obtain the results for conservative scattering. With two proper boundary conditons, the two integration constants, K and H, may be determined. The upward and downard fluxes in the context of the two-stream approximation are

$$F^+(\tau) = 2\pi\mu_1 I(\tau, \mu_1), \qquad F^-(\tau) = 2\pi\mu_1 I(\tau, -\mu_1). \qquad (3.3.5)$$

3.3.2 Eddington's approximation

We begin with the general approach of decomposing the equation of radiative transfer using the property of Legendre polynomials. In line with the Legendre polynomial expansion for the phase function denoted in Eq. (3.1.15), the scattered intensity may be expanded in terms of Legendre polynomials such that

$$I(\tau, \mu) = \sum_{\ell=0}^{N} I_\ell(\tau) P_\ell(\mu). \tag{3.3.6}$$

Using the orthogonal and recurrence properties of Legendre polynomials, Eq. (3.1.16) may be decomposed in N harmonics in the form

$$\frac{\ell}{2\ell - 1} \frac{dI_{\ell-1}}{d\tau} + \frac{\ell + 1}{2\ell + 3} \frac{dI_{\ell+1}}{d\tau}$$
$$= I_\ell \left(1 - \frac{\tilde{\omega}\tilde{\omega}_\ell}{2\ell + 1}\right) - \frac{\tilde{\omega}}{4\pi}\tilde{\omega}_\ell P_\ell(-\mu_0)F_\odot e^{-\tau/\mu_0},$$
$$\ell = 0, 1, 2, \ldots, N. \tag{3.3.7}$$

The method of solving the basic radiative transfer equation using the aforementioned procedure is referred to as the *spherical harmonic method* (Kourganoff, 1952). Numerical solutions to a set of differential equations may be carried out in the same way as they are in the discrete-ordinates method (Dave, 1975).

Eddington's approximation uses an approach similar to that of the two-stream approximation, and was originally used for studies of radiative equilibrium in stellar atmospheres (Eddington, 1916). Letting $N = 1$, the phase function and intensity expressions may be written as follows:

$$P(\mu, \mu') = 1 + 3g\mu\mu',$$
$$I(\tau, \mu) = I_0(\tau) + I_1(\tau)\mu, \qquad -1 \leq \mu \leq 1. \tag{3.3.8}$$

Subsequently, Eq. (3.3.7) reduces to a set of two simultaneous equations in the forms

$$\frac{dI_1}{d\tau} = 3(1 - \tilde{\omega})I_0 - \frac{3\tilde{\omega}}{4\pi}F_\odot e^{-\tau/\mu_0}, \tag{3.3.9a}$$

$$\frac{dI_0}{d\tau} = (1 - \tilde{\omega}g)I_1 + \frac{3\tilde{\omega}}{4\pi}g\mu_0 F_\odot e^{-\tau/\mu_0}. \tag{3.3.9b}$$

Differentiating Eq. (3.3.9b) with respect to τ and substituting the expression for $dI_1/d\tau$ from Eq. (3.3.9a) leads to

$$\frac{d^2 I_0}{d\tau^2} = k^2 I_0 - \chi e^{-\tau/\mu_0}, \tag{3.3.10}$$

where $\chi = 3\tilde{\omega}F_\odot(1 + g - \tilde{\omega}g)/4\pi$ and the eigenvalue is

$$k^2 = 3(1 - \tilde{\omega})(1 - \tilde{\omega}g). \qquad (3.3.11)$$

Here, the eigenvalues are exactly the same as they are for the two-stream approximation depicted in Eq. (3.3.4c). Equation (3.3.10) represents a well-known diffusion equation for radiative transfer. The diffusion approximation is particularly applicable for the radiation field in the deep domain of an optically thick layer.

Straightforward analyses yield the following solutions for the diffusion equation:

$$I_0 = Ke^{k\tau} + He^{-k\tau} + \psi e^{-\tau/\mu_0}, \qquad (3.3.12a)$$

where

$$\psi = \frac{3\tilde{\omega}}{4\pi}F_\odot[1 + g(1 - \tilde{\omega})]/(k^2 - 1/\mu_0^2).$$

Following a similar procedure, the solution for the second harmonic, I_1, is given by

$$I_1 = aKe^{k\tau} - aHe^{-k\tau} - \xi e^{-\tau/\mu_0}, \qquad (3.3.12b)$$

where $a^2 = 3(1 - \tilde{\omega})/(1 - \tilde{\omega}g)$, defined in the two-stream approximation [Eq. (3.3.4e)], and

$$\xi = \frac{3\tilde{\omega}}{4\pi}\frac{F_\odot}{\mu_0}\left[1 + 3g(1 - \tilde{\omega})\mu_0^2\right]/(k^2 - 1/\mu_0^2).$$

The integration constants, K and H, are to be determined from proper boundary conditions. Finally, the upward and downward fluxes are given by

$$\left.\begin{array}{c} F^+(\tau) \\ \\ F^-(\tau) \end{array}\right\} = 2\pi \int_0^{\pm 1}(I_0 + \mu I_1)\mu\,d\mu = \left\{\begin{array}{c} \pi\left(I_0 + \dfrac{2}{3}I_1\right) \\ \\ \pi\left(I_0 - \dfrac{2}{3}I_1\right). \end{array}\right. \qquad (3.3.13)$$

3.3.3 *Generalized two-stream equation*

Using the radiative transfer equation denoted in Eq. (3.1.16) and the upward and downward diffuse fluxes defined in Eq. (3.1.19b), we may write

$$\frac{1}{2\pi}\frac{dF^+(\tau)}{d\tau} = \int_0^1 I(\tau, \mu)\,d\mu - \frac{\tilde{\omega}}{2}\int_0^1\int_{-1}^1 I(\tau, \mu)P(\mu, \mu')\,d\mu'\,d\mu$$
$$- \frac{\tilde{\omega}}{4\pi}F_\odot e^{-\tau/\mu_0}\int_0^1 P(\mu, -\mu_0)\,d\mu, \qquad (3.3.14a)$$

$$\frac{1}{2\pi}\frac{dF^-(\tau)}{d\tau} = -\int_0^1 I(\tau, -\mu)\,d\mu + \frac{\tilde{\omega}}{2}\int_0^1\int_{-1}^1 I(\tau, \mu')P(-\mu, \mu')\,d\mu'\,d\mu$$
$$+ \frac{\tilde{\omega}}{4\pi}F_\odot e^{-\tau/\mu_0}\int_0^1 P(-\mu, -\mu_0)\,d\mu. \qquad (3.3.14b)$$

Table 3.1 Coefficients in two-stream approximations

Method	γ_1	γ_2	γ_3
Two-stream	$[1 - \tilde{\omega}(1 + g)/2]/\mu_1$	$\tilde{\omega}(1 - g)/2\mu_1$	$(1 - 3g\mu_1\mu_0)/2$
Eddington's	$[7 - (4 + 3g)\tilde{\omega}]/4$	$-[1 - (4 - 3g)\tilde{\omega}]/4$	$(2 - 3g\mu_0)/4$

The generalized two-stream approximation may be expressed by

$$\frac{dF^+(\tau)}{d\tau} = \gamma_1 F^+(\tau) - \gamma_2 F^-(\tau) - \gamma_3 \tilde{\omega} F_\odot e^{-\tau/\mu_0}, \qquad (3.3.15a)$$

$$\frac{dF^-(\tau)}{d\tau} = \gamma_2 F^+(\tau) - \gamma_1 F^-(\tau) + (1 - \gamma_3) \tilde{\omega} F_\odot e^{-\tau/\mu_0}. \qquad (3.3.15b)$$

The differential changes in upward and downward diffuse fluxes are directly related to the upward and downward diffuse fluxes, as well as the downward direct flux. The coefficients γ_i ($i = 1, 2, 3$) depend on the manner in which the intensity and phase function are approximated in Eq. (3.3.14). In the two-stream approximation, there are only upward and downward intensities in the directions μ_1 and $-\mu_1$ given by the Gauss quadrature formula, while the phase function is expanded in two terms in Legendre polynomials. In Eddington's approximation, both intensity and phase function are expanded in two polynomial terms. The coefficients γ_i can be directly derived from Eqs. (3.3.2a, b) and (3.3.9a, b), and are given in Table 3.1.

In Eq. (3.3.14), we let the last integral involving the phase function be

$$q = \frac{1}{2} \int_0^1 P(\mu, -\mu_0) \, d\mu. \qquad (3.3.16a)$$

Since the phase function is normalized to unity, we have

$$\frac{1}{2} \int_0^1 P(-\mu, -\mu_0) \, d\mu = 1 - q. \qquad (3.3.16b)$$

Equations (3.3.16a, b) can be evaluated exactly by numerical means. We may take $\gamma_3 = q$ in the two-stream approximation. This constitutes the modified two-stream approximation proposed by Liou (1973b) and Meador and Weaver (1980). The two-stream approximation yields negative albedo values for a thin atmosphere when $\gamma_3 < 0$ (i.e., $g > \mu_1/\mu_0$). This also occurs in Eddington's approximation when $g > 0.75/\mu_0$. These negative albedo values can be avoided by using q, the full phase function integration for the direct solar beam, denoted in Eq. (3.3.16b). The accuracy of the two-stream approximation has been discussed in Liou (1973a). The overall accuracy of the two-stream and Eddington's approximations can be improved by incorporating the δ-function adjustment for forward scattering. We will discuss forward scattering in Subsection 3.4. There are other two-stream approximations, such as those discussed by Zdunkowski et al. (1974) and Coakley and Chýlek (1975), who used hemispheric average intensities for the upward and downward diffuse fluxes, and by Meador and Weaver (1980), who used a combination of Eddington and δ-function methods to get the two-stream approximations. The accuracy of these methods is generally on the same order of magnitude as that of the δ–two-stream or δ-Eddington approximations (see Section 3.5).

The solutions for the equations of the generalized two-stream approximation expressed in Eq. (3.3.15a, b) are as follows:

$$F^+ = vKe^{k\tau} + uHe^{-k\tau} + \epsilon e^{-\tau/\mu_0}, \qquad (3.3.17a)$$

$$F^- = uKe^{k\tau} + vHe^{-k\tau} + \gamma e^{-\tau/\mu_0}, \qquad (3.3.17b)$$

where

$$k^2 = \gamma_1^2 - \gamma_2^2, \qquad (3.3.18a)$$

$$v = \frac{1}{2}[1 + (\gamma_1 - \gamma_2)/k], \qquad u = \frac{1}{2}[1 - (\gamma_1 - \gamma_2)/k], \quad (3.3.18b)$$

$$\epsilon = [\gamma_3(1/\mu_0 - \gamma_1) - \gamma_2(1 - \gamma_3)]\mu_0^2\tilde{\omega}F_\odot, \qquad (3.3.18c)$$

$$\gamma = -[(1 - \gamma_3)(1/\mu_0 + \gamma_1) + \gamma_2\gamma_3]\mu_0^2\tilde{\omega}F_\odot, \qquad (3.3.18d)$$

and where K and H are unknown coefficients to be determined from the boundary conditions.

3.4   Delta-function adjustment and similarity principle

The two-stream and Eddington methods for radiative transfer are good approximations for optically thick layers, but they produce inaccurate results for thin layers and when significant absorption is involved. The basic problem is that scattering by atmospheric particulates is highly peaked in the forward directions. Figure 3.5 illustrates the phase functions for cloud water droplets, aerosols, and molecules. The phase functions for cloud and aerosol particles are highly peaked in the forward direction. This is especially evident for cloud particles, for which the forward-scattered energy within $\sim 5°$ scattering angles produced by diffraction is four to five orders of magnitude greater than it is in the side and backward directions. The highly peaked diffraction pattern is typical for atmospheric particulates. It is clear that a two-term expansion in the phase function is far from adequate.

To incorporate the forward peak contribution in multiple scattering, we may consider an adjusted absorption and scattering atmosphere, such that the fraction of scattered energy residing in the forward peak, f, is removed from the scattering parameters: optical depth, τ; single-scattering albedo, $\tilde{\omega}$; and asymmetry factor, g. We use primes to represent the adjusted radiative parameters, as shown in Fig. 3.6. The optical (extinction) depth is the sum of the scattering (τ_s) and absorption

(τ_a) optical depths. The forward peak is produced by diffraction without the contribution of absorption. Thus the adjusted scattering and absorption optical depths must be

$$\tau_s' = (1 - f)\tau_s, \tag{3.4.1}$$

$$\tau_a' = \tau_a. \tag{3.4.2}$$

The total adjusted optical depth is

$$\tau' = \tau_s' + \tau_a' = (1 - f)\tau_s + \tau_a = \tau(1 - f\tilde{\omega}). \tag{3.4.3}$$

The adjusted single-scattering albedo is

$$\tilde{\omega}' = \frac{\tau_s'}{\tau'} = \frac{(1 - f)\tau_s}{(1 - f\tilde{\omega})\tau} = \frac{(1 - f)\tilde{\omega}}{1 - f\tilde{\omega}}. \tag{3.4.4}$$

Moreover, we multiply the asymmetry factor by the scattering optical depth to get the similarity equation

$$\tau_s'g' = \tau_s g - \tau_s f, \quad \text{that is,} \quad g' = \frac{g - f}{1 - f}, \tag{3.4.5}$$

where we note that the asymmetry factor in the forward peak is equal to unity. In the diffusion domain, the solution of the diffuse intensity is given by exponential functions with eigenvalues defined in Eq. (3.3.11). We may set the intensity solution in the adjusted atmosphere so that it is equivalent to that in the real atmosphere, in the form

$$k\tau = k'\tau'. \tag{3.4.6}$$

From Eqs. (3.4.3)–(3.4.6), the similarity relations for radiative transfer can be expressed in the forms

$$\frac{\tau}{\tau'} = \frac{k'}{k} = \frac{1 - \tilde{\omega}'}{1 - \tilde{\omega}} = \frac{\tilde{\omega}'(1 - g')}{\tilde{\omega}(1 - g)}. \tag{3.4.7a}$$

Using the expression for the eigenvalue defined in Eq. (3.3.4c), we also find the relation for the similarity parameter defined in Eq. (3.3.4e), as follows:

$$a = \left(\frac{1 - \tilde{\omega}}{1 - \omega g}\right)^{1/2} = \left(\frac{1 - \tilde{\omega}'}{1 - \tilde{\omega}'g'}\right)^{1/2}. \tag{3.4.7b}$$

The similarity principle can also be derived from the basic radiative transfer equation. We may begin with this equation in the form

$$\mu \frac{dI(\tau, \mu)}{d\tau} = I(\tau, \mu) - \frac{\tilde{\omega}}{2} \int_{-1}^{1} I(\tau, \mu')P(\mu, \mu') \, d\mu'. \tag{3.4.8}$$

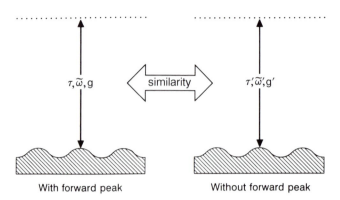

Fig. 3.6 Similarity principle for radiative transfer. The prime system represents adjustment radiative parameters such that the forward diffraction peak in scattering processes is removed.

From Eq. (3.3.8), the phase function in the limit of the two-stream and Eddington's approximations is given by $P(\mu, \mu') = 1 + 3g\mu\mu'$. However, the phase functions involving cloud and aerosol particles are highly peaked in the forward direction, and two-term expansions do not adequately account for the strong forward scattering. Let the fraction of the energy scattered in the forward direction ($\Theta = 0°$) be denoted by f. The normalized phase function may be expressed in terms of this value, as follows:

$$P(\mu, \mu') = 2f\delta(\mu - \mu') + (1 - f)(1 + 3g'\mu\mu'), \qquad (3.4.9)$$

where $\mu = \mu'$ when $\Theta = 0$, δ is the δ function, and g' denotes the scaled asymmetry factor. The phase function so defined is normalized to unity, and the asymmetry factor defined in Eq. (3.3.3) is given by

$$g = f + (1 - f)g'. \qquad (3.6.10)$$

The second moment of the phase function expansion is

$$\tilde{\omega}_2/5 = f. \qquad (3.6.11)$$

Thus the scaled asymmetry factor can be expressed by

$$g' = \frac{g - \tilde{\omega}_2/5}{1 - \tilde{\omega}_2/5}. \qquad (3.4.12)$$

Now, substituting Eq. (3.4.9) into Eq. (3.4.8), we obtain

$$\mu \frac{dI(\tau, \mu)}{d\tau} = I(\tau, \mu)(1 - \tilde{\omega}f) - \frac{\tilde{\omega}(1 - f)}{2}$$
$$\times \int_{-1}^{1} (1 + 3g'\mu\mu')I(\tau, \mu')\, d\mu'. \qquad (3.4.13)$$

Consequently, if we redefine the optical depth, single-scattering albedo, and phase function such that

$$\tau' = (1 - \tilde{\omega}f)\tau, \tag{3.4.14a}$$

$$\tilde{\omega}' = \frac{(1 - f)\tilde{\omega}}{1 - \tilde{\omega}f}, \tag{3.4.14b}$$

$$P'(\mu, \mu') = 1 + 3g'\mu\mu', \tag{3.4.14c}$$

Eq. (3.4.13) then becomes

$$\mu \frac{dI(\tau', \mu)}{d\tau'} = I(\tau', \mu) - \frac{\tilde{\omega}'}{2} \int_{-1}^{1} I(\tau', \mu')P'(\mu, \mu')\,d\mu'. \tag{3.4.15}$$

Equation (3.4.15) is exactly the same as Eq. (3.4.8), except that g, τ, and $\tilde{\omega}$ have been replaced by g', τ', and $\tilde{\omega}'$. By redefining the asymmetry factor, optical depth, and single-scattering albedo, the forward-scattering nature of the phase function is approximately accounted for in the basic radiative transfer equation. In essence, we have incorporated the second moment of the phase function expansion in the formulation of the radiative transfer equation. The "equivalence" between Eqs. (3.4.8) and (3.4.15) is the similarity principle that has been stated previously.

The phase functions for aerosol and cloud particles require involved scattering calculations, as will be discussed in Section 5.1. For many applications to radiative transfer in planetary atmospheres, an analytic expression for the phase function in terms of the asymmetry factor has been proposed:

$$P_{\mathrm{HG}}(\cos\Theta) = (1 - g^2)/(1 + g^2 - 2g\cos\Theta)^{3/2}$$
$$= \sum_{\ell=0}^{N}(2\ell + 1)g^{\ell}P_{\ell}(\cos\Theta). \tag{3.4.16}$$

This is referred to as the *Henyey–Greenstein phase function* (Henyey and Greenstein, 1941), which is adequate for scattering patterns that are not strongly peaked in the forward direction. Using this expression, the second moment for the phase function is given by

$$\tilde{\omega}_2/5 = f = g^2. \tag{3.4.17}$$

Thus, in the limit of the Henyey–Greenstein approximation, the forward fraction of the scattered light is now expressed in terms of the asymmetry factor. Subsequently, the scaled asymmetry factor, optical depth, and single-scattering albedo can now be expressed by

$$g' = \frac{g}{1 + g}, \qquad \tau' = (1 - \tilde{\omega}g^2)\tau, \qquad \tilde{\omega}' = \frac{(1 - g^2)\tilde{\omega}}{1 - \tilde{\omega}g^2}. \tag{3.4.18}$$

The similarity principle for radiative transfer was first stated by Sobolev (1975) for isotropic scattering. The general similarity relationships have been presented by van de Hulst (1980). The application of employing a Dirac δ function to approximate highly peaked forward scattering in radiative transfer has been discussed by a number of researchers, including Hansen (1969), Potter (1970), Joseph et al. (1976), and Wiscombe (1977).

The two-stream approximations are popular because they enable the analytic solutions for upward and downward fluxes to be derived, and the numerical computations for these fluxes to be efficiently performed. The incorporation of the delta-function adjustment to account for the strong forward scattering of large size parameters in the context of two-stream approximations has led to a significant improvement in the accuracy of radiative flux calculations. As pointed out previously, the δ adjustment provides a third term closure through the second moment of the phase function expansion. Schaller (1979) has illustrated that the δ–two-stream and δ–Eddington approximations have the same accuracy. King and Harshvardhan (1986) have undertaken a more comprehensive examination of the accuracy of various two-stream approximations. They have shown that relative errors of 15–20% could result for some values of optical depths, solar zenith angles and single-scattering albedos. In the next section, we will introduce the δ–four-stream approximation, which produces an accuracy within $\sim$5% in the radiative flux calculations.

3.5 δ–four-stream approximation for radiative transfer parameterizations

For atmospheric flux computations, Liou (1974) has proposed that the four-stream approximation could be of value. For this approximation, the solution for eigenvalues associated with the homogeneous part of the discretized equations can be derived analytically from the recurrence equation for eigenvalues. Thus, the computational time for the flux calculations does not significantly exceed that required for the two-stream approximation. Cuzzi et al. (1982) have carried out an examination of the four-stream approximation. Their findings indicate that the four-stream approximation, as well as the incorporation of the forward-peak adjustment in this approximation, does indeed have much to offer for flux calculations in terms of both accuracy and efficiency.

The four-stream approximation, as given in Liou (1974), is based on the general solution for the discrete-ordinates method for radiative transfer. In order to be able to understand the merit of the four-stream approximation, it is necessary to have some background in solving a set of differential equations based on Chandrasekhar's (1950) formulations. In particular, it is noted that the search for eigenvalues from the recurrence equation developed in the solution is both mathematically ambiguous and numerically troublesome, as pointed out in Subsection 3.2.1.

A systematic and independent development of the solution for this approximation has been presented by Liou et al. (1988). Specifically, this solution involves the computation of solar radiative fluxes using a relatively simple, convenient, and accurate method. Knowledge of the discrete-ordinates method for radiative transfer is desirable but not necessary. In addition, a wide range of accuracy checks for this approximation has been provided, including the δ adjustment to account for the forward diffraction peak based on the generalized similarity principle for radiative transfer.

Consider two radiative streams in the upper and lower hemispheres (i.e., let $n = 2$). At the same time, expand the scattering phase function into four terms (i.e., $N = 3$) in line with the four radiative streams. On the basis of Eq. (3.2.16), four first-order differential equations can then be written explicitly in matrix form:

$$\frac{d}{d\tau}\begin{bmatrix} I_{-2} \\ I_{-1} \\ I_1 \\ I_2 \end{bmatrix} = -\begin{bmatrix} -b_{2,2} & -b_{2,1} & -b_{2,-1} & -b_{2,-2} \\ -b_{1,2} & -b_{1,1} & -b_{1,-1} & -b_{1,-2} \\ b_{1,-2} & b_{1,-1} & b_{1,1} & b_{1,2} \\ b_{2,-2} & b_{2,-1} & b_{2,1} & b_{2,2} \end{bmatrix}\begin{bmatrix} I_{-2} \\ I_{-1} \\ I_1 \\ I_2 \end{bmatrix} - \begin{bmatrix} b_{-2,-0} \\ b_{-1,-0} \\ b_{1,-0} \\ b_{2,-0} \end{bmatrix} I_\odot,$$

(3.5.1)

where $I_\odot = I(\tau, -\mu_0)$ defined in Eq. (3.2.11). The 4×4 matrix represents the contribution of multiple scattering. Thus the derivative of the diffuse intensity at a specific quadrature angle is the weighted sum of the multiple-scattered intensity from all four quadrature angles. The last term represents the contribution of the unscattered component of the direct solar flux at position τ.

We proceed with a direct approach to find the eigenvalues and eigenvectors for Eq. (3.5.1). To do so, we define the sum and difference of the upward and downward intensities in the form

$$M_{1,2}^\pm = I_{1,2} \pm I_{-1,-2}.$$

(3.5.2)

From Eq. (3.5.1), we obtain the following four equations:

$$-\frac{dM_2^+}{d\tau} = b_{22}^- M_2^- + b_{21}^- M_1^- + b_2^+ I_\odot,$$

(3.5.3a)

$$-\frac{dM_2^-}{d\tau} = b_{22}^+ M_2^+ + b_{21}^+ M_1^+ + b_2^- I_\odot,$$

(3.5.3b)

$$-\frac{dM_1^+}{d\tau} = b_{12}^- M_2^- + b_{11}^- M_1^- + b_1^+ I_\odot,$$

(3.5.3c)

$$-\frac{dM_1^-}{d\tau} = b_{12}^+ M_2^+ + b_{11}^+ M_1^+ + b_1^- I_\odot,$$

(3.5.3d)

where the coefficients are defined by

$$\begin{aligned} b_{22}^\pm &= b_{2,2} \pm b_{2,-2}, & b_{21}^\pm &= b_{2,1} \pm b_{2,-1}, \\ b_{12}^\pm &= b_{1,2} \pm b_{1,-2}, & b_{11}^\pm &= b_{1,1} \pm b_{1,-1}, \\ b_2^\pm &= b_{2,-0} \pm b_{-2,-0}, & b_1^\pm &= b_{1,-0} \pm b_{-1,-0}, \end{aligned}$$

(3.5.4)

and $b_{i,j}$ have been defined in Eq. (3.2.13). Equations (3.5.3a–d) can be combined to yield

$$\frac{d^2}{d\tau^2} \begin{bmatrix} M_2^+ \\ M_1^+ \end{bmatrix} = \begin{bmatrix} a_{22} & a_{21} \\ a_{12} & a_{11} \end{bmatrix} \begin{bmatrix} M_2^+ \\ M_1^+ \end{bmatrix} + \begin{bmatrix} d_2 \\ d_1 \end{bmatrix} I_\odot, \qquad (3.5.5)$$

where

$$\begin{aligned}
a_{22} &= b_{22}^+ b_{22}^- + b_{12}^+ b_{21}^-, & a_{21} &= b_{22}^- b_{21}^+ + b_{21}^- b_{11}^+, \\
a_{12} &= b_{12}^- b_{22}^+ + b_{11}^- b_{12}^+, & a_{11} &= b_{12}^- b_{21}^+ + b_{11}^- b_{11}^+, \\
d_2 &= b_{22}^- b_2^- + b_{21}^- b_1^- + b_2^+/\mu_0, \\
d_1 &= b_{12}^- b_2^- + b_{11}^- b_1^- + b_1^+/\mu_0.
\end{aligned} \qquad (3.5.6)$$

Performing differential operations on Eq. (3.5.5) leads to

$$\frac{d^4 M_2^+}{d\tau^4} = b\frac{d^2 M_2^+}{d\tau^2} + cM_2^+ + \left(\frac{d_2}{\mu_0^2} + a_{21}d_1 - a_{11}d_2\right) I_\odot, \qquad (3.5.7a)$$

$$\frac{d^4 M_1^+}{d\tau^4} = b\frac{d^2 M_1^+}{d\tau^2} + cM_1^+ + \left(\frac{d_1}{\mu_0^2} + a_{12}d_2 - a_{22}d_1\right) I_\odot, \qquad (3.5.7b)$$

where the terms $b = a_{22} + a_{11}$ and $c = a_{21}a_{12} - a_{11}a_{22}$. The complete solution for M_2^+ (or M_1^+) is the sum of the solution for the homogeneous part of the fourth-order differential equation plus a particular solution. Thus,

$$\begin{bmatrix} M_2^+ \\ M_1^+ \end{bmatrix} = \sum_{j=-2}^{2} \begin{bmatrix} G_j \\ H_j \end{bmatrix} e^{-k_j \tau} + \begin{bmatrix} \eta_2 \\ \eta_1 \end{bmatrix} e^{-\tau/\mu_0}, \qquad (3.5.8)$$

where G_j and H_j are associated with eigenvectors, and η_2 and η_1 are results for the particular solutions. Considering the homogeneous part in Eq. (3.5.7a) and substituting the homogeneous solution for M_2^+ into this equation, we find

$$\sum_{j=-2}^{2} \left(k_j^4 - bk_j^2 - c\right) G_j e^{-k_j \tau} = 0. \qquad (3.5.9)$$

In order to have a nontrivial solution for M_2^+ (or M_1^+), we must have

$$f(k) = k^4 - bk^2 - c = 0. \qquad (3.5.10)$$

It follows that the eigenvalues are given by

$$k^2 = \left[b \pm (b^2 + 4c)^{1/2}\right]/2. \qquad (3.5.11)$$

From the definitions of b and c, we have $b^2 + 4c = (a_{11} - a_{22})^2 + 4a_{21}a_{12}$. The terms a_{21} and a_{12} can be expressed in terms of $c_{i,j}$ defined in Eq. (3.2.10).

Under the conditions that $0 < g < 1$ and $0 < \tilde{\omega} < 1$, we find that $a_{21} < 0$ and $a_{12} < 0$. This implies that $b^2 + 4c > 0$. Also, it can be shown that the term $c = -\prod_{\ell=0}^{3}(1 - \tilde{\omega}g^{\ell})/\mu_1^2\mu_2^2$. This term is less than zero, so that the eigenvalues are all real numbers. In the case of conservative scattering, $c = 0$. As a result, the two eigenvalues are zero. By substituting the particular solution for $M_{1,2}^+$ into Eqs. (3.5.7a, b), we obtain

$$\eta_2 = \frac{d_2/\mu_0^2 + a_{21}d_1 - a_{11}d_2}{f(1/\mu_0)} \frac{F_\odot}{2\pi}, \tag{3.5.12a}$$

$$\eta_1 = \frac{d_1/\mu_0^2 + a_{12}d_2 - a_{22}d_1}{f(1/\mu_0)} \frac{F_\odot}{2\pi} \tag{3.5.12b}$$

The function f in this equation has been defined in Eq. (3.5.10). Because G_j and H_j in Eq. (3.5.8) are defined after high-order differentiations, they are not mutually independent. We may determine their relationship from the homogeneous part of Eq. (3.5.5). A straightforward substitution yields

$$H_1 e^{-k_1\tau} + H_{-1}e^{k_1\tau} = A_1(G_1 e^{-k_1\tau} + G_{-1}e^{k_1\tau}), \tag{3.5.13a}$$

$$H_2 e^{-k_2\tau} + H_{-2}e^{k_2\tau} = A_2(G_2 e^{-k_2\tau} + G_{-2}e^{k_2\tau}), \tag{3.5.13b}$$

where $A_{1,2} = (k_{1,2}^2 - a_{22})/a_{21}$, and k_1 and k_2 are eigenvalues from Eq. (3.5.11).

Following the preceding procedures and analogous to Eq. (3.5.7), we may obtain expressions for $M_{1,2}$ in the form

$$\frac{d^4 M_2^-}{d\tau^4} = b'\frac{d^2 M_2^-}{d\tau^2} + c'M_2^- + \left(\frac{d_2'}{\mu_0^2} + a_{21}'d_1' - a_{11}'d_2'\right)I_\odot, \tag{3.5.14a}$$

$$\frac{d^4 M_1^-}{d\tau^4} = b'\frac{d^2 M_1^-}{d\tau^2} + c'M_1^- + \left(\frac{d_1'}{\mu_0^2} + a_{12}'d_2' - a_{22}'d_1'\right)I_\odot, \tag{3.5.14b}$$

where the primed coefficients can be obtained by replacing the superscripts $+$ and $-$ in Eq. (3.5.6) with $-$ and $+$, respectively. Also, we note that $b' = a_{22}' + a_{11}' = b$, and $c' = a_{21}'a_{12}' - a_{11}'a_{22}' = c$. The particular solutions for $M_{2,1}^-$ are

$$M_{2,1}^- = \eta_{2,1}'e^{-\tau/\mu_0}, \tag{3.5.15}$$

with

$$\eta_2' = \frac{d_2'/\mu_0^2 + a_{21}'d_1' - a_{11}'d_2'}{f(1/\mu_0)} \frac{F_\odot}{2\pi}, \tag{3.5.16a}$$

$$\eta_1' = \frac{d_1'/\mu_0^2 + a_{12}'d_2' - a_{22}'d_1'}{f(1/\mu_0)} \frac{F_\odot}{2\pi}. \tag{3.5.16b}$$

From Eqs. (3.5.3b, d), the homogeneous solutions for $M_{2,1}$ are given by

$$M_2^- = \frac{A_1 b_{21}^- - b_{11}^-}{a^-} k_1(-G_1 e^{-k_1 \tau} + G_{-1} e^{k_1 \tau})$$

$$+ \frac{A_2 b_{21}^- - b_{11}^-}{a^-} k_2(-G_2 e^{-k_2 \tau} + G_{-2} e^{k_2 \tau}), \qquad (3.5.17a)$$

$$M_1^- = \frac{b_{12}^- - A_1 b_{22}^-}{a^-} k_1(-G_1 e^{-k_1 \tau} + G_{-1} e^{k_1 \tau})$$

$$+ \frac{b_{12}^- - A_2 b_{22}^-}{a^-} k_2(-G_2 e^{-k_2 \tau} + G_{-2} e^{k_2 \tau}), \qquad (3.5.17b)$$

where $a^- = b_{22}^- b_{11}^- - b_{12}^- b_{21}^-$.

Finally, combining Eqs. (3.5.8), (3.5.13), and (3.5.17), the complete solutions for I_i ($i = -2, -1, 1, 2$) are given by

$$\begin{bmatrix} I_1 \\ I_{-1} \\ I_2 \\ I_{-2} \end{bmatrix} = \begin{bmatrix} \Phi_1^+ e_1^- & \Phi_1^- e_1^+ & \Phi_2^+ e_2^- & \Phi_2^- e_2^+ \\ \Phi_1^- e_1^- & \Phi_1^+ e_1^+ & \Phi_2^- e_2^- & \Phi_2^+ e_2^+ \\ \phi_1^+ e_1^- & \phi_1^- e_1^+ & \phi_2^+ e_2^- & \phi_2^- e_2^+ \\ \phi_1^- e_1^- & \phi_1^+ e_1^+ & \phi_2^- e_2^- & \phi_2^+ e_2^+ \end{bmatrix} \begin{bmatrix} G_1 \\ G_{-1} \\ G_2 \\ G_{-2} \end{bmatrix} + \begin{bmatrix} Z_1^+ \\ Z_1^- \\ Z_2^+ \\ Z_2^- \end{bmatrix} e^{-\tau/\mu_0},$$

$$(3.5.18)$$

where the elements $e_1^- = e^{-k_1 \tau}$, $e_1^+ = e^{k_1 \tau}$, $e_2^- = e^{-k_2 \tau}$, and $e_2^+ = e^{k_2 \tau}$, and the eigenvectors are:

$$\phi_{1,2}^\pm = \frac{1}{2}\left(1 \pm \frac{b_{11}^- - A_{1,2} b_{21}^-}{a^-} k_{1,2}\right), \qquad (3.5.19a)$$

$$\Phi_{1,2}^\pm = \frac{1}{2}\left(A_{1,2} \pm \frac{A_{1,2} b_{22}^- - b_{12}^-}{a^-} k_{1,2}\right). \qquad (3.5.19b)$$

In Eqs. (3.5.19a, b), $b_{ij}^\pm$ is defined by Eq. (3.5.4), with $b_{i,j}$ given in Eq. (3.2.13), and $k_{1,2}$ by eigenvalues of Eq. (3.5.11) with b and c defined below Eq. (3.5.7b). a^- and $A_{1,2}$ are defined by expressions below Eqs. (3.5.17b) and (3.5.13b), respectively, with a_{ij} given in Eq. (3.5.6). The Z functions are defined by

$$Z_{1,2}^\pm = \frac{1}{2}(\eta_{1,2} \pm \eta_{1,2}'), \qquad (3.5.19c)$$

where $\eta_{1,2}$ and $\eta_{1,2}'$ are defined by Eqs. (3.5.12) and (3.5.16). d_i and a_{ij} are given in Eq. (3.5.6), and $f(1/\mu_0)$ has the same expression as that in Eq. (3.5.10), except k is replaced by $1/\mu_0$. d_i' and a_{ij}' have the same expressions as those in Eq. (3.5.6) except that the superscripts $+$ and $-$ are replaced by $-$ and $+$, respectively. The coefficients G_j ($j = 1, 2, -1, -2$) are to be determined from radiation boundary conditions.

Consider a homogeneous layer characterized by an optical depth τ_* and assume that there is no diffuse radiation from the top and bottom of this layer; then the boundary conditions are

$$I_{-1,-2}(\tau = 0) = 0$$

$$I_{1,2}(\tau = \tau_*) = 0. \qquad (3.5.20)$$

The lower boundary condition can be modified to include surface albedo effects. Using these boundary conditions, G_j can be obtained by an inversion of the 4×4 matrix given in Eq. (3.5.18). The upward and total (diffuse plus direct) downward fluxes at a given level τ are given by

$$F^+(\tau) = 2\pi(a_1\mu_1 I_1 + a_2\mu_2 I_2), \qquad (3.5.21a)$$

$$F^-(\tau) = 2\pi(a_1\mu_1 I_{-1} + a_2\mu_2 I_{-2}) + \mu_0 F_\odot e^{-\tau/\mu_0}. \qquad (3.5.21b)$$

We may also apply the four-stream solutions to nonhomogeneous atmospheres in the manner presented in Section 3.7.

The regular Gauss quadratures and weights in the four-stream approximation are $\mu_1 = 0.3399810$, $\mu_2 = 0.8611363$, $a_1 = 0.6521452$, and $a_2 = 0.3478548$. When the isotropic surface reflection is included in this approximation or when it is applied to the thermal infrared radiative transfer involving isothermal emission, double Gauss quadratures and weights ($\mu_1 = 0.2113248$, $\mu_2 = 0.7886752$, and $a_1 = a_2 = 0.5$) offer some advantage in flux calculations because $\sum_i a_i\mu_i = 1/2$ in this case. In the case of conservative scattering, $\tilde{\omega} = 1$, $\phi_2^\pm = \Phi_2^\pm = 0.5$, the 4×4 matrix becomes 0 in Eq. (3.5.18). The solution for this equation does not exist. Direct formulation and solution from Eq. (3.5.1) are required by setting $\tilde{\omega} = 1$. However, we may use $\tilde{\omega} = 0.999999$ in numerical calculations and obtain the results for conservative scattering. In the case $\tilde{\omega} = 0$, the multiple-scattering term vanishes.

It is possible to incorporate a δ-function adjustment to account for the forward diffraction peak in the context of the four-stream approximation. In reference to Eq. (3.1.8), we may express the normalized phase function expansion by incorporating the δ-forward adjustment in the form

$$P_\delta(\cos\Theta) = 2f\delta(\cos\Theta - 1) + (1 - f)\sum_{\ell=0}^{N} \tilde{\omega}_\ell' P_\ell(\cos\Theta), \qquad (3.5.22)$$

where $\tilde{\omega}_\ell'$ is the adjusted coefficient in the phase function expansion. The forward peak coefficient f in the four-stream approximation can be evaluated by demanding that the next-highest-order coefficient in the prime expansion, $\tilde{\omega}_4'$, vanish. Setting $P(\cos\Theta) = P_\delta(\cos\Theta)$ and utilizing the orthogonal property of Legendre polynomials, we find

$$\tilde{\omega}_\ell' = [\tilde{\omega}_\ell - f(2\ell + 1)]/(1 - f). \qquad (3.5.23)$$

Letting $\tilde{\omega}_4' = 0$, we obtain $f = \tilde{\omega}_4/9$. Based on Eq. (3.5.23), $\tilde{\omega}_\ell'$ ($\ell = 0, 1, 2, 3$) can be evaluated from the expansion coefficients of the phase function, $\tilde{\omega}_\ell$ ($\ell = 0, 1, 2, 3, 4$).

The adjusted phase function from Eq. (3.5.22) is given by

$$P'(\cos \Theta) = \sum_{\ell=0}^{N} \tilde{\omega}_\ell' P_\ell(\cos \Theta). \tag{3.5.24}$$

This equation, together with Eqs. (3.4.14a, b), constitutes the generalized similarity principle for radiative transfer. That is, the removal of the forward diffraction peak in scattering processes using adjusted single- scattering parameters is "equivalent" to actual scattering processes.

The reflectance r and total transmittance t of the solar flux $\mu_0 F_\odot$ are defined in the forms

$$r(\mu_0) = F^+(0)/\mu_0 F_\odot, \tag{3.5.25a}$$

$$t(\mu_0) = F^-(\tau_*)/\mu_0 F_\odot. \tag{3.5.25b}$$

The accuracy of the δ–two-stream and δ–four-stream approximations is examined by comparing the approximate results with the "exact" values computed from the adding method for radiative transfer. Let the reflectance computed from the approximate and "exact" methods be denoted by $\hat{r}$ and r, respectively. Then the relative accuracy is defined by $(\Delta r/r)100\% = [(\hat{r} - r)/r]100\%$. Likewise, the relative accuracy of the total transmittance is defined by $(\Delta t/t)100\%$. The analytic Henyey–Greenstein phase function expanded in the asymmetry factor g was used in the computation [Eq. (3.4.16)].

Numerous asymmetry factors, single-scattering albedos, optical depths, and solar zenith angles were used in the computations. For presentation purposes, however, we select two single-scattering albedos of 1 and 0.8, optical depths from 0.1 to 50 (intervals of 0.1 from 0.1 to 1, 1 from 1 to 10, and 5 from 10 to 50), and cosines of the solar zenith angle from 0 (0.01) to 1 (intervals of 0.1). The asymmetry factor chosen for the graphic presentation is 0.75. To highlight the relative accuracy of the presentation, heavy shading is used for accuracy within 5%, while accuracy within 5–10% is denoted by light shading. White regions show errors greater than 10%.

Figure 3.7 shows the relative accuracy of the δ–two-stream (top graphs) and δ–four-stream (bottom graphs) approximations displayed in intervals of 0, 1, 2, 5, 10%, etc. The accuracy of the δ–two-stream approximation is comparable to that of the δ-Eddington approximation presented by King and Harshvardhan (1986). For conservative scattering, the reflection values produced by both approximations have low accuracy, on the order of 10 to 30% for $\mu_0 < 0.5$ and $\mu_0 > 0.9$ for $\tau < 1$. Errors greater than 10% occur for the total transmittance when $\mu_0 < 0.2$.

In general, reflectance and total transmittance values computed from the δ–four-stream approximation are accurate within about 5%, except for three small regions. For reflectance, 5–10% errors occur for $\mu_0 < 0.3$ and $0.6 < \mu_0 < 1$ when $\tau < 1$. For total transmittance, errors greater than 5% are produced for very high solar zenith angles ($\mu_0 < 0.1$). It is noted that these regions are associated with very small values. Thus, absolute errors are extremely small ($< 1\%$). In the case of $\tilde{\omega} = 0.8$, significant absorption could be built up for large optical depths and/or small solar zenith angles. The δ–two-stream (or δ-Eddington) approximation generally produces errors greater than 5–10%, as is evident from the graphic presentation. In particular, due to small transmittance values, errors of more than 50% may result in the case of large optical depths. The δ–four-stream approximation, on the other hand, has a relative accuracy (i.e., within about 5%) that is comparable to the case of conservative scattering. Errors of 5–10% occur only for very low solar zenith angles ($\mu_0 < 0.2$). Tables 3.2 and 3.3 present numerical results of reflectance and total transmittance computed from δ–two-stream, δ–four-stream, and doubling methods for $\tilde{\omega} = 1$ and 0.8.

In addition to the aforementioned results, computations have also been carried out using asymmetry factors of 0.7, 0.8, and 0.85 for the analytic Henyey–Greenstein phase function. The actual phase functions for cloud droplets were employed in accuracy checks, as were the surface albedos. The accuracy of the δ–four-stream approximation and, for that matter, the δ–two-stream or δ-Eddington approximation is not sensitive to small variations in the asymmetry factor and the detailed structure of the phase function. Also, variations in the surface albedo do not significantly alter the accuracy of the approximations.

Lastly, we have examined the accuracy of the δ–two-stream and δ–four-stream approximations in the case of Rayleigh scattering. Since $g = 0$ for Rayleigh atmospheres, there is no δ adjustment, and use of the two-stream method is equivalent to the isotropic scattering approximation. The four-stream approximation for flux calculations in Rayleigh atmospheres has an accuracy within about 3%.

For applications to the solar absorption bands, in which gaseous absorption in scattering atmospheres must be accounted for, the single-scattering albedo could be small. For this reason, we investigated the accuracy of the δ–two-stream and δ–four stream approximations using single-scattering albedos of 0.5 and 0.3 and keeping the other parameters the same as in Fig. 3.7. For cases involving large absorption, the reflectance values are generally very small. Thus we have presented the percentage of relative accuracy for absorptance, $(\Delta A/A)100\%$, where $A = 1 - r - t$, and total transmittance. Fig. 3.8 shows that the δ–two-stream approximation for absorption calculations produces adequate accuracy, which increases as $\tilde{\omega}$ decreases (i.e., absorptance increases). The δ–four-stream approximation has better accuracy than the δ–two-stream approximation, with errors for absorptance generally less than 2%. It is noted that as $\tilde{\omega}$ decreases, the effects of multiple-scattering on the flux calculations become less important. For total transmittance, errors from the

Table 3.2 Comparison of reflectance and total transmittance as computed by the δ–two-stream (a), δ–four-stream (b), and doubling (c) methods for conservative scattering; $\tilde{\omega} = 1$.

τ	μ_0	Reflectance					Transmittance				
		0.1	0.3	0.5	0.7	0.9	0.1	0.3	0.5	0.7	0.9
0.25	a	0.31216	0.12096	0.06384	0.03676	0.02101	0.68784	0.87904	0.93617	0.96324	0.97899
	b	0.38544	0.15451	0.07565	0.03972	0.02193	0.61457	0.84549	0.92436	0.96028	0.97807
	c	0.41610	0.15795	0.07179	0.03801	0.02249	0.58390	0.84205	0.92821	0.96200	0.97751
1.0	a	0.51352	0.32948	0.21009	0.13740	0.08956	0.48647	0.67050	0.78993	0.86260	0.91046
	b	0.58329	0.39015	0.24725	0.15534	0.09707	0.41670	0.60986	0.75277	0.84466	0.90293
	c	0.58148	0.38570	0.24048	0.15019	0.09672	0.41852	0.61430	0.75952	0.84981	0.90328
4.0	a	0.68564	0.59244	0.49891	0.41185	0.33578	0.31436	0.40756	0.50109	0.58814	0.66422
	b	0.72859	0.61975	0.52116	0.43006	0.34888	0.27140	0.38024	0.47883	0.56994	0.65111
	c	0.73254	0.61732	0.51932	0.42945	0.34822	0.26746	0.38268	0.48068	0.57055	0.65178
16.0	a	0.86858	0.82978	0.79098	0.75218	0.71340	0.13140	0.17020	0.20900	0.24780	0.28657
	b	0.87853	0.82988	0.78613	0.74539	0.70718	0.12145	0.17009	0.21384	0.25458	0.29278
	c	0.88103	0.82995	0.78658	0.74618	0.70721	0.11897	0.17005	0.21342	0.25382	0.29279
32.0	a	0.92599	0.90413	0.88228	0.86043	0.83858	0.07398	0.09582	0.11766	0.13951	0.16135
	b	0.93004	0.90202	0.87682	0.85335	0.83135	0.06993	0.09793	0.12313	0.14658	0.16858
	c	0.93151	0.90211	0.87714	0.85388	0.83145	0.06849	0.09789	0.12286	0.14612	0.16855

Table 3.3 Same as Table 3.2 except for the single-scattering albedo $\tilde{\omega}$ of 0.8.

τ	μ_0	Reflectance					Transmittance				
		0.1	0.3	0.5	0.7	0.9	0.1	0.3	0.5	0.7	0.9
0.25	a	0.21542	0.08867	0.04736	0.02733	0.01553	0.49230	0.76749	0.85858	0.90290	0.92904
	b	0.26907	0.10881	0.05290	0.02759	0.01540	0.45040	0.73848	0.84608	0.89840	0.92740
	c	0.28961	0.10686	0.04855	0.02590	0.01547	0.43017	0.73172	0.84756	0.89938	0.92669
1.0	a	0.29926	0.19772	0.12788	0.08327	0.05310	0.23476	0.40572	0.55685	0.65946	0.73071
	b	0.34599	0.21293	0.12880	0.07856	0.04963	0.19685	0.35288	0.51237	0.63149	0.71728
	c	0.35487	0.20714	0.12342	0.07622	0.04929	0.20556	0.35621	0.51606	0.63580	0.71772
4.0	a	0.32304	0.24597	0.18736	0.14125	0.10427	0.05288	0.07730	0.11690	0.17210	0.23407
	b	0.36061	0.24348	0.16937	0.12035	0.08906	0.04775	0.07364	0.10892	0.15813	0.21859
	c	0.37148	0.23862	0.16615	0.12036	0.08925	0.04539	0.07179	0.10718	0.15697	0.21953
16.0	a	0.32432	0.24785	0.19022	0.14525	0.10916	0.00015	0.00022	0.00032	0.00052	0.00098
	b	0.36145	0.24478	0.17131	0.12316	0.09281	0.00026	0.00040	0.00059	0.00086	0.00126
	c	0.37229	0.23990	0.16808	0.12315	0.09296	0.00027	0.00042	0.00062	0.00092	0.00139
32.0	a	0.32432	0.24785	0.19022	0.14525	0.10916	0.00000	0.00000	0.00000	0.00000	0.00000
	b	0.36145	0.24478	0.17131	0.12316	0.09281	0.00000	0.00000	0.00000	0.00000	0.00000
	c	0.37229	0.23990	0.16808	0.12315	0.09296	0.00000	0.00000	0.00000	0.00000	0.00000

δ–four-stream approximation are again within about 5%. Large relative errors can be produced by the δ–two-stream approximation when the transmittance values are small.

We have presented a simple and systematic formulation of the δ–four-stream approximation for solar flux calculations. While all approximate methods for radiative flux transfer have advantages and shortcomings in terms of their computational accuracy for different $\tilde{\omega}$, τ, and μ_0, this approximation can achieve relative accuracy within about 5% for all reasonable ranges of the single-scattering parameters at a given wavelength. For computations of solar fluxes covering the entire solar spectrum, the averaged accuracy should also be within about 5%. By virtue of the two intensity streams in the upper hemisphere and the two in the lower hemisphere, the δ–four-stream approximation has all the radiative characteristics inherent in the δ–two-stream approximation. The solution of this approximation, like various two-stream methods, is in analytic form so that the computer time involved is minimal. The method can be easily applied to nonhomogeneous atmospheres, as described in Section 3.7. For radiative transfer parameterizations in numerical models in which a single radiative transfer approximation is required, the δ–four-stream approximation would be an excellent method.

3.6   Principles of invariance and radiative flux transfer

The transfer of light beams in planetary atmospheres depends on the incoming and outgoing directions. If computations of flux are required, numerical integrations over the outgoing directions must be performed by virtue of the definition of flux. Numerical integrations require considerable computational effort. Hence, it is desirable to seek simplified and approximate expressions for the representation of a flux field. In this section, we wish to introduce the transfer of radiative flux based on the principles of invariance. It suffices to consider azimuthal-independent radiative transfer. Consider an atmosphere with an optical depth τ_*. The reflection function R and transmission function T are defined by

$$R(\mu, \mu_0) = \frac{\pi I_r(0, \mu)}{\mu_0 F_\odot}, \tag{3.6.1a}$$

$$T(\mu, \mu_0) = \frac{\pi I_t(\tau_*, -\mu)}{\mu_0 F_\odot}, \tag{3.6.1b}$$

$$T^{\text{dir}}(\mu_0) = \frac{\pi I_t^{\text{dir}}(\tau_*, -\mu_0)}{\mu_0 F_\odot} = e^{-\tau_*/\mu_0}, \tag{3.6.1c}$$

where I_r and I_t represent the reflected and transmitted intensities at the top and bottom of the atmosphere, respectively. The minus sign associated with μ indicates that the direction of the light beam is downward. The diffuse and direct (dir) components of the transmission function are separated in the definitions.

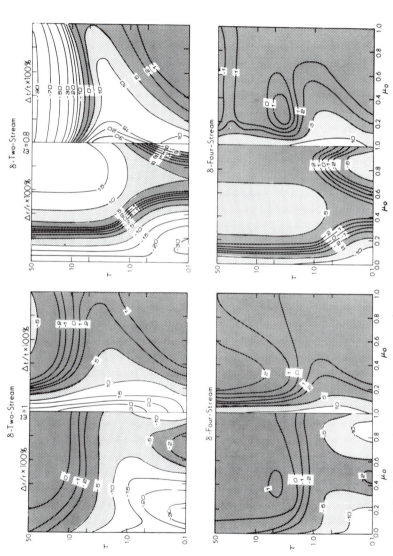

FIG. 3.7 Relative accuracy of the reflectance ($\hat{r}$) and total transmittance ($\hat{t}$) computed from the δ-two-stream (upper graphs) and δ-four-stream (lower graphs) approximations with respect to r and t derived from the adding method for radiative transfer. The relative accuracy is defined by $\Delta r/r = (\hat{r} - r)/r$ for reflectance and $\Delta t/t = (\hat{t} - t)/t$ for total transmittance. The results are shown in the domain of the optical depth τ and the cosine of the solar zenith angle μ_0, and expressed in terms of percentage. The heavy and light shadings denote errors within 5% and within 5–10%, respectively, while the white area represents errors greater than 10%. The left and right graphs are, respectively, for $\tilde{\omega} = 1$ (conservative scattering) and $\tilde{\omega} = 0.8$ (after Liou et al., 1988).

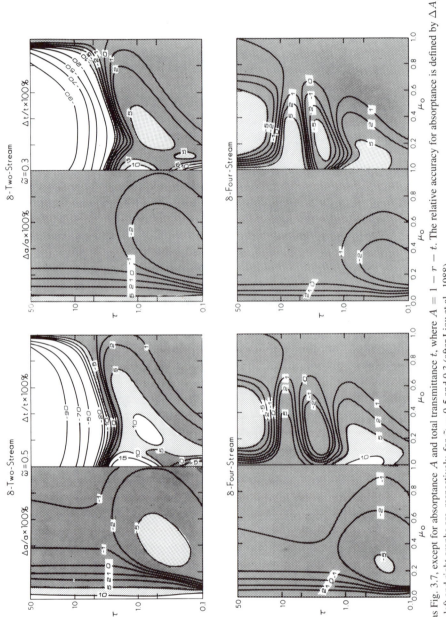

FIG. 3.8 Same as Fig. 3.7, except for absorptance A and total transmittance t, where $A = 1 - r - t$. The relative accuracy for absorptance is defined by $\Delta A/A = (\hat{A} - A)/A$. The left and right graphs are, respectively, for $\tilde{\omega} = 0.5$ and 0.3 (after Liou et al., 1988).

Next, we introduce the corresponding nondimensional parameters in flux forms. *Reflectance* (also referred to as *reflection* or *local* or *planetary albedo*) and *transmittance* (also referred to as *transmission*) are defined by

$$r(\mu_0) = \frac{F^+(0)}{\mu_0 F_\odot} = 2 \int_0^1 R(\mu, \mu_0)\mu \, d\mu, \tag{3.6.2a}$$

$$t(\mu_0) = \frac{F^-(\tau_*)}{\mu_0 F_\odot} = 2 \int_0^1 T(\mu, \mu_0)\mu \, d\mu, \tag{3.6.2b}$$

$$t^{\text{dir}}(\mu_0) = 2 \int_0^1 e^{-\tau_*/\mu_0}\mu \, d\mu = e^{-\tau_*/\mu_0}, \tag{3.6.2c}$$

where F^+ and F^- represent the diffuse upward and downward fluxes, respectively. These are obtained by integrating the upward and downward intensities over the upper and lower hemispheres. The total transmission $t(\mu_0)$ is therefore the sum of t and t^{dir}.

Finally, the global reflectance (or global albedo) $\bar{r}$ and global transmittance $\bar{t}$ may be defined in the forms

$$\bar{r} = \frac{f^+(0)}{\pi a_e^2 F_\odot} = 2 \int_0^1 r(\mu_0)\mu_0 \, d\mu_0, \tag{3.6.3a}$$

$$\bar{t} = \frac{f^-(\tau_*)}{\pi a_e^2 F_\odot} = 2 \int_0^1 t(\mu_0)\mu_0 \, d\mu_0, \tag{3.6.3b}$$

$$\bar{t}^{\text{dir}} = 2 \int_0^1 e^{-\tau_*/\mu_0}\mu_0 \, d\mu_0, \tag{3.6.3c}$$

where f^+ and f^- represent the total outgoing flux at the top and bottom of the atmosphere, respectively, a_e is the radius of the planet, and $\pi a_e^2 F_\odot$ represents the total incoming solar flux at TOA.

For a semi-infinite, plane-parallel atmosphere, the diffuse reflected intensity cannot be changed if a layer of finite optical depth, having the same optical properties as those of the original layer, is added (Ambartzumian, 1942). Based on this invariant principle, the reflection function at the top of a plane-parallel atmosphere can be expressed in terms of a known mathematical function, the so-called H function. More general principles of invariance for a finite, plane-parallel atmosphere have been developed by Chandrasekhar (1950), who used scattering and transmission functions in defining the four principles governing the reflection and transmission of a light beam in two layers. Liou (1980) has developed these four principles in terms of the conventional reflection and transmission functions defined in Subsection 3.2.2.

The four principles of invariance governing the reflection and transmission of light beam may be stated as follows:

1. The reflected intensity at any given optical depth level τ results from the reflection of (a) the attenuated solar flux and (b) the downward diffuse intensity at that level by the optical depth $\tau_* - \tau$.
2. The diffusely transmitted intensity at τ results from (a) the transmission of solar flux and (b) the reflection of the upward diffuse intensity above the level τ.
3. The reflected intensity at the top of the finite atmosphere ($\tau = 0$) is equivalent to (a) the reflection of solar flux plus (b) the direct and diffuse transmission of the upward diffuse intensity above the level τ.
4. The diffusely transmitted intensity at the bottom of a finite atmosphere ($\tau = \tau_*$) is equivalent to (a) the transmission of the attenuated solar flux at level τ plus (b) the direct and diffuse transmission of the downward diffuse intensity at the level τ by the optical depth $\tau_* - \tau$.

Using the definitions of the reflection and transmission functions in Eqs. (3.6.1a,b), letting $\tau = \tau_1$, and $\tau_* - \tau = \tau_2$, and defining the dimensionless upward and downward internal intensities by

$$U(\mu, \mu_0) = \frac{\pi I(\tau_1, \mu)}{\mu_0 F_\odot}, \tag{3.6.4a}$$

$$D(\mu, \mu_0) = \frac{\pi I(\tau_1, -\mu)}{\mu_0 F_\odot}, \tag{3.6.4b}$$

the four principles of invariance may be expressed in terms of the reflection and transmission functions as follows:

$$U(\mu, \mu_0) = R_2(\mu, \mu_0)e^{-\tau_1/\mu_0}$$
$$+ 2 \int_0^1 R_2(\mu, \mu')D(\mu', \mu_0)\mu' \, d\mu', \tag{3.6.4}$$

$$D(\mu, \mu_0) = T_1(\mu, \mu_0) + 2 \int_0^1 R_1(\mu, \mu'')U(\mu'', \mu_0)\mu'' \, d\mu'', \tag{3.6.5}$$

$$R_{12}(\mu, \mu_0) = R_1(\mu, \mu_0) + e^{-\tau_1/\mu_0}U(\mu, \mu_0)$$
$$+ 2 \int_0^1 T_1(\mu, \mu')U(\mu', \mu_0)\mu' \, d\mu', \tag{3.6.6}$$

$$T_{12}(\mu, \mu_0) = T_2(\mu, \mu_0)e^{-\tau_1/\mu_0} + e^{-\tau_2/\mu}D(\mu, \mu_0)$$
$$+ 2 \int_0^1 T_2(\mu, \mu')D(\mu', \mu_0)\mu' \, d\mu'. \tag{3.6.7}$$

The geometric configuration involving the basic variables is illustrated in Fig. 3.9. Although the preceding equations are written for azimuthal-independent cases, these equations may be modified for applications to general radiative transfer involving azimuthal terms and polarization effects by replacing μ with (μ, ϕ)

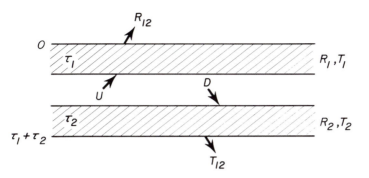

FIG. 3.9 Geometric configuration for the reflection and transmission functions defined in Eqs. (3.6.4)–(3.6.7) based on the principles of invariance for a finite atmosphere. For illustration purposes, we have defined $\tau = \tau_1$ and $\tau_* - \tau = \tau_2$ in the text.

and the diffuse intensity with the Stokes parameters defined in Subsection 5.1.2. Substituting Eq. (3.6.4) into Eq. (3.6.5) leads to

$$D(\mu, \mu_0) = T_1(\mu, \mu_0) + S_{12}(\mu, \mu_0)e^{-\tau_1/\mu_0}$$
$$+ 2 \int_0^1 S_{12}(\mu, \mu'')D(\mu'', \mu_0)\mu'' \, d\mu'', \qquad (3.6.8)$$

where

$$S_{12}(\mu, \mu'') = 2 \int_0^1 R_1(\mu, \mu')R_2(\mu', \mu'')\mu' \, d\mu' . \qquad (3.6.9)$$

Equations (3.6.4) and (3.6.6)–(3.6.9), which are postulated from the principles of invariance, are "equivalent" to the adding equations presented in Eqs. (3.2.44a–f) for the case involving radiation from above. The principles of invariance can also be formulated for the case involving radiation from below, and the resulting equations would be "equivalent" to the adding equations presented in Eqs. (3.2.46a–f).

We wish to derive a set of equations for the computation of fluxes. Analogous to the derivation of Eq. (3.6.8), substituting Eq. (3.6.5) into Eq. (3.6.4) yields

$$U(\mu, \mu_0) = R_2(\mu, \mu_0)e^{-\tau_1/\mu_0} + 2 \int_0^1 R_2(\mu, \mu')T_1(\mu', \mu_0)\mu' \, d\mu'$$
$$+ 2 \int_0^1 S_{12}(\mu, \mu'')U(\mu'', \mu_0)\mu'' \, d\mu''. \qquad (3.6.9a)$$

In order to obtain flux forms based on the principles of invariance, we define

$$y(\mu_0) = 2 \int_0^1 Y(\mu, \mu_0)\mu \, d\mu, \qquad (3.6.10)$$

where the notations $y(Y)$ can be $u(U)$, $d(D)$, $r_{12}(R_{12})$, $t_{12}(T_{12})$, or $s_{12}(S_{12})$. Carrying out solid-angle integrations on Eqs. (3.6.9a) and (3.6.5)–(3.6.7), we obtain

$$u(\mu_0) = r_2(\mu_0)e^{-\tau_1/\mu_0} + 2\int_0^1 r_2(\mu')T_1(\mu',\mu_0)\mu'\,d\mu'$$

$$+ 2\int_0^1 s_{12}(\mu'')U(\mu'',\mu_0)\mu''\,d\mu''\,, \qquad (3.6.11a)$$

$$d(\mu_0) = t_1(\mu_0) + 2\int_0^1 r_1(\mu')U(\mu',\mu_0)\mu'\,d\mu'\,, \qquad (3.6.11b)$$

$$r_{12}(\mu_0) = r_1(\mu_0) + 2\int_0^1 e^{-\tau_1/\mu}U(\mu,\mu_0)\mu\,d\mu$$

$$+ 2\int_0^1 t_1(\mu')U(\mu',\mu_0)\mu'\,d\mu'\,, \qquad (3.6.11c)$$

$$t_{12}(\mu_0) = t_2(\mu_0)e^{-\tau_1/\mu_0} + 2\int_0^1 e^{-\tau_2/\mu}D(\mu,\mu_0)\mu\,d\mu$$

$$+ 2\int_0^1 t_2(\mu')D(\mu',\mu_0)\mu'\,d\mu'\,. \qquad (3.6.11d)$$

Moreover, by using global values for reflectance and for direct and diffuse transmittance in these equations, viz.,

$$2\int_0^1 r_2(\mu')T_1(\mu',\mu_0)\mu'\,d\mu' \cong \bar{r}_2\,2\int_0^1 T_1(\mu',\mu_0)\mu'\,d\mu'$$

$$= \bar{r}_2 t_1(\mu_0), \qquad (3.6.12a)$$

$$2\int_0^1 e^{-\tau_1/\mu}U(\mu,\mu_0)\mu\,d\mu \cong e^{-\tau_1/\bar{\mu}}\,2\int_0^1 U(\mu,\mu_0)\mu\,d\mu$$

$$= e^{-\tau_1/\bar{\mu}}u(\mu_0), \qquad (3.6.12b)$$

where $1/\bar{\mu}$ denotes the diffusivity factor to be determined numerically, we have

$$u(\mu_0) \cong r_2(\mu_0)e^{-\tau_1/\mu_0} + \bar{r}_2 t_1(\mu_0) + \bar{s}_{12}u(\mu_0), \qquad (3.6.13a)$$

$$d(\mu_0) \cong t_1(\mu_0) + \bar{r}_1\mu(\mu_0), \qquad (3.6.13b)$$

$$r_{12}(\mu_0) \cong r_1(\mu_0) + e^{-\tau_1/\bar{\mu}}u(\mu_0) + \bar{t}_1 u(\mu_0), \qquad (3.6.13c)$$

$$t_{12}(\mu_0) \cong t_2(\mu_0)e^{-\tau_1/\mu_0} + e^{-\tau_2/\bar{\mu}}d(\mu_0) + \bar{t}_2 d(\mu_0), \qquad (3.6.13d)$$

where

$$\bar{s}_{12} = 2\int_0^1 r_1(\mu')r_2(\mu')\mu'\,d\mu' \cong \bar{r}_1\bar{r}_2\,. \qquad (3.6.14)$$

To finalize the iterative equations for reflectance and transmittance for a combined layer with an optical depth $(\tau_1 + \tau_2)$, we introduce a parameter referred to as the *upward generation function*, based on Eq. (3.6.13a), in the form

$$u(\mu_0) = \left[r_2(\mu_0)e^{-\tau_1/\mu_0} + \bar{r}_2 t_1(\mu_0)\right](1 - \bar{s}_{12})^{-1}\,. \qquad (3.6.15)$$

Further, we may define a number of total global transmittances as follows:

$$\bar{\bar{t}}_{12} = \bar{t}_{12} + \exp(-\tau_{1,2}/\bar{\mu}), \qquad (3.6.16a)$$

$$\bar{t}_{1,2} = t_{1,2}(\mu_0) + \exp(-\tau_{1,2}/\mu_0). \qquad (3.6.16b)$$

By adding the direct transmittance, $\exp[-(\tau_1 + \tau_2)/\mu_0)]$, to the diffuse transmittance in Eq. (3.6.13d), we obtain

$$r_{12}(\mu_0) = r_1(\mu_0) + \bar{\bar{t}}_1 u(\mu_0), \qquad (3.6.17)$$

$$\bar{t}_{12}(\mu_0) = \bar{t}_2(\mu_0)e^{-\tau_1/\mu_0} + \bar{\bar{t}}_2[t_1(\mu_0) + \bar{r}_1 u(\mu_0)]. \qquad (3.6.18)$$

Equations (3.6.15), (3.6.17), and (3.6.18) constitute a closed set of iterative equations for computing reflectance and transmittance for a combined layer. The physical meaning and configurations of these equations may be understood from Fig. 3.10. The reflectance of a combined layer is produced by (a) the reflectance of the first layer plus (b) the diffuse transmittance of the upward generation function, $u(\mu_0)$. The total transmittance of a combined layer is the result of (a) the transmittance of the direct transmittance component of the first layer and (b) the global diffuse transmittance of the diffuse transmittance plus the global diffuse reflectance of the upward generation function by the second layer. The upward generation function is the sum of (a) the reflectance of the direct transmittance of the first layer and (b) the global diffuse reflectance of the diffuse transmittance of the first layer by the second layer that undergoes multiple reflections.

In order to proceed with the computational procedures for flux, it is necessary to determine the reflectance and transmittance values as functions of the solar zenith angle for each layer. We may begin with a layer that is optically thin and use the single-scattering approximation given in Eqs. (3.2.31)–(3.2.34). Subsequently, we may perform zenith angle integrations to obtain reflectance, $r(\mu_0)$, and transmittance, $t(\mu_0)$. The global albedo $\bar{r}$ and global transmittance $\bar{t}$ may also be calculated. The forward diffraction peak may be incorporated in the computation through the δ-function adjustment discussed in Section 3.4.

The principles of invariance may be applied to a combination of a finite homogeneous layer with an optical depth of τ_* and a surface. Consider a Lambertian surface with an isotropic reflectance of r_s and zero transmittance. From Eqs. (3.6.17), the reflectance at the top of the layer is

$$r_*(\mu_0) = r_1(\mu_0) + \bar{\bar{t}}_1 u(\mu_0), \qquad (3.6.19)$$

where

$$u(\mu_0) = r_s \bar{t}_1(\mu_0)(1 - r_s \bar{r}_1)^{-1}. \qquad (3.6.20)$$

Based on Eq. (3.6.13b), the total transmission at the bottom of the layer (or at the surface) is given by

$$t_*(\mu_0) = \bar{t}_1(\mu_0) + \bar{r}_1 u(\mu_0). \qquad (3.6.21)$$

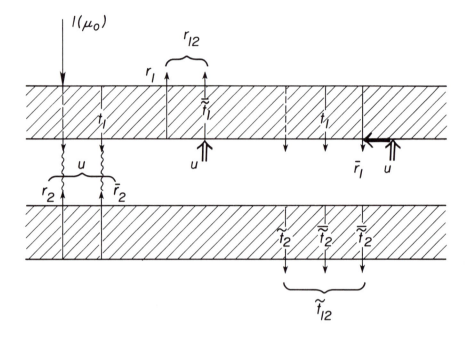

FIG. 3.10   Adding principles for reflectance, r_{12}, and total transmittance, $\tilde{t}_{12}$, according to the terms in Eqs. (3.6.17) and (3.6.18). The term u is defined in Eq. (3.6.15), $1(\mu_0)$ denotes unit solar flux, and the dashed lines in the diagram represent exponential attenuation. The wavy lines illustrate multiple reflections. The meanings of all other terms are explained in the text.

Equations (3.6.19)–(3.6.21) can also be derived from the adding principle for radiative transfer presented in Fig. 3.4.

3.7   Application of radiative transfer methods to nonhomogeneous atmospheres

One of the fundamental difficulties in radiative transfer involves accounting for the nonhomogeneous nature of the atmosphere. Figure 3.11 shows the profiles of molecular and aerosol number densities in the earth's atmosphere. Two typical aerosol concentrations are displayed. The clear condition has a visibility of $\sim$25 km, whereas the hazy condition represents an extreme aerosol concentration in an urban environment. These profiles illustrate that the atmosphere, even without clouds, is nonhomogeneous and cannot be represented by a single single-scattering albedo, $\tilde{\omega}$ and a phase function P. The radiative transfer equation for diffuse intensities must be modified to include variations in $\tilde{\omega}$ and P with optical depth. Using the basic radiative transfer equation denoted in Eq. (3.1.16), we may

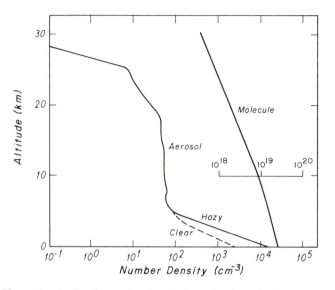

Fig. 3.11   The number density of aerosols and molecules as functions of altitude in a model atmosphere. Two aerosol concentrations, representing background (clear) and urban (hazy) conditions, are shown.

write

$$\mu \frac{dI(\tau,\mu)}{d\tau} = I(\tau,\mu) - \frac{\tilde{\omega}(\tau)}{2} \int_{-1}^{1} I(\tau,\mu')P(\tau;\mu,\mu')\,d\mu'$$
$$- \frac{\tilde{\omega}(\tau)}{4\pi}P(\tau;\mu,-\mu_0)F_\odot e^{-\tau/\mu_0}. \tag{3.7.1}$$

Since $\tilde{\omega}$ and P are functions of optical depth, analytic solutions for this equation are generally not possible. We may, however, devise a numerical procedure to compute the diffuse intensities in nonhomogeneous atmospheres.

3.7.1 Discrete-ordinates method

The discrete-ordinates method for radiative transfer can be applied to nonhomogeneous atmospheres by numerical approximations (Liou, 1975). For the present analysis, consider the azimuth-independent component. As illustrated in Fig. 3.12, the atmosphere may be divided into N homogeneous layers, each of which is characterized by a single- scattering albedo, a phase function, and an extinction coefficient (or optical depth). The solution for the azimuthally independent diffuse intensity, as given in Eq. (3.2.7), may be written for each individual layer ℓ in the form

$$I^{(\ell)}(\tau,\mu_i) = \sum_j L_j^{(\ell)}\phi_j^{(\ell)}(\mu_i)e^{-k_j^{(\ell)}\tau} + Z^{(\ell)}(\mu_i)e^{-\tau/\mu_0},$$
$$\ell = 1, 2, \ldots, N. \tag{3.7.2}$$

At TOA ($\tau = 0$), there is no downward diffuse flux, so that

$$I^{(1)}(0, -\mu_i) = 0. \tag{3.7.3}$$

Within the atmosphere, the upward and downward intensities must be continuous at the interface of each predivided layer. Thus we have

$$I^{(\ell)}(\tau_\ell, \mu_i) = I^{(\ell+1)}(\tau_\ell, \mu_i), \tag{3.7.4}$$

where τ_ℓ denotes the optical depth from TOA to the bottom of the ℓ layer. At the bottom of the atmosphere, surface reflectance must be accounted for. The reflection of sunlight at the surface depends significantly on its optical property with respect to the incident wavelength. For example, reflectance from vegetation and soil is highly wavelength dependent. Also, surface reflectance patterns are generally anisotropic. The interactions of the surface with solar radiation are extremely important in determining the solar flux available at the surface and, hence, the surface temperature, especially over land. However, for the computation of solar fluxes in the atmosphere, it is appropriate to use a Lambertian surface. Let the surface albedo be r_s. Then the upward diffuse intensities that are reflected from the surface may be expressed by

$$I^{(N)}(\tau_N, +\mu_i) = \frac{r_s}{\pi} \left[F^-(\tau_N) + \mu_0 F_\odot e^{-\tau_N/\mu_0} \right], \tag{3.7.5}$$

where the downward diffuse flux reaching the surface is

$$F^-(\tau_N) = 2\pi \sum_{i=1}^{N} I^{(N)}(\tau_N, -\mu_i) a_i \mu_i . \tag{3.7.6}$$

Matching the boundary and continuity conditions that are required for the diffuse intensities, we obtain the following set of equations for the determination of the unknown coefficients:

$$\sum_j L_j^{(1)} \phi_j^{(1)}(-\mu_i) = -Z^{(1)}(-\mu_i), \quad i = 1, \ldots, n, \tag{3.7.7}$$

$$\sum_j \left[L_j^{(\ell)} \gamma_j^{(\ell)}(\mu_i) + L_j^{(\ell+1)} \delta_j^{(\ell+1)}(\mu_i) \right] = -^{(\ell)} \eta^{(\ell+1)}(\mu_i),$$

$$i = -n, \ldots, -1, 1, \ldots, n, \quad \ell = 1, 2, \ldots, N, \tag{3.7.8}$$

$$\sum_j L_j^{(N)} \beta_j^{(N)}(+\mu_i) = -\epsilon^{(N)}(+\mu_i), \quad i = 1, \ldots, n, \tag{3.7.9}$$

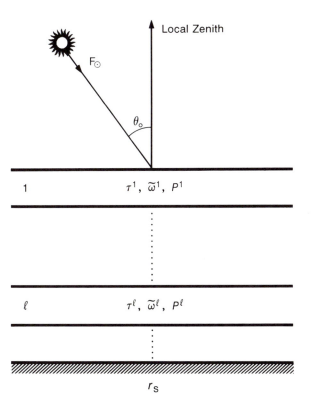

FIG. 3.12 A nonhomogeneous atmosphere is divided into ℓ homogeneous layers with respect to the single–scattering albedo $\tilde{\omega}$, phase function P, and optical depth τ. $F_\odot$ denotes the solar flux at TOA, θ_0 is the solar zenith angle, and r_s is the surface albedo.

where

$$\gamma_j^{(\ell)}(\mu_i) = \phi_j^{(\ell)}(\mu_i)e^{-k_j^{(\ell)}\tau_\ell} , \tag{3.7.10}$$

$$\delta_j^{(\ell+1)}(\mu_i) = -\phi_j^{(\ell+1)}(\mu_i)e^{-k_j^{(\ell-1)}\tau_\ell} , \tag{3.7.11}$$

$$^{(\ell)}\eta^{(\ell+1)}(\mu_i) = \left[z^{(\ell)}(\mu_i) - z^{(\ell+1)}(\mu_i)\right]e^{-\tau_\ell/\mu_0} , \tag{3.7.12}$$

$$\beta_j^{(N)}(+\mu_i) = \left(\phi_j^{(N)}(+\mu_i) - 2r_s\sum_{i=1}^{n}\phi_j^{(N)}(-\mu_i)a_i\mu_i\right)$$
$$\times e^{-k_j^{(N)}\tau_N} , \tag{3.7.13}$$

$$\epsilon_j^{(N)}(+\mu_i) = \left(Z^{(N)}(+\mu_i) - 2r_s\sum_{i=1}^{n}Z^{(N)}(-\mu_i)a_i\mu_i - \frac{r_s}{\pi}\mu_0F_\odot\right)$$
$$\times e^{-\tau_N/\mu_0} . \tag{3.7.14}$$

Thus we have $N \times 2n$ equations for the determination of $N \times 2n$ unknown coefficients, $L_j^{(\ell)}$. Equations (3.7.7)–(3.7.9) may be expressed in terms of matrix representations in the form

$$\phi \mathbf{L} = \chi. \tag{3.7.15}$$

These matrices are given by

$$\mathbf{L} = \begin{bmatrix} L_{-n}^{(1)} \\ \vdots \\ L_n^{(1)} \\ L_{-n}^{(2)} \\ \vdots \\ L_n^{(2)} \\ \vdots \\ L_{-n}^{(N)} \\ \vdots \\ L_n^{(N)} \end{bmatrix}, \tag{3.7.16}$$

$$\chi = \begin{bmatrix} Z^{(1)}(-\mu_n) \\ \vdots \\ Z^{(1)}(-\mu_1) \\ {}^{(1)}\eta^{(2)}(-\mu_n) \\ \vdots \\ {}^{(1)}\eta^{(2)}(+\mu_n) \\ \vdots \\ \epsilon^{(N)}(+\mu_1) \\ \vdots \\ \epsilon^{(N)}(+\mu_n) \end{bmatrix}, \tag{3.7.17}$$

and

$$
\phi =
\begin{bmatrix}
\phi^{(1)}_{-n}(-\mu_n) & \cdots & \phi^{(1)}_n(-\mu_n) & 0 & \cdots & 0 & 0 & \cdots & 0 \\
\vdots & & \vdots & \vdots & & \vdots & \vdots & & \vdots \\
\phi^{(1)}_{-n}(-\mu_1) & \cdots & \phi^{(1)}_n(-\mu_1) & 0 & \cdots & 0 & 0 & \cdots & 0 \\
\gamma^{(1)}_{-n}(-\mu_n) & \cdots & \gamma^{(1)}_n(-\mu_n) & \delta^{(2)}_{-n}(-\mu_n) & \cdots & \delta^{(2)}_n(-\mu_n) & 0 & \cdots & 0 \\
\vdots & & \vdots & \vdots & & \vdots & \vdots & & \vdots \\
\gamma^{(1)}_{-n}(\mu_n) & \cdots & \gamma^{(1)}_n(\mu_n) & \delta^{(2)}_{-n}(\mu_n) & \cdots & \delta^{(2)}_n(\mu_n) & 0 & \cdots & 0 \\
\vdots & & \vdots & \vdots & & \vdots & & & \vdots \\
0 & \cdots & 0 & 0 & \cdots & 0 & \beta^{(N)}_{-n}(\mu_1) & \cdots & \beta^{(N)}_n(\mu_1) \\
\vdots & & \vdots & \vdots & & \vdots & \vdots & & \vdots \\
0 & \cdots & 0 & 0 & \cdots & 0 & \beta^{(N)}_{-n}(\mu_n) & \cdots & \beta^{(N)}_n(\mu_n)
\end{bmatrix}
$$

$$(3.7.18)$$

In the context of a two-stream approximation, we may write

$$
\begin{bmatrix}
x & x & & & & & \\
x & x & x & x & & & \\
x & x & x & x & & & \\
& & x & x & x & x & \\
& & x & x & x & x & \\
& & & & \cdot & \cdot & \cdot & \cdot \\
& & & & \cdot & \cdot & \cdot & \cdot \\
& & & & & & x & x
\end{bmatrix}
\begin{bmatrix}
L^{(1)}_{-1} \\
L^{(1)}_{1} \\
L^{(2)}_{-1} \\
\cdot \\
\cdot \\
\cdot \\
L^{(N)}_{-1} \\
L^{(N)}_{1}
\end{bmatrix}
=
\begin{bmatrix}
x \\
x \\
x \\
x \\
x \\
\cdot \\
\cdot \\
x
\end{bmatrix},
\qquad (3.7.19)
$$

where the xs are elements that can be found in Eqs. (3.3.5a,b), the blanks denote zeros, and we have set $L_{-1} = K$, and $L_1 = H$ in these equations. Likewise, a matrix formulation may be set up for the four-stream approximation described in Section 3.5 for applications to nonhomogeneous atmospheres. Applications of the general discrete-ordinates method to nonhomogeneous atmospheres for intensity computations have been illustrated by Stamnes (1986).

3.7.2 *Adding method*

The adding principle may be applied to nonhomogeneous atmospheres by using the following numerical procedures:

1. The atmosphere may be divided into N homogeneous layers, each of which is characterized by a single-scattering albedo, phase function, and optical depth, as shown in the previous section. Let R_ℓ and T_ℓ ($\ell = 1, 2, \ldots, N$) denote the reflection and transmission functions for each

homogeneous layer. Since homogeneous layers are considered, we have $R_\ell^* = R_\ell$ and $T_\ell^* = T_\ell$. R_ℓ and T_ℓ may be obtained from the doubling method described previously. The surface is considered to be a layer whose reflection function is R_{N+1} and whose transmission function is $T_{N+1} = 0$. If the surface is approximated as a Lambertian reflector, then R_{N+1} is the surface albedo r_s, which ranges from 0 to 1.

2. As shown in Fig. 3.13, the layers are added one at a time from TOA downward to obtain $R_{1,\ell}$ and $T_{1,\ell}$ for ℓ from 2 to $(N + 1)$, and $R_{1,\ell}^*$, and $T_{1,\ell}^*$ for ℓ from 2 to N. For example, $R_{1,\ell}$ is the reflection function for composite layers 1 through ℓ, with the lower part of the atmosphere and surface absent.

3. The layers are then added one at a time from the surface upward to obtain $R_{\ell+1,N+1}$ and $T_{\ell+1,N+1}$ for ℓ from $(N - 1)$ to 1.

4. We now consider the composite layers $(1, \ell)$ and $(\ell+1, N+1)$ and use the adding equations for the internal intensities denoted in Eqs. (3.2.44c,d) to obtain

$$D = T_{1,\ell} + ST_{1,\ell} + S\exp(-\tau_{1,\ell}/\mu_0), \tag{3.7.20a}$$

$$U = R_{\ell+1,N+1}D + R_{\ell+1,N+1}\exp(-\tau_{1,\ell}/\mu_0), \tag{3.7.20b}$$

where

$$S = Q(1 - Q)^{-1}, \tag{3.7.21}$$

$$Q = R_{1,\ell}^* R_{\ell+1,N+1}, \tag{3.7.22}$$

and $\tau_{1,\ell}$ is the optical depth from TOA to the bottom of the layer.

To obtain the upward and downward fluxes at the interface between the ℓ and $\ell + 1$ layers, angular integrations are performed. It is necessary to consider only the azimuth-independent condition, so that

$$F^+ = \mu_0 F_\odot \left(2\int_0^1 U(\mu, \mu_0)\mu\,d\mu\right), \tag{3.7.23}$$

$$F_{\text{dif}}^- = \mu_0 F_\odot \left(2\int_0^1 D(\mu, \mu_0)\mu\,d\mu\right). \tag{3.7.24}$$

The downward direct solar flux is

$$F_{\text{dir}}^- = \mu_0 F_\odot \exp(-\tau_{1,\ell}/\mu_0). \tag{3.7.25}$$

Thus the net flux is

$$F = F^+ - (F_{\text{dif}}^- + F_{\text{dir}}^-). \tag{3.7.26}$$

The heating rate for a given layer can be evaluated from the divergence of the net fluxes.

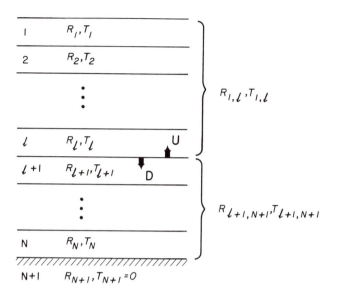

Fig. 3.13   Diagram illustrating the computation of internal intensities based on the adding principle for radiative transfer.

3.8   Absorption of solar flux in scattering atmospheres, spectral integration, and parameterization

3.8.1   *Spectral integration*

In the discussion of various methodologies and approximations for radiative transfer, we have assumed that the transfer of radiation is monochromatic. In order to obtain atmospheric solar heating rates and fluxes at TOA and the surface, spectral integration must be performed.

Molecules scatter sunlight. These scattering processes are important in the ultraviolet (uv) and visible regions. The scattering of sunlight by cloud and aerosol particles is generally coupled with absorption. There are several gases, chiefly ozone and water vapor, that absorb solar radiation. The details of gaseous absorption in the solar spectrum will be discussed in the next subsection. To determine the distribution of solar fluxes, it is necessary to combine scattering and absorption by particulates, scattering by molecules, and gaseous absorption. In this section, we present a procedure based on which spectral integration may be performed.

In principle, if the absorption coefficients, in terms of the line position, intensity, and half-width, are known, line-by-line computations including the multiple-scattering contribution may be performed. However, this is computationally formidable even for a small spectral interval. We may approach the incorporation of gaseous absorption in a scattering atmosphere on the basis of the k-distribution method discussed in Section 2.6. Consider a wavelength interval

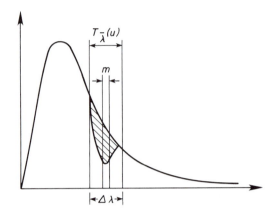

FIG. 3.14　Graphical representation of the incorporation of gaseous absorption in multiple-scattering processes involving a spectral interval $\Delta\lambda$. $T_{\bar{\lambda}}(u)$ denotes the spectral transmittance, and m is the index for the sub-spectral interval.

$\Delta\lambda$; the spectral transmittance may be defined in the form

$$T_{\bar{\lambda}}(u) = \int_{\Delta\lambda} e^{-k_\lambda u}\, \frac{d\lambda}{\Delta\lambda}. \qquad (3.8.1a)$$

As illustrated in Section 2.6, $T_{\bar{\lambda}}(u)$ can be replaced by a finite number of terms in the form

$$T_{\bar{\lambda}}(u) = \sum_{m=1}^{M} \Delta g_m\, e^{-k_m u}, \qquad (3.8.1b)$$

where Δg_m and k_m represent the weights and equivalent absorption coefficients, respectively, and M is the total number of subintervals for the spectral band $\Delta\lambda$. A graphic representation of the subintervals is displayed in Fig. 3.14.

　　An alternative approach to incorporating the gaseous absorption in multiple-scattering processes utilizes the distribution of photon path length to account for the amount of absorber along the path length when scattering events are present. Introduce the probability distribution $p(u)$, such that $p(u)\, du$ is the probability that a photon, which contributes to the conservative scattering intensity I_c has traveled a path length between u and $u + du$. Then the monochromatic scattered intensity is related to I_c in the form

$$I_\lambda(\mu, \phi; \mu_0, \phi_0) = I_c(\mu, \phi; \mu_0, \phi_0) \int_0^\infty p(u) e^{-k_\lambda u}\, du. \qquad (3.8.2a)$$

Integration over a finite spectral interval $\Delta\lambda$ leads to

$$I_{\bar{\lambda}}(\mu, \phi; \mu_0, \phi_0) = I_c(\mu, \phi; \mu_0, \phi_0) \int_0^\infty p(u) T_{\bar{\lambda}}(u)\, du$$

$$= I_c T_{\bar{\lambda}}(\langle u \rangle), \qquad (3.8.2b)$$

where $\langle u \rangle$ denotes the mean amount of absorber along the photon path. The photon path length approach requires a radiative transfer method to solve for I_c and a separate program to obtain $p(u)$ and $\langle u \rangle$. Usually $p(u)$ is determined by Monte Carlo techniques, but in some cases inverse Laplace transform methods may be applied to derive analytic solutions for $p(u)$ (Fouquart and Bonnel, 1980).

In the following we shall describe a general procedure for spectral integration involving both absorption and scattering due to gases and particulates. Let the scattering and absorption optical depths for a given layer consisting of particles and/or molecules be denoted by $\Delta\tau_s$ and $\Delta\tau_a$, respectively, and let $\Delta\tau = \Delta\tau_s + \Delta\tau_a$. If the gaseous path length for this layer is denoted by Δu, then the total optical depth for a combination of particles/molecules and absorbing gases in a subspectral interval m is given by

$$\Delta\tau_m = \Delta\tau + k_m \, \Delta u. \qquad (3.8.3)$$

Here, we may safely assume that the scattering and absorption properties of particles and/or molecules remain constant in a small spectral band. The single-scattering albedo for this subspectral interval may be written

$$\tilde{\omega}_m = \frac{\Delta\tau_s}{\Delta\tau + k_m \, \Delta u}. \qquad (3.8.4)$$

The phase function due to particles/molecules does not change since the absorbing gases do not contribute significantly to scattering. Consequently, the asymmetry factor will remain the same as in the case with only particles/molecules.

Consider a layer containing a mixture of Rayleigh molecules, Mie particles, and absorbing gases. The total optical depth for such a mixture is

$$\Delta\tau_m = \Delta\tau^R + \Delta\tau^M + k_m \, \Delta u, \qquad (3.8.5)$$

where $\Delta\tau^R$ and $\Delta\tau^M$ represent the optical depths of Rayleigh molecules and Mie particles, respectively, and $\Delta\tau^M = \Delta\tau_s^M + \Delta\tau_a^M$, with $\Delta\tau_s^M$ and $\Delta\tau_a^M$ being the scattering and absorption optical depths of Mie particles, respectively. Thus the single-scattering albedo and the phase function may be written in the forms

$$\tilde{\omega}_m = \frac{\Delta\tau^R + \Delta\tau_s^M}{\Delta\tau^R + \Delta\tau^M + k_m \, \Delta u}, \qquad (3.8.6)$$

$$P(\Theta) = \frac{\Delta\tau_s^M P^M(\Theta) + \Delta\tau^R P^R(\Theta)}{\Delta\tau_s^M + \Delta\tau^R}, \qquad (3.8.7)$$

where P^M and P^R denote the phase functions for Mie particles and Rayleigh molecules, respectively. The asymmetry factor for Rayleigh molecules is zero.

Thus, for a combination of Mie particles and Rayleigh molecules, the asymmetry factor is given by

$$g = \frac{\Delta\tau_s^M}{\Delta\tau_s^M + \Delta\tau^R} g^M. \tag{3.8.8}$$

Note that $P(\Theta)$ and g may be assumed to be constant over the spectral band.

Once the single-scattering properties have been defined for m subspectral intervals, monochromatic radiative transfer calculations may be carried out m times. The solar flux for a spectral interval may then be expressed by

$$F_{\bar{\lambda}} = \int_{\Delta\lambda} F_\lambda \frac{d\lambda}{\Delta\lambda} = \sum_m F_m \Delta g_m. \tag{3.8.9}$$

If the solar spectrum can be divided into n spectral intervals ($n = 1, 2, \ldots, N$), then the total solar flux is

$$F = \sum_n F_{\bar{\lambda},n} \Delta w_n, \tag{3.8.10}$$

where Δw_n is the fractional solar flux for the n spectral interval.

3.8.2 *Solar absorption spectrum of atmospheric gases*

Absorption of solar radiation in the atmosphere is primarily due to atomic and molecular oxygen and nitrogen, ozone, water vapor, and carbon dioxide. Other minor gases such as nitrous oxide, nitric oxide, carbon monoxide, and methane also exhibit absorption spectra in the solar region. Absorption spectra due to electronic transitions of molecular and atomic oxygen and nitrogen and ozone occur chiefly in the uv region, while those due to vibrational and rotational transitions of triatomic molecules such as H_2O, CO_2, and O_3 are found in the ir region, as discussed in Section 2.2.5. Absorption of solar energy in the visible region is very minor. As solar radiation penetrates the earth's atmosphere, most of its energy in the uv region is absorbed by oxygen and nitrogen species in the upper atmosphere. A large portion of solar energy in the near-ir region is absorbed by water vapor in the troposphere (see Fig. 3.1).

3.8.2.1   Water vapor

The ν_2 fundamental band of water vapor is centered at 6.25 μm, as discussed in Subsection 2.2.5.2. The ν_1 and ν_3 fundamentals of water vapor produce bands at $3657.05\,\text{cm}^{-1}$ (2.74 μm) and $3755.92\,\text{cm}^{-1}$ (2.66 μm), respectively. These two bands are close to one another and combine to form a band referred to as the 2.7 μm band. The $2\nu_2$ band is centered at $3161.60\,\text{cm}^{-1}$ (3.2μm band) in the tail of the solar spectrum. In addition, the solar spectrum contains a large number of overtone and combination bands, which arise from ground-state transitions. At the near-ir region, these bands absorb a significant amount of solar flux in the lower atmosphere, as is evident in the solar absorption spectrum. They are centered at 0.94, 1.1, 1.38, and 1.87 μm and are commonly identified in groups by the Greek letters (ρ, σ, τ), ϕ, ψ, and ω, respectively. Although the overtone and combination bands centered at 0.72 and 0.82 μm are relatively weak, their contributions to the solar heating of the atmosphere appear to be not insignificant (Kratz and Cess, 1985; Chou, 1986). The line-by-line data for the water vapor absorption bands in the solar spectrum are available from the HITRAN (Rothman et al., 1987).

3.8.2.2   Carbon dioxide

Carbon dioxide exhibits a number of overtone and combination bands in the solar region. The 2.0, 1.6, and 1.4 μm bands are so weak that, for all practical purposes, they can be neglected in solar absorption calculations. The 2.7 μm band of CO_2, which overlaps with the 2.7 μm band of water vapor, is somewhat stronger and contributes to solar absorption in the lower stratosphere. The ν_3 fundamental band of carbon dioxide centered at 4.3 μm is not important in the solar region as far as the absorption of solar flux is concerned.

3.8.2.3   Ozone

The absorption of ozone in the solar spectral region is due to electronic transitions. The strongest ozone bands are the *Hartley bands*, which cover the region from 2000 to 3000 Å and are centered at 2553 Å. The absorption of solar flux in these ozone bands takes place primarily in the upper stratosphere and in the mesosphere. The weak bands between 3000 and 3600 Å have more structure and are called the *Huggins bands*. Ozone also shows weak absorption bands in the visible and near-ir regions from about 4400 to 11,800 Å. These bands are referred to as *Chappuis bands*. The absorption coefficients in these bands are slightly dependent upon temperature. They have been measured by a number of earlier researchers (Inn and Tanaka, 1953; Vigroux, 1953). The ozone absorption coefficients for a temperature of $-44°$ C are listed in Table 3.4. The fractional solar flux associated with each spectral interval is also given in this table.

3.8.2.4   Oxygen and nitrogen

The uv absorption spectrum of molecular oxygen begins at about 2600 Å and continues down to the shorter wavelengths. The bands between 2600 and 2000 Å, referred to as the Herzberg bands, are weak and of little importance in the absorption of solar radiation because of an overlap with much stronger ozone bands in this spectral region. Nevertheless, the Herzberg bands are considered to be of significance in the formation of ozone. Adjacent to the Herzberg bands are the strong Schumann–Runge band system and continuum, which begin at 2000 Å and continue down to $\sim$1250 Å. Several other bands exist between 1250 and 1000 Å. Of particular interest is the strong Lyman-α line at 1216 Å, located in one of the

Table 3.4 Ozone absorption coefficients and fractional solar fluxes[a]

$\Delta\lambda$ (μm)	i	$k_i(-44°\text{ C})$	w_i[a]
0.20–0.21	1	9.8	1.24×10^{-4}
0.21–0.22	2	27	2.97×10^{-4}
0.22–0.23	3	75	4.59×10^{-4}
0.23–0.24	4	164	4.59×10^{-4}
0.24–0.25	5	254	5.14×10^{-4}
0.25–0.26	6	290	7.55×10^{-4}
0.26–0.27	7	241	1.35×10^{-3}
0.27–0.28	8	145	1.59×10^{-3}
0.28–0.30	9	33.7	6.46×10^{-3}
0.30–0.32	10	2.8	1.01×10^{-2}
0.32–0.34	11	0.16	1.50×10^{-2}
0.34–0.35	12	0.014	7.95×10^{-3}
0.45–0.50	13	0.0107	7.46×10^{-2}
0.50–0.55	14	0.055	6.78×10^{-2}
0.55–0.60	15	0.11	6.30×10^{-2}
0.60–0.65	16	0.09	5.87×10^{-2}
0.65–0.70	17	0.038	5.33×10^{-2}
0.70–0.80	18	0.015	9.14×10^{-2}

[a] Based on the fractional solar fluxes presented by Thekaekara (1974, 1976).

windows of the O_2 bands. The regions below about 1000 Å contain very strong O_2 bands, which are known as the Hopfield bands.

Molecular oxygen has three absorption bands in the visible region of the solar spectrum. These are the A band centered at 0.762 μm, the B band at 0.688 μm, and the γ band at 0.628 μm. These bands are produced by a magnetic dipole electronic transition and are associated with vibrational transitions. The intensities of these bands are very weak. However, because their positions are near the peak of the solar spectrum, absorption due to molecular oxygen in the earth's atmosphere cannot be neglected.

The absorption spectrum of molecular nitrogen begins at 1450 Å. The region from 1450 to 1000 Å contains the Lyman-Birge-Hopfield bands, which consist of narrow and sharp lines. From 1000 to 800 Å, the absorption spectrum of N_2 is occupied by the Tanaka–Worley bands. These bands are very complicated and the absorption coefficients are highly variable. Below 800 Å, the absorption spectrum of N_2 is generally made up of the ionization continuum.

Because of the absorption of solar uv radiation, some oxygen and nitrogen molecules in the upper atmosphere undergo photochemical dissociation and are dissociated into atomic oxygen and nitrogen. Atomic nitrogen exhibits an absorption spectrum from ~10 to 1000 Å. Although atomic nitrogen probably is not abundant enough to be a significant absorber, it may play an important role in the absorption of uv radiation in the thermosphere. Atomic oxygen also shows an absorption continuum in the region of 10 to 1000 Å. Absorption cross sections of O_2, N_2, O, N, and O_3 have been determined from measurements.

3.8.3 Transfer of broadband solar flux in the atmosphere

First consider a nonscattering atmosphere. From Eq. (3.1.20) for monochromatic direct solar flux, the total direct downward solar flux may be written

$$F_s(z) = \int_0^\infty \mu_0 F_\odot \exp\left(\frac{-k_\lambda u(z)}{\mu_0}\right) d\lambda, \qquad (3.8.11)$$

where $k_\lambda u(z)$ represents the optical depth, k_λ is the absorption coefficient, and the absorbing gaseous path length is defined by

$$u(z) = \int_z^{z_\infty} \rho_a(z')\, dz', \qquad (3.8.12)$$

where ρ_a denotes the density of the absorbing gas and z_∞ denotes the height at TOA. The total solar flux at TOA can be obtained from an integration over the entire solar spectrum as follows:

$$S = \int_0^\infty F_\odot\, d\lambda. \qquad (3.8.13)$$

Monochromatic absorptance may be expressed by

$$A_\lambda(u/\mu_0) = 1 - \exp(-k_\lambda u/\mu_0). \qquad (3.8.14)$$

We may define broadband solar absorptance in the height coordinate in which the monochromatic absorptance is weighed by the solar flux so that

$$A(z) = \frac{1}{S} \int_0^\infty F_\odot A_\lambda(u/\mu_0)\, d\lambda. \qquad (3.8.15)$$

It follows that Eq. (3.8.11) may be rewritten in the form

$$F_s(z) = \mu_0 S[1 - A(z)]. \qquad (3.8.16)$$

Thus the solar flux, at a given level z in a nonscattering atmosphere, is directly related to the broadband solar absorptance defined in Eq. (3.8.15). If we let $w_\lambda = F_\odot/S$, broadband solar absorptance may be rewritten in terms of spectral absorptance A_i, in the form

$$A(z) = \int_0^\infty A_\lambda(u/\mu_0) w_\lambda\, d\lambda \cong \sum_i A_i(u/\mu_0) w_i\, \Delta\lambda_i. \qquad (3.8.17)$$

Spectral absorptance may be obtained from known absorption coefficients for various absorbers in the solar spectrum. In view of the solar absorption spectrum, two spectral intervals corresponding to O_3 and (H_2O, CO_2) absorption may be separated for the computations of solar absorption in the stratosphere and troposphere. We may write

$$A(z) = \int_0^\lambda A_\lambda(u/\mu_0) w_\lambda\, d\lambda + \int_\lambda^\infty A_\lambda(u/\mu_0) w_\lambda\, d\lambda$$

Table 3.5 Empirical constants for the H_2O and CO_2 bands and fractional solar fluxes[a]

λ (μm)	i	C_i	D_i	K_i	$\Delta\nu_i$ (cm^{-1})	w_i[a]
H_2O band						
0.72	1a	16	31	7	1375	0.0694
0.82	1b	9	20	5	1525	0.0801
0.94	1	−135	230	125	1400	0.1346
1.1	2	−292	345	180	1000	0.0892
1.38	3	202	460	198	1500	0.1021
1.87	4	127	232	144	1100	0.0622
2.7	5	337	246	150	1000	0.0300
3.2	6	−144	295	151	540	0.0218
CO_2 band						
2.7	5	−137	77	68	320	—

[a] Based on the fractional fluxes presented by Thekaekara (1974, 1976).

$$= \sum_i A_i(u_0/\mu_0)w_i\,\Delta\lambda_i + \left(\sum_i A_i(u_w/\mu_0)w_i\,\Delta\lambda_i + \epsilon A_5(u_c/\mu_0)\right)$$

$$= A(u_0/\mu_0) + \left[A(u_w/\mu_0) + \epsilon A_5(u_c/\mu_0)\right], \qquad (3.8.18)$$

where u_o, u_w, and u_c denote the path lengths for O_3 (or O_2), H_2O and CO_2, respectively; ϵ is a correction factor for the overlap of H_2O and CO_2 at the 2.7 μm band; and the index i denotes the number of subspectral intervals employed in the calculation.

3.8.3.1 Water vapor and carbon dioxide

Absorptance data for the H_2O and CO_2 bands are available from laboratory measurements (Howard et al., 1956). Based on these data, Liou and Sasamori (1975) have derived an empirical equation for spectral absorptance in the form

$$A_i(u) = \frac{1}{\Delta\nu_i}\left[C_i + D_i\log_{10}\left(up^{\eta_i} + \chi_{0i}\right)\right], \qquad (3.8.19)$$

where $\Delta\nu_i$ is the bandwidth, $\chi_{0i} = 10^{-C_i/D_i}$, $\eta_i = K_i/D_i$, and C_i, D_i and K_i are empirical constants. This parameterized equation is applicable to both strong and weak absorption for the 0.94, 1.1, 1.37, 1.87, 2.7, and 3.2 μm H_2O bands. When $u \to \infty$, $A_i(u)$ approaches the strong-line limits that are determined from the experiments. When $u \to 0$, $A_i(u) \to 0$. The subscripts i (=1–6) are used to denote these H_2O bands. In addition, band absorptance for 0.72 and 0.82 μm is determined from line-by-line data using Eq. (3.8.19). Table 3.5 lists the relevant constants and the fractional solar fluxes in these bands.

At the 2.7 μm band ($i = 5$), band absorptance due to H_2O and CO_2 can be written

$$A_5(u_w, u_c) = A_5(u_w) + \epsilon\, A_5(u_c), \qquad (3.8.20)$$

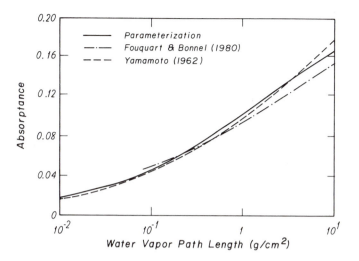

FIG. 3.15   Broadband water vapor absorptance as a function of path length at standard temperature and pressure from various sources, including the parameterization presented in the text.

where the overlap correction depends on water vapor absorption and is empirically determined from laboratory measurements in the form

$$\epsilon = a - bA_5(u_w), \tag{3.8.21}$$

where $a = 0.75$ and $b = 0.48$ for $A_5(u_w)\,\Delta\nu_5 \geq 300$ cm^{-1}, and $a = 1$ and $b = 1.12$ for $A_5(u_w)\,\Delta\nu_5 \leq 300$ cm^{-1}.

Using Eq. (3.8.19), the solar broadband absorptance for H$_2$O, $A(u_w)$, as a function of the H$_2$O path length u is illustrated in Fig. 3.15. Also shown are results derived by Yamamoto (1962) in an earlier study using laboratory data, and by Fouquart and Bonnel (1980), who used line-by-line data in the calculation. The use of laboratory data appears to overestimate the broadband absorptance for $u > 1\,\mathrm{g\,cm^{-2}}$.

Band absorptances for the H$_2$O bands can also be computed from line-by-line data using band models (Kratz and Cess, 1985) and the k-distribution method (Chou, 1986). We shall discuss the k-distribution method for applications to solar radiative transfer. The transmittance for a spectral interval $\Delta\lambda_i$ can be expressed in terms of the absorption coefficient in the form

$$T_i(u) = \int_{-\infty}^{\infty} e^{-ku}\, h_i(k)\, d\log k, \tag{3.8.22a}$$

where $h_i(k)$ is the probability density function in the logarithmic scale. This function is normalized by

$$\int_{-\infty}^{\infty} h_i(k)\, d\log k = \int_0^1 dg_i = 1, \tag{3.8.22b}$$

where the differential of the cumulative probability function is $dg_i = h_i(k)\,d\log k$. From Eq. (3.8.11), the spectral solar flux may be expressed by

$$F_{\tilde{\nu}}(u) = \mu_0 S \sum_i T_i(u/\mu_0)\,w_i\,\Delta\nu_i$$

$$= \mu_0 S \int_{-\infty}^{\infty} e^{-ku/\mu_0}\,h(k)\,d\log k, \qquad (3.8.23)$$

with $h(k) = \sum_i w_i \Delta\nu_i h_i(k)$. Chou (1986) has computed $h(k)$ for the solar absorption bands based on line-by-line data. The spectral solar fluxes used in the calculations were based on the data given by Labs and Neckel (1968) using a solar constant value of 1365 W m^{-2}. Listed in Table 3.6 are the flux-weighted k-distribution functions at $p = 300$ mb and $T = 240$ K. The units for the absorption coefficient k are g^{-1}cm^2 and W m^{-2} for the weighted probability distribution function. The number of k and $h(k)$ may be reduced by simple summations. The optimal numbers of the equivalent absorption coefficients k and the weight, $dg = h(k)\,d\log k$, for flux and heating rate calculations may be determined by numerical experimentation. A parameterization of the broadband absorptance for H$_2$O based on the aforementioned line-by-line approach has been developed by means of numerical fittings and is given in the form

$$A(u_w) = f_w\left\{1 - \exp\left[-11.5u_w/(1 + 10.5u_w + 64u_w^{0.59})\right]\right\}, \qquad (3.8.24)$$

where $f_w = 0.5343$, representing the fraction of solar flux in the spectral range 2600–14500 cm^{-1} where H$_2$O exhibits absorption lines. Solar heating rates below about 200 mb without scattering contributions computed from this parameterized equation are accurate within a few percent of the results derived from a more exact monochromatic calculation (< 0.1 K d^{-1}).

In order to apply the band absorptance for H$_2$O to nonhomogeneous atmospheres, a pressure correction must be performed. In the solar region, it suffices to use the scaling approximation for the path length discussed in Subsection 2.5.1, in the form

$$\tilde{u} = \int_0^u \left(\frac{p}{p_r}\right)^n du. \qquad (3.8.25)$$

There appears to be no formal physical foundation for the selection of n. The best value to achieve the most accurate flux and heating rate calculations can be derived by numerical experimentation (in comparison with the results computed from the line-by-line approach). Based on line-by-line calculations, Chou (1986) found that in the pressure scaling for H$_2$O absorption, the reference pressure may be placed at the upper troposphere where absorption in the line center is usually saturated. Radiative transfer in the H$_2$O bands in the troposphere is governed by the absorption in the wing regions. A reference pressure of 300 mb and a scaling index n of 0.8 have been suggested. This scaling index appears to work well for all

H_2O bands. If the laboratory data described previously are used in the calculation, $n = \eta = K/D$, which varies slightly within the absorption band. The reference pressure p_r may be set at 1013 mb.

3.8.3.2 Ozone

The ozone absorption coefficients, k_i, have been determined from measurements. For each wavelength interval, absorptance may be written

$$A_i(u_0) = 1 - \exp(-k_i u_0). \tag{3.8.26a}$$

Total ozone absorptance is then

$$A(u_0) = \sum_i A_i(u_0) \, w_i \, \Delta\nu_i. \tag{3.8.26b}$$

Based on the fitting of computational results, Lacis and Hansen (1974) have derived parameterization expressions for the ozone absorptance in the uv and visible regions. For the Hartley and Huggins bands, the absorptance is given by

$$A^{uv}(u_0) = \frac{1.082 u_0}{(1 + 138.6 u_0)^{0.805}} + \frac{0.0658 u_0}{1 + (103.6 u_0)^3}, \tag{3.8.27}$$

where u_0 is in units of cm (STP). This expression has a maximum error of $\leq 0.5\,\%$ in the interval $10^{-4} \leq u_0 \leq 1$ cm. For the Chappuis band, the parameterization for the absorptance is

$$A^{vis}(u_0) = \frac{0.02118 \, u_0}{1 + 0.042 \, u_0 + 0.000323 \, u_0^2}. \tag{3.8.28}$$

This expression has an accuracy of about four decimal places in the interval $10^{-4} \leq u_0 \leq 1$ cm. The total absorptance due to ozone is then the sum of the preceding two absorptance values. Although the parameterization equations are derived by numerical fittings, these equations appear to be adequate and reliable for the computation of solar heating rates due to ozone without scattering contributions.

3.8.3.3 Oxygen

Absorption of solar flux by O_2 in the visible region is important in the upper atmosphere and could affect the solar flux available at the surface and in the tropopause. Line-by-line information for O_2 is available, and a precise computation for the band absorptance can be performed. Based on the line-by-line calculations using the Voigt line shape, a parameterization of the band absorptance due to O_2 in the visible region has been presented by Kiehl and Yamamouchi (1985) based

Table 3.6 The flux-weighted k-distribution function, $h(k)$, at $p = 300$ mb and $T = 240$ K. The units are $\mathrm{g^{-1}\,cm^2}$ for the absorption coefficient k and $\mathrm{W\,m^{-2}}$ for the distribution function[a]

$\log_{10} k/\lambda\,(\mu m)$	2.7/3.2	1.87	1.38	1.1	0.94	0.82	0.72	Total
−4.00	6.02	97.91	106.78	73.60	196.98	184.03	115.66	780.98
−3.75	2.00	18.26	16.78	7.57	17.68	25.81	15.68	103.78
−3.50	2.06	19.56	16.45	8.33	16.13	27.74	17.05	107.31
−3.25	3.09	21.58	16.95	11.40	15.91	29.30	26.72	124.95
−3.00	4.31	19.94	14.81	15.87	16.67	30.74	29.28	131.63
−2.75	5.19	17.73	14.74	21.28	21.97	29.40	30.49	140.80
−2.50	6.80	17.28	16.13	26.40	28.14	26.52	30.29	151.56
−2.25	7.30	16.67	15.64	25.43	29.58	21.78	28.77	145.17
−2.00	8.07	15.00	13.37	25.11	32.89	17.63	26.46	138.53
−1.75	8.62	12.82	16.08	25.88	35.65	13.62	21.13	133.80
−1.50	9.37	11.35	18.40	24.27	33.00	10.50	17.20	124.10
−1.25	10.00	11.83	20.08	24.14	29.68	8.02	12.09	115.85
−1.00	9.91	10.54	23.28	23.39	26.18	5.93	8.52	107.74
−0.75	9.98	9.49	23.62	21.73	22.64	4.10	6.03	97.58
−0.50	9.76	10.62	24.13	17.77	17.08	2.66	4.37	86.38
−0.25	8.89	11.88	23.67	13.56	12.99	1.62	2.98	75.59
0.00	8.04	11.50	24.08	10.14	9.69	1.10	1.97	66.52
0.25	7.28	12.21	21.07	7.44	6.88	0.81	1.13	56.81
0.50	7.40	11.29	18.87	5.59	5.00	0.52	0.58	49.25
0.75	7.58	9.19	16.73	3.95	3.61	0.30	0.36	41.73
1.00	8.17	7.47	13.30	2.75	2.40	0.09	0.10	34.28
1.25	7.21	6.04	10.03	1.76	1.54			26.60
1.50	7.07	4.45	7.62	1.20	0.90			21.24
1.75	5.88	3.45	5.21	0.90	0.69			16.13
2.00	4.56	2.35	3.71	0.74	0.32			11.69
2.25	3.67	1.85	2.71	0.51	0.23			8.97
2.50	2.72	1.27	1.93	0.25	0.03			6.20
2.75	1.86	0.90	1.20	0.09				4.04
3.00	1.42	0.68	0.82	0.01				2.93
3.25	0.97	0.48	0.57					2.02
3.50	0.67	0.31	0.27					1.25
3.75	0.46	0.15	0.16					0.77
4.00	0.32	0.05	0.03					0.40
4.25	0.27							0.27
4.50	0.23							0.23

[a] After Chou (1986).

on numerical fitting. There are three isotopes for the oxygen atom, denoted by i ($= 1, 2, 3$). The absorptance for each O_2 band may be parameterized in the form

$$A_j = 2A_o(\bar{T})\,\ell n\left(1 + \sum_{i=1}^{3} \frac{U_{ij}}{\left[4 + U_{ij}\left(1 + 1/\beta_{ij}\right)\right]^{1/2}}\right), \qquad j = 1, 2, 3,$$

(3.8.29)

Table 3.7 Band parameters for the oxygen bands in the visible[a]

Band	Band center (μm)	Band Strength (cm^{-2} atm^{-1} STP)	Solar Flux $\left(\mathrm{W\,m^{-2}\,(cm^{-1})^{-1}}\right)$
A	0.762	5.30×10^{-3}	6.980×10^{-2}
B	0.688	4.08×10^{-4}	6.642×10^{-2}
γ	0.6280	1.52×10^{-5}	6.225×10^{-2}

Isotope	Abundance q	Mean line spacing $\delta(\mathrm{cm}^{-1})$
$^{16}O^{16}O$	0.997580	2.8290
$^{16}O^{17}O$	0.000373	1.4145
$^{16}O^{18}O$	0.002039	1.4145

[a] After Kiehl and Yamamouchi (1985).

where $\bar{T}$ is the pressure-weighted temperature along the path and the dimensionless path length and mean line parameter are, respectively, defined by

$$U_{ij} = q_i \int_z^{z\infty} \frac{S_{ij}(T)}{A_o(T)}\, p_o\, dz, \tag{3.8.30a}$$

$$\beta_{ij} = \frac{1}{U_{ij}} \int_z^{z\infty} \frac{4\alpha(T)}{\delta_i}\, \frac{p}{p_r}\, dU_{ij}(z). \tag{3.8.30b}$$

In these equations, S_{ij} denotes the band strength in cm^{-2} atm^{-1}; q_i is the isotope fraction; $p_o = 0.21p$, the partial pressure of O_2, where p is the total atmospheric pressure; p_r is the reference pressure of one atmosphere; δ_i is the mean line spacing; and α is the mean line half-width given by

$$\alpha(T) = 0.05516(300/T)^{1/2} \text{ (cm}^{-1}). \tag{3.8.30c}$$

The effective bandwidth parameter is evaluated as follows:

$$A_o(T) = 46.31(T/300)^{1/2} \text{ (cm}^{-1}). \tag{3.8.30d}$$

The relevant parameters for the calculation of O_2 band absorptance are listed in Table 3.7.

3.8.4 Inclusion of Rayleigh scattering and surface reflection

In order to include the contribution of scattering due to molecules in the calculation of solar fluxes and heating rates due to various gases, we must first briefly present the theory of Rayleigh scattering. Our objective here will be to establish expressions for the scattering cross section and phase function for molecules.

The scattered intensity I for molecules or particles may be expressed in terms of the incident intensity I_0, scattering cross section σ_s, and phase function $P(\Theta)$, in the form (more detailed discussion is given in Subsection 5.1.2)

$$I(\Theta) = I_0 \frac{\sigma_s}{R^2} \frac{P(\Theta)}{4\pi}, \qquad (3.8.31)$$

where R denotes the distance. The term σ_s/R^2 represents the effective solid angle for single-scattering processes.

Based on the scattering theory developed by Rayleigh in 1871, the scattering cross section for molecules is given by

$$\sigma_s = \frac{8\pi^3 (m_r^2 - 1)^2}{3\lambda^4 N_s^2} f(\delta_*), \qquad (3.8.32)$$

where m_r is the real part of the refractive index, N_s is the total number of molecules per unit volume ($\cong 2.55 \times 10^{19}$ cm^{-3}), and the anisotropic correction term, $f(\delta_*) = (6 + 3\delta_*)/(6 - 7\delta_*)$, with the anisotropic factor $\delta_* = 0.035$. The normalized phase function for Rayleigh scattering is

$$P(\Theta) = \frac{3}{4}(1 + \cos^2 \Theta). \qquad (3.8.33)$$

The real part of the refractive index may be approximately fitted by

$$(m_r - 1) \times 10^8 \cong 6.43 \times 10^3 + \frac{2.95 \times 10^6}{146 - \lambda^{-2}} + \frac{2.55 \times 10^4}{41 - \lambda^{-2}}, \qquad (3.8.34)$$

where λ is in units of μm.

Heating due to the absorption of ozone molecules depends significantly on Rayleigh scattering and surface reflection. It is appropriate to assume that the Rayleigh scattering layer is primarily confined to the lower atmosphere between 0 and ∼10 km. Based on multiple-scattering calculations, reflection by the Rayleigh layer, r_a, as a function of the cosine of the solar zenith angle, may be parameterized in the form

$$r_a(\mu_0) = a/(1 + b\mu_0), \qquad (3.8.35)$$

where the empirical coefficients $a \cong 0.219$ and $b \cong 0.816$. The global reflection due to Rayleigh scattering is then

$$\bar{r}_a = 2 \int_0^1 r_a(\mu_0)\mu_0 \, d\mu_0 = 0.144. \qquad (3.8.36)$$

Assuming that the underlying surface has a Lambertian albedo of r_s, the combined reflection due to the Rayleigh layer and the surface, taking into account multiple reflections between these two layers, is given by

$$r(\mu_0) = r_a(\mu_0) + \frac{[1 - r_a(\mu_0)](1 - \bar{r}_a) r_s}{(1 - \bar{r}_a r_s)}. \qquad (3.8.37)$$

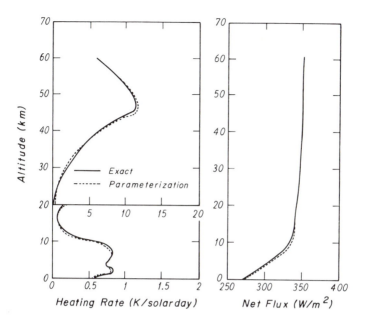

FIG. 3.16   Comparisons of solar heating rates and net fluxes computed from the flux parameterization scheme with results from a more exact radiation program. The heating rate is presented in terms of degrees per solar day (12 hours).

The contribution of surface reflection and Rayleigh scattering produces an upward flux. Hence, Eq. (3.8.16) for the net solar flux must be modified so that

$$F_S(z) = \mu_0 S [1 - A(z)] - \mu_0 S [1 - A(z_b)] r(\mu_0)$$
$$\times [1 - A (u'(z - z_b)/\bar{\mu})], \tag{3.8.38}$$

where the second term on the right-hand side represents the effective upward flux, which, when combined with the downward flux in a nonscattering atmosphere, will give the correct net solar flux. In this equation, z_b may be set at 10 km for ozone heating rate calculations, and the ozone path length u' should be calculated with respect to the reference height z_b from below this height. The diffusivity factor $1/\bar{\mu}$ is ~1.9 for ozone applications. For water vapor absorption, largely confined to below 10 km, we may set $z_b = r_a(\mu_0) = 0$ and take $1/\bar{\mu}$ of ~1.66.

Based on the preceding analysis, the broadband solar heating rate may be written in the form

$$\left(\frac{\partial T}{\partial t}\right)_S = -\frac{1}{\rho C_p} \frac{dF(z)}{dz} = \frac{\mu_0 S}{\rho C_p} \left(\frac{dA(z)}{dz} - [1 - A(z_b)] r(\mu_0) \frac{dA(u'/\bar{\mu})}{dz}\right).$$
$$\tag{3.8.39}$$

From Eq. (3.8.18) we have

$$\frac{dA(z)}{dz} = \rho_o \frac{dA(u_o/\mu_0)}{du_o} + \rho_w \frac{dA(u_w/\mu_0)}{du_w} + \frac{d}{dz}[\epsilon A_5(u_c/\mu_0)]. \qquad (3.8.40)$$

Moreover, using Eqs. (3.8.26a) and (3.8.19), we find

$$\frac{dA(u_o/\mu_0)}{du_o} = \frac{1}{\mu_0} \sum_i w_i k_i \, \Delta\nu_i \exp\left(-k_i u_o/\mu_0\right), \qquad (3.8.41)$$

and

$$\frac{dA(u_w/\mu_0)}{du_w} = \frac{\log_{10} e}{\mu_0} \sum_i w_i D_i \left(\frac{\tilde{p}}{\mu_0} + \chi_{0i}\right)^{-1} \frac{d\tilde{p}}{du_w}, \qquad (3.8.42)$$

where the reduced pressure is given by

$$\tilde{p} = \int_0^{u_w} p^\eta \, du. \qquad (3.8.43)$$

The term $dA_5(u_c/\mu_0)/du_c$ has the same analytic form as in Eq. (3.8.42), except no summation is involved.

    Figure 3.16 shows the comparisons of the solar net fluxes and heating rates computed from the preceding parameterization scheme and a more exact and comprehensive radiative transfer program using the discrete-ordinates method. A daytime fraction of 0.5, along with the values shown in the diagram, were used in the computation. Agreement between these two methods is within $\sim$0.02 K per solar day (12 hours) for solar heating rates and within 5 W m^{-2} for net fluxes. Below $\sim$10 km, solar heating is basically produced by absorption of H_2O. Between $\sim$10–15 km, absorption of CO_2 contributes to within $\sim$0.1 K per solar day. Above 15 km, significant solar heating is produced exclusively by O_3 absorption. The net solar flux ranges from $\sim$350 W m^{-2} in the atmosphere to $\sim$270 W m^{-2} at the surface. We have also performed a series of comparisons between the parameterization method and more exact programs involving different atmospheric conditions and solar zenith angles. The accuracy of the parameterization method is similar to that cited above.

REFERENCES

Ambartzumian, V. A., 1942: A new method for computing light scattering in turbid media. *Izv. Akad. Nauk SSSR, Ser. Geogr. i Geofiz.*, **3**, 97–104.

Asano, S., 1975: On the discrete-ordinates method for the radiative transfer. *J. Meteor. Soc. Japan*, 92–95.

Chandrasekhar, S., 1950: *Radiative Transfer*. Oxford University Press, Oxford, 393 pp.

Chou, M. D., 1986: Atmospheric solar heating rate in the water vapor bands. *J. Climate Appl. Meteor.*, **25**, 1532–1542.

Coakley, J. A., Jr., and P. Chýlek, 1975: The two-stream approximation in radiative transfer: Including the angle of the incident radiation. *J. Atmos. Sci.*, **32**, 409–418.

Cuzzi, J. N., T. P. Ackerman, and L. C. Helmle, 1982: The delta–four-stream approximation for radiative flux transfer. *J. Atmos. Sci.*, **39**, 917–925.

Dave, J. V., 1975: A direct solution of the spherical harmonics approximation to the radiative transfer equation for an arbitrary solar elevation. Part I: Theory. *J. Atmos. Sci.*, **32**, 790–798.

Eddington, A. S., 1916: On the radiative equilibrium of the stars. *Mon. Not. Roy. Astronom. Soc.*, **77**, 16–35.

Fouquart, Y., and B. Bonnel, 1980: Computations of solar heating of the Earth's atmosphere: A new parameterization. *Contrib. Atmos. Phys.*, **53**, 35–62.

Hansen, J. E., 1969: Radiative transfer by doubling very thin layers. *Astrophys. J.*, **155**, 565–573.

Henyey, L. C., and J. L. Greenstein, 1941: Diffuse radiation in the galaxy. *Astrophys. J.*, **93**, 70–83.

Herman, B. M., and F. F. Abraham, 1960: A note on the two-stream theory of radiational transfer through clouds. *J. Meteor.*, **17**, 471–473.

Hovenier, J.W., 1969: Symmetry relationships for scattering of polarized light in a slab of randomly oriented particles. *J. Atmos. Sci.*, **26**, 488–499.

Howard, J. N., D. L. Burch, and D. Williams, 1956: Near-infrared transmission through synthetic atmospheres. *J. Opt. Soc. Amer.*, **46**, 186–190.

IMSL User's Manual, 1987: *Math/Library*, IMSL, Houston, Texas, 1151 pp.

Inn, E. C. Y., and Y. Tanaka, 1953: Absorption coefficient of ozone in the ultraviolet and visible regions. *J. Opt. Soc. Amer.*, **43**, 870–873.

Joseph, J. H., W. J. Wiscombe, and J. A. Weinman, 1976: The delta-Eddington approximation for radiative flux transfer. *J. Atmos. Sci.*, **33**, 2452–2459.

Kiehl, J. T., and T. Yamamouchi, 1985: A parameterization for absorption due to the A, B and γ oxygen bands. *Tellus*, **37B**, 1–6.

King, M. D., and Harshvardhan, 1986: Comparative accuracy of selected multiple scattering approximations. *J. Atmos. Sci.*, **43**, 784–801.

Kourganoff, V., 1952: *Basic Methods in Transfer Problems.* Oxford at the Clarendon Press, London and New York, 281 pp.

Kneizys, F., E. Shettle, L. Abreu, J. Chetwynd, G. Anderson, W. Gallery, J. Selby, and S. Clough, 1988: *Users guide to LOWTRAN 7*. AFGL-TR-88-0177, Environmental Research Papers, No. 1010, Air Force Geophysics Laboratory, Hanscom AFB, Massachusetts.

Kratz, D. P., and R. D. Cess, 1985: Solar absorption by atmospheric water vapor: A comparison of radiation models. *Tellus*, **37B**, 53–63.

Labs, D., and H. Neckel, 1968: The radiation of the solar photosphere from 2000 Å to 100 μ. *Z. Astrophys.*, 69, 1–73.

Lacis, A. A., and J. E. Hansen, 1974: A parameterization for the absorption of solar radiation in the earth's atmosphere. *J. Atmos. Sci.*, **31**, 118–133.

Liou, K. N., 1973a: A numerical experiment on Chandrasekhar's discrete-ordinates method for radiative transfer: Application to cloudy and hazy atmospheres. *J. Atmos. Sci.*, **30**, 1303–1326.

Liou, K. N., 1973b: Transfer of solar irradiance through cirrus cloud layers. *J. Geophys. Res.*, **78**, 1409–1418.

Liou, K. N., 1974: Analytic two-stream and four-stream solutions for radiative transfer. *J. Atmos. Sci.*, **31**, 1473–1475.

Liou, K. N., 1975: Applications of the discrete-ordinate method for radiative transfer to inhomogeneous aerosol atmospheres. *J. Geophys. Res.*, **80**, 3434–3440.

Liou, K. N., 1980: *An Introduction to Atmospheric Radiation.* Academic Press, New York, 392 pp.

Liou, K. N., and T. Sasamori, 1975: On the transfer of solar radiation in aerosol atmospheres. *J. Atmos. Sci.*, **32**, 2166–2177.

Liou, K. N., Q. Fu, and T. P. Ackerman, 1988: A simple formulation of the delta–four-stream approximation for radiative transfer parameterizations. *J. Atmos. Sci.*, **45**, 1940–1947.

Meador, W. E., and W. R. Weaver, 1980: Two stream approximations to radiative transfer in planetary atmospheres: A unified description of existing methods and a new improvement. *J. Atmos. Sci.*, **37**, 630–643.

Neiburger, M., 1949: Reflection, absorption, and transmission of insolation by stratus cloud. *J. Meteor.*, **6**, 98–104.

Peebles, G. H., and M. S. Plesset, 1951: Transmission of gamma rays through large thicknesses of heavy materials. *Phys. Rev.*, **81**, 430–439.

Potter, J. F., 1970: The delta-function approximation in radiative transfer theory. *J. Atmos. Sci.*, **27**, 943–949.

Rothman, L. S., R. R. Gamache, A. Goldman, L. R. Brown, R. A. Toth, H. M. Pickett, R. L. Poynter, J.-M. Flaud, C. Camy-Peyret, A. Barbe, N. Husson, C. P. Rinsland, and M. A. H. Smith, 1987: The HITRAN database: 1986 edition. *Appl. Opt.*, 26, 4058–4097.

Sagan, C., and J. B. Pollack, 1967: Anisotropic nonconservative scattering and the clouds of Venus. *J. Geophys. Res.*, **72**, 469–477.

Schaller, E., 1979: A delta–two-stream approximation in radiative flux calculations. *Contrib. Atmos. Phys.*, **52**, 17–26.

Schuster, A., 1905: Radiation through a foggy atmosphere. *Astrophys. J.*, **21**, 1–22.

Sobolev, V. V., 1975: *Light Scattering in Planetary Atmospheres.* Translated by W. M. Irvine. Pergamon Press, Oxford, 254 pp.

Stamnes, K., 1986: The theory of multiple scattering of radiation in plane parallel atmospheres. *Rev. Geophys.*, **24**, 299–310.

Stamnes, K., and H. Dale, 1981: A new look at the discrete-ordinate method for radiative transfer calculations in anisotropically scattering atmospheres. II. Intensity computations. *J. Atmos. Sci.*, **38**, 2696–2706.

Stamnes, K., and R. A. Swanson, 1981: A new look at the discrete-ordinate method for radiative transfer calculations in anisotropically scattering atmospheres. *J. Atmos. Sci.*, **38**, 387–399.

Stokes, G. G., 1862: On the intensity of the light reflected from or transmitted through a pile of plates. *Proc. Roy. Soc. London*, **11**, 545–556.

Thekaekara, M. P., 1974: Extraterrestrial solar spectrum, 3000–6100 Å at 1 Å intervals. *Appl. Opt.*, **13**, 518–522.

Thekaekara, M. P., 1976: Solar irradiance: Total and spectral and its possible variations. *Appl. Opt.*, **15**, 915–920.

van de Hulst, H. C., 1980: *Multiple Light Scattering. Tables, Formulas, and Applications*, Vols. 1 and 2. Academic Press, New York, 739 pp.

Vigroux, E., 1953: Contribution a l'étude expérimentale de l'absorption de l'ozone. *Ann. Phys.*, **8**, 709–762.

Wiscombe, W., 1977: The delta-M method: Rapid yet accurate radiative flux calculations for strongly asymmetric phase functions. *J. Atmos. Sci.*, **34**, 1408–1422.

Yamamoto, G., 1962: Direct absorption of solar radiation by atmospheric water vapor, carbon dioxide and molecular oxygen. *J. Atmos. Sci.*, **19**, 182–188.

Zdunkowski, W. G., G. J. Korb, and C. T. Davis, 1974: Radiative transfer in model clouds of variable and height constant liquid water content as computed by approximate and exact methods. *Beitr. Phys. Atmos.*, **47**, 157–186.

4

THEORY, OBSERVATION, AND MODELING OF CLOUD PROCESSES IN THE ATMOSPHERE

Clouds are conventionally classified in terms of their position and appearance in the atmosphere. Clouds with base heights above ~ 6 km are designated as high clouds, a category that includes cirrus (Ci), cirrostratus (Cs), and cirrocumulus (Cc). The group of middle clouds with base heights between ~ 2 and 6 km consists of altocumulus (Ac) and altostratus (As). Stratus (St), stratocumulus (Sc), and nimbostratus (Ns) constitute the group of low clouds with base heights below ~ 2 km. Cumulus (Cu) and cumulonimbus (Cb) belong to the classification of vertically developed clouds.

The importance of clouds in climate and weather processes has been recognized through a number of observational and modeling studies. First, the radiation budgets at the top of the atmosphere, as observed from satellites, are closely related to the cloud field. Second, a small change in the cloud parameters may significantly amplify or offset climatic temperature perturbations due to external radiative forcing, such as that due to the increase in CO_2 and other greenhouse gases. Third, certain kinds of clouds, such as stratus, are primarily responsible for the reflection of sunlight (solar albedo effect), while other kinds, such as semitransparent high cirrus, emit significant infrared radiation (ir greenhouse effect). Clearly, the stability or instability of the climate system would depend on the role that clouds play in climatic perturbations. Clouds are also important physical elements in numerical weather prediction. Yet clouds remain one of the least understood components of the weather and climate systems.

We begin the discussion of clouds with a global view from satellites. Subsequently, we look at some aspects of cloud climatology and examine the microphysical properties of various cloud types. We then develop basic equations that govern cloud and moisture fields, including parameterizations for the sources and sinks, and present a discussion of the transport of moisture and heat by eddies. Subjects associated with stratiform cloud models and cumulus convection are further discussed, with a view that they are closely associated with radiative processes in the atmosphere. Stratiform cloud formation schemes associated with large-scale models will be discussed in Section 7.4.

4.1 A global view of clouds from satellites

Clouds are global in nature, as is evident from satellite cloud pictures. We first describe the formation and dissipation of cirrus clouds based on Geostationary Operational Environmental Satellite (GOES) ir pictures.

4.1.1 *Cirrus clouds*

Shown in Fig. 4.1 is a full disk ir picture taken at 2345 GMT, February 23, 1984 (Liou, 1986). Warmer areas are darker, while cooler areas are lighter. Because temperature normally decreases with height in the troposphere, the whitest areas are assumed to be high clouds. Two conditions minimize the possibility that the white areas may be associated with cold surface temperatures: First, the picture was taken over the Northern Hemisphere during the later winter, when the polar regions would generally be outside the frame of the satellite camera; second, the picture was taken during daytime.

Cirrus clouds are indeed globally distributed, being present at all latitudes and without respect to land or sea or season of the year. These clouds are undergoing continuous changes in area coverage, thickness, texture, and position. The most striking cirriform cloud feature shown in Fig. 4.1 is the large spiral-comma-shaped pattern west of the Washington State coastline. This system is associated with a major surface cyclone located to the northeast of the cloud center. There are lower clouds and precipitation associated with the large-scale rising motion under much of this high-cloud canopy.

To the north of this cloud band, over the northern-most areas of the Pacific, another bright cirriform area is associated with a complex of three surface lows and associated frontal systems. Further to the west, the leading edge of a large cirriform cloud mass is moving into the ir picture. This cloud mass represents a major storm developing off the coast of Japan.

The most impressive area of cirriform cloudiness in the entire photograph is seen over the Pacific Ocean between Hawaii and Mexico. This pattern is associated with a huge, although rather weak, trough aloft that is linked to the subtropical jet stream that curves southward near about 35° N, 160° W. The brightness of these cirriform clouds, coupled with light grey areas (representing middle clouds) below, hints strongly of an active zone of weather activity. Further east, broken, largely transverse bands of cirriform clouds are spreading eastward into Mexico.

The equatorial area in Fig. 4.1 is characterized by strong, predominantly diurnal convection over western South America with a large production of anvil cirrus and a zone of Intertropical Convergence Zone (ITCZ) that extends across the Pacific at 10° S. In the central Pacific, a collection of mesoscale and synoptic scale

FIG. 4.1   Full disc ir picture at 2345 GMT February 23, 1984, illustrating the globally distributed cirrus clouds.

clusters of cumulonimbus, some embedded in areas of middle clouds, is producing the brightest (coldest) cirriform cloudiness in the picture. The area that is moving slowly westward is characteristic of many disturbances in the ITCZ.

In the Southern Hemisphere, the dominant cloudiness is due to the cirriform clouds associated with the strong cold front that extends north-northwestward from an occluded front anchored in an intense low centered at 57° S, 140° W. Also, some spiral shaped cirrus and middle clouds are shown in the vicinity of 31° S, 104° W; these are associated with a low-pressure system that has been cut off from the westerlies. On the western edge of the picture, a weak cold front with a thin cirrus band approaches New Zealand. To the north of this front, a band of cirrus stretching north-northwestward from 30° S, 165° E is associated with a surface low at 19° S, 158° E and a shear line aloft.

Having described the principal cirrus cloud systems at 2345 GMT February 23, 1984, we shall now briefly discuss the changes in these systems during the ensuing 48 hours. The cirriform cloud pattern west of Washington State, as depicted

Fig. 4.2   Same as Fig. 4.1, except for February 25, 1984.

in Fig. 4.1, changes from a well-defined spiral with the original surface low to a disorganized, blotchy mass of less bright clouds in the midst of the dissipation–reformation process, and finally to a redevelopment of brighter, more organized masses coming together with a major storm development, as shown in Fig. 4.2.

A huge cirriform spiral, characteristic of extratropical cyclones, is seen in the north Pacific. The thin line of cirrus on the poleward side of the cold frontal band stretches from 47° N, 135° W to 35° N, 154° W. This line occurs adjacent to the polar jet stream that cuts across the frontal zone near the west coast of central British Columbia. Another major storm, which boasts a huge cirriform frontal band and a spiral center at 43° N, 161° E, is also evident in Fig. 4.2. This is the same storm that was just coming into view from the west in Fig. 4.1.

The principal equatorial feature in Fig. 4.2 is the ITCZ near 5° S stretching across the Pacific Ocean. This zone is much better defined in Fig. 4.2 because it is composed of at least six synoptic-scale areas of bright cumulonimbus cloud clusters that generally move east to west. In the Southern Hemisphere there has been relatively little change in the 48 hour period between Figs. 4.1 and 4.2 due to

FIG. 4.3   A GOES visible image for 1615 GMT June 17, 1976, showing extended and persistent stratus and stratocumulus clouds (after Moeng and Arakawa, 1980).

the fact that the 500 mb flow pattern was initially a Rex-type blocking situation. The cirrus cloud band near 55° S, 175° W in Fig. 4.2 is associated with a strong, new cold front beginning to sweep northeastward.

In light of the preceding description, the formation, maintenance, and dissipation of cirrus clouds are principally associated with large-scale synoptic features and disturbances. In the tropics, these phenomena are related to deep-cumulus outflows. Thin cirrus has not been identified due to the limitations of ir remperature techniques and may or may not be present.

4.1.2   *Subtropical stratiform and tropical cumuliform clouds*

As seen from satellite ir pictures, large areas of subtropical anticyclones over the north Pacific, south Pacific, and south Atlantic oceans are covered by clouds. These regions are characterized by general subsidence and relatively cold sea surface temperatures. Clouds that are produced in these regions are predominantly the low stratiform type with warm dry air aloft. These clouds are generally dense enough to form a horizontally continuous and persistent layer of stratus or stratocumulus clouds, especially near the coastal areas. Fig. 4.3 is a GOES visible image taken at 1615 GMT June 17, 1976, illustrating extended and persistent cloudiness near the west coast of North America during the Northern Hemisphere summer (Moeng and Arakawa, 1980). These stratus clouds fill the upper portion of the planetary boundary layer and are capped by a sharp inversion layer.

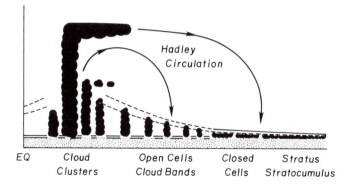

FIG. 4.4   A schematic cross-section diagram illustrating the formation of subtropical stratiform and tropical cumuliform clouds. The planetary boundary layer is represented by the stippled area. The solid lines represent the Hadley circulation. The dashed lines above the cumulus clouds represent the trade wind inversion (after Arakawa, 1975).

In the large areas extending westward and equatorward from the coast, the cloudiness is dominated by scattered, shallow cumulus clouds of small fractional cloud cover and is organized into mesoscale cellular convection. The frequent occurrence of mesoscale cellular convection over large areas of the oceans has been noted from satellites. Two types of cellular organization occur. So-called *closed cells* are characterized by approximately polygonal areas of stratiform clouds surrounded by clear air. *Open cells* are characterized by approximately polygonal clear areas surrounded by cumuliform clouds. The typical sizes of the cells range from about 20 to 100 km. Thus, these cells can only be observed from satellites.

Mesoscale cellular convection is also frequently observed in the tropics. There are more open cells than closed cells in the trade wind regions, because heating by the underlying oceans is strong. When the vertical wind shear is sufficiently strong and is associated with a synoptic-scale disturbance, convection may be organized into cloud bands or cloud clusters with a horizontal dimension on the order of several hundred kilometers. The frequency of cloud clusters is highest in, or close to, the ITCZ. A typical cloud cluster produces high-level cirrus outflow as illustrated in the previous subsection. The majority of cloud clusters propagate zonally and steadily, and are related to a dynamic coupling with long-lived planetary wave disturbances. A schematic cross-section diagram illustrating the formation of subtropical stratiform and tropical cumuliform clouds is shown in Fig. 4.4. From right to left, the sea surface temperature increases and subsidence decreases. The planetary boundary layer is represented by the stippled area with continuous and discontinuous double lines on the top. The dashed lines above the cumulus clouds represent the trade wind inversion.

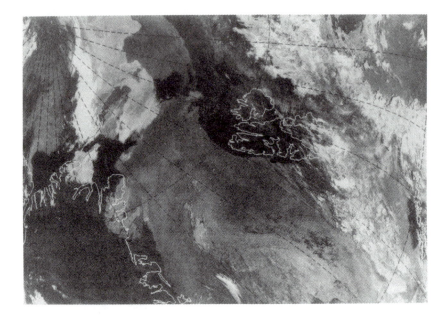

FIG. 4.5   Satellite picture of Arctic stratus clouds taken at 1018 UTC June 26, 1984, using the difference of the equivalent brightness temperatures of NOAA-6 AVHRR channel 3 (3.55–3.93 μm) and channel 4 (10.5–11.5 μm). The area of stratus is indicated by the nearly homogeneous gray color between Svalbard (right side) and Greenland (left side) (after Finger and Wendling, 1990).

4.1.3   Arctic stratus

One of the important climatic features in the Arctic is the persistence of extensive layers of stratiform clouds in the polar oceans, particularly during the melting season. Arctic stratus clouds tend to occur in well-defined layers separated by clear interstices, with individual cloud layers 300–500 m thick. Because of the temperature configurations of the surface and clouds, specific remote sensing techniques are required to map the cloud regions from satellites. Fig. 4.5 shows a satellite picture of Arctic stratus in June, using the difference of the equivalent brightness temperatures of NOAA 6 Advanced Very High Resolution Radiometer (AVHRR) channels 3 and 4 (Finger and Wendling, 1990). The structure of stratus clouds is related to three factors: the large-scale transports of sensible and latent heat into the Arctic Basin; boundary-layer turbulence; and the microphysical and optical properties of the liquid water drops (See section 4.7.1 for further discussion). Vowinckel and Orvig (1970) have shown that the frequency of stratiform clouds during July exceeds 70%, while the total mean cloud cover during this period exceeds 90% over the polar oceans. The change in cloud cover from winter to summer is remarkable. Over a period of about 4 weeks, cloud cover increases from a winter amount of 40–60% to a summer amount of 70–90%.

4.2   Some aspects of cloud climatology

An accurate global cloud climatology is essential to the development of global climate modeling; it can serve both as a source of input data for models that prescribe
cloudiness and as a means of validating models that predict cloud formation. In
recognition of the importance of cloud climatology data with regard to climate
problems, the International Satellite Cloud Climatology Project (ISCCP) was established, commencing in 1984, to provide a 5 year global cloud cover climatology
from satellite observations (Schiffer and Rossow, 1983). At this point, we shall
describe the global cloud climatology that is currently available. In discussions
hereafter, the terms *cloud cover, cloud amount, cloudiness*, and *fractional cloud
coverage* will be used interchangeably.

The most widely cited cloud climatology in the past has been that of London
(1957). This climatology was compiled from surface cloud observations in the
1930s and 1940s. Its primary merit is that it provides the zonally averaged cloud
parameters in terms of cloud cover amount, type, and height for the Northern
Hemisphere. In this climatology, clouds are divided into six types: high clouds
(Ci, Cs), middle clouds (As, Ac), low clouds (St, Sc), cumulus, cumulonimbus,
and nimbostratus. Sasamori et al. (1972) have provided the fractional cloud cover
including cirrus clouds for the Southern Hemisphere. Figure 4.6 shows the cloud
base heights for the six types as functions of latitude as well as the mean annual
cloud cover and thickness of each type. It is noted that the cirrus base heights
given in Fig. 4.6 vary with latitude but their thicknesses are fixed at 1.7 km. This
is probably due to the limitations of the surface observations.

In recent years, the cloud climatology data analyzed from the Air Force three-
dimensional nephanalysis (3DNEPH) cloud data base have been presented by a
number of researchers (Gordon et al., 1984; Henderson-Sellers, 1986; Koenig
et al., 1987; Stowe et al., 1988). The 3DNEPH integrates satellite and conventional cloud information to produce a high-resolution global cloud analysis, including high, middle, low and total cloud cover. Figure 4.7 shows the January and
July, 1979, total cloud climatology for the Northern Hemisphere as derived from
3DNEPH data. The most striking features in the January, 1979, cloud climatology
are the almost continuous belt of low cloud amounts associated with the subtropical
high, and the regions of high cloud amounts south (ITCZ) and north (mid-latitude
storm tracks) of the subtropical high. There is considerable longitudinal variation
in the total cloud amounts in the equatorial regions. The cloud amount maxima
along the east coasts of the United States and China are associated with major
storm tracks. The July cloud climatology has the following noticeable features:
a well-defined ITCZ in the Atlantic, in the western and eastern Pacific and over
equatorial Africa, and the tropical and subtropical highs, located south and north
of the ITCZ in the Atlantic and eastern Pacific. In January a total cloud amount
minimum is located over the eastern Pacific off Central America. However, in

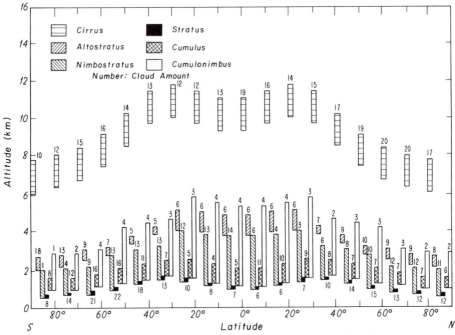

FIG. 4.6   Zonally averaged climatology of cloud type, cover, and thickness for both Northern and Southern Hemispheres at 10° latitudinal intervals.

July this region shows a total cloud amount maximum. The enhanced cloudiness over the north Atlantic, the north Pacific, northern Europe and the northwestern U.S.S.R. is related to the mid-latitude storm tracks.

Hughes (1984) has undertaken a comprehensive review of past cloud climatologies based on conventional surface observations, satellite data, and satellite-derived nephanalyses before 1984. The uncertainties and inaccuracies associated with our present knowledge of global cloud distributions have been carefully described. In order to have a global view of the cloud climatology, we present some zonally averaged data. Zonal averages involve a high degree of spatial compression in the longitudinal direction, and adjacent longitudinal profiles may be extremely different.

Figure 4.8 shows the meridional profiles of zonally averaged cloud amounts for January, July and the annual condition as presented by Brooks (1927), London (1957), van Loon (1972), and Berlyand and Strokina (1980). The results from London and van Loon are for the Northern and Southern Hemispheres, respectively. The overall trends for both January and July are very similar for these climatologies, despite significant variations among them. The near equatorial maximum associated with the ITCZ lies between 0–10° N in July and 0–10° S in January. The minimum, associated with the subtropical anticyclones and desert areas, occurs in both hemispheres at ~15° in the winter and at ~25° in the summer. The

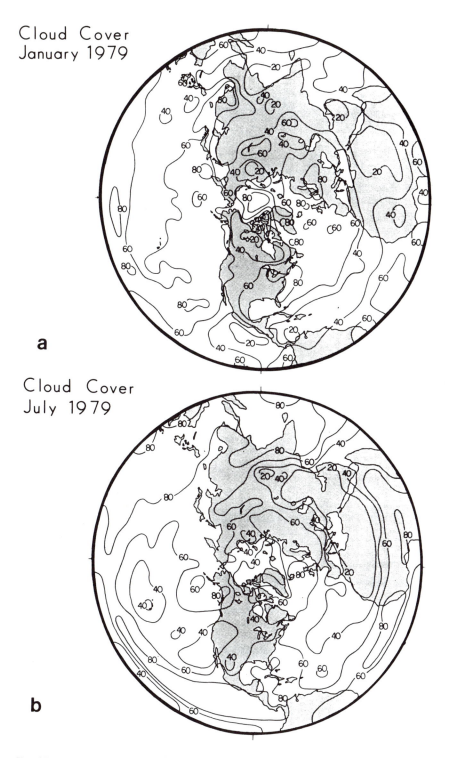

Cloud Cover
January 1979

a

Cloud Cover
July 1979

b

Fig. 4.7 The average daily (a) January and (b) July 3DNEPH total cloud amounts for the Northern Hemisphere (after Koenig et al., 1987).

cloud amount tends to decrease poleward of 60° S in both January and July, but increases in July poleward of 60° N. The most cloudy latitude band in January is 60–70° S and in July, 80–90° N. The latter area of cloudiness is largely associated with Arctic stratus, as pointed out in Subsection 4.1.3. The cloud amount values presented by Berlyand and Strokina (1980) are consistently higher at most latitudes, as shown in the annual condition, probably because only daytime observations were used in the construction of the cloud climatology. Surprisingly, the analysis of cloud climatology carried out by Brooks in 1927, using limited surface observations, produced realistic cloud cover features. It is anticipated that a more accurate cloud cover climatology will be constructed from the ISCCP data mentioned previously.

Most of the cloud climatologies presented in the past represent total cloud amount or cloud cover. The data presented by London (1957), as well as the 3DNEPH data described previously, contain information on the cloud type. Gordon et al. (1984) have carried out two different monthly mean analyses of low, middle, and high clouds amounts for January, 1977, and July, 1979, using both 3DNEPH and surface observations of clouds. For the first time, the layer cloud amounts are presented in a manner consistent with a general circulation model. The positions of the clouds are defined in terms of season and latitude, as depicted in Fig. 4.9, where $\sigma = p/p_*$, with p_* being the surface pressure. Table 4.1 lists the mean cloud amount in terms of cloud type for the Northern and Southern Hemisphere and the global means. There are significant deviations between the high cloud results derived by Gordon et al. and those derived by London, although both used surface cloud observations. The discrepancies are probably due to an enhancement of the surface network since 1954, as well as to natural variability. During January, 1977, there is more low cloudiness derived from surface observations than from 3DNEPH. But during July, 1979, the reverse is shown. 3DNEPH shows more high cloudiness in the Northern Hemisphere than do the results derived from surface observations. 3DNEPH total cloud cover is also substantially larger.

In this discussion, we have presented several types of cloud climatologies to illustrate the characteristic distribution of clouds. It is evident that a unique and agreed global cloud climatology does not exist at present. Future efforts are required to narrow down the uncertainties and inaccuracies in the processing of cloud data derived from satellite and/or surface observations.

4.3   Microphysical properties of various cloud types

4.3.1   *Stratiform and cumuliform clouds*

Between the 1940s and 1960s there have been extensive observational studies conducted on the droplet size distribution and liquid water content (LWC) for various stratus and cumulus clouds. Some results have been presented in a number

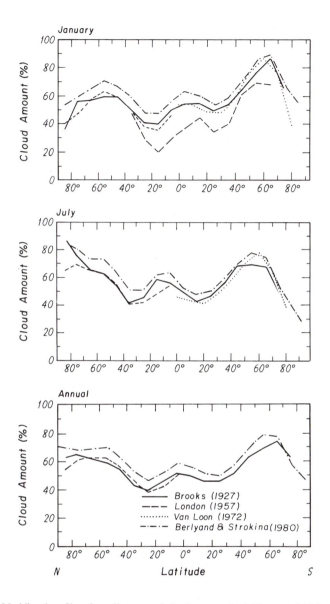

FIG. 4.8   Meridional profiles of zonally averaged cloud amounts for (a) January, (b) July, and (c) annual conditions.

of cloud physics textbooks (Mason, 1971; Pruppacher and Klett, 1978). In this section we attempt to summarize the observed microphysical cloud data that are pertinent to a discussion of cloud modeling, radiative transfer, radiation budgets, and climate. The microphysical properties of cirrus clouds will be discussed in the next subsection.

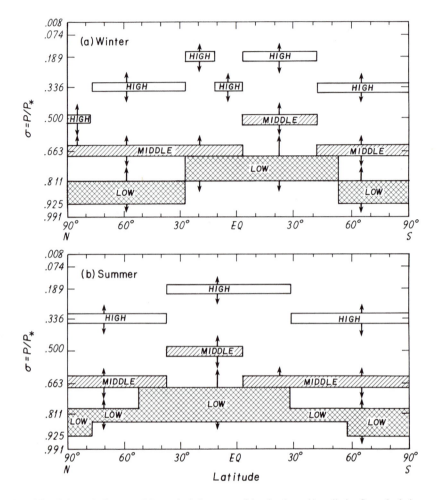

FIG. 4.9   Schematic diagram of the vertical placement of the clouds used in radiative flux calculations: (top) winter climatology, (bottom) summer climatology (after Gordon et al., 1984).

Diem (1948) observed from aircraft the droplet size distributions of six cloud types, using a photomicrographic technique and oil-coated slides for exposure. These clouds included altostratus, nimbostratus, and stratus (over land). The bases of these clouds ranged from about 1520 to 2740 m, while their tops varied from about 1830 to 3960 m. The typical cloud thickness was about 1825 m. Neiburger (1949) measured the droplet sizes in California stratus (over the oceans). Small slides coated with a thin film of lampblack were exposed to the airstream, and the droplet diameter was determined from the rings left in the soot. Zaitsev (1950) utilized the photomicrographic technique to determine the droplet size distribution in fair weather cumulus and cumulus congestus. The droplet size distribution and LWC varied with height in cumulus congestus. Borovikov et al. (1961) presented

Table 4.1 Mean cloud amount based on 3DNEPH (for the Northern Hemisphere), surface observations of clouds (SFCOBS) and London's results[a]

Case	Layer	Mean	3DNEPH	SFCOBS	London
January, 1977	High	GBL		0.109	0.187
		NH	0.218	0.119	0.174
		SH		0.099	0.199
	Middle	GBL		0.338	0.339
		NH	0.196	0.119	0.084
		SH		0.120	0.083
	Low	GBL		0.338	0.339
		NH	0.249	0.286	0.329
		SH		0.389	0.350
	Total	GBL		0.489	0.504
		NH	0.503	0.455	0.488
		SH		0.522	0.520
July, 1979	High	GBL		0.094	0.183
		NH	0.187	0.121	0.198
		SH		0.067	0.169
	Middle	GBL		0.166	0.083
		NH	0.251	0.127	0.083
		SH		0.105	0.082
	Low	GBL		0.357	0.339
		NH	0.376	0.338	0.352
		SH		0.375	0.326
	Total	GBL		0.491	0.501
		NH	0.587	0.493	0.520
		SH		0.490	0.483

[a] After Gordon et al. (1984).

an extensive series of droplet size measurements in stratiform clouds in various parts of Russia under a variety of climatic conditions. Weickmann and aufm Kampe (1953) performed an extensive series of measurements of droplet size and LWC in cumuliform clouds (Cu, Cu congestus, Cb) over New Jersey and Florida using slides coated with castor oil and the in situ photographic technique, which take into account the evaporation of droplets before photographing. Battan and Reitan (1957) reported the mean droplet size distribution in Cu and Cu congestus over the central United States. Squires (1958) measured the droplet characteristics of continental, transitional and maritime cumuli, orographic clouds and dark stratus in the vicinity of Hawaii; he used the data to investigate raindrop growth, by means of slides coated with a layer of magnesium oxide. Durbin (1959) obtained 150 droplet samples from cumulus clouds with heights ranging from ~230 to 2130 m using the impaction method. Averaged spectra of the droplet diameter were found to have a maximum at ~8 μm. In some cases, a secondary maximum at ~15 μm was also found. Singleton and Smith (1960) obtained the droplet sizes for low stratiform clouds with thicknesses varying from ~210 to 2250 m and reported that the thicker a layer of cloud, the broader the mean droplet spectrum. Warner

(1969) observed the droplet size distribution in cumulus clouds near the eastern Australian coast and found that the bimodal nature of these clouds increases with height above the cloud base. This effect is due, in part, to mixing between the cloud and its environment. Extensive observations of the droplet size distribution for stratiform and cumuliform clouds were also carried out in China (Gu, 1980). The droplet characteristics of Ac have evidently not been adequately observed. The droplet size distribution diagram presented by aufm Kampe and Weickmann (1957) appears to be the only source of information for Ac.

It is evident from the preceding discussion that the droplet size distributions of cumulus clouds and low stratiform clouds have been comprehensively measured all over the world for a period of more than 20 years. Cloud droplet spectra may be characterized by the function $n(r)$, in units of number per volume per radius r. Thus $n(r)dr$ represents the number of droplets per unit volume with radii in the interval $(r, r + dr)$. As observed by the aforementioned researchers, droplet size distributions generally vary with the position in a cloud and with time at a given location. Statistical variability also arises from the randomness of droplet sampling locations. Droplet size distributions obtained from aircraft observations usually represent results for averaged conditions over a large volume of clouds.

A summary of the representative mean droplet size distributions for various cloud types is displayed in Table 4.2. In this table, N denotes the droplet number concentration, r_m is the mode radius (i.e., the radius corresponding to the peak of the $n(r)\,dr$ curve), and Δr is the radius range. The droplet number concentration is given by

$$N = \int_{\Delta r} n(r)\,dr. \tag{4.3.1}$$

LWC is related to the droplet size distribution by

$$\text{LWC} = \frac{4}{3}\pi\rho_\ell \int_{\Delta r} r^3 n(r)\,dr. \tag{4.3.2}$$

where ρ_ℓ is the density of liquid water.

Figure 4.10 shows the representative droplet size distributions for the cloud types identified in Table 4.2. In the presentation of the droplet size distributions, we have smoothed the original data and normalized each distribution so that the integration of $n(r)$ over Δr yields approximately the total number density N. Droplet size distributions for low stratiform clouds are displayed in Fig. 4.10(a). Stratus over land consists of droplets of large radius and has a greater LWC than stratus and stratocumulus over the oceans. The middle cloud type has relatively narrow droplet size distribution, as shown in Fig. 4.10(b). There are a variety of droplet size distributions for convective clouds. These are illustrated in Fig. 4.10(c). The fair weather cumulus has a narrow droplet size distribution with the largest radius cut off at $\sim20\,\mu$m. The cumulus congestus shows a double maxima and has droplets with radii up to $\sim40\,\mu$m. The cumulonimbus has a much broader droplet

Table 4.2 Characteristics of the droplet size distribution for various cloud types

	Cloud Type	Investigator	N (cm^{-3})	r_m (μm)	Δr (μm)	LWC $(g\,m^{-3})$
Low clouds	St I (ocean)	Neiburger	464	3.5	0–16	0.24
	St II (land)	Diem	260	4.5	0–20	0.44
	Sc	Diem	350	4.0	0–12	0.09
	Ns	Diem	330	4.0	0–20	0.40
Middle clouds	As	Diem	450	4.5	0–13	0.41
	Ac	aufm Kampe and Weickmann	—	5.0	0–12	—
Cumulus	Cu (fair weather)	Battan and Reitan	293	4.0	0–20	0.33
	Cu (congestus)	Durbin	207	3.5	0–40	0.66
	Cb	Weickmann and aufm Kampe	72	5.0	0–70	2.50

size distribution that extends to a radius of $70\,\mu m$. The droplet concentration for cumulonimbus is low, with a value of $\sim 72\,cm^{-3}$. The nimbostratus also shows double maxima, but the largest radius does not exceed $\sim 20\,\mu m$.

Several mathematical expressions for the cloud droplet size distributions have

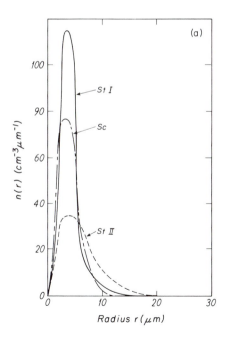

FIG. 4.10(a) Droplet size distributions of stratocumulus and stratus over land (St II) and the oceans (St I).

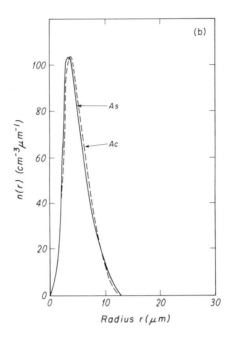

FIG. 4.10(b)   Droplet size distributions of altostratus and altocumulus.

been suggested for the purpose of scattering and absorption calculations. In connection with this, the modified gamma distribution developed by Deirmendjian (1969) is quite useful. This size distribution is a generalization of that proposed by Borovikov et al. (1961) for clouds and can be written in the form

$$n(r) = N \frac{6^6}{5!} \frac{1}{r_m} \left(\frac{r}{r_m} \right)^6 e^{-6r/r_m}. \tag{4.3.3}$$

The cloud model with $r_m = 4\,\mu\text{m}$ fits well with observed mean droplet size distributions for fair weather cumulus. A linear combination of two modified gamma distributions can be used to fit the observed droplet size spectra that have bimodal distributions. For example, the composite distribution with two mode radii at 4 and 7 μm fits the observed droplet size distribution for Cu congestus, while modal radii at 4 and 10 μm fit that for Ns.

4.3.2   Cirrus clouds

Cirrus clouds are normally located in the upper troposphere where observations are difficult to make. Weickmann (1945) was the first cloud physicist to undertake the measurement of the composition of cirrus. Using the technique of direct in-flight samplings by means of oil-covered slides and, subsequently, microphotographs,

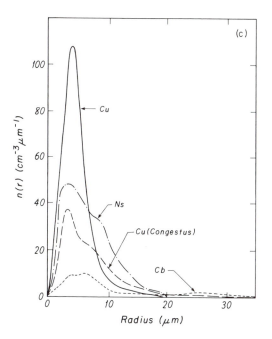

FIG. 4.10(c) Droplet size distributions of fair weather cumulus, nimbostratus, cumulus congestus, and cumulonimbus.

he showed that cirrus clouds are predominantly composed of columnar crystals. In cirrostratus, crystals have a length of $\sim$100 μm and a width of $\sim$40 μm, which vary with temperature, duration of supersaturation, and concentration. The concentration of ice crystals in cirrostratus clouds is $\sim$0.1 cm^{-3}. In cirrocumulus, ice crystals are in the form of bundles of incompletely built columns. On average, the bundles are $\sim$200–300 μm in length with a width of $\sim$50–100 μm. The ice water content (IWC) in cirrus clouds is on the order of 0.01 g m^{-3}. On the basis of snowflake replicas collected throughout the world, Schaefer (1951) reported that cirrus clouds result from spontaneous nucleation and that the common crystal types in cirrus are hexagonal plates and columns, and irregular or asymmetric crystals.

Evidently, there was a lack of investigation of the composition of cirrus clouds during the 1950s and 1960s. It was not until the early 1970s that the study of the microstructure of cirrus clouds was carried out (Heymsfield and Knollenberg, 1972). The optical-array spectrometers designed by Knollenberg (1970) were used to make continuous particle size distribution measurements in cirrus. The mean crystal length was found to be $\sim$600–1000 μm with a concentration of $\sim$0.01–0.025 cm^{-3} and an IWC of $\sim$0.15–0.25 g m^{-3}. The predominant ice crystal habits were columns, bullet rosettes (75%) and plates (25%), as illustrated in Fig. 4.11.

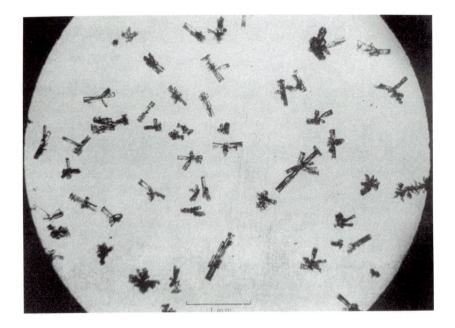

FIG. 4.11   Photograph of cirrostratus crystals captured in silicone oil (after Heymsfield and Knollen-berg, 1972).

Hobbs et al. (1975) carried out a series of measurements to explore the distribution and mass concentration of ice in cirrus clouds. An airborne optical ice particle counter was used to measure the concentration of ice particles, and a continuous particle sampler with a metal foil impactor was employed to collect and replicate the cloud particles. The crystal shapes associated with cirrostratus, cirrocumulus, and cirrus were found to be columns, plates, and bullets with sizes ranging from about 100 to 1000 μm. Many crystals with sizes smaller than 100 μm could be missed by the sampling techniques. The IWC observed was from $\sim$0.006 to 0.3 g m^{-3}. Cirrus clouds frequently appear to be associated with upper level troughs.

A series of in situ measurements of the microphysical properties of cirrus clouds was also undertaken by Heymsfield (1975a, 1977). The predominant ice crystal types for cirrus uncinus and cirrostratus were polycrystalline bullet rosettes, single bullets, banded columns, and plates. The IWC was found to be less than 0.02 g m^{-3}. There is strong evidence for the temperature dependence of particle sizes and concentrations. For cirrus uncinus, there is a bimodal distribution with a second maximum concentration peak at about the 500 μm region. Moreover, based on this data set, there are quite a few relatively small crystals ($\sim$20–50 μm) existing in cirrus clouds that have not been previously reported. Heymsfield (1977) also presented comprehensive ice crystal data collected from stratiform ice clouds

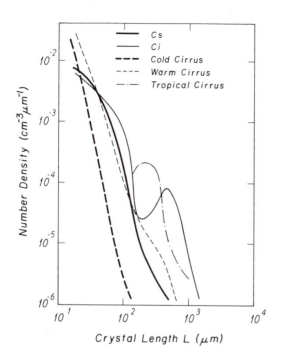

FIG. 4.12   Representative ice crystal size distributions for cirrus clouds.

associated with warm frontal overrunning systems, warm frontal occlusions, closed lows aloft, and the jet stream. IWC and the mean length of the ice crystal may be correlated with temperature. Heymsfield and Platt (1984) classified the ice crystal size distribution according to warm ($-30°$ C) and cold ($-50°$ C) cirrus clouds. Griffith et al. (1980) carried out aircraft observations of the thermal infrared radiative properties and microphysics of tropical cirrus clouds. The ice crystal size distributions reported by these authors are displayed in Fig. 4.12.

Varley et al. (1980) reported observations of the typical habit and distribution of ice crystals found in relatively thin cirrus clouds. Three instruments were employed, in addition to the replicas for the identification of the shape and size of ice crystals in cirrus clouds. These included a scatter probe that was developed for the sizing of spherical water droplets from 2 to $30\,\mu$m and Knollenberg's one-dimensional (26–$312\,\mu$m, 400–$4700\,\mu$m) and two-dimensional (25–$800\,\mu$m, 200–$6400\,\mu$m) probes. The predominant crystal form was found to be bullet rosettes with a few plates and columns. The IWC for thin cirrus and cirrostratus are, respectively, on the order of 10^{-3} and 10^{-2} g m^{-3}. The measurements revealed that a large number of small ice crystals on the order of 20–$50\,\mu$m exist in cirrus clouds, confirming Heymsfield's observations. In addition, observations indicated that some thin cirrus clouds are directly correlated with high pressure systems. A

Table 4.3 Aircraft observations of the composition and structure of cirrus clouds

Cloud type	Investigator	Synoptic condition	Composition
Cirrostratus, cirrocumulus	Weickmann	—	Column, bundle of columns $L\sim100\text{--}300\,\mu m$ IWC $\sim0.01\,g\,m^{-3}$
Cirrus uncinus, cirrostratus, anvil	Heymsfield and Knollenberg	—	Bullet rosette, column (75%), plate (25%) $L\sim600\text{--}1000\,\mu m$ IWC $\sim0.15\text{--}0.25\,g\,m^{-3}$
Cirrus, cirrostratus ($\sim6\text{--}7\,km$)	Hobbs et al.	Upper-level trough, frontal system	Bullet, column, plate $L\sim100\text{--}700\,\mu m$ IWC $\sim0.01\text{--}0.1\,g\,m^{-3}$
Cirrus uncinus	Heymsfield	Temp. $\sim-19\text{--}-58°$ C, strong wind shear	Bullet rosette, column, plate $L\sim20\text{--}2000\,\mu m$ IWC $\sim0.15\text{--}3\,g\,m^{-3}$
Cirrostratus			$L\sim20\text{--}500\,\mu m$ IWC $\sim0.01\text{--}0.15\,g\,m^{-3}$
Stratiform ice clouds	Heymsfield	Temp. $\sim-10\text{--}-60°$ C, frontal system jet stream	Bullet rosette, column, thick plate $L\sim300\text{--}600\,\mu m$ IWC $\sim0.001\text{--}1\,g\,m^{-3}$
Thin cirrus, cirrostratus ($\sim8\text{--}9\,km$)	Varley et al.	Upper level trough, high-pressure system	Bullet rosette, column, plate, $L\sim20\text{--}2000\,\mu m$ IWC $\sim0.001\text{--}0.05\,g\,m^{-3}$

summary of the above mentioned aircraft observations is presented in Table 4.3.

Heymsfield and Platt (1984) have grouped the observed data for the ice crystal distribution as a function of maximum dimension and temperature for a 5° C interval from −20° to −60° C. Based on these results, it appears that temperature is the predominant factor controlling ice crystal size distribution and IWC. The observed mean IWC as a function of temperature, shown in Fig. 4.13, may be parameterized in the form (Liou, 1986)

$$ln(\text{IWC}) = -7.6 + 4\exp[-0.2443\times10^{-3}(|T|-20)^{2.455}], \qquad |T| > 20. \quad (4.3.4)$$

The crystal size distribution $n(L)$ in cirrus clouds can be represented by two equations in the forms (Heymsfield and Platt, 1984)

$$n(L) = \begin{cases} A_1 L^{B_1}(\text{IWC}), & L \le L_o \\ A_2 L^{B_2}(\text{IWC}), & L > L_o, \end{cases} \quad (4.3.5)$$

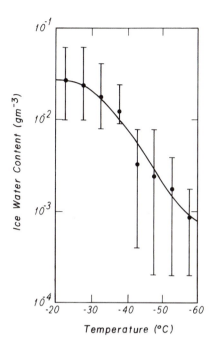

Fig. 4.13 IWC as a function of temperature. The solid curve represents the best fit to the data points.

where L is the maximum crystal dimension (or the length) in μm, IWC is in units of g m^{-3}, $n(L)$ is in units of m^{-3} μm^{-1}, and B_1 and B_2 are the slopes of the curves. The terms A_1, A_2, and L_o are defined by

$$A_1 = \frac{N_{100}/\text{IWC}}{(100)^{B_1}} \tag{4.3.6a}$$

$$A_2 = \frac{N_{1000}/\text{IWC}}{(1000)^{B_2}} \tag{4.3.6b}$$

$$L_o = \left(\frac{A_2}{A_1}\right)^{1/(B_1-B_2)}, \tag{4.3.6c}$$

where N_{100} and N_{1000} denote the ice crystal concentrations at 100 and 1000 μm. Based on the aircraft observations of ice crystal size distributions, the values of B_1, B_2, N_{100}/IWC and N_{1000}/IWC can be related to temperature in the range of $-20°$ to $-60°$ C. Using the curve fitting technique, we find

$$B_1 = \sum_{n=0}^{3} a_n T^n \tag{4.3.7a}$$

$$B_2 = \sum_{n=0}^{4} b_n T^n \tag{4.3.7b}$$

Table 4.4 Empirical coefficients for the parameters defining the ice crystal size distribution

n	a_n	b_n	c_n	d_n	e_n
0	-1.1430×10^1	1.8940×10^1	-1.0159×10^1	1.6764×10^1	1.5508×10^2
1	-7.3892×10^{-1}	2.7658×10^0	-1.4538×10^0	-1.5072×10^{-1}	1.8377×10^1
2	-1.8647×10^{-2}	1.2833×10^{-1}	-1.3511×10^{-2}	-1.9713×10^{-2}	8.5312×10^{-1}
3	-1.4045×10^{-4}	2.7750×10^{-3}	1.1318×10^{-3}	-3.5051×10^{-4}	1.6879×10^{-2}
4	0	2.2994×10^{-5}	2.2360×10^{-5}	-1.6727×10^{-6}	1.1873×10^{-4}

$$\ell n(N_{100}/\text{IWC}) = \begin{cases} \sum_{n=0}^4 c_n T^n, & T \geq -37.5^\circ \text{ C} \\ \sum_{n=0}^4 d_n T^n, & T < -37.5^\circ \text{ C} \end{cases} \qquad (4.3.7c)$$

$$\frac{N_{1000}}{\text{IWC}} = \sum_{n=0}^4 e_n T^n. \qquad (4.3.7d)$$

The empirical coefficients are listed in Table 4.4. It follows that once the temperature is given, the ice crystal size distribution and IWC may be computed. The preceding parameterized equations generally have accuracies within about 5%.

With respect to the aspect ratio, defined as the ratio of width to length, of ice crystals, Auer and Veal (1970) have presented a comprehensive analysis based on available laboratory and field data for various types of ice particles. Heymsfield (1972) has made an effort to relate the major and minor axes of ice crystals occurring in cirrus clouds. A number of empirical equations were derived to correlate the length (L) and width (D) of bullet (b) and column (c) types for temperatures below about -20° C. These are, respectively, given by (L and D are in units of mm)

$$D_b = \begin{cases} 0.25 L_b^{0.79}, & L \leq 0.3 \text{ mm} \\ 0.19 L_b^{0.53}, & L > 0.3 \text{ mm} \end{cases} \qquad (4.3.8a)$$

$$D_c = \begin{cases} 0.5 L_c, & L \leq 0.3 \text{ mm} \\ 0.2 L_c^{0.41}, & L > 0.3 \text{ mm}. \end{cases} \qquad (4.3.8b)$$

The spatial orientation of ice crystals in cirrus clouds is a significant factor with respect to light scattering and radiative transfer, and must be addressed. Jayaweera and Mason (1965) have studied the behavior of freely falling cylinders in a viscous fluid and found that if the ratio of diameter to length is less than unity, cylinders will fall with their long axes horizontal. Observations by Ono (1969) of natural clouds have shown that columnar and plate crystals fall with their major axes oriented horizontally. Based on lidar backscattering measurements, Platt et al. (1978) have demonstrated that in cirrus clouds at approximately -15° C ice crystals are predominantly plates oriented horizontally. Platt et al. found that the return signals of vertically pointed laser beams from ice crystals largely retain the polarization state of the incident energy and that for this to occur the plates must be perpendicular to the laser beams.

In recent years, there have been suggestions that cirrus clouds in the tropics and midlatitudes may contain a substantial number of ice crystals less than 20 μm that cannot be detected with conventional imaging probes. These suggestions have been made on the basis of indirect radiometer, satellite, and lidar observations during the First ISCCP Regional Experiment (FIRE), which was carried out in Wisconsin, October–November, 1986 (Sassen et al., 1989; Ackerman et al., 1990). Platt et al. (1989) have also shown some evidence of small ice crystals based on lidar and radiometer observations of tropical cirrus. If a large number of small ice crystals of less than 20 μm do indeed exist in cirrus clouds, they would have significant effects on the transfer of solar and ir radiation.

One final note is in order: A type of ice crystal clouds, referred to as *polar stratospheric clouds* (PSCs), have been observed in the polar stratosphere between $\sim$15 and 20 km. These clouds have frequently been detected over the Arctic and Antarctic by the Stratospheric Aerosol Measurement II (SAMII) payload on board Nimbus 7 during the winter months when the ambient temperature falls below $\sim$195 K (McCormick and Trepte, 1987). The condensation of both water vapor and nitric acid (HNO_3) and the subsequent formation of HNO_3 trihydrates serve as nuclei for ice crystal growth. The particle sizes are normally on the order of 1 μm, although larger sizes of $\sim$50 μm have also been detected using Formvar crystal replication techniques (Goodman et al., 1989). Larger PSC crystals are predominantly solid and hollow columns. PSCs are important from the standpoint of both chemical processes and radiation budgets. Chemical reactions occur on the crystal surface resulting from the release of chlorine, leading to the potential catalytic destruction of ozone. Large PSC crystals may irreversibly remove H_2O and HNO_3 from higher to lower levels and therefore affect the formation of PSCs. Because of high locations and significant temperature contrast between the surface and clouds, PSCs, similar to high tropical anvil cirrus, will absorb the warmer ir radiation emitted from the surface and reduce radiative cooling . (See Subsection 5.2.4.2 for further discussion.)

4.4   Development of basic equations for moisture and cloud fields

In order to develop a set of governing equations for moisture, cloud and temperature fields, first an averaging operator (denoted with an overbar) either in time or space, or both, is defined such that

$$\bar{\chi} = \frac{1}{t(s)} \int_0^{t(s)} \chi \, dt(s), \qquad (4.4.1)$$

where t denotes time, s space, and χ can be any atmospheric variable. If the deviation from the mean for χ is denoted by χ', then

$$\chi = \bar{\chi} + \chi'. \qquad (4.4.2)$$

It follows that $\bar{\chi}' = 0$.

4.4.1   Equations of state, continuity, and motion

The equation of state for air relates pressure, density, and temperature, and may be expressed by

$$p = \rho RT, \tag{4.4.3}$$

where R is the gas constant for air. Letting $p = \bar{p} + p'$, $\rho = \bar{\rho} + \rho'$, and $T = \bar{T} + T'$, we find

$$\bar{p} + p' = R(\bar{\rho} + \rho')(\bar{T} + T'). \tag{4.4.4}$$

By applying averaging procedures, the following is obtained:

$$\bar{p} = R\bar{\rho}\bar{T}\left(1 + \frac{\overline{\rho'T'}}{\bar{\rho}\bar{T}}\right). \tag{4.4.5}$$

In the earth's atmosphere, $|\rho'/\bar{\rho}|$ and $|T'/\bar{T}|$ are usually less than $\sim 10^{-2}$ for synoptic-scale systems. Thus, the last term in Eq. (4.4.5) may be neglected so that

$$\bar{p} \approx R\bar{\rho}\bar{T}. \tag{4.4.6}$$

Dividing Eq. (4.4.4) by Eq. (4.4.6), we obtain

$$\frac{p'}{\bar{p}} \approx \frac{\rho'}{\bar{\rho}} + \frac{T'}{\bar{T}}. \tag{4.4.7}$$

Under the condition that the pressure fluctuations in the earth's atmosphere may be neglected (e.g., the Boussinesq approximation under which the air density is treated as a constant, except when it is coupled with gravity in the buoyancy term), we have

$$|\rho'/\bar{\rho}| \approx |T'/\bar{T}|. \tag{4.4.8}$$

Variations in the density or temperature range from $\sim 10^{-4}$ to $\sim 10^{-2}$.

The equation of continuity is a consequence of the conservation of air mass and can be expressed by

$$\frac{\partial \rho}{\partial t} = -\nabla \cdot (\rho \mathbf{v}). \tag{4.4.9a}$$

Using the definition of the total derivative, d/dt, and the vector operation, an alternate form for Eq. (4.4.9a) is

$$-\frac{1}{\rho}\frac{d\rho}{dt} = \nabla \cdot \mathbf{v}. \tag{4.4.9b}$$

The air density fluctuations, ρ', are small in comparison with the mean value, $\bar{\rho}$; that is, $\rho'/\bar{\rho} \ll 1$, as indicated previously. To a good approximation, we may set $\rho = \bar{\rho}$. Thus, by applying the averaging procedure to Eq. (4.4.9a), we obtain

$$\frac{\partial \bar{\rho}}{\partial t} \cong -\nabla \cdot (\bar{\rho}\bar{\mathbf{v}}). \tag{4.4.10}$$

On the basis of Newton's second law of motion for a rotational fluid, the acceleration must be balanced by the sum of the forces. The principal forces per unit mass in atmospheric motions are the pressure force, gravitation $\mathbf{g}$, the friction force $\mathbf{F}$, and the force produced by the rotation of the fluid. In vector form, we may write

$$\frac{d\mathbf{v}}{dt} = \frac{1}{\rho}\nabla p - 2\mathbf{\Omega} \times \mathbf{v} + \mathbf{g} + \mathbf{F}, \tag{4.4.11}$$

where the first and second terms on the right-hand side represent the pressure gradient and Coriolis forces, respectively. In Cartesian coordinates, we have (neglecting the friction force)

$$\frac{du}{dt} = \frac{\partial u}{\partial t} + \mathbf{v} \cdot \nabla u = fv - \frac{1}{\rho}\frac{\partial p}{\partial x}, \tag{4.4.12}$$

$$\frac{dv}{dt} = \frac{\partial v}{\partial t} + \mathbf{v} \cdot \nabla v = -fu - \frac{1}{\rho}\frac{\partial p}{\partial y}, \tag{4.4.13}$$

$$\frac{dw}{dt} = \frac{\partial w}{\partial t} + \mathbf{v} \cdot \nabla w = -g - \frac{1}{\rho}\frac{\partial p}{\partial z}, \tag{4.4.14}$$

where f is the Coriolis parameter. When the vertical scale of disturbances is much smaller than the horizontal scale, the vertical acceleration may be neglected so that

$$0 = -\frac{1}{\rho}\frac{\partial p}{\partial z} - g. \tag{4.4.15}$$

This is the hydrostatic equation.

By combining the equation of continuity with Eqs. (4.4.12)–(4.4.14), we obtain

$$\frac{\partial}{\partial t}(\rho u) + \nabla \cdot (\rho \mathbf{v} u) = \rho fv - \frac{\partial p}{\partial x}, \tag{4.4.16}$$

$$\frac{\partial}{\partial t}(\rho v) + \nabla \cdot (\rho \mathbf{v} v) = -\rho fu - \frac{\partial p}{\partial y}, \tag{4.4.17a}$$

$$\frac{\partial}{\partial t}(\rho w) + \nabla \cdot (\rho \mathbf{v} w) = -\rho g - \frac{\partial p}{\partial z}. \tag{4.4.17b}$$

Carrying out the averaging procedure yields

$$\frac{\partial}{\partial t}(\bar{\rho}\bar{u}) + \nabla \cdot \bar{\rho}\left(\bar{\mathbf{v}}\bar{u} + \overline{\mathbf{v}'u'}\right) = \bar{\rho}f\bar{v} - \frac{\partial \bar{p}}{\partial x}. \tag{4.4.18}$$

$$\frac{\partial}{\partial t}(\bar{\rho}\bar{v}) + \nabla \cdot \bar{\rho}\left(\bar{\mathbf{v}}\bar{v} + \overline{\mathbf{v}'v'}\right) = -\bar{\rho}f\bar{u} - \frac{\partial \bar{p}}{\partial y}, \tag{4.4.19a}$$

$$\frac{\partial}{\partial t}(\bar{\rho}\bar{w}) + \nabla \cdot \bar{\rho}\left(\bar{\mathbf{v}}\bar{w} + \overline{\mathbf{v}'w'}\right) = -(\bar{\rho} + \rho')g - \frac{\partial \bar{p}}{\partial z}. \tag{4.4.19b}$$

By utilizing the equation of continuity in its averaged form, we obtain

$$\frac{d\bar{u}}{dt} = \frac{\partial \bar{u}}{\partial t} + \bar{\mathbf{v}} \cdot \nabla \bar{u} = -\frac{1}{\bar{\rho}}\nabla \cdot \bar{\rho}\overline{\mathbf{v}'u'} + f\bar{v} - \frac{1}{\bar{\rho}}\frac{\partial \bar{p}}{\partial x}, \tag{4.4.20}$$

$$\frac{d\bar{v}}{dt} = \frac{\partial \bar{v}}{\partial t} + \bar{\mathbf{v}} \cdot \nabla \bar{v} = -\frac{1}{\bar{\rho}}\nabla \cdot \bar{\rho}\overline{\mathbf{v}'v'} - f\bar{u} - \frac{1}{\bar{\rho}}\frac{\partial \bar{p}}{\partial y}, \tag{4.4.21}$$

$$\frac{d\bar{w}}{dt} = \frac{\partial \bar{w}}{\partial t} + \bar{\mathbf{v}} \cdot \nabla \bar{w} = -\frac{1}{\bar{\rho}}\nabla \cdot \bar{\rho}\overline{\mathbf{v}'w'} - \frac{\bar{\rho} + \rho'}{\bar{\rho}}g - \frac{1}{\bar{\rho}}\frac{\partial \bar{p}}{\partial z}. \tag{4.4.22a}$$

In Eq. (4.4.22a) we have retained the density perturbation ρ' when it is coupled with gravity in the buoyancy terms of the vertical momentum equation. This is referred to as the *Boussinesq approximation*. For large-scale hydrostatic equilibrium, the mean pressure gradient force must be balanced by the gravitational attraction force so that

$$0 = -\frac{1}{\bar{\rho}}\frac{\partial \bar{p}}{\partial z} - g. \tag{4.4.22b}$$

In Eq. (4.4.20), the term $\bar{\rho}\overline{\mathbf{v}'u'}$ $(\overline{u'u'}, \overline{v'u'}, \overline{w'u'})$ represents the eddy fluxes of x momentum in the x, y, and z directions, respectively. In Eqs. (4.4.21) and (4.4.22a), the terms $\bar{\rho}\overline{\mathbf{v}'v'}$ and $\bar{\rho}\overline{\mathbf{v}'w'}$ denote the eddy fluxes of y and z momenta in the x, y, and z directions, respectively. These terms are referred to as *eddy stresses*.

4.4.2   *Moisture and liquid water content fields*

Let the density of water vapor be ρ_v. On the basis of the principle of conservation of mass, the time rate of change in water vapor per unit volume should be equal to the divergence of water vapor fluxes and the potential sources of water vapor, namely, condensation and/or evaporation. Thus, we write

$$\frac{\partial \rho_v}{\partial t} = -\nabla \cdot (\mathbf{v}\rho_v) + S_v, \tag{4.4.23}$$

where S_v denotes the source and sink of water vapor density. The specific humidity (or approximate mixing ratio) is defined by

$$q = \rho_v/\rho. \tag{4.4.24}$$

Replacing ρ_v with ρq, Eq. (4.4.23) becomes

$$\frac{\partial(\rho q)}{\partial t} = -\nabla \cdot (\mathbf{v}\rho q) + S_v. \tag{4.4.25}$$

After performing the averaging procedures, and subject to the assumption that $\rho = \bar{\rho}$, Eq. (4.4.25) may be rewritten in the form

$$\frac{\partial}{\partial t}(\bar{\rho}\bar{q}) = -\nabla \cdot \bar{\rho}\left(\bar{\mathbf{v}}\bar{q} + \overline{\mathbf{v}'q'}\right) + \bar{S}_v. \qquad (4.4.26)$$

Combining the equation of continuity in Eq. (4.4.10) with Eq. (4.4.26) yields

$$\frac{d\bar{q}}{dt} = \frac{\partial \bar{q}}{\partial t} + \mathbf{v} \cdot \nabla \bar{q} = -\frac{1}{\bar{\rho}}\nabla \cdot \bar{\rho}\overline{\mathbf{v}'q'} + \bar{S}_v/\bar{\rho}. \qquad (4.4.27)$$

The first term on the right-hand side of Eq. (4.4.27) represents the divergence of water vapor eddy fluxes.

Next, consider a small volume within a cloud that consists of a sample of cloud droplets and/or ice crystals, and let its LWC be ρ_m (e.g., $\mathrm{g\,m^{-3}}$). Again, by virtue of the principle of mass conservation, we have

$$\frac{\partial \rho_m}{\partial t} = -\nabla \cdot (\mathbf{v}_m \rho_m) + S_m, \qquad (4.4.28)$$

where S_m represents the source and sink of LWC and $\mathbf{v}_m$ denotes the vector velocity associated with LWC. The LWC mixing ratio is defined by

$$q_m = \frac{\rho_m}{\rho}. \qquad (4.4.29)$$

Using q_m, Eq. (4.4.28) becomes

$$\frac{\partial}{\partial t}(\rho q_m) = -\nabla \cdot (\rho \mathbf{v}_m q_m) + S_m. \qquad (4.4.30)$$

Carrying out averaging procedures yields

$$\frac{\partial}{\partial t}(\bar{\rho}\bar{q}_m) = -\nabla \cdot \bar{\rho}\left(\bar{\mathbf{v}}_m \bar{q}_m + \overline{\mathbf{v}'_m q'_m}\right) + \bar{S}_m. \qquad (4.4.31)$$

The velocity components in clouds generally differ from those in clear columns. However, we shall prove that, as a good approximation, the horizontal velocity in clouds and clear columns may be considered to be the same even for very large particles. We shall consider a spherical raindrop. The horizontal resisting force, f_r, on a raindrop must be directly proportional to the difference in the horizontal velocity involving the air and raindrop, and the raindrop diameter (Byers, 1965). Thus,

$$f_r = 3\pi\mu D(u - u_m)\frac{C_D \mathrm{Re}}{24}, \qquad (4.4.32)$$

where D is the raindrop diameter, μ is the viscosity of air, C_D is the drag coefficient, Re is the Reynolds number, and u and u_m are the horizontal velocities of the air and the raindrop, respectively. The horizontal resisting force must be equal to the mass of the raindrop multiplied by its horizontal deceleration, viz.,

$$\frac{4}{3}\pi \left(\frac{D}{2}\right)^3 \rho_\ell \frac{du_m}{dt} = 3\pi\mu D(u - u_m)\frac{C_D Re}{24}, \qquad (4.4.33)$$

where ρ_ℓ denotes the density of liquid water. Consider a raindrop with a diameter of ~ 0.48 cm. For this diameter, we find $C_D Re/24 \sim 77$, and $\mu \sim 1.8 \times 10^{-4}$ g cm^{-1} s^{-1} (Kessler, 1969). Thus the horizontal acceleration in clouds may be approximated by

$$\frac{du_m}{dt} \approx 1.09(u - u_m). \qquad (4.4.34)$$

The horizontal velocity difference, $(u - u_m)$, may be treated as a constant, so that

$$\frac{du_m}{dt} = \frac{du}{dt} = \frac{\partial u}{\partial z}\frac{dz}{dt}. \qquad (4.4.35)$$

The fall velocity of a raindrop with a diameter of 0.48 cm, $dz/dt \approx 9.07$ m s^{-1}. Thus we have

$$u - u_m \approx 832 \frac{\partial u}{\partial z}. \qquad (4.4.36)$$

Given a wind shear $\partial u/\partial z$ of 0.02 s^{-1} (20 m s^{-1} km^{-1}), which is a very large value, $u - u_m = 0.1664$ m s^{-1}. This value is less than 1% of the representative wind speeds in the air. Thus, for all practical purposes, the horizontal velocity of cloud droplets and/or ice crystals may be assumed to be the same as that of air (i.e., $\mathbf{v}_{m2} \approx \mathbf{v}_2$).

Let the fall velocity of liquid water be $\bar{w}_m$. Equation (4.4.31) may then be written in the z coordinate in the form

$$\frac{\partial}{\partial t}(\bar{\rho}\bar{q}_m) \cong -\nabla_2 \cdot (\bar{\rho}\bar{\mathbf{v}}_2\bar{q}_m) - \frac{\partial}{\partial z}[\bar{\rho}(\bar{w}_m + \bar{w})\bar{q}_m]$$
$$- \nabla \cdot \bar{\rho}\,\overline{\mathbf{v}'_m q'_m} + \bar{S}_m. \qquad (4.4.37)$$

Inserting the equation of continuity, denoted by Eq. (4.4.10), into Eq. (4.4.37) leads to

$$\frac{d\bar{q}_m}{dt} \cong \frac{\partial \bar{q}_m}{\partial t} + \mathbf{v} \cdot \nabla \bar{q}_m = -\frac{1}{\bar{\rho}}\frac{\partial}{\partial z}(\bar{\rho}\bar{w}_m\bar{q}_m)$$
$$- \frac{1}{\bar{\rho}}\nabla \cdot \bar{\rho}\,\overline{\mathbf{v}'_m q'_m} + \frac{\bar{S}_m}{\bar{\rho}}. \qquad (4.4.38)$$

The first term on the right-hand side of Eq. (4.4.38) represents the flux of LWC associated with the fall velocity $\bar{w}_m$. This equation is valid for both cloud droplets and raindrops.

4.4.3   *Thermodynamic heat transfer*

During the phase transition of water vapor molecules, heat exchanges take place. Therefore, variations in the moisture field will affect the temperature field. To develop a consistent equation for temperature changes, we begin with the first law of thermodynamics for moist air in the form

$$C_p \frac{dT}{dt} - \alpha \frac{dp}{dt} = -L \frac{dq}{dt} + Q_R, \tag{4.4.39}$$

where Q_R represents radiative heat exchange. The first term on the right-hand side of Eq. (4.4.39) denotes the latent heat transfer, with L the latent heat per unit mass. In Eq. (4.4.39), we have neglected heat exchanges produced by molecular conduction and kinematic motion, because they are not important in the present discussion.

Using the equation of state and the hydrostatic equation for air in Eqs. (4.4.3) and (4.4.15), respectively, the second term in Eq. (4.4.39) may be rewritten, leading to

$$C_p \frac{dT}{dt} + g \frac{dz}{dt} + L \frac{dq}{dt} = Q_R. \tag{4.4.40}$$

On the basis of Eq. (4.4.40) derived from the first law of thermodynamics, we define a conserved quantity, called the *moist static energy*, in the form

$$E = C_p T + gz + Lq. \tag{4.4.41}$$

Thus, we have

$$\frac{dE}{dt} = \frac{\partial E}{\partial t} + \mathbf{v} \cdot \nabla E = Q_R. \tag{4.4.42}$$

Equation (4.4.42) shows that the total change in moist static energy must be balanced by radiative heat exchanges.

Combining Eq. (4.4.9a), the equation of continuity, and Eq. (4.4.42) yields

$$\frac{\partial}{\partial t}(\rho E) + \nabla \cdot (\rho \mathbf{v} E) = \rho Q_R. \tag{4.4.43}$$

We then carry out the averaging procedure to obtain

$$\frac{\partial}{\partial t}(\bar{\rho}\bar{E}) + \nabla \cdot \bar{\rho}\left(\bar{\mathbf{v}}\bar{E} + \overline{\mathbf{v}'E'}\right) = \bar{\rho}\bar{Q}_R. \tag{4.4.44}$$

Multiplying Eq. (4.4.10) by $\bar{E}$ and subtracting this equation from Eq. (4.4.44), we obtain

$$\frac{d\bar{E}}{dt} = \frac{\partial \bar{E}}{\partial t} + \bar{\mathbf{v}} \cdot \nabla \bar{E} = -\frac{1}{\bar{\rho}} \nabla \cdot \bar{\rho}\overline{\mathbf{v}'E'} + \bar{Q}_R. \tag{4.4.45}$$

On substituting Eq. (4.4.41) for moist static energy into Eq. (4.4.45), we find

$$\frac{d\bar{T}}{dt} + \gamma_d \bar{w} + \frac{L}{C_p}\frac{d\bar{q}}{dt} = -\frac{1}{\bar{\rho}C_p}\nabla \cdot \bar{\rho}\left[\overline{\mathbf{v}'(C_pT' + gz')} + L\overline{\mathbf{v}'q'}\right] + \frac{\bar{Q}_R}{C_p}, \quad (4.4.46)$$

where the adiabatic lapse rate $\gamma_d = g/C_p$, and $\bar{w} = d\bar{z}/dt$. Multiplying the water vapor mixing ratio equation (4.4.27) by L/C_p and subtracting this equation from Eq. (4.4.46) yields

$$\frac{d\bar{T}}{dt} = \frac{\partial\bar{T}}{\partial t} + \bar{\mathbf{v}} \cdot \nabla\bar{T} = -\gamma_d\bar{w} - \frac{1}{\bar{\rho}C_p}\bar{\nabla} \cdot \overline{\rho\mathbf{v}'(C_pT' + gz')}$$
$$+ \bar{Q}/C_p + \bar{Q}_R/C_p, \quad (4.4.47)$$

where we have defined the condensational heating term

$$\bar{Q} = -L\bar{S}_v/\bar{\rho}. \quad (4.4.48)$$

In terms of the potential temperature, we begin with its definition in the form

$$\Theta = T\left(\frac{p_0}{p}\right)^\kappa, \quad (4.4.49)$$

where $p_0 = 1000\,\text{mb}$ and $\kappa = R/C_p$. Since

$$\frac{1}{\Theta}\frac{d\Theta}{dt} = \frac{1}{T}\left(\frac{dT}{dt} - \frac{1}{\rho C_p}\frac{dp}{dt}\right), \quad (4.4.50)$$

the first law of thermodynamics for moist air may be expressed by

$$\frac{d\Theta}{dt} = \frac{\partial\Theta}{\partial t} + \mathbf{v} \cdot \nabla\Theta = \frac{\Theta}{T}\frac{1}{C_p}(Q + Q_R), \quad (4.4.51)$$

where the heating due to condensation and/or evaporation, Q, is set as $-L\,dq/dt$. Combining this equation with the equation of continuity leads to

$$\frac{\partial}{\partial t}(\rho\Theta) + \nabla \cdot (\rho\mathbf{v}\Theta) = \left(\frac{p_0}{p}\right)^\kappa\frac{\rho}{C_p}(Q + Q_R). \quad (4.4.52)$$

Since the variation in pressure is small, we may set $\bar{p} = p$. Using the same procedures described previously for the temperature field, we find

$$\frac{d\bar{\Theta}}{dt} = \frac{\partial\bar{\Theta}}{\partial t} + \bar{\mathbf{v}} \cdot \nabla\bar{\Theta} = -\frac{1}{\bar{\rho}}\nabla \cdot \overline{\rho\mathbf{v}'\Theta'} + \left(\frac{p_0}{\bar{p}}\right)\frac{\bar{Q} + \bar{Q}_R}{C_p}. \quad (4.4.53)$$

In addition, a thermodynamic parameter called the *equivalent potential temperature* may be defined from the first law of thermodynamics. Consider a saturated parcel undergoing pseudoadiabatic ascent during which $Q_R = 0$. Thus the first law of thermodynamics via Eqs. (4.4.39) and (4.4.50) may be written in the form

$$\frac{d\ell n\Theta}{dt} = -\frac{L}{TC_p}\frac{dq}{dt}. \tag{4.4.54}$$

Since the rate of change in temperature of the condensed water is much smaller than that in specific humidity associated with latent heat release during condensation, we may write

$$\frac{d\ell n\Theta}{dt} \cong -\frac{d}{dt}\left(\frac{Lq}{C_pT}\right). \tag{4.4.55}$$

Equation (4.4.55) would apply exactly for a reversible ascent. After integrating this equation from the initial state (Θ, q) to a state where $q = 0$, and letting $\Theta_e = \Theta$ when $q = 0$, we obtain

$$\Theta_e = \Theta \exp\left(\frac{Lq}{C_pT}\right). \tag{4.4.56}$$

Equation (4.4.56) defines the so-called equivalent potential temperature Θ_e. Hence, Eq. (4.4.54) may be rewritten in the form

$$\frac{d\ell n\Theta_e}{dt} = 0 \text{ (adiabatic).} \tag{4.4.57}$$

The equivalent potential temperature is conserved for a parcel undergoing pseudoadiabatic changes of state. It is also conserved in the dry adiabatic processes for an unsaturated parcel if the temperature used in Eq. (4.4.56) is the temperature that the parcel would have if it were expanded adiabatically to saturation.

Returning to Eq. (4.4.39), the first law of thermodynamics (including the contribution from radiative flux exchanges) may be expressed in terms of the equivalent potential temperature in the form

$$\frac{d\Theta_e}{dt} = \frac{\partial\Theta_e}{\partial t} + \mathbf{v}\cdot\nabla\Theta_e = \frac{\Theta_e}{TC_p}Q_R. \tag{4.4.58}$$

Using the equation of continuity, we have

$$\frac{\partial}{\partial t}(\rho\Theta_e) + \nabla\cdot(\rho\mathbf{v}\Theta_e) = \frac{\rho\Theta_e}{TC_p}Q_R. \tag{4.4.59}$$

As before, carrying out the averaging procedure yields

$$\frac{d\bar{\Theta}_e}{dt} = \frac{\partial\bar{\Theta}_e}{\partial t} + \mathbf{v}\cdot\nabla\bar{\Theta}_e \cong -\frac{1}{\bar{\rho}}\nabla\cdot\overline{\rho\mathbf{v}'\Theta'_e} + \frac{\bar{\Theta}_e}{\bar{T}C_p}\bar{Q}_R, \tag{4.4.60}$$

where we have assumed that $\Theta_e Q_R/(\bar{T} + T') \approx \Theta_e Q_R/\bar{T}$ and $\overline{\Theta'_e Q'_R} \approx 0$.

4.4.4   Sources and sinks of cloud and moisture fields

To complete the governing equations for the cloud field, it is necessary to consider precipitation processes. Let the subscripts mi and mr denote the IWC mixing ratio for ice particles and LWC mixing ratio for raindrops, respectively. A general set of equations governing the moisture and cloud fields may be written as follows:

$$\frac{d\bar{q}}{dt} = -\frac{1}{\bar{\rho}}\nabla \cdot \overline{\bar{\rho}\mathbf{v}'q'} + \frac{\bar{S}_v}{\bar{\rho}}, \tag{4.4.61}$$

$$\frac{d\bar{q}_m}{dt} = -\frac{1}{\bar{\rho}}\frac{\partial}{\partial z}(\bar{\rho}\bar{w}_m\bar{q}_m) - \frac{1}{\bar{\rho}}\nabla \cdot \bar{\rho}\overline{\mathbf{v}'q'_m} + \frac{\bar{S}_m}{\bar{\rho}}, \tag{4.4.62}$$

$$\frac{d\bar{q}_{mi}}{dt} = -\frac{1}{\bar{\rho}}\frac{\partial}{\partial z}(\bar{\rho}\bar{w}_{mi}\bar{q}_{mi}) - \frac{1}{\bar{\rho}}\nabla \cdot \bar{\rho}\overline{\mathbf{v}'q'_{mi}} + \frac{\bar{S}_{mi}}{\bar{\rho}} \tag{4.4.63}$$

$$\frac{d\bar{q}_{mr}}{dt} = -\frac{1}{\bar{\rho}}\frac{\partial}{\partial z}(\bar{\rho}\bar{w}_{mr}\bar{q}_{mr}) - \frac{1}{\bar{\rho}}\nabla \cdot \bar{\rho}\overline{\mathbf{v}'q'_{mr}} + \frac{\bar{S}_{mr}}{\bar{\rho}}. \tag{4.4.64}$$

All the terms in Eqs. (4.4.63) and (4.4.64) are self-explanatory. This set of equations is composed of the so-called bulk equations for water vapor, LWC (or IWC) for cloud particles, precipitating ice crystals, and raindrops in terms of the nondimensional mixing ratio.

The potential sources and sinks in each equation may be understood from the cloud microphysical processes that occur in a cumulonimbus cloud, as illustrated in Fig. 4.14. These processes include, but are not limited to, condensation of water vapor into cloud droplets (S_1); conversion of cloud droplets to large raindrops by means of accretion (autoconversion) (S_2); glaciation of raindrops to snowflakes (S_3); deposition of vapor to ice (S_4); melting of ice crystals (S_5); evaporation of cloud droplets (S_6); evaporation of raindrops (S_7); sublimation of ice crystals (S_8); and sublimation of ice crystals in the process of melting (S_9). Having defined these cloud microphysical processes, and noting that each process is a discrete event, the sources and sinks depicted in Eqs. (4.4.61)–(4.4.64) may be expressed by

$$\bar{S}_v = -\bar{S}_1 - \bar{S}_4 + \bar{S}_6 + \bar{S}_7 + \bar{S}_8 + \bar{S}_9, \tag{4.4.65}$$

$$\bar{S}_m = \bar{S}_1 - \bar{S}_2 - \bar{S}_6, \tag{4.4.66}$$

$$\bar{S}_{mi} = \bar{S}_3 + \bar{S}_4 - \bar{S}_5 - \bar{S}_8 - \bar{S}_9, \tag{4.4.67}$$

$$\bar{S}_{mr} = \bar{S}_2 - \bar{S}_3 + \bar{S}_5 - \bar{S}_7. \tag{4.4.68}$$

The conservation of mass requires that

$$\bar{S}_v + \bar{S}_m + \bar{S}_{mi} + \bar{S}_{mr} = 0. \tag{4.4.69}$$

The total latent heat release is

$$\bar{Q} = [\, L(\bar{S}_1 - \bar{S}_6 - \bar{S}_7) + L_s(\bar{S}_4 - \bar{S}_8 - \bar{S}_9)$$
$$+ L_f(\bar{S}_3 - \bar{S}_5)\,]/\bar{\rho}, \tag{4.4.70}$$

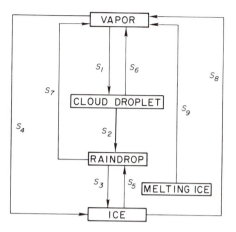

FIG. 4.14   Some aspects of cloud microphysical processes in a cumulonimbus cloud. The notations are: S_1, condensation; S_2, autoconversion; S_3, glaciation; S_4, deposition; S_5, melting; S_6, cloud droplet evaporation; S_7, raindrop evaporation; S_8, sublimation; and S_9, sublimation involving melting ice.

where L, L_s and L_f denote latent heat transformation between vapor and liquid water, vapor and ice, and liquid water and ice, respectively. In order to solve Eqs. (4.4.61)-(4.4.64), all the sources and sinks must be expressed in terms of $\bar{q}, \bar{q}_m, \bar{q}_{mi}$ and $\bar{q}_{mr}$. Thus the detailed particle size distributions cannot be accounted for, but must be parameterized.

The total moisture equation may be written by summing Eqs. (4.4.61)-(4.4.64). Letting $\bar{q}_t = \bar{q} + \bar{q}_m + \bar{q}_{mi} + \bar{q}_{mr}$, we find

$$\frac{d\bar{q}_t}{dt} = \frac{\partial \bar{q}_t}{\partial t} + \mathbf{v} \cdot \nabla \bar{q}_t = -\frac{1}{\bar{\rho}} \nabla \cdot \overline{\bar{\rho} \mathbf{v}' q_t'}$$
$$- \frac{1}{\bar{\rho}} \frac{\partial}{\partial z} (\bar{\rho} \bar{w}_m \bar{q}_m) - \frac{1}{\bar{\rho}} \frac{\partial}{\partial z} (\bar{\rho} \bar{w}_{mi} \bar{q}_{mi}) - \frac{1}{\bar{\rho}} \frac{\partial}{\partial z} (\bar{\rho} \bar{w}_{mr} \bar{q}_{mr}). \qquad (4.4.71)$$

This equation is general and can be simplified to accommodate the formation of different cloud conditions. For example, for nonprecipitating stratus, $\bar{q}_{mi} = \bar{q}_{mr} = 0$, whereas for cirrus clouds, we may let the IWC mixing ratio for ice crystals be $\bar{q}_{mi}$ and set $\bar{q}_m = \bar{q}_{mr} = 0$. To solve the equation, the eddy terms must also be expressed in terms of mean quantities.

4.5   Parameterization of sources and sinks of moisture

We wish to relate microphysical processes to the bulk quantities in Eqs. (4.4.61)–(4.4.64). Let the mass of a cloud droplet or raindrop, with a radius r, be m, and the droplet size distribution in units of number per volume per length be denoted by $n(r)$, which is confined to a minimum and maximum radius range of (r_1, r_2). Then the bulk LWC for that droplet size distribution may be expressed by

$$\bar{\rho}_m = \bar{\rho}\bar{q}_m = \int_{r_1}^{r_2} n(r)m(r) \, dr, \qquad (4.5.1)$$

where $\bar{q}_m$ is the LWC mixing ratio defined previously, $\bar{\rho}$ is the density of air, which is approximately constant with respect to time, and m is the mass of a cloud droplet, which is equal to $\rho_\ell 4\pi r^3/3$. Variations in droplet size distribution are much smaller than those in mass. Thus the time rate of change of the bulk LWC may be written

$$\frac{d\bar{\rho}_m}{dt} = \bar{\rho}\frac{d\bar{q}_m}{dt} \cong \int_{r_1}^{r_2} n(r) \frac{dm(r)}{dt} \, dr. \qquad (4.5.2)$$

Another important parameter is the bulk terminal velocity for precipitation particles, which is defined as the ratio of the vertical flux of precipitation particles to LWC. The vertical flux of precipitation particles is given by

$$\bar{F}_m = \int_{r_1}^{r_2} n(r)m(r)w(r) \, dr, \qquad (4.5.3)$$

where w denotes the terminal velocity for individual particles. Thus the bulk terminal velocity is

$$\bar{w}_m = \bar{F}_m/\bar{\rho}_m. \qquad (4.5.4)$$

In order to obtain $\bar{q}_m$ and $\bar{w}_m$, we must have the particle size distribution and terminal velocity for individual particles. Below, we present expressions for $n(r)$ and $w(r)$ for raindrops and snowflakes in connection with the derivation of parameterized equations for various microphysical processes.

1. Liquid phase. The cloud droplet size distribution can be approximated from observations (see Section 4.3 for a more detailed discussion). The terminal velocity for cloud droplets with radii less than $\sim 50\,\mu$m is governed by the Stokes law, and is proportional to the square of the droplet radius, viz.,

$$w(r) = k_1 r^2, \qquad r < 50\,\mu\text{m}, \qquad (4.5.5)$$

where the constant of proportionality, $k_1 = 1.19 \times 10^6\,\text{cm}^{-1}\,\text{s}^{-1}$.

Raindrop size distributions have been measured at the surface in terms of the rainfall rate. Based on measurements, Marshall and Palmer (1948) have suggested that droplet size distributions can be fitted by an appropriate negative exponential form, given by

$$n(D) = n_0 e^{-\Lambda D}, \qquad (4.5.6)$$

where the diameter, $D = 2r$, $n(D)\,dD$ represents the number of droplets per unit volume with a diameter between D and $D + dD$ in units of cm. The slope factor Λ

was found to depend only on the rainfall rate $\mathcal{R}$, which is measured in mm/h and is given by $\Lambda = 41\mathcal{R}^{-0.21}$, and the intercept parameter, $n_0 = 0.08\,\mathrm{cm}^{-4}$. Diameter, rather than radius, is conventionally used for precipitation particles that could be nonspherical.

The most comprehensive observational data available on the fall velocity of raindrops have been given by Gunn and Kinzer (1949). Based on the measured data, Liu and Orville (1969) have proposed the following form for the terminal velocity of raindrops:

$$w_r = aD^b(\bar{\rho}_0/\bar{\rho})^{1/2}, \qquad (4.5.7)$$

where $\bar{\rho}_0$ is a reference air density of $1.2 \times 10^{-3}\,\mathrm{g\,cm}^{-3}$, $a = 2115\,\mathrm{cm}^{1-b}\,\mathrm{s}^{-1}$, and $b = 0.8$. The square-root factor, which involves air density, allows for increasing fall speeds with increasing altitude. For raindrops in the diameter interval, $0.12 < D < 0.4\,\mathrm{cm}$, a value of 0.5 for b has also been suggested (Rogers, 1979).

2. Ice phase. The size distributions for snowflakes have been measured at the ground by Gunn and Marshall (1958). Analogous to the Marshall–Palmer distribution for raindrops, they proposed the following exponential form

$$n_i(D) = n_{0i}e^{-\Lambda_i D}, \qquad (4.5.8)$$

where D denotes the diameter of snowflakes. According to the analyses performed by Sekhon and Srivastava (1970), $\Lambda_i = 22.9\mathcal{R}^{-0.45}$ and $n_{0i} = 2.5 \times 10^{-2}\mathcal{R}^{-0.94}$. However, it has been suggested that the intercept parameter n_{0i}, has a constant value of $0.03\,\mathrm{cm}^{-4}$ (Lin et al., 1983).

The terminal velocities for graupel-like, hexagonal snowflakes have been suggested by Locatelli and Hobbs (1974) in the form

$$w_i = cD^d(\bar{\rho}_0/\bar{\rho})^{1/2}, \qquad (4.5.9)$$

with $c = 152.93\,\mathrm{cm}^{1-d}\mathrm{s}^{-1}$ and $d = 0.25$.

In view of the mathematical forms for the preceding particle size distribution and terminal velocity, Eqs. (4.5.1)–(4.5.3) involve the gamma function as follows:

$$\int_0^\infty D^n e^{-\Lambda D}\, dD = \frac{\Gamma(n+1)}{\Lambda^{n+1}}, \qquad (4.5.10)$$

where, without sacrificing the accuracy of the integration, we have set the integration limits at (D_1, D_2) to $(0, \infty)$.

From the definition of the bulk LWC in Eq. (4.5.1), and using the particle size distributions denoted in Eqs. (4.5.6) and (4.5.8), we obtain

$$\bar{\rho}_{mr} = \bar{\rho}\bar{q}_{mr} = \pi n_0 \rho_\ell \Lambda^{-4}, \qquad (4.5.11a)$$

$$\bar{\rho}_{mi} = \bar{\rho}\bar{q}_{mi} = \pi n_{0i} \rho_i \Lambda_i^{-4}, \qquad (4.5.11b)$$

where ρ_ℓ and ρ_i are the density of liquid water and snowflakes, respectively. For practical purposes, $\rho_\ell \cong 1\,\mathrm{g\,cm^{-3}}$ and $\rho_i \cong 0.1\,\mathrm{g\,cm^{-3}}$ (note that for graupels, the density varies from 0.7 to $0.9\,\mathrm{g\,cm^{-3}}$). These two equations relate the LWC mixing ratio to the slope factor. Moreover, using the particle size distribution and terminal velocity denoted above, as well as the expressions in Eq. (4.5.11), we obtain the bulk terminal velocities for raindrops and snowflakes in the form

$$\bar{w}_{mr} = \frac{a}{6}(\bar{\rho}_0/\bar{\rho})^{1/2}\Gamma(4+b)\left[\pi n_0 \rho_\ell/(\bar{\rho}\bar{q}_{mr})\right]^{-b/4}, \qquad (4.5.12a)$$

$$\bar{w}_{mi} = \frac{c}{6}(\bar{\rho}_0/\bar{\rho})^{1/2}\Gamma(4+d)\left[\pi n_0 \rho_i/(\bar{\rho}\bar{q}_{mi})\right]^{-d/4}. \qquad (4.5.12b)$$

Below we shall derive parameterized expressions for condensation, precipitation, evaporation, freezing, melting, deposition, sublimation, and Bergeron processes for potential use in numerical models.

4.5.1 *Condensation*

Based on the steady-state, one-dimensional diffusion equation for water vapor and latent heat transport, the condensation growth equation for a water droplet with a radius r is given by (Pruppacher and Klett, 1978)

$$r\frac{dr}{dt} = \frac{S-1}{\rho_\ell(A+B)}, \qquad (4.5.13)$$

where the ambient saturation ratio $S = \bar{q}/\bar{q}_s(T)$, and A and B are coefficients associated with temperature and pressure, given by

$$A = \frac{L^2}{R_v K T^2}, \qquad B = \frac{R_v T}{\tilde{D}e_s(T)}, \qquad (4.5.14)$$

where L is the latent heat per unit mass for condensation, R_v is the gas constant for water vapor, e_s is the saturation vapor pressure, which is a function of temperature only, and K and $\tilde{D}$ are, respectively, the thermal conductivity for air and diffusion coefficients for water vapor. The values of K and $\tilde{D}$ are listed in Table 4.5. Using K and $\tilde{D}$ that correspond to $0°$ C, we find $A = 5.579 \times 10^2/T^2$ and $B = 2.040 \times 10^4 T/e_s(T)$, both in units of $\mathrm{cm\,s\,g^{-1}}$.

Since the mass of a spherical droplet is $m = \rho_\ell 4\pi r^3/3$, Eq. (4.5.13) can be rewritten in the form

$$\frac{dm(r)}{dt} = \frac{4\pi r}{\bar{q}_s(A+B)}(\bar{q}-\bar{q}_s). \qquad (4.5.15)$$

Substituting this expression into Eq. (4.5.2), we obtain

$$\left(\frac{d\bar{q}_m}{dt}\right)_{\mathrm{CON}} = \frac{\bar{S}_1}{\bar{\rho}} = \frac{1}{\Delta t_1}(\bar{q}-\bar{q}_s), \qquad \bar{q} > \bar{q}_s, \qquad (4.5.16)$$

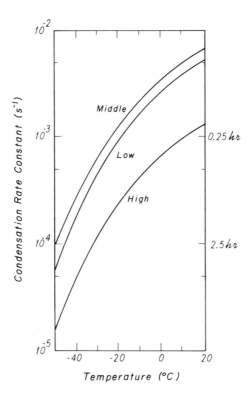

FIG. 4.15   The condensation rate constant, $1/\Delta t_1$, as a function of temperature for three cloud types.

where the reciprocal of the time scale, Δt_1, or the condensation rate constant, for the condensation process is

$$\frac{1}{\Delta t_1} = \frac{4\pi}{\bar{\rho}\bar{q}_s(A + B)} \int_{r_1}^{r_2} n(r)r \, dr. \qquad (4.5.17)$$

Figure 4.15 illustrates $1/\Delta t_1$ as a function of temperature using the observed particle size distributions for stratus, altostratus, and cirrus clouds, discussed in Section 4.3. The condensation rate constants of 10^{-3} and 10^{-4} s^{-1} correspond to time constants of $\sim$0.25 and 2.5 h, respectively. The rate constant increases with temperature. Middle clouds have the largest condensation rate because of their high particle concentration.

4.5.2   *Precipitation*

The precipitation process converts cloud droplets to raindrops. Assume that a collector droplet with radius R has a collision efficiency E with respect to a droplet with radius r. Let the respective terminal velocities for the R and r droplets be

Table 4.5 Values of the diffusion coefficient $\tilde{D}$ and coefficient of thermal conductivity K

$T(°C)$	$K(\mathrm{erg\,cm^{-1}s^{-1}deg^{-1}})$	$\tilde{D}(\mathrm{cm^2\,s^{-1}})$ [a]
−20	2.28×10^3	0.197
−10	2.36×10^3	0.211
0	2.43×10^3	0.226
10	2.50×10^3	0.241
20	2.57×10^3	0.257
30	2.64×10^3	0.273
40	2.70×10^3	0.289

[a] The tabulated values of $\tilde{D}$ are for a pressure of 1000 mb. To obtain $\tilde{D}$ for an arbitrary pressure p(mb), the tabulated values should be multiplied by $(1000/p)$ (from *Smithsonian Meteorological Tables*, 1958).

$w(R)$ and $w(r)$. The total volume of sweeps by the collector droplet per unit time is then $\pi(R + r)^2[w(R) - w(r)]$. In addition, the differential LWC for droplets with a size distribution $n(r)$ is $m(r)n(r)\,dr$. Thus the time rate of change of the total mass for the collector droplet may be expressed by

$$\frac{dm(R)}{dt} = \int_0^R \pi(R + r)^2 \left[w(R) - w(r)\right] E(R,r)m(r)n(r)\,dr. \qquad (4.5.18)$$

For collision to be effective, we may assume that $r \ll R$. Assuming an average collision efficiency of $\bar{E}$ and using the definition of LWC, we have

$$\frac{dm(R)}{dt} \cong \pi R^2 \bar{E} w(R) \bar{\rho}_m. \qquad (4.5.19)$$

Moreover, using Eq. (4.5.2) for the bulk LWC, we find

$$\left(\frac{d\bar{q}_{mr}}{dt}\right)_{\mathrm{PRE}} = \bar{q}_m \bar{E} \int_{R_1}^{R_2} \pi R^2 w(R)n(R)\,dR. \qquad (4.5.20)$$

The initial stage of the conversion of cloud droplets to raindrops is associated with the size distribution and terminal velocity for cloud droplets. Substituting Eq. (4.5.5) for the terminal velocity of cloud droplets into Eq. (4.5.20), we obtain

$$\left(\frac{d\bar{q}_{mr}}{dt}\right)_{\mathrm{PRE}}^{(1)} = \frac{1}{\Delta t_2}\bar{q}_m, \qquad (4.5.21)$$

where the reciprocal time scale is

$$\frac{1}{\Delta t_2} = \pi \bar{E} k_1 \int_{R_1}^{R_2} R^4 n(R)\,dR. \qquad (4.5.22a)$$

We may define a mean radius in the form

$$\bar{R} = \left(\frac{1}{N}\int_{R_1}^{R_2} R^4 n(R)\,dR\right)^{1/4}, \qquad (4.5.22b)$$

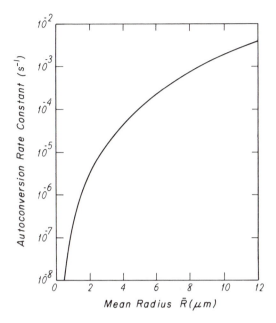

FIG. 4.16 The autoconversion rate constant, $1/\Delta t_2$, as a function of the mean particle radius.

where the total number density is given by

$$N = \int_{R_1}^{R_2} n(R)\,dR. \tag{4.5.22c}$$

Thus,

$$\frac{1}{\Delta t_2} \cong \pi \bar{E} k_1 \bar{R}^4 N. \tag{4.5.22d}$$

In order to have an estimate of the reciprocal time scale for collision processes, referred to as *autoconversion*, we may take $\bar{E} = 1$ and $N = 100\,\text{cm}^{-3}$. Using the constant k_1, given in Eq. (4.5.5), the autoconversion rate, $1/\Delta t_2$, as a function of the mean radius $\bar{R}$ is shown in Fig. 4.16. It is clear that $1/\Delta t_2$ depends significantly on $\bar{R}$. The variation in $1/\Delta t_2$ for $\bar{R}$ from 2 to 10 μm is four orders of magnitude. The dependence of $\bar{R}^4$ on the autoconversion rate is the most important factor in the initiation of precipitation.

The second stage for precipitation processes involves large raindrops. Using Eqs. (4.5.6) and (4.5.7) for the raindrop size distribution and the terminal velocity for individual raindrops, and utilizing the bulk LWC expression in Eq. (4.5.11a), Eq. (4.5.20) may be expressed by

$$\left(\frac{d\bar{q}_{mr}}{dt}\right)^{(2)}_{\text{PRE}} = k_2 \bar{q}_m \bar{q}_{mr}^{(3+b)/4}, \tag{4.5.23}$$

where

$$k_2 = \frac{\pi}{4} \bar{E} a n_0 \Gamma(3+b)(\bar{\rho}_0/\bar{\rho})(\pi n_0 \rho_\ell/\bar{\rho})^{-(3+b)/4}$$
$$\cong 8.02 \times 10^{-5} \bar{\rho}^{0.45} \ (\text{s}^{-1}). \tag{4.5.24}$$

This value is obtained by assuming that the mean collision efficiency, $\bar{E} = 1$. Other constants have been introduced previously.

Combining the aforementioned two processes for precipitation, we have

$$\left(\frac{d\bar{q}_{mr}}{dt}\right)_{\text{PRE}} = \frac{\bar{S}_2}{\bar{\rho}} = \frac{1}{\Delta t_2}\bar{q}_m + k_2 \bar{q}_m \bar{q}_{mr}^{(3+b)/4}. \tag{4.5.25}$$

Modeling studies for cumulus clouds and thunderstorms by Ogura and Takahashi (1971) used only the first term on the right-hand side of Eq. (4.5.25). Kessler (1969) has developed a similar equation, except that the exponent is given by 7/8.

For large-scale cloud modeling, Sundqvist (1978) has proposed the following form for precipitation:

$$\frac{\bar{S}_2}{\bar{\rho}} = \frac{1}{\Delta t_2}\bar{q}_m \left\{1 - \exp\left[-(\bar{q}_m/\bar{q}_{m0})^2\right]\right\}. \tag{4.5.26}$$

where $\bar{q}_{m0}$ is a reference LWC for cloud droplets. This expression takes into account the likelihood that large cloud droplets (i.e., large $\bar{q}_m$) would produce significant precipitation on a shorter time scale, whereas small cloud droplets would require a much longer time scale to precipitate. For stratiform clouds, $\Delta t_2 \cong 10^4$ s and $\bar{q}_{m0} \cong 0.3 \times 10^{-3}$. For convective clouds, $\Delta t_2 \cong 10^3$ s and $\bar{q}_{m0} \cong 10^{-3}$. Sundqvist et al. (1989) have modified the reciprocal time scale, $1/\Delta t_2$, as a function of the rate of precipitation so that the collection by raindrops, described by the second term on the right-hand side of Eq. (4.5.25), may be realistically simulated.

4.5.3 Evaporation of cloud droplets and raindrops

In cases when the mixing ratio of water vapor for the ambient air is smaller than the saturation mixing ratio within the cloud, evaporation may take place. This is the reverse of the condensation process denoted in Eq. (4.5.16). Thus we have

$$\left(\frac{d\bar{q}_m}{dt}\right)_{\text{EVAC}} = \frac{\bar{S}_6}{\bar{\rho}} = -\frac{1}{\Delta t_1}(\bar{q} - \bar{q}_s), \qquad \bar{q} < \bar{q}_s. \tag{4.5.27}$$

The time rate of change in mass for raindrops due to evaporation can be expressed by an equation analogous to Eq. (4.5.15) for condensation. However, a ventilation factor must be included. Based on laboratory experiments, the air flow past a droplet increases the evaporation by a factor that is dependent on the square root of the Reynolds number. It has been suggested that the ventilation factor Ve could best be represented by

$$Ve = \alpha + \beta F Re^{1/2}, \tag{4.5.28}$$

where F is a function of the Reynolds number, Re. For Re > 100, corresponding to raindrops of radius greater than about 0.025 cm, F is very close to one. However, for smaller sizes where Re < 100, F varies significantly with Re. The value of F rises to a peak of about 2.2 when Re $\cong 2.1$ ($r \cong 60\,\mu$m), then drops off to zero when Re < 1.0 ($r < 40\,\mu$m). The coefficients $\alpha = 1$ and $\beta = 0.22$ (Kinzer and Gunn, 1951). Beard and Pruppacher (1971) have performed comprehensive evaporation experiments and statistical fittings, and proposed that $\alpha = 0.78$, $\beta = 0.31$, and $F = S_c^{1/3}$, where the Schmidt number, $S_c = \mu/(\rho\tilde{D})$, with μ being the dynamic viscosity of air and $\tilde{D}$ the diffusivity of water vapor in air. The Reynolds number can be expressed in terms of the droplet diameter D and terminal velocity w in the form Re $= \bar{\rho}wD/\mu$. Using the terminal velocity expression for raindrops in Eq. (4.5.7), we find

$$Re = \frac{\bar{\rho}}{\mu}(\bar{\rho}_0/\bar{\rho})^{1/2}aD^{1+b}. \tag{4.5.29}$$

The ventilation factor may then be expressed by

$$Ve = \alpha + \beta^* D^{(1+b)/2}, \tag{4.5.30a}$$

where

$$\beta^* = \beta S_c^{1/3}(\bar{\rho}/\mu)^{1/2}(\bar{\rho}_0/\bar{\rho})^{1/4}a^{1/2}. \tag{4.5.30b}$$

Taking the ventilation effects into account and using Eq. (4.5.15), the time rate of change in mass for raindrops due to evaporation may be written in the form

$$\frac{dm(D)}{dt} = -\frac{2\pi D}{A+B}\frac{(\bar{q}-\bar{q}_s)}{\bar{q}_s}Ve, \qquad q < q_s. \tag{4.5.31}$$

Substituting this expression into Eq. (4.5.2) for the time rate of change in the bulk LWC mixing ratio, and utilizing the LWC expression in Eq. (4.5.11a), we obtain

$$\left(\frac{d\bar{q}_{mr}}{dt}\right)_{EVA} = \frac{\bar{S}_7}{\bar{\rho}} = -\frac{1}{A+B}\frac{\bar{q}-\bar{q}_s}{\bar{q}_s}\left(\alpha'\bar{q}_{mr}^{1/2} + \beta'\bar{q}_{mr}^{(5+b)/8}\right), \tag{4.5.32}$$

where

$$\alpha' = 2\pi n_0\alpha(\pi\rho_\ell n_0/\bar{\rho})^{-1/2}\bar{\rho}^{-1} \cong 0.78\bar{\rho}^{-0.5}, \tag{4.5.32a}$$

$$\beta' = 2\pi n_0\beta^*\Gamma\left((5+b)/2\right)(\pi\rho_\ell n_0/\bar{\rho})^{-(5+b)/8}\bar{\rho}^{-1}$$
$$\cong 46\bar{\rho}^{-0.358}. \tag{4.5.32b}$$

4.5.4   *Freezing and melting*

The freezing of cloud droplets is a sink term for the LWC of cloud droplets at temperatures below $0°$ C. There appears to be no satisfactory theory to describe the time rate of change of the bulk LWC for the freezing processes involving raindrops. It is generally assumed that freezing takes place when the air temperature is below $0°$ C and that the time rate of increase for the IWC mixing ratio may be written

$$\left(\frac{d\bar{q}_{mi}}{dt}\right)_{\text{FRE}} = \frac{\bar{S}_3}{\bar{\rho}} = \frac{1}{\Delta t_3}\bar{q}_m, \qquad T < 0°\text{ C}. \tag{4.5.33}$$

The estimate of $1/\Delta t_3$ is purely empirical. Values in the range of 0–0.05 s^{-1} have been suggested (Ogura and Takahashi, 1971).

The time rate of melting for snow or hail is a result of conduction and convection of heat to the particle surface, and of diffusion of the latent heat of condensation. It may be expressed by (Mason, 1971)

$$L_f \frac{dm}{dt} = -4\pi r V e \left[K(T - T_0) + L\tilde{D}(\rho_v - \rho_{vs})\right], \tag{4.5.34}$$

where L_f is the latent heat of fusion, T_0 is the surface temperature of ice particles, ρ_v is the environmental vapor density, ρ_{vs} is the saturation vapor density, and other notations have been defined previously. The first and second terms on the right-hand side represent, respectively, the conduction of heat through air and the condensation of heat onto the surface of ice particles. By assuming saturation conditions and a small temperature range of about $10°$ C, we have $\rho_v - \rho_{vs} \cong \chi(T - T_0)$ from the equation of state, where χ is a constant. Thus the time rate of change in mass for melting processes may be written

$$\frac{dm}{dt} = -\frac{2\pi}{L_f}(K + L\tilde{D}\chi)(T - T_0)D\left(\alpha + \beta_1^* D^{(1+d)/2}\right). \tag{4.5.35a}$$

Here, we have used the ventilation factor derived for raindrops in Eq. (4.5.30), except that a and b have been replaced by c and d for ice conditions given in Eq. (4.5.9) and

$$\beta_1^* = \beta S_c^{1/3}(\bar{\rho}/\mu)^{1/2}(\bar{\rho}_0/\bar{\rho})^{1/4}c^{1/2}. \tag{4.5.35b}$$

By substituting Eq. (4.5.35) into Eq. (4.5.2) and using the size distribution for snowflakes in Eq. (4.5.8), we find

$$\left(\frac{d\bar{q}_{mr}}{dt}\right)_{\text{MEL}} = \frac{\bar{S}_5}{\bar{\rho}} = -\frac{2\pi}{L_f}(K + L\tilde{D}\chi)(T - T_0)$$
$$\times \left(\alpha'' \bar{q}_{mi}^{1/2} + \beta'' \bar{q}_{mi}^{(5+d)/8}\right), \tag{4.5.36a}$$

where

$$\alpha'' = \alpha(\pi \rho_i n_{0i}/\bar{\rho})^{-1/2} \cong 1.51 \bar{\rho}^{-0.5}, \tag{4.5.36b}$$

$$\beta'' = \beta_1^* \Gamma\left((5+d)/2\right)(\pi \rho_1 n_{0i}/\bar{\rho})^{-(5+d)/8} \cong 38.89 \bar{\rho}^{-0.427}. \tag{4.5.36c}$$

4.5.5 Deposition and sublimation

The deposition growth of nonspherical ice crystals may be formulated by utilizing the concept of electrostatic current. The growth equation is similar to that for water droplets denoted in Eq. (4.5.15) and may be expressed by

$$\frac{dm}{dt} = 4\pi C \frac{S_i - 1}{\rho_i(A' + B')} Ve, \tag{4.5.37}$$

where C is the capacity of a nonspherical ice particle; the ice saturation ratio, $S_i = \bar{q}/\bar{q}_{is}$, where $\bar{q}_{is}$ is the ice saturation mixing ratio; and the coefficients

$$A' = \frac{L_s^2}{KR_vT^2}, \qquad B' = \frac{R_vT}{\tilde{D}e_{is}(T)}, \tag{4.5.38}$$

where L_s is the latent heat per unit mass for deposition, and e_{is} is the saturation vapor pressure over ice. The ventilation factor for ice particles may be assumed to have the same form as that for raindrops.

Following the same procedure as for the evaporation of raindrops, using the size distribution for snowflakes denoted in Eq. (4.5.8), and assuming that ice particles are spherical, we find

$$\left(\frac{d\bar{q}_{mi}}{dt}\right)_{\text{DEP}} = \frac{\bar{S}_4}{\bar{\rho}} = \frac{2\pi}{A' + B'} \frac{\bar{q} - \bar{q}_{is}}{\bar{q}_{is}} \left(\alpha'' \bar{q}_{mi}^{1/2} + \beta'' \bar{q}_{mi}^{(5+d)/8}\right), \tag{4.5.39}$$

where α'' and β'' have been defined in Subsection 4.5.4.

Sublimation occurs if the air is subsaturated with respect to ice. Using Eq. (4.5.39), we have

$$\left(\frac{d\bar{q}_{mi}}{dt}\right)_{\text{SUB}} = \frac{\bar{S}_8}{\bar{\rho}} = -\frac{2\pi}{A' + B'} \frac{\bar{q} - \bar{q}_{is}}{\bar{q}_{is}}$$
$$\times \left(\alpha'' \bar{q}_{mi}^{1/2} + \beta'' \bar{q}_{mi}^{(5+d)/8}\right), \qquad q < q_{is}. \tag{4.5.40}$$

When the ice particle surface is wet, the sublimation equation will take the same form as Eq. (4.5.40), except that the saturation vapor pressure is with respect to liquid water. Thus,

$$\frac{\bar{S}_9}{\bar{\rho}} = -\frac{2\pi}{A' + B''} \frac{\bar{q} - \bar{q}_s}{\bar{q}_s} \left(\alpha'' \bar{q}_{mi}^{1/2} + \beta'' \bar{q}_{mi}^{(5+d)/8}\right), \tag{4.5.41}$$

where $B'' = R_v T/[\tilde{D}e_s(T)]$. The preceding parameterized equations express the sources and sinks for vapor, cloud LWC, rain LWC, and IWC denoted in Eqs. (4.4.65)–(4.4.68) in terms of the nondimensional mixing ratios.

4.5.6 Bergeron process

When ice crystals and supercooled water droplets coexist in a cloud, ice crystals will grow at the expense of water droplets because of the vapor pressure difference between the two. The processes involving phase changes between liquid water and ice can be described by the second law of thermodynamics. Based on the conservation of the Gibbs function during phase changes, the differential change of the saturation vapor pressure and temperature can be shown to be governed by the Clausius–Clapeyron equation in the form

$$\frac{de_*}{dT} = \frac{L}{R_v} \frac{e_*}{T^2}, \tag{4.5.42}$$

where e_* is the saturation vapor pressure and L is the latent heat involving vapor and water. At the triple point, vapor, ice and water may coexist. Let the vapor pressure at this point be e_0; it follows that the vapor pressure difference between ice and water below the freezing temperature may be expressed by

$$e_s - e_{is} = e_0 \left[\exp\left(\frac{L}{R_v T_0 T}(T - T_0) \right) - \exp\left(\frac{L_s}{R_v T_0 T}(T - T_0) \right) \right], \tag{4.5.43}$$

where e_s and e_{is} denote the saturation vapor pressure over water and ice, respectively, and L_s is the latent heat involving vapor and ice. The depositional growth of ice crystals, referred to as the *Bergeron process*, is effective and may be considered to be the most significant process by which small ice crystals grow to large sizes before triggering collision and coalescence.

The formation of ice crystals on active ice nuclei by heterogeneous nucleation and the growth of ice crystals by deposition of vapor are phenomena that occur primarily at temperatures between 0 and $-40°$ C. Below about $-40°$ C, homogeneous nucleation dominates. The activated ice-forming nuclei may be related to temperature as follows (Koenig and Murray, 1976):

$$N = a_1 \exp\left\{-a_2 \max\left[(T - T_*), T_0\right]\right\}, \tag{4.5.44}$$

where $a_1 = 1$, $a_2 = 0.5758$, $T_0 = 0°$ C, and T_* is a threshold temperature, which may be taken to be $-20°$C. The rate of the formation of ice crystals due to heterogeneous nucleation is given by $m_0 \, dN/dt$, where m_0 is the initial mass of nucleated ice crystals ($\cong 10^{-14}$ kg). The growth rate of an ice crystal is governed by the diffusion process. The time rate of change of mass is proportional to mass itself and may be approximated by

$$\frac{dm}{dt} = a_3 m^{a_4}, \tag{4.5.45}$$

where a_3 and a_4 are empirical constants that are dependent on temperature.

Using a bulk IWC definition similar to that expressed in Eq. (4.5.2), that is, $\bar{\rho}\bar{q}_{mi} \cong Nm$, and including heterogeneous nucleation, we have

$$\left(\frac{d\bar{q}_{mi}}{dt}\right)_{BEG} = m_0 \frac{dN}{dt} + a_3 \bar{\rho}^{a_4-1} N^{1-a_4} \bar{q}_{mi}^{a_4}, \tag{4.5.46}$$

where the activated ice nuclei concentration is given in Eq. (4.5.44).

The preceding cloud microphysical processes should be sufficient for incorporation in large-scale stratiform cloud models. However, for application to the modeling of local convective clouds (see, e.g., Fig. 4.24), it appears that we must consider additional classes of ice particles. Five classes of hydrometeors, including water droplets, ice crystals, snowflakes, raindrops, and graupel/hail have been accounted for in the parameterization of cloud microphysical processes by Lin et al. (1983). Figure 4.17 illustrates the interactions of these microphysical processes, which are grouped as follows. First, autoconversion is associated with ice crystal aggregation to form snowflakes, with graupel formation through snow crystal aggregations, and with precipitation autoconversion. Parameterizations of these processes are similar to those discussed in Subsection 4.5.2, but empirical adjustments on the coefficients may differ somewhat. Second, evaporation/sublimation processes are related to the sublimation of snowflakes and graupel, to the depositional growth of cloud ice, and to the evaporation of raindrops (see Subsections 4.5.3 and 4.5.5 for similar analyses). Third, the accretion process relates to the formation of snowflakes, raindrops, and graupel from water droplets and ice crystals; to the formation of raindrops from ice crystals, water droplets, and snowflakes; and to the formation of graupel from ice crystals, water droplets, snowflakes, and raindrops. In Subsection 4.5.2, we have developed the equation for accretion involving raindrops and cloud droplets. Accretion involving ice and ice, or ice and water, can be parameterized using the same physical principle, except that empirical adjustments must be done for exchange constants. Fourth, freezing/melting processes involve the freezing of raindrops to form graupel and the melting of snowflakes and graupels to form raindrops (see Subsection 4.5.4 for similar discussions). Finally, the Bergeron process is associated with the depositional formation of snowflakes from supercooled water droplets. This important process has been described in some detail in Subsection 4.5.6.

4.6 Transport of heat and moisture by eddies

In Section 4.4, we have developed a set of thermodynamic and dynamic equations for the mean quantities associated with cloud fields. These equations involve the eddy flux transport of momentum, heat, and moisture. From Eqs. (4.4.20), (4.4.27),

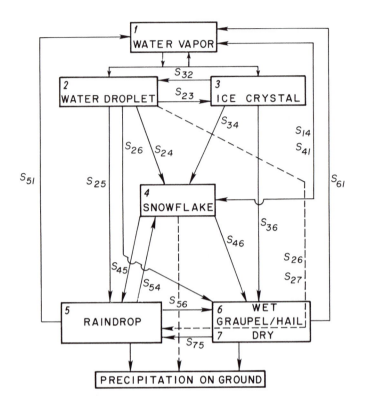

Fig. 4.17 Interactions of cloud microphysical processes involving water vapor, water droplets, ice crystals, snowflakes, raindrops, and graupel/hail. The notations are defined as follows: S_{51}, evaporation of raindrops; S_{61}, sublimation of graupel; S_{14}, depositional growth of snowflakes; S_{41}, sublimation of snowflakes; S_{32}, melting of ice crystals to form water droplets; S_{23}, depositional growth of ice crystals at the expense of water droplets (Bergeron process) plus homogeneous freezing of water droplets to form raindrops; S_{25}, autoconversion of water droplets to form raindrops plus accretion of water droplets by raindrops; S_{27}, accretion of water droplets by graupel; S_{26}, wet growth of graupel; S_{24}, accretion of water droplets by snowflakes ($T < T_0$) plus the Bergeron process; S_{36}, accretion of ice crystals by graupel; S_{34}, accretion of ice crystals by snowflakes plus autoconversion of ice crystal to form snowflakes; S_{45}, melting of snowflakes to form raindrops; S_{54}, accretion of raindrops by snowflakes and ice crystals to produce snowflakes or graupel; S_{75}, melting of graupel to form raindrops, S_{56}, freezing of raindrops to form graupel plus accretion of raindrops by graupel (after Lin et al., 1983).

and (4.4.47) and using Cartesian coordinates, we may decompose these transports into vertical and horizontal components as follows:

$$F^u = -\frac{1}{\bar{\rho}} \frac{\partial}{\partial z} \bar{\rho} \,\overline{w'u'} - \frac{1}{\bar{\rho}} \nabla_2 \cdot \bar{\rho} \,\overline{\mathbf{v}_2' u'}, \qquad (4.6.1)$$

$$F^v = -\frac{1}{\bar{\rho}} \frac{\partial}{\partial z} \bar{\rho} \,\overline{w'v'} - \frac{1}{\bar{\rho}} \nabla_2 \cdot \bar{\rho} \,\overline{\mathbf{v}_2' v'}, \qquad (4.6.2)$$

$$F^T = -\frac{1}{\bar{\rho}C_p}\frac{\partial}{\partial z}\bar{\rho}\overline{w'(C_pT' + gz')}$$

$$-\frac{1}{\bar{\rho}C_p}\nabla_2 \cdot \bar{\rho}\overline{\mathbf{v}_2'(C_pT' + gz')}, \tag{4.6.3}$$

$$F^q = -\frac{1}{\bar{\rho}}\frac{\partial}{\partial z}\bar{\rho}\overline{w'q'} - \frac{1}{\bar{\rho}}\nabla_2 \cdot \bar{\rho}\overline{\mathbf{v}_2'q'}. \tag{4.6.4}$$

Note that Eq. (4.6.3) can also be expressed in terms of the potential temperature denoted in Eq. (4.4.53). Moreover, the eddy transport of LWC from Eq. (4.4.38) may be written

$$F^c = -\frac{1}{\bar{\rho}}\frac{\partial}{\partial z}\bar{\rho}\overline{w'q_m'} - \frac{1}{\bar{\rho}}\nabla_2 \cdot \bar{\rho}\overline{\mathbf{v}_2'q_m'}. \tag{4.6.5}$$

Below, we present the means by which parameterizations of eddy fluxes may be performed. We begin with eddy transports in the vertical.

The motions that facilitate the transfer of heat, humidity, and momentum have a random character. Random three-dimensional velocity fluctuations, which are not predictable in detail, are aptly described by the term *turbulence*. Because of the random nature of turbulent flow, statistical treatments are used in the analyses. It is assumed that all concentrations of the properties of a fluid can be decomposed into slowly varying, average values and rapidly varying fluctuations around the averages. Let χ denote the concentration of the property per unit mass; then the average value at a given point in space may be expressed by Eq. (4.4.1). The fluctuating part is then given by $\chi' = \chi - \bar{\chi}$ and $\overline{\chi'} = 0$.

We now consider the vertical transport of χ. At any instant, the rate of vertical transport is expressed by $\rho w\chi$, where ρ denotes the density of air and w is the vertical velocity. Let $\rho = \bar{\rho} + \rho'$ and $w = \bar{w} + w'$; then consider the net average rate of flux vertical transport in the form

$$F_\chi = \overline{\rho w\chi} = \overline{(\bar{\rho} + \rho')(\bar{w} + w')(\bar{\chi} + \chi')}$$
$$= \bar{\rho}\bar{w}\bar{\chi} + \bar{\rho}\overline{w'\chi'} + \bar{w}\overline{\rho'\chi'} + \bar{\chi}\overline{\rho'w'} + \overline{\rho'w'\chi'}. \tag{4.6.6}$$

If the total mass below a given point is constant in the average, the average net vertical transport of air must be zero. That is, $\overline{\rho w} = \bar{\rho}\bar{w} + \overline{\rho'w'} = 0$. Moreover, $\bar{\rho}\overline{w'\chi'} \gg \bar{\chi}\overline{\rho'w'}$ and $\overline{\rho'w'\chi'}$ so that the turbulent flux of χ in the vertical is approximately given by

$$F_\chi = \overline{\rho w\chi} \approx \bar{\rho}\overline{w'\chi'}, \tag{4.6.7}$$

where χ can be the temperature, humidity, or velocity.

In order to evaluate the $\overline{w'\chi'}$ term, the so-called mixing length model has been utilized to describe the property of turbulent transfer. According to the mixing length theory proposed by Taylor (1915) and Prandtl (1925), the separate masses of a fluid, which are called *eddies*, are supposed to spring into existence in some undefined way and then, after moving unchanged over a certain path length, become indistinguishable from the surrounding fluid. Based on this argument, the lifetime of an eddy is analogous to that of a molecule or atom between successive collisions. The concept of a mixing action in the fluid is implicit in this theory, and the path length of the eddy, analogous to the mean free path of a molecule or atom, is referred to as the *mixing length*. Now consider an eddy that is homogeneous with respect to χ and imagine that this eddy passes through a horizontal reference surface after traversing a distance ℓ' above or below its surface. In accordance with the mixing length hypothesis, a parcel of fluid that is displaced vertically will carry the mean property $\bar{\chi}$ of its original level for a characteristic distance ℓ'. This vertical displacement will then create a turbulent fluctuation in the property χ'. If χ is conserved during the vertical displacement, the magnitude of χ' will depend on ℓ' and the gradient of the mean property such that

$$\chi' = -\ell' \frac{\partial \bar{\chi}}{\partial z}, \tag{4.6.8}$$

where ℓ' is positive when the eddy moves from below and negative when it comes from above. Inserting Eq. (4.6.8) into Eq. (4.6.7), we obtain

$$F_\chi = -\bar{\rho} \overline{w'\ell'} \frac{\partial \bar{\chi}}{\partial z}. \tag{4.6.9}$$

If we define a turbulent transfer coefficient in the form

$$K_\chi = \overline{w'\ell'}, \tag{4.6.10}$$

then the eddy flux of the property may be written

$$F_\chi = \bar{\rho} \overline{w'\chi'} = -\bar{\rho} K_\chi \frac{\partial \bar{\chi}}{\partial z}. \tag{4.6.11}$$

In order to estimate w' in Eq. (4.6.10) in terms of mean fields, the vertical stability of the atmosphere is assumed to be nearly neutral (adiabatic lapse rate) so that buoyancy effects are small. Under this condition, the horizontal scale of the eddies should be comparable to the vertical scale; that is $|w'| \approx |u'|$. Thus, from Eq. (4.6.8), we may set

$$w' = \ell' \left| \frac{\partial \bar{u}}{\partial z} \right|. \tag{4.6.12}$$

The absolute value sign indicates that w' is always positive. The turbulent transfer coefficient defined in Eq. (4.6.10) is then given by

$$K_\chi = \overline{\ell'^2} \left| \frac{\partial \bar{u}}{\partial z} \right|. \tag{4.6.13}$$

We now require an assumption concerning $\overline{\ell'^2}$. Very close to the surface, eddies are restricted in their vertical motions by the surface. Higher above the surface, eddies have more freedom in the vertical direction. This led Prandtl to postulate that $\ell' \approx z$.

In considering the turbulent transfer of the horizontal velocity component, u, the eddy momentum flux, after using Eqs. (4.6.11) and (4.6.13), may be expressed by

$$F_m = \bar{\rho}\,\overline{w'u'} = -\bar{\rho}K_m\frac{\partial\bar{u}}{\partial z} = -\bar{\rho}\overline{\ell'^2}\left(\frac{\partial\bar{u}}{\partial z}\right)^2. \tag{4.6.14}$$

Define the friction velocity, u_*, in terms of the von Karman constant, k, in the form

$$u_* = \left|\frac{F_m}{\bar{\rho}}\right|_{z=0}^{1/2} = kz\frac{\partial\bar{u}}{\partial z}. \tag{4.6.15}$$

The frictional velocity is introduced as a convenient scaling velocity. Thus we find

$$\bar{u} = \frac{u_*}{k}\,\ell n\,\frac{z}{z_0}, \tag{4.6.16}$$

which specifies that $\bar{u}$ vanishes at a height z_0, the so-called *roughness length*, above the surface. Equation (4.6.16) represents the logarithmic velocity profile that appears to be valid under conditions of horizontal uniformity, steady state, and neutral static stability. The von Karman constant has been determined in wind tunnels as well as in the atmosphere and is estimated to be in the range from 0.35 to 0.4. The roughness length z_0 is a variable that depends on surface characteristics, and varies from 10^{-5} m for smooth surfaces to 0.1 m for very rough terrain (Fleagle and Businger, 1980). From the preceding analysis, the transfer coefficient for momentum, referred to as the *eddy viscosity*, may be expressed by

$$K_m = k^2 z^2\frac{\partial\bar{u}}{\partial z} = kzu_*. \tag{4.6.17}$$

In reference to Eq. (4.6.13), we have $\overline{\ell'^2} = (kz)^2$. The vertical eddy momentum flux is a function of static stability in the boundary layer. The manner in which it can be parameterized depends on the closure scheme imposed.

The sensible heat flux in the vertical may be written

$$F_h = \bar{\rho}C_p\overline{w'T'} + \bar{\rho}g\,\overline{w'z'}. \tag{4.6.18}$$

Temperature is not a conserved parameter. Variations of the temperature in the vertical direction within a parcel are intimately coupled with changes in the potential energy of the parcel, gz. The latent heat flux may be defined in the form

$$F_q = \bar{\rho}L\overline{w'q'}. \tag{4.6.19}$$

On the basis of the mixing length concept introduced previously, Eqs. (4.6.18) and (4.6.19) may be expressed by

$$F_h = -\bar{\rho} C_p K_h \left(\frac{\partial \bar{T}}{\partial z} + \gamma_d \right), \qquad (4.6.20)$$

$$F_q = -\bar{\rho} L K_q \frac{\partial \bar{q}}{\partial z}. \qquad (4.6.21)$$

where K_h and K_q are, respectively, referred to as the *eddy thermal diffusion coefficient* and *eddy diffusion coefficient*, and γ_d is the dry adiabatic lapse rate. There is substantial experimental evidence that the profiles of potential temperature and specific humidity are similar over land. However, over water, additional experimental information is needed. For practical applications, it is generally assumed that $K_q \cong K_h$.

We shall now briefly describe the manner in which K_m and K_h may be estimated in realistic atmospheres. The air near the earth's surface is generally either statically stable due to temperature inversions or statically unstable due to superadiabatic lapse rates. These conditions are referred to as *diabatic*. Under diabatic conditions, Obukhov (1946) has postulated that the vertical wind shear in the surface layer, denoted in Eq. (4.6.15), could be modified so that

$$\frac{\partial \bar{u}}{\partial z} = \frac{u_*}{kz} \phi_m(\xi), \qquad (4.6.22)$$

where ϕ_m is an empirical function referred to as the *dimensionless wind shear*, which is a function a dimensionless length, $\xi = z/L$, and the Obukhov length is defined by

$$L = \frac{\bar{T}}{gk} \frac{u_*^2}{\overline{w'\Theta'}/u_*}. \qquad (4.6.23)$$

This length is related to the ratio of the generation of mechanical turbulence by eddies to the production of turbulent kinetic energy by buoyancy forces. The Obukhov length may be interpreted as the height at which mechanical and thermal productions of turbulence are of equal importance. Based on Eqs. (4.6.17) and (4.6.22), the eddy viscosity for diabatic conditions may be expressed by

$$K_m = kzu_*/\phi_m. \qquad (4.6.24)$$

It is now necessary to find the dimensionless wind shear, ϕ_m. Under diabatic conditions, the buoyancy force per unit mass acting on an eddy, whose density differs from its surroundings by ρ', is $(-\rho'/\bar{\rho})g$. If the eddy has traveled a vertical length ℓ' and arrived at the reference height, then the buoyancy force has undertaken specific work, given by $(-\rho'/\bar{\rho})g\ell'$. The energy associated with the work transfers directly to the vertical component of turbulent velocity. For the eddy momentum flux, a modification of Eq. (4.6.14) may be written in the form

$$\overline{w'u'} \propto \overline{\ell'^2} \left(\frac{\partial \bar{u}}{\partial z}\right)^2 - \alpha' \frac{g}{\bar{\rho}} \overline{\rho'\ell'}, \tag{4.6.25}$$

where α' is an empirical coefficient relating the work done by the buoyancy force to that done by the vertical wind shear. Based on the Boussinesq approximation, we can show that $-\rho'/\bar{\rho} \cong \Theta'/\bar{\Theta} \cong T'/\bar{T}$. Using the mixing length theory, $\Theta' = -\ell' \, \partial\bar{\Theta}/\partial z$. Considering $k_m^2 \propto \overline{w'\ell'u'\ell'}$, it can be shown that

$$\phi_m = (1 - \alpha Ri)^{-1/4}, \tag{4.6.26}$$

where α replaces α' to account for the statistical operations, and the gradient Richardson number is

$$Ri = \frac{g}{\bar{T}} \frac{\partial\bar{\Theta}/\partial z}{(\partial\bar{u}/\partial z)^2}. \tag{4.6.27}$$

This number is a measure of the stability of the air and is derived based on dimensional argument. In the boundary layer, $\partial\bar{\Theta}/\partial z \cong \partial\bar{T}/\partial z + \gamma_d$, and $\bar{T} \cong \bar{\Theta}$. If $Ri = 0$, $\phi_m = 1$, implying adiabatic conditions. It is still necessary to determine α in order to obtain ϕ_m. In experiments where both the profiles and fluxes are measured, ϕ_m and Ri from Eqs. (4.6.22) and (4.6.27) can be evaluated and the empirical coefficient α can be determined.

In a manner similar to Eq. (4.6.22), the eddy thermal diffusion coefficient may be expressed by

$$K_h = kzu_*/\phi_h, \tag{4.6.28}$$

where ϕ_h represents modifications to the temperature profile in diabatic conditions. From Eqs. (4.6.24) and (4.6.28), we have

$$\frac{K_h}{K_m} = \frac{\phi_m}{\phi_h}. \tag{4.6.29}$$

Businger et al. (1971) have carried out comprehensive measurements of heat and momentum fluxes for a wide range of stability conditions, and suggest the following expressions for ϕ_m and ϕ_h:

$$\phi_m = \begin{cases} (1 - 15\xi)^{-1/4}, & \text{unstable } (\xi < 0) \\ (1 + 4.7\xi), & \text{stable } (\xi > 0) \end{cases} \tag{4.6.30}$$

$$\phi_h = \begin{cases} 0.74(1 - 9\xi)^{-1/2}, & \text{unstable } (\xi < 0) \\ (0.74 + 4.7\xi), & \text{stable } (\xi > 0). \end{cases} \tag{4.6.31}$$

Graphic illustrations of these relationships are displayed in Fig. 4.18. Using Eq. (4.6.23) and noting $\overline{w'\Theta'} = -K_h \partial\bar{\Theta}/\partial z$, the dimensionless height ξ can be related to the Richardson number in the form

$$\xi = \frac{z}{L} = Ri \frac{\phi_m^2}{\phi_h}. \tag{4.6.32}$$

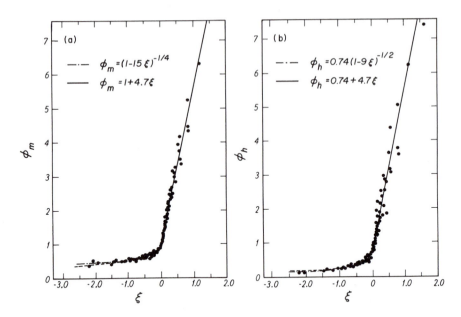

FIG. 4.18 Experimental results of (a) dimensionless wind shear ϕ_m, defined in Eq. (4.6.22), and (b) temperature gradients ϕ_h, defined in Eq. (4.6.28) as functions of a dimensionless length $\xi = z/L$, where L is the Obukhov length (after Businger et al., 1971).

From Eqs. (4.6.30) and (4.6.31), a unique relationship between ξ and Ri can be established. Based on the observational data, the frictional velocity, defined in Eq. (4.6.16) for neutral conditions, may also be modified to include diabatic effects as follows:

$$u_* = k\bar{u} / \left[\ell n(z/z_0) - \psi_m(\xi) \right], \qquad (4.6.33)$$

where ψ_m is given in the form (Businger, 1973)

$$\psi_m = \begin{cases} 2\,\ell n[(1 + \phi_m^{-1})/2] + \ell n[(1 + \phi_m^{-2})/2] - 2\tan^{-1}(\phi_m^{-1} + \pi/2), & \xi < 0 \\ -4.7\xi, & \xi > 0. \end{cases}$$
$$(4.6.34)$$

Using the preceding equations, K_m and K_h may be computed once the mean wind shear, $\partial \bar{u}/\partial z$, and potential temperature gradient, $\partial \bar{\Theta}/\partial z$, are given.

As a first order approximation, the dimensionless height may be expressed by

$$\xi \propto \left(\frac{kz}{u_*} \right)^2 \frac{g}{\bar{T}} \frac{\partial \bar{\Theta}}{\partial z}. \qquad (4.6.35)$$

Using Eq. (4.6.31) for unstable conditions and noting that $|-9\xi| \geq 1$, the eddy thermal diffusion coefficient from Eq. (4.6.28) may be written

$$K_h \propto (kz)^2 \left(\frac{g}{\bar{T}} \frac{\partial \bar{\Theta}}{\partial z} \right)^{1/2}. \qquad (4.6.36)$$

This is the similarity equation for the eddy thermal diffusion coefficient derived by Priestly (1959).

The preceding discussion outlines the so-called first-order closure method. This method is based on the mixing length, or K theory. For applications to diabatic atmospheres, the theory depends heavily on the empirical functions derived from experiments.

Now consider a three-dimensional eddy. Applying the mixing length theory, the eddy flux transport of momentum may be written in Cartesian coordinates as follows:

$$F^u = -\frac{1}{\bar{\rho}} \nabla \cdot \bar{\rho} \overline{\mathbf{v}'u'}$$
$$= \frac{1}{\bar{\rho}} \left(\frac{\partial}{\partial x} \bar{\rho} K_{mx} \frac{\partial \bar{u}}{\partial x} + \frac{\partial}{\partial y} \bar{\rho} K_{my} \frac{\partial \bar{u}}{\partial y} + \frac{\partial}{\partial z} \bar{\rho} K_{mz} \frac{\partial \bar{u}}{\partial z} \right). \quad (4.6.37)$$

Similar expressions may be written for heat and moisture transports. If we assume that the exchange processes in space are about the same in three directions, we may set $K_{mx} = K_{my} = K_{mz} = K_m$. In this case, we have

$$F^u \cong \frac{1}{\bar{\rho}} \nabla \cdot \bar{\rho} K_m \nabla \bar{u}. \quad (4.6.38)$$

Further, if the eddy viscosity is assumed to be constant, we have

$$F^u \cong K_m \nabla^2 \bar{u}. \quad (4.6.39a)$$

This is the diffusion approximation. Using this approximation, the eddy transport of heat, moisture, and cloud LWC may be expressed by

$$F^\Theta \cong K_h \nabla^2 \bar{\Theta}, \quad (4.6.39b)$$
$$F^q \cong K_q \nabla^2 \bar{q}, \quad (4.6.39c)$$
$$F^c \cong K_c \nabla^2 \bar{q}_m. \quad (4.6.39d)$$

Estimates of the exchange coefficients rely heavily on empirical approaches.

More recently, various turbulence closure models, the so-called second-order closure methods, have been developed with increasing sophistication. Basic equations for eddy fluxes in the models are derived from the original governing equations (see, e.g., Mellor and Yamada, 1974). A third-order closure has also been developed in connection with the development of a cumulus cloud model (Krueger, 1988). Several cloud modeling efforts that will be discussed later use the second-order closure scheme to compute eddy fluxes.

4.7 Stratiform cloud models

We shall confine our present discussion to the modeling of two types of stratiform clouds that are critically important to the question of clouds in climate: low-level stratus and high-level cirrus. Some of the fundamentals presented herein may also be applicable to middle-level water and/or mixed clouds.

Understanding of the formation and dissipation of stratiform clouds requires appropriate models. We first describe the basic equations for velocity, potential temperature, specific humidity, LWC mixing ratio, and continuity that have been introduced in Eqs. (4.4.21), (4.4.45), (4.4.53), (4.4.27), (4.4.38), and (4.4.9), respectively. It appears adequate to consider two-dimensional space and utilize the Boussinesq approximation for the modeling of stratiform clouds. Based on this approximation, air density is treated as a constant, except when it is coupled with gravity in the buoyancy term of the vertical momentum equation. This is valid for motions in which the vertical scale is less than the atmospheric scale height ($\sim$8 km). Under the preceding conditions, the governing equations may be written in the forms

$$\frac{\partial \bar{v}}{\partial t} = -\bar{v}\frac{\partial \bar{v}}{\partial y} - \bar{w}\frac{\partial \bar{v}}{\partial z} - \frac{1}{\bar{\rho}}\frac{\partial \bar{p}}{\partial y} - \frac{\partial}{\partial y}\overline{v'^2} - \frac{\partial}{\partial z}\overline{v'w'}, \tag{4.7.1}$$

$$\frac{\partial \bar{w}}{\partial t} = -\bar{v}\frac{\partial \bar{w}}{\partial y} - \bar{w}\frac{\partial \bar{w}}{\partial z} - \frac{1}{\bar{\rho}}\frac{\partial \bar{p}}{\partial z} + gB - \frac{\partial}{\partial y}\overline{v'w'} - \frac{\partial}{\partial z}\overline{w'^2}, \tag{4.7.2}$$

$$\frac{\partial \bar{E}}{\partial t} = -\bar{v}\frac{\partial \bar{E}}{\partial y} - \bar{w}\frac{\partial \bar{E}}{\partial z} - \left(\frac{\partial}{\partial y}\overline{v'E'} + \frac{\partial}{\partial z}\overline{w'E'}\right) + \bar{Q}_R, \tag{4.7.3a}$$

$$\frac{\partial \bar{\Theta}}{\partial t} = -\bar{v}\frac{\partial \bar{\Theta}}{\partial y} - \bar{w}\frac{\partial \bar{\Theta}}{\partial z} - \left(\frac{\partial}{\partial y}\overline{v'\Theta'} + \frac{\partial}{\partial z}\overline{w'\Theta'}\right)$$
$$+ \frac{\pi^{-1}}{C_p}(\bar{Q} + \bar{Q}_R), \tag{4.7.3b}$$

$$\frac{\partial \bar{q}}{\partial t} = -\bar{v}\frac{\partial \bar{q}}{\partial y} - \bar{w}\frac{\partial \bar{q}}{\partial z} - \left(\frac{\partial}{\partial y}\overline{v'q'} + \frac{\partial}{\partial z}\overline{w'q'}\right) + \frac{\bar{S}_v}{\bar{\rho}}, \tag{4.7.4}$$

$$\frac{\partial \bar{q}_m}{\partial t} = -\bar{v}\frac{\partial \bar{q}_m}{\partial y} - \bar{w}\frac{\partial \bar{q}_m}{\partial z} - \frac{\partial}{\partial z}(\bar{w}_m\bar{q}_m)$$
$$- \left(\frac{\partial}{\partial y}\overline{v'q'_m} + \frac{\partial}{\partial z}\overline{w'q'_m}\right) + \frac{\bar{S}_m}{\bar{\rho}}, \tag{4.7.5}$$

$$0 = \frac{\partial \bar{v}}{\partial y} + \frac{\partial \bar{w}}{\partial z}, \tag{4.7.6}$$

where $B = \rho/\bar{\rho}$, $\bar{\rho}$ in this case represents a reference air density, and $\pi^{-1} = (p_0/\bar{p})^\kappa$, with $p_0 = 1000$ mb. The buoyancy term can be represented by the potential temperature. Including the contribution of moisture, this term may be expressed by

$$B = \Theta/\bar{\Theta} + c\bar{q} - \bar{q}_m, \tag{4.7.7}$$

where the coefficient, $c \cong 0.61 = 1/\varepsilon - 1$, with $\varepsilon = 0.622$, which is the ratio of the molecular weight of water vapor to that of dry air. Inclusion of the humidity increases the buoyancy force. However, if condensation occurs, this force is reduced. The effect of humidity may be absorbed in the virtual temperature, defined by $T_v = (1 + 0.61\bar{q})\bar{T}$. In Eqs. (4.7.3a, b), the thermodynamic equation is written in terms of both moist static energy and potential temperature.

The wind fields are normally solved by introducing the stream function ψ in the form

$$\bar{v} = -\frac{\partial\bar{\psi}}{\partial z}, \qquad w = \frac{\partial\bar{\psi}}{\partial y}. \tag{4.7.8}$$

If we define the horizontal zonal component of vorticity in the form

$$\xi = \frac{\partial\bar{w}}{\partial y} - \frac{\partial\bar{v}}{\partial z} = \nabla^2\bar{\psi}, \tag{4.7.9}$$

where the Laplacian operator, $\nabla^2 = \partial^2/\partial y^2 + \partial^2/\partial z^2$, Eqs. (4.7.1) and (4.7.2) can be combined to give

$$\frac{\partial\bar{\xi}}{\partial t} = J(\bar{\xi}, \bar{\psi}) + g\frac{\partial B}{\partial y} + \left(\frac{\partial^2}{\partial y\,\partial z}\overline{v'^2} - \frac{\partial^2}{\partial y^2}\overline{v'w'}\right)$$
$$+ \left(\frac{\partial^2}{\partial z^2}\overline{w'v'} - \frac{\partial^2}{\partial y\partial z}\overline{w'^2}\right), \tag{4.7.10}$$

where the Jacobian is

$$J(\bar{\xi}, \bar{\psi}) = \frac{\partial\bar{\xi}}{\partial y}\frac{\partial\bar{\psi}}{\partial z} - \frac{\partial\bar{\xi}}{\partial z}\frac{\partial\bar{\psi}}{\partial y}. \tag{4.7.11}$$

Moreover, referring to Eq. (4.4.71), Eqs. (4.7.4) and (4.7.5) can be combined to give

$$\frac{\partial\bar{q}_t}{\partial t} = -\bar{v}\frac{\partial\bar{q}_t}{\partial y} - \bar{w}\frac{\partial\bar{q}_t}{\partial z}$$
$$- \left(\frac{\partial}{\partial y}\overline{v'q_t'} + \frac{\partial}{\partial z}\overline{w'q_t'}\right) - \frac{\partial}{\partial z}(\bar{w}_m\bar{q}_m). \tag{4.7.12}$$

The governing equations may be solved if the initial and boundary conditions are given. The source and sink terms for moisture, fall velocity of cloud particles, and eddy terms must be expressed in terms of the basic model parameters.

4.7.1   Stratus clouds

1. Marine Stratus. As discussed in Subsection 4.1.2, during the Northern Hemisphere summer, extended stratus clouds occur and persist over the western coast of North America. They appear in the southeastern part of the Pacific marine subtropical highs and are associated with large-scale subsidence and cold sea surface temperatures. In addition, marine stratus and stratocumulus clouds also cover a large portion of the eastern Atlantic Ocean and a small portion of the western

Indian Ocean. It is clear that extended marine stratiform clouds are significant climatic features. These clouds are characterized by a nearly adiabatic, well-mixed subcloud layer and a saturation adiabatic layer capped by a strong temperature inversion. Lilly (1968) has constructed a model of the cloud-topped boundary layers to understand the stability of persistent stratocumulus. Evaporative cooling accompanying entrainment at cloud top could lead to cloud breakup. Since the appearance of Lilly's study, many modeling studies have been undertaken to understand the role of radiative cooling, entrainment, and other dynamic factors in the behavior and structure of marine stratiform clouds. These studies include simple mixed-layer models, high-order closure models, and large-eddy simulation models. A review of these models has been given by Cotton and Anthes (1989).

Moeng and Arakawa (1980) have developed a two-dimensional model using a second order turbulence closure scheme to investigate the physical processes involved in the formation and dissipation of stratus clouds off the California coast. The model's vertical extent is from the surface to 3 km, while the horizontal extent is 1000 km. The horizontal axis is oriented along the observed surface wind direction from northeast to southwest. Because the horizontal grid size is generally much larger than the vertical grid size, horizontal turbulent diffusion can be disregarded. The fall velocity of cloud droplets in stratus clouds ($< 10\,\mu$m) is small and can also be neglected, as is the diffusion term, $\partial^2 \overline{w'^2}/\partial y\,\partial z$. The radiative heating term, $\bar{\rho}Q_R$, can be expressed by $-\partial\bar{F}/\partial z$. Thus, from Eqs. (4.7.10), (4.7.3a), (4.7.12), and (4.7.6), the model equations governing the mean vorticity, moist static energy and total moisture fields may be written in the forms

$$\frac{\partial\bar{\xi}}{\partial t} = J(\bar{\xi}, \bar{\psi}) + g\,\frac{\partial B}{\partial y} + \frac{\partial^2}{\partial z^2}\,\overline{v'w'}, \tag{4.7.13}$$

$$\frac{\partial\bar{E}}{\partial t} = -\bar{v}\,\frac{\partial\bar{E}}{\partial y} - \bar{w}\,\frac{\partial\bar{E}}{\partial z} - \frac{\partial}{\partial z}\,\overline{w'E'} - \frac{1}{\bar{\rho}}\,\frac{\partial\bar{F}}{\partial z}, \tag{4.7.14}$$

$$\frac{\partial\bar{q}_t}{\partial t} = -\bar{v}\,\frac{\partial\bar{q}_t}{\partial y} - \bar{w}\,\frac{\partial\bar{q}_t}{\partial z} - \frac{\partial}{\partial z}\,\overline{w'q'_t}, \tag{4.7.15}$$

where $\bar{q}_t = \bar{q} + \bar{q}_m$, is the sum of the water vapor and LWC mixing ratios. The B term in Eq. (4.7.7) may be expressed in terms of the virtual dry static energy in the form

$$B \cong \left[(C_pT + gz) + C_p\bar{T}(0.61\bar{q} - \bar{q}_m)\right]/C_p\bar{T}. \tag{4.7.16}$$

Moeng and Arakawa used a broadband emissivity approach and a second-order turbulence closure scheme that included water vapor saturation in the stratus model. Figure 4.19 shows a schematic diagram derived from model results of the breakup process for stratus clouds. It is inferred that breakup of the cloud layer occurs when entrained, unsaturated air from the inversion layer is cooled by evaporation and becomes denser than the surrounding cloud air. Mixed parcels that are negatively buoyant can be produced with respect to unmixed clouds.

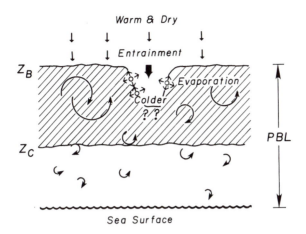

FIG. 4.19   A schematic diagram illustrating the mechanisms involved in the instability process for stratus clouds (after Moeng and Arakawa, 1980).

This may occur when cloud air is advected into a region of relatively warm sea surface temperatures and relatively weak large-scale subsidence. The instability destroys the horizontally uniform stratus cloud layer and reduces the cloud cover. Later studies suggest that there is only a weak correlation between observed cloud breakup and the existence of conditions that could generate negatively buoyant mixed parcels (Kuo and Schubert, 1988). The search for additional mechanisms involving stratocumulus transitions is a subject of contemporary research that requires the support of observations.

2. Arctic stratus. The persistence of extensive layers of stratiform clouds over the polar regions during the melting season has been illustrated in Subsection 4.1.3. Using a moist Boussinesq approximation and a fixed pressure field, Herman and Goody (1976) have developed a simple, equivalent one-dimensional model to understand the physical processes involved in the formation and dissipation of arctic stratus. The basic model equations for the equivalent potential temperature, total water mixing ratio, and horizontal velocities in a semi-two-dimensional context may be expressed in the forms

$$\frac{\delta \Theta_e}{\delta t} = \frac{\bar{\Theta}_e}{C_p \bar{T}} \bar{Q}_R - \frac{\partial}{\partial z} \overline{w' \Theta'_e}, \tag{4.7.17}$$

$$\frac{\delta \bar{q}_t}{\delta t} = -\frac{\partial}{\partial z} \overline{w' q'_t} - \bar{w}_m \frac{\partial \bar{q}_m}{\partial z}, \tag{4.7.18}$$

$$\frac{\delta \bar{u}}{\delta t} = f\bar{v} - \frac{\partial}{\partial z} \overline{w' u'}, \tag{4.7.19}$$

$$\frac{\delta \bar{v}}{\delta t} = -f\bar{u} - \frac{\partial}{\partial z} \overline{w' v'}, \tag{4.7.20}$$

where Θ_e is the equivalent potential temperature defined in Eq. (4.4.56) and the downstream derivative, $\delta/\delta t = U_0 \, \partial/\partial x$, relates the change in time to the change experienced over the corresponding travel distance x moving with a constant geostrophic current U_0. Herman and Goody used the eddy diffusion coefficient for momentum, heat, and moisture, based on the mixing length approach for vertical eddy transport terms. The solar and thermal ir radiative transfer parameterization schemes were based on the two-stream approximation and broadband emissivity approach. The surface temperature is assumed to be constant ($0°$ C), and the diurnal cycle is neglected. This model was used to explain the major observed properties of summertime Arctic stratus, including the growth of the cloud top and the development of layers. Stratus clouds are formed when moist air from the surrounding land masses is modified as it passes over melting pack-ice. The commonly observed layering is a result of the trapping of solar radiation inside a cloud that is opaque to thermal radiation.

Herman and Goody used the simplified model to examine the radiative and turbulent processes within arctic stratus. Figure 4.20 illustrates the regions associated with the dominant terms in Eq. (4.7.17) for the local rate of change in equivalent potential temperature. The terms heating and cooling refer to the divergence of the turbulent flux under stable and unstable conditions, respectively. The solar and ir terms refer to the divergence of net solar and thermal ir fluxes. The ir cooling zone in the upper left-hand corner represents the cooling of upper regions of the boundary layer by direct ir flux exchange with space. Diffusive and ir cooling lower the boundary layer temperature in about a day. Equilibrium between the ir boundary exchange and diffusion is rapidly established close to the surface. Once cloud droplets have formed, the absorption properties of the droplets become important in the radiative region. The upper cloud layer becomes unstable due to the intense ir loss from the cloud top, and a radiative–convective balance is established in the cloud layer. At the cloud top, convective warming is approximately equal to radiative cooling, while in the cloud interior convective cooling balances the heating due to solar absorption. A region of intense radiative heating forms within the cloud through a greenhouse mechanism. After about 3 days, however, the heating and cooling terms become small, and several steady-state zones are established. Two radiative equilibrium zones are located between the radiative–diffusive and radiative–convective zones.

Using the data collected during the Arctic Stratus Experiment in June, 1980, Curry (1986) has studied the interactions involving turbulence, radiation and cloud microphysics in arctic stratus clouds. These clouds are not maintained by surface moisture fluxes because of the static stability of the region. Convection within the clouds is promoted by the ir radiative cooling, in line with the theoretical findings by Herman and Goody. Turbulent entrainment is confined to above ~ 50 m below the mean cloud top and does not directly affect the evolution of the droplet spectra. LWC production is closely related to radiative cooling within the cloud, as indicated by observed dynamic and water fluxes as well as cloud microphysical parameters.

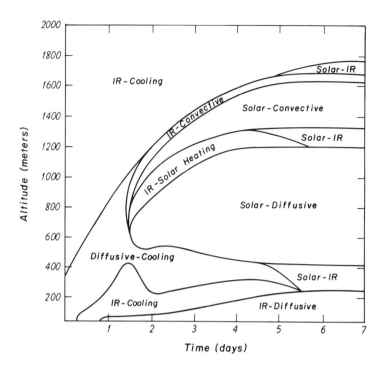

Fig. 4.20 Components of the radiative and turbulent regions computed for a case involving warm air flowing over a colder surface (after Herman and Goody, 1976).

4.7.2 *Cirrus clouds*

In Subsection 4.1.1, we have illustrated that cirrus cloud systems are normally associated with synoptic and mesoscale disturbances and are partly related to deep cumulus outflows. Day-to-day experience reveals that cirrus clouds have a fibrous appearance. Most extensive cirrus sheets are associated with frontal systems, or are formed as widespread anvil clouds remain after cumulonimbus has dissipated. The most pronounced cirrus bands frequently occur in southwesterly flows adjacent to the warm side of the jet stream core (Conover, 1960). This cloud region stretches over a narrow area 45–65 km wide and several hundred kilometers long. Cirrus bands often parallel certain high-level temperature discontinuities.

Ludlam (1948) first attempted to explain the formation of cirrus uncinus based on the principle of thermodynamics. Since ice clouds are generally formed at saturation humidities, with respect to water, and are therefore at considerable supersaturation with respect to ice, ice particles grow rapidly on a relatively sparse number of effective nuclei and achieve high fall velocities. Because of wind shear, the typical form of cirrus uncinus appears as a trail of precipitation. Furthermore, Ludlam (1956) has described a qualitative mechanism for the continuous generation of ice crystals in localized regions. Crystal growth in ice crystal trails warms the surrounding air by releasing significant amounts of latent heat. Convective bubbles of warmed air then rise out of the trails and form fresh clouds after ascending a few hundred meters adjacent to the head of the trail.

The mechanisms for the formation of cirrus uncinus clouds in terms of their shape and extent have been investigated by Yagi et al. (1968). The movement and elongation of cirrus uncinus are due to the prevailing vertical wind shear at the cloud level. The horizontal component of comma-shaped cirrus could not be explained without considering the temporal and spatial variations in the vertical wind shear at the cloud level. Harimaya (1968) has constructed a microphysical model, coupled with assumed horizontal winds, to compute the shape of cirrus uncinus. The resulting computations suggest that the shape of cirrus uncinus depends primarily on the mass of ice particles and the vertical wind shear. Although this model did not include an interactive dynamic formation, it was, nevertheless, a pioneering modeling attempt for cirrus clouds. Yagi (1969) has further examined the detailed synoptic conditions at the cirrus cloud level. At the cirrus cloud top, there is a layer with a dry adiabatic lapse rate and vertical wind shear, while above the cloud, there is a stable layer that prevents further cloud growth. In the case of cirrus uncinus, the trail occurs in a stable layer, whereas in the case of spissatus and fiblatus clouds, this lower layer is quite unstable. It appears that the static stability in the lower cloud layer plays a significant role in the shape and extent of cirrus clouds.

Based on comprehensive aircraft measurements of the temperatures, horizontal wind velocities, and particle spectra at different altitudes, Heymsfield (1975b) has suggested several plausible physical mechanisms for the formation of cirrus uncinus clouds. For cirrus uncinus oriented perpendicular to the direction of the winds, there appears to be a layer of lifting along which convective cells develop. For isolated cirrus uncinus, a stable layer seems to exist below the cirrus head region, where wave motions generate disturbances from which convection occurs. Two different mechanisms are apparent for the maintenance of "long-lasting" cirrus uncinus clouds (on the order of several hours). One is associated with local cooling in the trail of the cloud due to ice crystal evaporation. The other is the production of new cells due to the convergence with positive wind shear and divergence with negative wind shear that are associated with the downdraft of an existing generating cell. A conceptual model of a cirrus uncinus cloud is illustrated in Fig. 4.21.

As discussed in Section 4.1, satellite imagery suggests that large portions of the tropics are covered by extensive cirrus systems which vary spatially and temporally. Cirrus formations associated with mesoscale complexes may have large vertical extents with bases near 5–6 km and tops between 12 and 16 km. They evolve during the life cycle of the complexes and are modulated by large-scale disturbances. At various stages, the cirrus system may involve a mid-level deck with an overlaying thinner anvil near the tropopause, a single deep deck,

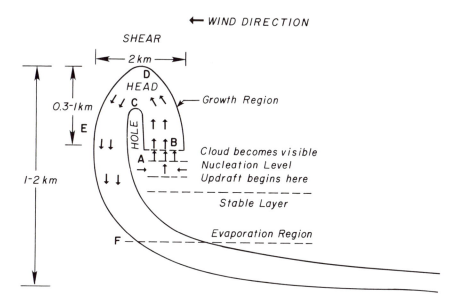

FIG. 4.21   A conceptual model of the cirrus uncinus cloud (after Heymsfield, 1975b).

or a single elevated deck in the upper atmosphere (Houze et al. 1981). Tropical anvils are relatively stable and long-lived. Unlike water clouds, these high anvils contain irregular ice crystals with low concentrations and could be optically thin and nonblack. The influence of cirrus clouds on the atmospheric radiation field depends on their solar and thermal ir radiative properties that, in turn, are modulated by cloud compositions and positions. Cloud–radiation feedback is important in the formation, maintenance, and dissipation of tropical anvils produced by strong diurnal convection in the equatorial region.

Outflow cirrus clouds from tropical towers appear to be maintained in a convectively active state by radiative flux gradients within the clouds (Danielson, 1982). These clouds are warmed and lifted into the stratosphere by net radiative heating. Danielson has suggested that extended anvils would become radiatively destabilized by cooling at tops and warming at bases. The resulting increase in the lapse rate within the anvils would drive convective fluxes by providing an upward flux of water vapor. The additional moisture would promote rapid ice crystal growth and fallout, hence serving as a dehydration mechanism for the tropical stratosphere. Ackerman et al. (1988) have computed heating rates in typical tropical anvils and found that the difference between heating rates at the cloud bottom and top ranges from 30 to 200 K d^{-1}, leading to convective instability in the anvils. Radiative heating could have important consequences for upward mass transport in the tropics. Using the radiative heating rates presented by Ackerman et al. (1988), Lilly (1988) has analyzed the dynamic mechanism of the formation of cirrus anvils using a mixed layer model, and has shown that destabilization of the layer could be produced by the strong radiative heating gradient. Although the preceding studies have shed some light on the dynamics of tropical anvils, a comprehensive understanding of their evolution and the role of radiative heating during their evolution requires an analysis of results produced by an appropriate cloud model.

For applications to the modeling of cirrus clouds, the mixing length theory introduced in Section 4.6 may be used, as a first-order approximation, for the eddy terms. On the basis of the preceding discussion for stratiform cloud models, the following system of equations involving vorticity, potential temperature, specific humidity, and the IWC mixing ratio $\bar{q}_{mi}$, may be used to model the formation and dissipation of cirrus clouds:

$$\frac{\partial \bar{\xi}}{\partial t} = J(\bar{\xi}, \bar{\psi}) + K_m \nabla^2 \bar{\xi} + g \frac{\partial B}{\partial y}, \tag{4.7.21}$$

$$\frac{\partial \bar{\Theta}}{\partial t} = J(\bar{\Theta}, \bar{\psi}) + K_h \nabla^2 \bar{\Theta} + \frac{\pi^{-1}}{C_p}(\bar{Q} + \bar{Q}_R), \tag{4.7.22}$$

$$\frac{\partial \bar{q}}{\partial t} = J(\bar{q}, \bar{\psi}) + K_q \nabla^2 \bar{q} + \frac{\bar{S}_v}{\bar{\rho}}, \tag{4.7.23}$$

$$\frac{\partial \bar{q}_{mi}}{\partial t} = J(\bar{q}_{mi}, \bar{\psi}) + K_c \nabla^2 \bar{q}_{mi} + \frac{\bar{S}_{mi}}{\bar{\rho}} - \frac{\partial}{\partial z}(\bar{q}_{mi}\bar{w}_{mi}). \tag{4.7.24}$$

In Eqs. (4.7.22)–(4.7.24) the Jacobian J is similar to that defined in Eq. (4.7.11), except that $\bar{\xi}$ is replaced by $\bar{\Theta}$, $\bar{q}$, and q_{mi}, and K_c is the eddy diffusivity for ice particles. The terms $\bar{S}_v$ and $\bar{S}_{mi}$ that denote sources and sinks for water vapor and ice particles have been defined in Subsection 4.4.4. Other notations have also been defined previously.

A two-dimensional cirrus cloud model has been developed by Starr and Cox (1985) in which a specified initial state is assumed to be horizontally uniform, cloud free, and at rest, except for constant vertical motion. The vertical and horizontal scales are taken to be 3.1 and 6.3 km, respectively. The direction of the y axis is normal to some uniform horizontal current. The eddy diffusivities for momentum, heat and water vapor are assumed to be the same, with a value of $1\,\mathrm{m^2\,s^{-1}}$. The eddy diffusivity for ice crystals is taken to be $0.5\,\mathrm{m^2\,s^{-1}}$. Parameterizations have been developed for the fall velocity of ice particles, deposition of vapor to ice, sublimation of ice to vapor, and broadband radiative processes. The model has been used to simulate the IWC mixing ratio for thin cirrus cloud layers and to investigate the quantitative role of dynamics, microphysics and radiation in the life cycle of the high-level clouds. Figure 4.22 illustrates the vertical profile of the horizontally averaged heating rates resulting from phase changes of ice and radiative flux exchanges for three time periods simulated from the cirrus cloud model. The terminal velocity of ice crystals becomes considerably larger than the

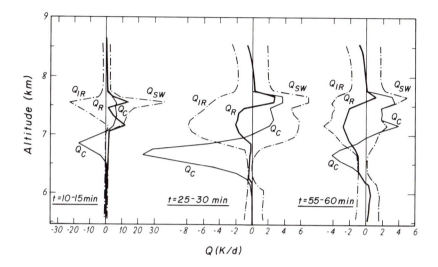

FIG. 4.22 Vertical profiles of horizontally averaged heating rates due to phase changes involving ice (Q_c), and ir (Q_{ir}), solar (Q_s), and net (Q_R) radiative processes over three time periods during a simulation of a thin cirrus cloud layer (after Starr and Cox, 1985).

initial vertical velocity in the model simulation. As a result, ice falls out of the cloud and the vapor sublimates onto it. The cooling associated with this sublimation has a maximum below the cloud base. The positive heating rate in the cloud is produced by the generation of ice crystals. The heating rates produced by solar and ir flux exchanges are comparable to the sublimational heating at all stages of the cloud evolution. Solar heating at the cloud top is offset by strong ir cooling. A net cooling of $-2\,\mathrm{K\,d^{-1}}$ is shown throughout most of the cloud layer and is significantly influenced by IWC generated from the model. Radiation and IWC feedbacks are critical in the formation and dissipation of cirrus clouds.

4.8   Convection in the atmosphere and cumulus cloud models

4.8.1   *Convective adjustment*

Convective adjustment is a numerical scheme that has been used to account for the convective nature of the atmosphere. To introduce this scheme, we shall begin with the first law of thermodynamics, denoted by Eq. (4.4.40), and consider this law in the context of one-dimensional space. The local rate of temperature change may be expressed in terms of static stability and radiative flux divergence in the form:

$$\rho C_p \frac{\partial T}{\partial t} = \rho C_p w(\gamma - \gamma_c) + \rho Q_R, \qquad (4.8.1)$$

where the vertical velocity, $w = dz/dt$, the atmospheric lapse rate, $\gamma = -\partial T/\partial z$, and

$$\gamma_c = \gamma_d + \frac{L}{C_p}\frac{dq_s}{dz}. \tag{4.8.2}$$

The dry adiabatic lapse rate, $\gamma_d = g/C_p$, and the last term in Eq. (4.8.2) represent a modification of the dry adiabatic lapse rate due to the saturation specific humidity gradient. Since specific humidity generally decreases with height, it has a negative value. The value of γ_c depends on the atmospheric humidity profile and varies from the tropics to the polar regions.

The local rate of temperature change, as shown in Eq. (4.8.1), depends on two factors. The first is due to the convective nature of the atmosphere and is governed by atmospheric stability, while the second is caused by the radiative flux exchange in the atmosphere. For the latter case, let the radiative flux divergence in the vertical, ρQ_R, be denoted by $-\partial F/\partial z$. For the former case, the term is associated with the divergence of the convective flux and may be written

$$-\frac{\partial F_v}{\partial z} = \rho C_p w(\gamma - \gamma_c). \tag{4.8.3}$$

Convective flux is related to small- and large-scale mean and eddy motions. If $\gamma \leq \gamma_c$, there will be no transport of convective flux. We may separate convective and radiative contributions to temperature perturbations as follows:

$$\rho C_p \frac{\partial T}{\partial t} = \rho C_p \left[\left(\frac{\partial T}{\partial t}\right)_{CON} + \left(\frac{\partial T}{\partial t}\right)_{RAD} \right] = -\frac{\partial F_v}{\partial z} - \frac{\partial F}{\partial z}. \tag{4.8.4}$$

Since a large portion of solar flux is absorbed at the surface, the air immediately above it is mostly unstable. In a one-dimensional context, the vertical transport of sensible and latent heat fluxes from the surface to the atmosphere is principally due to eddy motion. While it is possible to utilize the K theory to estimate the vertical transport of sensible and latent eddy fluxes, as demonstrated in Section 4.6, the procedure becomes rather complicated. The convective adjustment scheme, which is based on the concept of static stability, was first proposed by Manabe and Wetherald (1967) in connection with climatic sensitivity experiments involving various radiative forcings.

In the convective adjustment, the atmosphere is divided into layers that include nonconvection, convection without contact with the surface, and convection in contact with the surface. For a nonconvective layer, there would be no variation in the vertical eddy flux so that

$$-\frac{\partial F_v}{\partial z} = \rho C_p \left(\frac{\partial T}{\partial t}\right)_{CON} = 0, \tag{4.8.5}$$

and no adjustment is required. In a convective layer that is not in contact with the surface, the vertical temperature profile is numerically adjusted by using a critical lapse rate γ_c subject to the condition that the total potential energy is conserved within the layer. That is,

$$\int_{z_b}^{z_t} \rho C_p \left(\frac{\partial T}{\partial t} \right)_{\text{CON}} dz = -\int_{z_b}^{z_t} \frac{\partial F_v}{\partial z} dz = F_v(z_b) - F_v(z_t) = 0, \quad (4.8.6)$$

where z_t and z_b denote the top and bottom heights, respectively, of the unstable layer. This implies that convection develops when the atmospheric lapse rate exceeds the critical lapse rate. Convection then transports heat upward until the critical lapse rate is established, resulting in a redistribution of temperature with the total energy conserved. When a convective layer is in contact with the surface, the heat flux from the surface must be considered. Thus

$$\int_{0}^{z_t} \rho C_p \left(\frac{\partial T}{\partial t} \right)_{\text{CON}} dz = -\int_{0}^{z_t} \frac{\partial F_v}{\partial z} dz = F_v(0) - F_v(z_t) = F^+(0), \quad (4.8.7)$$

where $F^+(0)$ is the net radiative flux from the surface.

On the basis of Eqs. (4.8.5)–(4.8.7), an iterative procedure may be constructed for the computation of $(\partial T / \partial t)$. This procedure begins at the surface and scans the layers above progressively and iteratively until all layers of the supercritical lapse rate have been eliminated. This process is repeated at each time step. For a one-dimensional model, atmospheric temperatures are first constructed based on the balance between the radiative heating and cooling rates. Then the surface temperature is calculated from the balance of the solar and ir net fluxes at the surface. Since solar fluxes heat the surface, there will be net upward fluxes, which are distributed in the layer above the surface according to Eq. (4.8.7). Adjusting the temperature in steps upward and using Eq. (4.8.6) whenever the layer is convectively unstable eliminates all the supercritical lapse rates. The convective adjustment scheme is outlined in the following.

1. Compute $T_k^{(0)}$ for $k = 2, 3, 4, \ldots, N$ by using

$$T_k^{(0)} = T_k^\tau + \left(\frac{\partial T}{\partial t} \right)^\tau \Delta t, \quad (4.8.8)$$

where the superscript (n) $(n = 0, 1, 2, \ldots)$ denotes the iterative step during the τth time step, and T_k is the temperature at the kth layer. The subscript $k = 1$ denotes the surface, and $k = N$ the top level. The meaning of superscript (n) and subscript k is graphically displayed in Fig. 4.23.

2. Calculate the radiative equilibrium temperature for the surface $(k = 1)$ by using

$$T_1^{(0)} = \left(\frac{F_{s,1} + F_{ir,1}^\tau}{\sigma} \right)^{1/4}, \quad (4.8.9)$$

where $F_{s,1}$ and $F_{ir,1}^\tau$ are the net solar flux and downward longwave flux at the surface, respectively, and σ is the Stefan–Boltzmann constant. To a good approximation, the net solar flux in the atmosphere may be computed independent of the temperature.

3. Compute the critical temperature difference for each layer as follows:

$$\text{LRC}_k = (z_k - z_{k-1})\gamma_c, \qquad (4.8.10)$$

where z_k is the altitude of the kth level.

4. Calculate $T_1^{(1)}$ and $T_2^{(1)}$ such that they satisfy

$$\frac{C_p}{g}\Delta p_2 \left(T_2^{(1)} - T_2^{(0)}\right) = \sigma\Delta t \left[\left(T_1^{(0)}\right)^4 - \left(\left(T_1^{(1)}\right)\right)^4\right], \qquad (4.8.11a)$$

$$T_2^{(1)} = T_1^{(1)} - \text{LRC}_2, \qquad (4.8.11b)$$

where $\Delta p_2 = p_1 - p_2$. The term on the right-hand side of Eq. (4.8.11a) represents the net upward radiative flux. After eliminating $T_1^{(1)}$ in Eqs. (4.8.11a) and (4.8.11b), $T_2^{(1)}$ may be solved from the following equations using Newton's Method (see, e.g., Burden et al., 1978):

$$\sigma\Delta t \left(T_2^{(1)} + \text{LRC}_2\right)^4 + \frac{C_p}{g}\Delta p_2 T_2^{(1)} - \frac{C_p}{g}\Delta p_2 T_2^{(0)} - \sigma\Delta t \left(T_1^{(0)}\right)^4 = 0. \qquad (4.8.12)$$

5. If $T_2^{(1)} - T_3^{(0)} > \text{LRC}_3$, that is, if the layer is unstable, $T_1^{(2)}$, $T_2^{(2)}$, and $T_3^{(1)}$ are computed so that they satisfy

$$\frac{C_p}{g}\left[\Delta p_2\left(T_2^{(2)} - T_2^{(1)}\right) + \Delta p_3 \left(T_3^{(1)} - T_3^{(0)}\right)\right]$$
$$= \sigma\Delta t\left[\left(T_1^{(1)}\right)^4 - \left(T_1^{(2)}\right)^4\right], \qquad (4.8.13a)$$

$$T_3^{(1)} = T_2^{(2)} - \text{LRC}_3, \qquad (4.8.13b)$$

$$T_1^{(2)} = T_2^{(2)} + \text{LRC}_2, \qquad (4.8.13c)$$

where $\Delta p_3 = p_2 - p_3$. These equations are also solved using Newton's Method. If $T_2^{(1)} - T_3^{(0)} \le \text{LRC}_3$, the layer is stable. We then set $T_2^{(2)}$ and $T_3^{(1)}$ by using

$$T_2^{(2)} = T_2^{(1)}, \qquad T_3^{(1)} = T_3^{(0)}. \qquad (4.8.14)$$

6. For levels $k = 3, 4, 5, \ldots, N$, it is necessary to find $T_{k+1}^{(1)}$ and $T_k^{(2)}$. If $T_k^{(1)} - T_{k+1}^{(0)} > \text{LRC}_{k+1}$, we use

$$\frac{C_p}{g}\left[\Delta p_{k+1}\left(T_{k+1}^{(1)} - T_{k+1}^{(0)}\right) + \Delta p_k\left(T_k^{(2)} - T_k^{(1)}\right)\right] = 0, \qquad (4.8.15a)$$

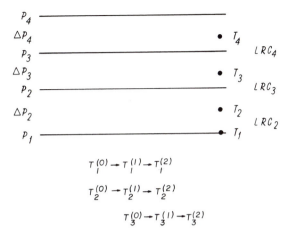

FIG. 4.23 Graphic representation of the temperature iteration in the convective adjustment scheme. The superscript (n) $(n = 0, 1, 2, \ldots)$ denotes the iterative step. The subscripts 1 and $(2, 3, \ldots)$ denote the surface and atmospheric layers, respectively. The term LRC_k $(k = 2, 3, \ldots)$ represents the critical temperature difference.

and

$$T_k^{(2)} - T_{k+1}^{(1)} = \mathrm{LRC}_{k+1}, \qquad k \geq 3. \tag{4.8.15b}$$

Substituting Eq. (4.8.15b) into Eq. (4.8.15a) and solving for $T_k^{(2)}$, we obtain

$$T_k^{(2)} = \frac{\Delta p_{k+1}\left(T_{k+1}^{(0)} + \mathrm{LRC}_{k+1}\right) + \Delta p_k T_k^{(1)}}{\Delta p_k + \Delta p_{k+1}}. \tag{4.8.15c}$$

$T_{k+1}^{(1)}$ is then solved by substituting $T_k^{(2)}$ back into Eq. (4.8.15b). If $T_k^{(1)} - T_{k+1}^{(0)} < \mathrm{LRC}_{k+1}$, we set

$$T_k^{(2)} = T_k^{(1)}, \qquad T_{k+1}^{(1)} = T_{k+1}^{(0)}. \tag{4.8.16}$$

7. The replacement $T_k^{(2)} \rightarrow T_k^{(0)}$ is made for $k = 1, 2, 3, \ldots, N$.
8. Repeat steps 4 through 7 until all layers of the supercritical lapse rate have been eliminated.

The convective adjustment scheme depends on the assignment of the critical lapse rate γ_c. For applications to prediction models, two values have been used. The first is the dry adiabatic lapse rate, and the case in which it is used is called *dry convection*. Moisture is not considered. The second is the moist adiabatic lapse rate denoted in Eq. (4.8.2), and the case in which it is used is referred to as *moist convection*. This moist rate depends on the vertical gradient of the saturation specific humidity. If the atmospheric lapse rate $\gamma \geq \gamma_c$ and the specific humidity $q \geq q_c$, where q_c denotes a preset critical specific humidity, the air is unstable and condensation occurs. The convective adjustment scheme will eliminate all layers of the supercritical lapse rate, allowing prediction experiments to be carried out. In addition, the total precipitation may also be computed, assuming that all condensates rain out instantly, in the form

$$\tilde{P} = -\int_0^{p_*} [q(p) - q_c(p)] \frac{dp}{g}, \qquad (4.8.17)$$

where the critical specific humidity is related to the saturation specific humidity via $q_c = h_c q_s(T)$, and where h_c (≤ 1) is the critical relative humidity, which is empirically determined so that condensation and precipitation may occur when the relative humidity is less than 100%. There appears to be no theoretical foundation for the assignment of h_c.

For applications to one-dimensional climate models, the critical lapse rate γ_c, is usually assumed to be $6.5\,\mathrm{K\,km^{-1}}$ for the globally averaged condition. This number is based on the fact that the climatological atmospheric temperature profile in the troposphere has a lapse rate close to this value. However, in climate temperature perturbation experiments it is important to incorporate critical lapse rates, which vary with the moisture content in the experiment. This will allow moisture feedback effects on the temperature profile to be taken into account (see Section 7.1 for further discussion on this subject).

4.8.2   Cumulus convection: single-cloud model

A significant portion of the large-scale disturbances in the tropics as well as the general circulation of the atmosphere are driven by latent heat release, which takes place primarily in deep cumulus towers. The horizontal space and time scales of these cumulus systems are two or more orders of magnitude smaller than those of synoptic-scale systems, though their vertical scales are similar. Because the horizontal space scale of these cumulus clouds is much smaller than the grid scale used in representing large-scale flows in numerical models, their activities can only be incorporated into the large-scale equations through parameterizations.

Observations reveal that deep cumulus towers and their effects on the environment vary from cloud to cloud. However, some common features may be seen in Fig. 4.24. Condensation occurs in the updraft. In the downdraft, evaporation takes place and is most significant below the cloud base, where the evaporation of falling raindrops produces cooling. Evaporation and/or sublimation of ice particles also takes place at the cloud top when the cloud air is detrained into the environment, and along the cloud edges through turbulent mixing with the environment. Cloud droplet freezing occurs at the level above the freezing point. It is clear that the net latent heat release to the environment at any level depends on

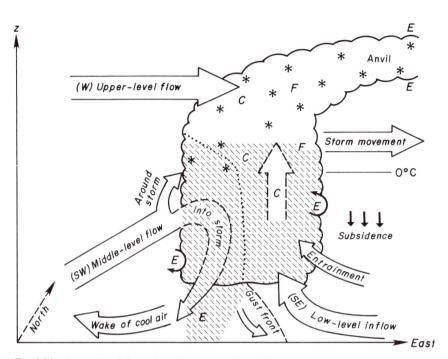

Fig. 4.24 Dynamic and thermodynamic representations of a mature thunderstorm. C refers to condensation, E is evaporation, and F is freezing. Letters in arrows refer to wind direction (after Anthes et al., 1982).

the differences between condensation and evaporation and between freezing and melting at that level. Furthermore, the integrated latent heat during the life cycle of deep cumulus is related to the precipitation reaching the ground. However, the vertical distribution of the latent heat released in a sample of cumuli could have an important effect on the development of synoptic disturbances. In view of the general features depicted in Fig. 4.24, condensation and freezing exceed evaporation and sublimation in the upper portion of the cloud. Consequently, the latent heat warms and dries the air.

In order to obtain an appropriate profile of the heating rate that is due to condensation (and freezing), cumulus convection may be assumed to occur in regions of deep layers of conditional instability and low-level convergence. The distributions of temperature and moisture in clouds follow the moist adiabatic process to the appropriate condensation level. The cloud base height is determined from the lifting condensation level of rising surface air, and the cloud top extends approximately to a level where the temperature, determined by the moist adiabatic process, equals the environmental temperature.

In the Kuo-type parameterization approach (Kuo, 1965), cumulus clouds are assumed to exist for only a short time and are dissolved by mixing with the

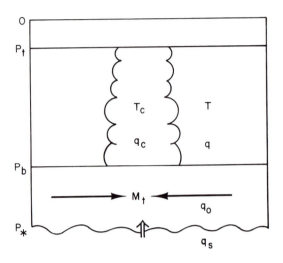

FIG. 4.25   An illustration of cumulus convection. The term M_t represents the total rate of moisture increase per unit of horizontal area in a cumulus cloud; q_s and q_o are the specific humidities for the surface and for the air immediately above the surface, respectively; T_c and q_c are the temperature and specific humidity within the cloud, respectively; and other notations are defined in the text.

environmental air at the cloud top. In this way, the heat and moisture of the cloud air are imparted to the environment. The total rate of moisture increase per unit horizontal area in a cumulus cloud is a result of the horizontal convergence and evaporation from the surface, as shown in Fig. 4.25:

$$M_t = - \int_0^{p_*} \nabla_2 \cdot (q\mathbf{v}_2) \frac{dp}{g} + C_D \rho_0 |\mathbf{v}_0| (q_s - q_0), \tag{4.8.18}$$

where p_* is the surface pressure; q_s is the saturation specific humidity corresponding to the surface temperature T_s; q_0 is the specific humidity for air with density ρ_0, immediately above the surface; $\mathbf{v}_0$ is the horizontal velocity immediately above the ocean; and C_D is the drag coefficient. The total amount of water vapor needed to generate a certain cloud cover over a unit area is denoted by W, representing the sum of the vapor flux associated with condensation, which raises the cloud temperature from T to T_c. The associated specific humidity of the cloud column increases from q to q_c. Thus we write

$$W = \frac{C_p}{L} \int_{p_t}^{p_b} (T_c - T) \frac{dp}{g} + \int_{p_t}^{p_b} (q_c - q) \frac{dp}{g}. \tag{4.8.19}$$

For deep cumulus, the bottom pressure $p_b \cong p_*$, and the top pressure $p_t \cong 0$. The time scale during which the cloud temperature and specific humidity increase from T to T_c and q to q_c is inversely proportional to the ratio of the total rate of moisture increase M_t to the total amount of water vapor W. The average heating rate profile produced by cumulus convection must be proportional to the temperature increase and inversely proportional to the time scale for condensation. Thus we may write

$$Q_c = K_c C_p (T_c - T), \tag{4.8.20}$$

where $K_c = M_t / W$, which is, in essence, a reciprocal time scale for the release of condensational heating in the cloud to the environment.

The preceding discussion outlines the Kuo scheme for cumulus convection, which was originally designed for the modeling of tropical cyclones. This parameterization scheme gives a fairly realistic vertical distribution of latent heat when compared with observational data. However, it underestimates the total amount of latent heat released, and hence, the precipitation. Also, the scheme is deficient in its estimate of the effects cumulus clouds have on the large-scale moisture field. This deficiency is due to the fact that the scheme does not consider the upward transport of water vapor by shallow clouds from below the cloud base. A modified scheme has been suggested (Kuo, 1974), in which it is assumed that the total latent heat release is proportional to $(1 - b)M_t$. The part, bM_t, represents the portion of moisture supply that is stored in the air. The values of b are much smaller than 1. The heating rate profile may then be computed from

$$Q_c = (1 - b)M_t L(T_c - T)/\langle T_c - T \rangle, \tag{4.8.21}$$

where the ensemble average is

$$\langle T_c - T \rangle = \int_{p_t}^{p_b} (T_c - T) \frac{dp}{g}.$$

The specification of the parameter b is critical for the production of realistic heating profiles. Although a number of methods have been proposed for specifying b, the most frequently used method has been developed by Anthes (1977). In this method, the partitioning of convective heating and moistening is assumed to depend on the mean saturation deficit of the entire cloud layer. The following parameterized form has been proposed:

$$b = \begin{cases} \left(\frac{1 - \langle h \rangle}{1 - h_c} \right)^n, & \langle h \rangle > h_c \\ 1, & \langle h \rangle \le h_c, \end{cases} \tag{4.8.21a}$$

where n is a certain constant, h_c is a threshold relative humidity below which no clouds are formed, and the ensemble average of the relative humidity for the cloud layer is given by

$$\langle h \rangle = \frac{1}{p_b - p_t} \int_{p_t}^{p_b} \frac{q}{q_s(T)} dp, \tag{4.8.21b}$$

The constant n is a tunable parameter, as is the threshold relative humidity. The best agreement between simulated and observed rainfall rates is found for values of n between 2 and 3, and h_c between 0.25 and 0.5.

The total precipitation due to cumulus convection over a time period Δt is given by

$$\tilde{P} = \int_{\Delta t} \int_0^{p_*} \frac{Q_c}{L} \frac{dp}{g} \, dt. \qquad (4.8.22)$$

To find T_c, we use the first law of thermodynamics for moist air in the form

$$-\frac{L}{C_p} \frac{dq_s}{dz} = \frac{dT}{dz} + \frac{g}{C_p}. \qquad (4.8.23)$$

Moreover, using the definition of $q_s (= e_s/p)$, where e_s is the saturation vapor pressure, the equation of state, the hydrostatic equation, and the Clausius–Clapeyron equation denoted in Eq. (4.5.42), we find

$$\frac{dq_s}{dz} = \frac{q_s L}{R_v T^2} \frac{dT}{dz} + \frac{q_s g}{RT}. \qquad (4.8.24)$$

Combining Eqs. (4.8.23) and (4.8.24) leads to

$$\gamma_s = -\left(\frac{\partial T}{\partial z}\right) = \gamma_d \left(1 + \frac{L q_s}{RT}\right) \Big/ \left(1 + \frac{q_s L^2}{C_p R_v T^2}\right). \qquad (4.8.25)$$

To apply Eq. (4.8.25) to the cloud region, we replace T with T_c and q_s with q_c. In the pressure coordinate we find

$$\frac{\partial \ln T_c}{\partial \ln p} = \frac{R}{C_p} \frac{1 + L q_c/RT_c}{1 + L^2 q_c/R_v C_p T_c^2}, \qquad (4.8.26)$$

where q_c may be related to q_s via a threshold relative humidity for cumulus convection, h_c. As a first approximation, we may set $q_c = h_c q_s$. Since $T_c(p_b) = T(p_b)$, upward stepwise calculations then give the cloud temperature at the model layer.

4.8.3 Cumulus convection: multicloud system

Cumulus convection is related to condensation processes, turbulent transport of heat and moisture, and the modification of radiative processes in clouds. Horizontal eddy terms are generally small when compared with the vertical flux term and may be neglected. Based on the analyses presented in Section 4.4, the basic averaged equations for potential temperature [Eq. (4.4.53)], specific humidity [Eq. (4.4.27)], and moist static energy [Eq. (4.4.45)] may be expressed in the forms

$$\frac{\partial \bar{\Theta}}{\partial t} + \bar{\mathbf{v}}_2 \cdot \nabla_2 \bar{\Theta} + \bar{w} \frac{\partial \bar{\Theta}}{\partial z} = -\frac{1}{\bar{\rho}} \frac{\partial}{\partial z} (\bar{\rho} \, \overline{w' \Theta'}) + \frac{\pi (\bar{Q} + \bar{Q}_R)}{C_p}, \qquad (4.8.27)$$

$$\frac{\partial \bar{q}}{\partial t} + \bar{\mathbf{v}}_2 \cdot \nabla_2 \bar{q} + \bar{w} \frac{\partial \bar{q}}{\partial z} = -\frac{1}{\bar{\rho}} \frac{\partial}{\partial z} (\bar{\rho} \, \overline{w'q'}) + \frac{\bar{S}_v}{\bar{\rho}}, \tag{4.8.28}$$

$$\frac{\partial \bar{E}}{\partial t} + \bar{\mathbf{v}}_2 \cdot \nabla_2 \bar{E} + \bar{w} \frac{\partial \bar{E}}{\partial z} = -\frac{1}{\bar{\rho}} \frac{\partial}{\partial z} (\bar{\rho} \, \overline{w'E'}) + \bar{Q}_R, \tag{4.8.29}$$

where

$$\bar{Q} = -L\bar{S}_v/\bar{\rho},$$
$$\bar{S}_v/\bar{\rho} = \left[-\bar{S}_1 + (\bar{S}_6 + \bar{S}_7) \right]/\bar{\rho}. \tag{4.8.30}$$

In Eq. (4.8.30), $\bar{S}_1$ denotes condensation and $\bar{S}_6$ and $\bar{S}_7$ are the evaporation of cloud droplets and raindrops, respectively. We may also define dry static energy

$$s = C_p T + gz \tag{4.8.31}$$

to obtain the following budget equation:

$$\frac{\partial \bar{s}}{\partial t} + \bar{\mathbf{v}} \cdot \nabla_2 \bar{s} + \bar{w} \frac{\partial \bar{s}}{\partial z} = -\frac{1}{\bar{\rho}} \frac{\partial}{\partial z} (\bar{\rho} \, \overline{w's'}) + \bar{Q}_R - \frac{L\bar{S}_v}{\bar{\rho}}. \tag{4.8.32}$$

Referring to Eqs. (4.8.31) and (4.8.28) for dry static energy and specific humidity, and utilizing the equation of continuity denoted in Eq. (4.4.10), we have

$$\frac{\partial}{\partial t} (\bar{\rho}\bar{s}) + \nabla_2 \cdot (\bar{\rho}\bar{\mathbf{v}}_2 \bar{s}) + \frac{\partial}{\partial z} (\bar{\rho}\bar{w}\bar{s}) = -\frac{\partial}{\partial z} (\bar{\rho} \, \overline{w's'}) + \bar{\rho}\bar{Q}_R - L\bar{S}_v = \bar{\rho}Q_1, \tag{4.8.33}$$

$$\frac{\partial}{\partial t} (\bar{\rho}\bar{q}) + \nabla_2 \cdot (\bar{\rho}\bar{\mathbf{v}}_2 \bar{q}) + \frac{\partial}{\partial z} (\bar{\rho}\bar{w}\bar{q}) = -\frac{\partial}{\partial z} (\bar{\rho} \, \overline{w'q'}) + \bar{S}_v = \frac{\bar{\rho}Q_2}{L}, \tag{4.8.34}$$

where Q_1 and Q_2 are referred to as the *apparent heating source* and *apparent moisture sink*, respectively (Yanai et al., 1973). In the parametrization problem, the two tendency terms and Q_1 and Q_2 in Eqs. (4.8.33) and (4.8.34) are unknowns. Additional equations are needed to close the system.

Consider a grid area where multiple cumulus clouds occur at some level between the cloud base and the highest cloud top, as shown in Fig. 4.26. Let M_i and σ_i denote the mass flux for an individual cumulus and the fractional cloud cover, respectively, and set $M_c = \sum_i M_i$ and $\sigma_c = \sum_i \sigma_i$. On the basis of the mass conservation principle, we may express the eddy terms for dry static energy in the form

$$\bar{\rho} \, \overline{w's'} = \sum_i M_i s_i + (-M_c)\tilde{s} = \sum_i M_i (s_i - \tilde{s}), \tag{4.8.35a}$$

where $\tilde{s}$ denotes the dry static energy for the cloud environment. Likewise, we have

$$\bar{\rho} \, \overline{w'q'} = \sum_i M_i (q_i - \tilde{q}). \tag{4.8.35b}$$

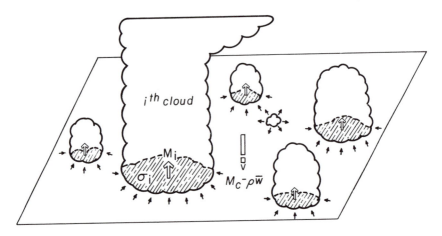

FIG. 4.26   An ensemble of cumulus clouds in a unit horizontal area at some level between the cloud base and the highest cloud top. The taller clouds are shown penetrating this level and entraining environmental air. A cloud which has lost buoyancy is shown detraining cloud air into the environment, where M_c is the cumulus mass flux (after Arakawa and Schubert, 1974).

Moreover, the averaged dry static energy and specific humidity may be expressed in terms of the following linear averages:

$$\bar{s} = (1 - \sigma_c)\tilde{s} + \sum_i \sigma_i s_i, \tag{4.8.36a}$$

$$\bar{q} = (1 - \sigma_c)\tilde{q} + \sum_i \sigma_i q_i. \tag{4.8.36b}$$

Assume that the total cloud cover is much less than unity and that entrainment occurs at all levels except near the cloud top where detrainment and evaporation of cloud liquid water take place. It may be shown that

$$\bar{\rho}Q_1 = M_c \frac{\partial \bar{s}}{\partial z} + D(\hat{s} - L\hat{q}_m - \bar{s}) + \bar{\rho}\bar{Q}_R, \tag{4.8.37a}$$

$$-\bar{\rho}Q_2 = M_c L \frac{\partial \bar{q}}{\partial z} + DL(\hat{q} + \hat{q}_m - \bar{q}), \tag{4.8.37b}$$

where the detrainment $D = -\sum_i \partial M_i / \partial z$ and the caret denotes the cloud top value. To obtain Q_1 and Q_2, the cumulus mass flux M_c and detrainment D must be evaluated by parameterization (Arakawa and Schubert, 1974). The Arakawa and Schubert scheme differs conceptually from the Kuo scheme discussed in Subsection 4.8.2. In the former, it is the resulting subsidence, due to the convective mass flux, that is responsible for the heating.

Spectral cumulus cloud ensemble models have been introduced to compute Q_1 and Q_2 as functions of height for given large-scale conditions (Ooyama, 1971;

Arakawa and Schubert, 1974). A spectral ensemble model has been proposed such that

$$M_c(z) = \int_0^{\lambda_{max}} M_B(\lambda)\eta(\lambda, z)\, d\lambda, \qquad (4.8.38a)$$

$$D(z) = -M_B(\hat{\lambda})\eta(\hat{\lambda}, z)\frac{d\hat{\lambda}}{dz}, \qquad (4.8.38b)$$

where M_B is the cloud-base subensemble mass flux, λ is the fractional rate of entrainment, which is used to identify the cloud type, λ_{max} is the maximum value of λ, $\hat{\lambda}$ is the value of λ for the clouds detraining at level z, and η is the cloud mass flux normalized at the cloud base that satisfies $\partial\eta/\partial z = \lambda\eta$. A measure of the ability of the large-scale environment to generate cumulus clouds is given by the cloud work function defined by

$$A(\lambda) = \int_{z_b}^{z_t} \eta(\lambda, z)\frac{T_c(z, \lambda) - \bar{T}(z)}{\bar{T}(z)}g\, dz. \qquad (4.8.39)$$

This function is related to the buoyancy force inside the clouds of type λ and therefore is the rate of kinetic energy generation per unit cloud mass flux at the cloud base level. In order to obtain the cloud-base subensemble mass flux, a constraint must be used. The closure assumption is that the cloud work function is invariant for each cloud type if the time variation in the large-scale forcing is sufficiently slow. Under this assumption, we have

$$\frac{dA(\lambda)}{dt} = \left(\frac{dA(\lambda)}{dt}\right)_{Cu} + \left(\frac{dA(\lambda)}{dt}\right)_{Large-scale} \cong 0. \qquad (4.8.40)$$

Arakawa and Schubert showed that

$$\int_0^{\lambda_{max}} K(\lambda, \lambda')M_B(\lambda')\, d\lambda' \cong F(\lambda), \qquad (4.8.41)$$

where K is a kernel that can be determined from the large-scale temperature and specific humidity, and F is the large-scale forcing that may include radiative, turbulent, and advective processes. This equation is used to determine the cloud-base mass flux M_B. On the basis of the preceding analysis, the apparent heating source and moisture sink defined in Eqs. (4.8.37a) and (4.8.37b) may be evaluated.

   The apparent heating source and moisture sink may be diagnostically computed from Eqs. (4.8.33) and (4.3.34) if spatially smoothed values of velocity components, dry static energy s, and specific humidity q are determined from observed data, and if time sequences of $\bar{s}$ and $\bar{q}$ are available. Figure 4.27 shows the mean vertical profiles of $Q_1 - Q_R$ and Q_2 from the data obtained during the GARP Atlantic Tropical Experiment (GATE) August–September, 1974 (Arakawa and Chen,

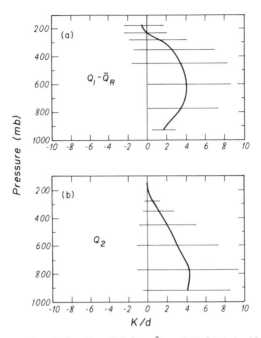

FIG. 4.27   Time-averaged vertical profiles of (a) $Q_1 - \bar{Q}_R$ and (b) Q_2, derived from GATE data. The terms Q_1, $\bar{Q}_R$, and Q_2 denote the apparent heating source, radiative heating, and apparent moisture sink defined in Eqs. (4.8.33) and (4.8.34). The bars represent standard deviations. To obtain the units Kd^{-1}, Q_1, Q_2, and $\bar{Q}_R$ are normalized by C_p (after Arakawa and Chen, 1986).

1986). There are significant variations in these two components. On average, the moisture sink has large values in the lower troposphere and the $Q_1 - Q_R$ component has a maximum near $\sim$600 mb. This component shows negative values near the cloud top, revealing significant cooling produced by ir flux exchanges between the cloud and the atmosphere above. Arakawa and Chen have presented cumulus parameterization as a closure problem and have shown that large uncertainties still exist in the selection of closure assumptions for cumulus parameterization. Reliable observations are required to verify and improve cumulus parameterizations.

REFERENCES

Ackerman, S. A., W. L. Smith, J. D. Spinhirne, and H. E. Revercomb, 1990: The 27–28 October 1986 FIRE IFO cirrus case study: Spectral properties of cirrus clouds in the 8–12 μm window. *Mon. Wea. Rev.*, **118**, 2377–2388.

Ackerman, T. P., K. N. Liou, F. P. J. Valero, and L. Pfister, 1988: Heating rates in tropical anvils. *J. Atmos. Sci.*, **45**, 1606–1623.

Anthes, R. A., 1977: A cumulus parameterization scheme utilizing a one-dimensional cloud model. *Mon. Wea. Rev.*, **105**, 270–286.

Anthes, R. A., H. D. Orville and D. J. Raymond, 1982: Mathematical modeling of convection. In *The Thunderstorm: A Social, Scientific and Technology Documentary*, Vol. 2, Thunderstorm Morphology and Dynamics, E. Kessler, Ed. U.S. Department of Commerce, pp. 495–579.

Arakawa, A., 1975: Modelling clouds and cloud processes for use in climate models. GARP Publ. Ser. No. 16, WMO, pp. 183–197.

Arakawa, A., and W. H. Schubert, 1974: Interaction of a cumulus cloud ensemble with the large-scale environment. Part I. *J. Atmos. Sci.*, **31**, 674–701.

Arakawa, A., and J. M. Chen, 1986: Closure assumptions in the cumulus parameterization problem. In *Short- and Medium-Range Numerical Weather Prediction*, WMO/IUGG NWP Symposium, Tokyo, pp. 107–131.

Auer, A. H., Jr., and D. L. Veal, 1970: The dimension of ice crystals in natural clouds. *J. Atmos. Sci.*, **27**, 919–926.

aufm Kampe, H. J., and H. K. Weickmann, 1957: Physics of clouds. In *Meteor. Monogr.*, Vol. 3, Amer. Meteor. Soc., pp. 182–225.

Battan, L. J., and C. H. Reitan, 1957: Droplet size measurement in convective clouds. In *Artificial Stimulation of Rain*, H. K. Weickmann, Ed., Pergamon Press, London, pp. 184–191.

Beard, K. V., and H. R. Pruppacher, 1971: A wind tunnel investigation of the rate of evaporation of small water drops falling at terminal velocity in air. *J. Atmos. Sci.*, **28**, 1455–1464.

Beryland, T. G., and L. A. Strokina, 1980: Global distribution of total cloud amount (in Russian). Gidrometeoizdata, Leningrad, 71 pp. (English translation by S. Warren).

Borovikov, A. M., I. I. Gaivoronskii, E. G. Zak, V. V. Kostarev, I. P. Mazin, V. E. Minervin, A.Kh. Khrgian, and S.M. Shmeter, 1961: *Cloud Physics*. Gidrometeoro-logischeskoe Izdatelstvo, Leningrad, 392 pp. (English translation by Isreal Program for Scientific Translations, Jeruselum, U.S. Dept. of Commerce, Washington, D.C., 1963).

Brooks, C. E. P., 1927: The mean cloudiness over the earth. *Mem. Roy. Meteor. Soc.*, **1**, 127–138.

Burden, R. L., J. D. Faires, and A.C. Reynolds, 1978: *Numerical Analysis*. Prindle, Weber & Schmidt, Boston, 579 pp.

Businger, J. A., 1973: Turbulent transfer in the atmospheric surface layer. In *Workshop on Micrometeorology*, D. Haugen, Ed., Amer. Meteor. Soc., pp. 67–100.

Businger, J. A., J. C. Wyngaard, Y. Izumi, and E. F. Bradley, 1971: Flux-profile relationships in the atmospheric surface layer. *J. Atmos. Sci.*, **28**, 181–189.

Byers, H. R., 1965: *Elements of Cloud Physics*. University of Chicago Press, 191 pp.

Conover, J., 1960: Cirrus patterns and related air motions near the jet stream as derived by photography. *J. Meteor.*, **17**, 532–546.

Cotton, W., and R. Anthes, 1989: *Storm and Cloud Dynamics*. Academic Press, San Diego, 880 pp.

Curry, J., 1986: Interactions among turbulence, radiation and microphysics in Arctic stratus clouds. *J. Atmos. Sci.*, **43**, 90–106.

Danielson, E. F., 1982: A dehydration mechanism for the stratosphere. *Geophys. Rev. lett.*, **9**, 605–608.

Deirmendjian, D., 1969: *Electromagnetic Scattering on Spherical Polydispersions*. Elsevier, New York, 291 pp.

Diem, M., 1948: Messungen der Frosse von Wolkenelementen II. *Meteor. Rundschau*, **1**, 261–273.

Durbin, W.G., 1959: Droplet sampling in cumulus clouds. *Tellus*, **11**, 202–215.

Finger, J. E., and P. Wendling, 1990: Turbulence structure of Arctic stratus clouds derived from measurements and calculations. *J. Atmos. Sci.*, **47**, 1351–1373

Fleagle, R. G., and J. A. Businger, 1980: *An Introduction to Atmospheric Physics*. 2nd Ed., Academic Press, New York, 432 pp.

Goodman, J., O. B. Toon, R. F. Pueschel, K. G. Snetsinger, and S. Verma, 1989: Antarctic stratospheric ice crystals. *J. Geophys. Res.*, **94**, 16,449–16,457.

Gordon, C. T., R. D. Hovanec, and W. F. Stern, 1984: Analyses of monthly mean cloudiness and their influence upon model-diagnosed radiative fluxes. *J. Geophys. Res.*, **89**, 4713–4738.

Griffith, K. T., S. K. Cox, and R. G. Knollenberg, 1980: Infrared radiative properties of tropical cirrus clouds inferred from aircraft measurements. *J. Atmos. Sci.*, **37**, 1077–1087.

Gu, Z., 1980: *Physical Principle for Clouds and Precipitation*. Science Press, Beijing, China, 219 pp.

Gunn, K. L. S., and J. S. Marshall, 1958: The distribution with size of aggregate snowflakes. J. Meteor., **15**, 452–461.

Gunn, R., and G .D. Kinzer, 1949: The terminal velocity of fall for water droplets in stagnant air. *J. Meteor.*, **6**, 243–248.

Harimaya, T., 1968: On the shape of cirrus uncinus clouds: A numerical computation. Studies of cirrus clouds: Part III. *J. Meteor. Soc. Japan*, **46**, 272–279.

Henderson-Sellers, A., 1986: Layer cloud amounts for January and July 1979 from 3D-Nephanalysis. *J. Climate Appl. Meteor.*, **25**, 118–132.

Herman, G. and R. Goody, 1976: Formation and persistence of summertime arctic stratus clouds. *J. Atmos. Sci.*, **33**, 1537–1553.

Heymsfield, A. J., 1972: Ice crystal terminal velocities. *J. Atmos. Sci.*, **29**, 1348–1356.

Heymsfield, A. J., 1975a: Cirrus uncinus generating cells and the evolution of cirriform clouds. Part I: Aircraft observations of the growth of the ice phase. *J. Atmos. Sci.*, **32**, 799–808.

Heymsfield, A. J., 1975b: Cirrus uncinus generating cells and the evolution of cirriform clouds. Part II. The structure and circulations of the cirrus uncinus generating head. *J. Atmos. Sci.*, **32**, 809–819.

Heymsfield, A. J., 1977: Precipitation development in stratiform ice clouds: A microphysical and dynamical study. *J. Atmos. Sci.*, **34**, 367–381.

Heymsfield, A. J., and R. G. Knollenberg, 1972: Properties of cirrus generating cells. *J. Atmos. Sci.*, **29**, 1358–1366.

Heymsfield, A.J., and C.M.R. Platt, 1984: A parameterization of the particle size spectrum of ice clouds in terms of the ambient temperature and the ice water content. *J. Atmos. Sci.*, **41**, 846–855.

Hobbs, P. V., L. F. Radke, and D. G. Atkinson, 1975: *Airborne Measurements and Observations in Cirrus Clouds.* AFCRL-TR-75-0249, Air Force Geophysics Laboratory, Hanscom AFB, 117 pp.

Houze, R. A. Jr., S. G. Geotis, F. D. Marks, and A. K. West, 1981: Winter monsoon convection in the vicinity of North Borneo. Part I. Structure and time variation of the clouds and precipitation. *Mon. Wea. Rev.*, **109**, 1595–1614.

Hughes, N. A., 1984: Global cloud climatologies: A historical review. *J. Climate Appl. Meteor.*, **23**, 724–751.

Jayaweera, K., and B. J. Mason, 1965: The behavior of freely falling cylinders and cones in a viscous fluid. *J. Fluid Mech.*, **22**, 709–720.

Kessler, E., 1969: *On the Distribution and Continuity of Water Substance in Atmospheric Circulations. Meteor. Monogr.*, No. 10, Amer. Meteor. Soc., Boston, 84 pp.

Kinzer, G. D., and R. Gunn, 1951: The evaporation, temperature and thermal relaxation time of freely falling water drops. *J. Meteor.*, **8**, 71–83.

Knollenberg, R. G., 1970: The optical array: an alternative to scattering or extinction for airborne particle size determination. *J. Appl. Meteor.*, **9**, 86–103.

Koenig, G., K. N. Liou, and M. Griffin, 1987: An investigation of cloud/radiation interactions using three-dimensional nephanalysis and earth radiation budget data bases. *J. Geophys. Res.*, **92**, 5540–5554.

Koenig, L. R., and F. Murray, 1976: Ice-bearing cumulus cloud evolution: Numerical simulation and general comparison against observations. *J. Appl. Meteor.*, **15**, 747–762.

Krueger, S., 1988: Numerical simulation of tropical cumulus clouds and their interactions with the subcloud layer. *J. Atmos. Sci.*, **45**, 2221–2250.

Kuo, H. C., and W. H. Schubert, 1988: Stability of cloud-topped boundary layers. *Quart. J. Roy. Meteor. Soc.*, **114**, 887–916.

Kuo, H. L., 1965: On formation and intensification of tropical cyclones through latent heat release by cumulus convection. *J. Atmos. Sci.*, **22**, 40–63.

Kuo, H. L., 1974: Further studies of the parameterization of the influence of cumulus convection on large-scale flow. *J. Atmos. Sci.*, **31**, 1232–1240.

Lilly, D. K., 1968: Models of cloud topped mixed layers under a strong inversion. *Quart. J. Roy. Meteor. Soc.*, **94**, 292–309.

Lilly, D. K., 1988: Cirrus outflow dynamics. *J. Atmos. Sci.*, **45**, 1594–1605.

Lin, Y. L., R. D. Farley, and H. D. Orville, 1983: Bulk parameterization of the snow field in a cloud model. *J. Climate Appl. Meteor.*, **22**, 1065–1092.

Liou, K. N., 1986: Influence of cirrus clouds on weather and climate processes: A global perspective. *Mon. Wea. Rev.*, **114**, 1167–1199.

Liu, J. Y., and H. D. Orville, 1969: Numerical modeling of precipitation and cloud shadow effects on mountain-induced cumuli. *J. Atmos. Sci.*, **26**, 1283–1298.

Locatelli, J. D., and P. V. Hobbs, 1974: Fall speeds and masses of solid precipitation particles. *J. Geophys. Res.*, **79**, 2185–2197.

London, J., 1957:*A Study of the Atmospheric Heat Balance.* Final Report, Contract AF19(122)-165, Dept. of Meteor. and Oceanogr., New York Univ. (ASTIA 117227, Air Force Geophysics Laboratory, Hanscom AFB), 99 pp.

Ludlam, F. H., 1948: The forms of ice clouds, *Quart. J. Roy. Meteor. Soc.*, **74**, 39–56.

Ludlam, F. H., 1956: The forms of ice clouds, II. *Quart. J. Roy. Meteor. Soc.*, **82**, 257–265.

Manabe, S., and R. T. Wetherald, 1967: Thermal equilibrium of the atmosphere with a given distribution of relative humidity. *J. Atmos. Sci.*, **24**, 241–259.

Marshall, J. S., and W. Palmer, 1948: The distribution of raindrops with size. *J. Meteor.*, **5**, 165–166.

Mason, B. J., 1971: *The Physics of Clouds.* 2nd Ed., Oxford University Press, Oxford, 671 pp.

McCormick, M. P., and C. R. Trepte, 1987: Polar stratospheric optical depth observed between 1978 and 1985. *J. Geophys. Res.*, **92**, 4297–4306.

Mellor, G. L., and T. Yamada, 1974: A hierarchy of turbulence closure models for planetary boundary layers. *J. Atmos. Sci.*, **31**, 1791–1806.

Moeng, C. H. and A. Arakawa, 1980: A numerical study of a marine subtropical stratus cloud layer and its stability. *J. Atmos. Sci.*, **37**, 2661–2676.

Neiburger, M., 1949: Reflection, absorption and transmission of insolation by stratus clouds. *J. Meteor.*, **6**, 98–104.

Obukhov, A. M., 1946: Turbulence in an atmosphere with nonuniform temperature. *Tr. Inst. Teoret. Geofiz.*, Akad. Nauk. SSSR No. 1. (English translation in *Boundary Layer Meteor.*, **2**, 7–29, 1971).

Ogura, Y., and T. Takahashi, 1971: Numerical simulation of the life cycle of a thunderstorm cell. *Mon. Wea. Rev.*, **99**, 895–911.

Ono, A., 1969: The shape and riming properties of ice crystals in natural clouds. *J. Atmos. Sci.*, **26**, 138–147.

Ooyama, K., 1971: A theory on parameterization of cumulus convection. *J. Meteor. Soc. Japan*, **49**, 744–756.

Platt, C. M. R., N. L. Abshire, and G. T. McNice, 1978: Some microphysical properties of an ice cloud from lidar observation of horizontally oriented crystals. *J. Appl. Meteor.*, **17**, 1220–1224.

Platt, C. M. R., J. D. Spinhirne, and W. D. Hart, 1989: Optical and microphysical properties of a cold cirrus cloud: Evidence for regions of small ice particles. *J. Geophy. Res.*, **94**, 11,151–11,164.

Prandtl, L., 1925: Bericht über Untersuchungen zur ausgebildeten Turbulenz. *Z. Angew. Math. Mech.*, *5*, 136–139.

Priestly, C. H. B., 1959: *Turbulent Transfer in the Lower Atmosphere*. University of Chicago Press, Chicago, 130 pp.

Pruppacher, H., and J. D. Klett, 1978: *Microphysics of Clouds and Precipitation*. D. Reidel, Dordrecht, 714 pp.

Rogers, R. R., 1979: *A Short Course in Cloud Physics*. 2nd Ed., Pergamon Press, Oxford, 235 pp.

Sasamori, T., J. London, and D. V. Hoyt, 1972: Radiation budget of the Southern Hemisphere. In *Meteorology of the Southern Hemisphere*, C. W. Newton, Ed., *Meteor. Monogr.*, No. 13, pp. 9–24.

Sassen, K., D. O'C. Starr, and T. Uttal, 1989: Mesoscale and microscale structure of cirrus clouds: Three case studies. *J. Atmos. Sci.*, **46**, 371–396.

Schaefer, V. J., 1951: Snow and its relationship to experimental meteorology. In *Compendium of Meteorology*, T.F. Malone, Ed., Amer. Meteor. Soc., Boston, pp. 221–234.

Schiffer, R. A., and W. B. Rossow, 1983: The international satellite cloud climatology project (ISCCP): The first project of the world climate research programme. *Bull. Amer. Meteor. Soc.*, **64**, 779–784.

Sekhon, R. S., and R. C. Srivastava, 1970: Snow size spectra and radar reflectivity. *J. Atmos. Sci.*, **27**, 299–307.

Singleton, F., and D. F. Smith, 1960: Some observations of drop size distributions in low layer clouds. *Quart. J. Roy. Meteor. Soc.*, **86**, 454–467.

Smith, W. L., Jr., P. F. Hein, and S. Cox, 1990: The 27–28 October 1986 FIRE IFO cirrus case study: In situ observations of radiation and dynamic properties of a cirrus cloud layer. *Mon. Wea. Rev.*, **118**, 2389–2401.

Smithsonian Meteorological Tables, 1958: R. J. List, Ed., Smithsonian Institution, Washington, D.C., 527 pp.

Squires, P., 1958: The microstructure and colloidal stability of warm clouds. I. The Relation between structure and stability. *Tellus*, **10**, 256–261.

Starr, D. O'C., and S. K. Cox, 1985: Cirrus Clouds. Part I: Cirrus cloud model. *J. Atmos. Sci.*, **42**, 2663–2681.

Stowe, L. L., C. G. Wellemeyer, T. F. Eck, H. Y. Yeh, and the NIMBUS 7 Cloud Data Processing Team, 1988: NIMBUS 7 global cloud climatology. Part I. Algorithms and validation. *J. Climate*, **1**, 445–470.

Sundqvist, H., 1978: A parameterization scheme for non-convective condensation including prediction of cloud water content. *Quart. J. Roy. Meteor. Soc.*, **104**, 677–690.

Sundqvist, H., E. Berge, and J. E. Kristjansson, 1989: Condensation and cloud parameterization studies with a mesoscale numerical weather prediction model.

Mon. Wea. Rev., **117**, 1641–1657.

Taylor, G. I., 1915: Eddy motion in the atmosphere. *Phil. Trans. Roy. Soc.*, A, **215**, 1–26.

van Loon, H., 1972: Cloudiness and precipitation in the Southern Hemisphere. In *Meteorology of the Southern Hemisphere*, C.W. Newton, Ed., *Meteor. Monogr.*, No. 13, pp. 101–112.

Varley, D. J., I. D. Cohen, and A. A. Barnes, 1980: *Cirrus Particle Distribution Study, Part VII.* AFGL-TR-80-0324, Air Force Geophysics Laboratory, Hanscom AFB, Massachusetts, 82 pp.

Vowinckel, E., and S. Orvig, 1970: The climate of the north polar basin. In *World Survey of Climatology*, Vol. 14, *Climates of Polar Regions*, S. Orvig, Ed., Elsevier, Amsterdam, pp. 129–252.

Warner, J., 1969: The microstructure of cumulus cloud. Part I. General features of the droplet spectrum. *J. Atmos. Sci.*, **26**, 1049–1282.

Weickmann, H. K., 1945: Formen und Bildung atmosphärisher Eiskristalle. *Beitr. Phys. Atmos.*, **28**, 12–52.

Weickmann, H. K., and H. J. aufm Kampe, 1953: Physical properties of cumulus clouds. *J. Meteor.*, **10**, 204–211.

Yagi, T., 1969: On the relation between the shape of cirrus clouds and the static stability of the cloud level. Studies of cirrus clouds: Part IV. *J. Meteor. Soc. Japan*, **47**, 59–64.

Yagi, T., T. Harimaya, and C. Magono, 1968: On the shape and movement of cirrus uncinus clouds by the trigonometric method utilizing stereophotographs. Studies of cirrus clouds: Part I. *J. Meteor. Soc. Japan*, **46**, 266–271.

Yanai, M., S. Esbensen, and J. Chu, 1973: Determination of bulk properties of tropical cloud clusters from large-scale heat and moisture budgets. *J. Atmos. Sci.*, **30**, 611–627.

Zaitsev, V. A., 1950: Vodnost' i raspredelnie Kapel'v Kuchevykh Oblakakh. (Water content and distribution of drops in cumulus clouds.) *Trudy Glav. Geofiz. Observat.*, **19**, 122–132.

5

RADIATIVE TRANSFER IN CLOUDS

Clouds reflect, absorb, and transmit solar radiation. The amount of solar flux reflected, absorbed, and transmitted by clouds is a function of the optical depth, the geometry governing the sun, and the direction of detection. Under the plane-parallel assumption, a number of the radiative transfer methodologies that have been introduced in Chapter 3 can be used to determine the radiative properties of clouds. As shown in various sections in that chapter, the fundamental radiative transfer equation has been developed for an ensemble of molecules and/or particulates. In the case of clouds, the basic scattering and absorption properties of cloud particles are determined by particle size distribution. Clouds can also reflect and transmit the thermal infrared (ir) radiation emitted from the surface and the atmosphere and, at the same time, emit ir radiation according to the temperature structure within them. Although Chapter 2 provides a comprehensive discussion of the transfer of ir radiation in clear atmospheres, the general effects of scattering by cloud particles on ir radiation transfer have not been addressed.

In this chapter, we present all aspects of radiative transfer in clouds in a systematic manner, beginning with the single-scattering, absorption, and polarization properties of water droplets and ice crystals. Solutions for the scattering of light by spherical and nonspherical particles will be presented without the detailed mathematical deductions of the related scattering theories. The typical radiative properties of clouds covering various solar and ir spectral regions are described, as are broadband radiative properties and parameterizations. Where possible, available aircraft observations of cloud radiation fields are illustrated in order to provide an independent check with theory. Finally, a discussion is presented, at the fundamental level, on topics relating to radiative transfer involving finite clouds and anisotropic medium associated with horizontally oriented ice crystals.

In order to develop practical retrieval methodologies for the inference of cloud optical depth and of some measure of cloud parameters from space, we must simplify the radiative transfer solutions and, most importantly, reduce the complexity of computer algorithms. The general principles for radiative transfer in clouds described in this chapter may be of some value for applications to remote sensing of clouds from satellites.

5.1   Scattering and absorption properties of cloud particles

5.1.1   *Refractive indices of ice and water*

Theoretical computations of the radiative properties of water and ice clouds require knowledge of the laboratory measurements of the complex refractive indices of pure water and ice as functions of wavelength. The complex refractive index $m(\lambda) = m_r(\lambda) - im_i(\lambda)$, where λ is the wavelength in a vacuum, m_r is the real refractive index, and m_i is the imaginary refractive index. The real refractive index determines the phase speed of the electromagnetic wave, while the imaginary refractive index is related to the absorption coefficient k_λ through $k_\lambda = 4\pi m_i/\lambda$.

Ice is a uniaxial, doubly refracting, and optically positive crystal. Its refractive index generally varies in two directions for the extraordinary and ordinary waves. The refractive indices of ice for the extraordinary wave, n_e, and the ordinary wave, n_o, at a temperature of $-3°$ C for the visible wave numbers (24,716–14,154 cm^{-1}) differ only in the third decimal place (Hobbs, 1974). For example, at a wave number of 20,342 cm^{-1}, $n_e = 1.3140$ and $n_o = 1.3126$. Thus, for all practical purposes it suffices to use the refractive index for the extraordinary wave in atmospheric scattering and absorption calculations.

The real and imaginary parts of the refractive index for ice and water have been comprehensively reviewed and tabulated by Irvine and Pollack (1968), based on the available laboratory measurements of reflection and transmission. Bertie et al. (1969) have measured the absorptance and reflectance of a film of ice in the range of 8000 to 30 cm^{-1} at a temperature of $-173°$ C and have derived the real and imaginary refractive indices. Schaaf and Williams (1973) have performed similar measurements for ice at $-7°$ C in the wave-number range of 5000 to 300 cm^{-1}. Seki et al. (1981) have measured the reflectance of a single crystal of hexagonal ice at a temperature of 80 K in the wavelength range of 0.044 to 0.207 μm in the ultraviolet. A more up-to-date review of the refractive index of ice has been given by Warren (1984). The refractive indices of water for a wavelength range of 0.2 to 200 μm at a temperature of 20° C have been compiled by Hale and Querry (1973).

The real and imaginary refractive indices for ice and water as functions of wavelength, using the results presented by Warren (1984) and Hale and Querry (1973), are displayed in Fig. 5.1. The real refractive indices of water exhibit great deviations from those of ice for wavelengths greater than about 10 μm. The imaginary refractive index of ice is much more complex, with values ranging from about 10^{-6} at 0.9 μm to about 0.75 at 3 μm. Ice exhibits relatively strong absorption at about 1.6 μm, where water shows a minimum. On the other hand, water has much larger absorption than ice in the wavelength regions from about 20 to 30 μm and for wavelengths greater than about 100 μm. The imaginary refractive indices for ice and water vary rapidly in the solar and ir spectra. Because of these rapid variations, care must be taken in light scattering and absorption calculations covering the entire solar and ir spectral regions.

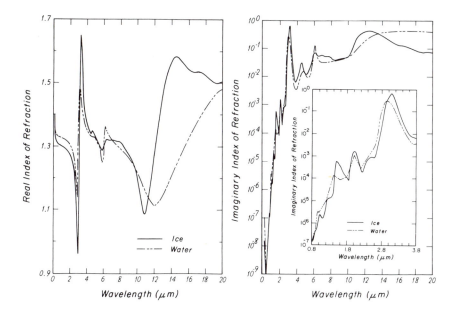

FIG. 5.1   Real and imaginary refractive indices of ice and water in the solar and ir spectral regions.

The spectral refractive indices of ice and water are fundamental parameters that determine the relative scattering and absorption properties of cloud particles due to a single-scattering event for a given size parameter. The size parameter α is defined as the ratio of particle circumference to incident wavelength. In the case of a sphere with a radius $r, \alpha = 2\pi r / \lambda$. If there is no absorption, values of the real part are responsible for the scattering process. When absorption is involved, the amount of energy scattered and absorbed depends on both the real and imaginary parts, as well as on the size parameter. In order to compute the solar and thermal ir fluxes that are reflected, absorbed, and transmitted through cloud layers, accurate values of real and imaginary refractive indices are required. These values are equally important in the development of remote sounding techniques for the discrimination of ice and water, and the determination of sizes and shapes of ice and water particles.

5.1.2   *Representation of a light beam*

Electromagnetic waves are characterized by certain polarization configurations that are described by the vibration of the electric vector and by the phase difference between the two components of this vector. These components are commonly denoted as E_ℓ and E_r, the electric fields parallel (ℓ) and perpendicular (r) to a reference plane. This reference plane is commonly denoted as the plane containing the incident and scattered directions and is referred to as the *scattering plane*.

Electric fields are complex, oscillating functions and may be expressed by

$$E_\ell = a_\ell \exp[-i(\xi + \delta_\ell)], \tag{5.1.1a}$$

$$E_r = a_r \exp[-i(\xi + \delta_r)], \tag{5.1.1b}$$

where a_ℓ and a_r are amplitudes, δ_ℓ and δ_r are phases, $\xi = kz - \omega t, k = 2\pi/\lambda, \omega$ is the circular frequency, and $i = \sqrt{-1}$. From these two equations, we can show that the electric fields are defined by the equation of an ellipse.

An electromagnetic wave can be represented by the amplitudes of the two electric components and their phase difference. Stokes in 1852 proposed four quantities, now known as the *Stokes parameters*, to represent the complete electromagnetic properties of a light beam. The Stokes parameters are real values defined by

$$I = E_\ell E_\ell^* + E_r E_r^* = a_\ell^2 + a_r^2, \tag{5.1.2a}$$

$$Q = E_\ell E_\ell^* - E_r E_r^* = a_\ell^2 - a_r^2, \tag{5.1.2b}$$

$$U = E_\ell E_r^* + E_r E_\ell^* = 2a_\ell a_r \cos \delta, \tag{5.1.2c}$$

$$V = i(E_r E_\ell^* - E_\ell E_r^*) = 2a_\ell a_r \sin \delta, \tag{5.1.2d}$$

where the superscript * denotes the complex conjugate and the phase difference $\delta = \delta_r - \delta_\ell$.

The Stokes parameters can be expressed in terms of the geometry governing an ellipse. Let β denote an angle whose tangent is the ratio of the axes of the ellipse traced by the endpoint of the electric vector, as displayed in Fig. 5.1a. If the major and minor axes of the ellipse are given by a and b, respectively, then $\tan \beta = \pm b/a$. Also let χ be the orientation angle between the major axis of the ellipse and the ℓ direction. When the plane waves are time harmonics, we may express the electric field vectors along the ℓ and r directions in terms of amplitude and phase using the cosine representation in the forms

$$E_\ell = a_\ell \cos(\xi + \delta_\ell), \tag{5.1.3a}$$

$$E_r = a_r \cos(\xi + \delta_r). \tag{5.1.3b}$$

Let x and y denote the directions along the major and minor axes, respectively. Then the electric fields in the $x - y$ plane may be written

$$\begin{pmatrix} E_x \\ E_y \end{pmatrix} = \begin{pmatrix} \cos \chi & \sin \chi \\ -\sin \chi & \cos \chi \end{pmatrix} \begin{pmatrix} E_\ell \\ E_r \end{pmatrix}, \tag{5.1.3c}$$

where E_x and E_y may also be expressed in terms of amplitudes (a, b) and an arbitrary phase δ_0 using cosine and sine representations such that they satisfy the elliptical equation in the forms

$$E_x = a \cos(\xi + \delta_0), \tag{5.1.3d}$$

$$E_y = \pm b \cos(\xi + \delta_0). \tag{5.1.3e}$$

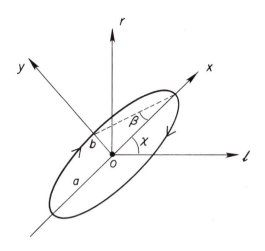

FIG. 5.1(a)   Geometric representation of elliptical polarization of a light beam. a and b are the lengths of the semimajor and -minor axes, respectively, χ is the orientation angle between the $O\ell$ and OX axes, and β is the ellipticity angle whose tangent is the ratio of the ellipse traced by the endpoint of the electric vector, i.e., $\tan \beta = \pm b/a$, where $+$ and $-$ stand for the right- and left-handed polarization, respectively.

After eliminating the propagation constant and all the phases using Eqs. (5.1.3a–e), the Stokes parameters can be written in terms of the total intensity, and the ellipticity and orientation angles in the forms

$$I = I_\ell + I_r, \tag{5.1.4a}$$

$$Q = I_\ell - I_r = I \cos 2\beta \cos 2\chi, \tag{5.1.4b}$$

$$U = I \cos 2\beta \sin 2\chi, \tag{5.1.4c}$$

$$V = I \sin 2\beta. \tag{5.1.4d}$$

A light beam is composed of many simple waves and is generally character-ized by partially elliptical polarization. If the phase differences between the two electric components are $0°$ or an integer order of $180°$ (i.e., $\beta = 0$), the light beam is linearly polarized ($V = 0$). If, on the other hand, the amplitudes of the two electric components are the same and their phase differences are an odd integer order of $90°$, then the light beam is circularly polarized ($Q = U = 0$). When the ellipticity angle β is positive or negative, the circular polarization is said to be right- or left-handed. Natural light, such as sunlight, is characterized by having the same amplitudes in the two electric components and random phase differences. In this case, $Q = U = V = 0$, and the light beam is unpolarized. The degree of polarization of a light beam is defined by

$$PO = (Q^2 + U^2 + V^2)/I. \tag{5.1.5a}$$

If ellipticity is not considered, we may define the degree of linear polarization in the form

$$LP = -Q/I = -(I_\ell - I_r)/(I_\ell + I_r). \tag{5.1.5b}$$

Consider a particle of arbitrary shape and size. The scattered electric field at a distance R from the particle must be related to the two components of the incident electric field (E_ℓ^0, E_r^0). In the far field, the two-by-two amplitude matrix that transforms the incident electric vector into the scattered electric vector may be written in the form

$$\begin{bmatrix} E_\ell \\ E_r \end{bmatrix} = \frac{\exp(-ikR + ikz)}{ikR} \begin{bmatrix} S_2 & S_3 \\ S_4 & S_1 \end{bmatrix} \begin{bmatrix} E_\ell^0 \\ E_r^0 \end{bmatrix}, \tag{5.1.5c}$$

where z is the vertical direction in the Cartesian coordinates and $S_j (j = 1, 2, 3, 4)$ are the amplitude functions.

In terms of the Stokes parameters defined in Eqs. (5.1.2a–d), we find

$$\begin{bmatrix} I \\ Q \\ U \\ V \end{bmatrix} = \frac{\mathbf{F}}{k^2 R^2} \begin{bmatrix} I_0 \\ Q_0 \\ U_0 \\ V_0 \end{bmatrix}, \tag{5.1.5d}$$

where the subscript 0 denotes the incident beam and the transformation matrix is given by

$$\mathbf{F} = \begin{bmatrix} \frac{1}{2}(M_2 + M_3 + M_4 + M_1) & \frac{1}{2}(M_2 - M_3 + M_4 - M_1) & S_{23} + S_{41} & -D_{23} - D_{41} \\ \frac{1}{2}(M_2 + M_3 - M_4 - M_1) & \frac{1}{2}(M_2 - M_3 - M_4 + M_1) & S_{23} - S_{41} & -D_{23} + D_{41} \\ S_{24} + S_{31} & S_{24} - S_{31} & S_{21} + S_{34} & -D_{21} + D_{34} \\ D_{24} + D_{31} & D_{24} - D_{31} & D_{21} + D_{34} & S_{21} - S_{34} \end{bmatrix}. \tag{5.1.5}$$

Expressions for the matrix elements can be derived from the definition of Stokes parameters in terms of the electric fields and are given by

$$M_k = |S_k|^2, \tag{5.1.6a}$$

$$S_{kj} = S_{jk} = (S_j S_k^* + S_k S_j^*)/2, \tag{5.1.6b}$$

$$-D_{kj} = D_{jk} = (S_j S_k^* - S_k S_j^*)i/2, \qquad j, k = 1, 2, 3, 4. \tag{5.1.6c}$$

The preceding matrix elements are real numbers. In radiative transfer, it is conventional to define the scattering phase matrix, $\mathbf{P}$, such that its first element is normalized to unity as follows:

$$\int_0^{2\pi} \int_0^\pi \frac{P_{11}(\Theta)}{4\pi} \sin\Theta \, d\Theta \, d\Phi = 1, \tag{5.1.7}$$

where Θ and Φ denote the scattering and azimuthal angles, respectively.

Next, we define the scattering cross section. The scattering cross section of a scatterer represents the amount of incident flux that is removed from the original direction due to a single-scattering event such that this flux is distributed isotropically throughout the area of a sphere whose radius is R and whose center is the scatterer. The scattering cross section is related to the first element of the scattering phase matrix in the form

$$\sigma_s = \frac{1}{k^2} \int_0^{2\pi} \int_0^{\pi} \left(\frac{1}{2} \sum_{k=1}^{4} M_k \right) \sin \Theta \, d\Theta \, d\Phi. \tag{5.1.8}$$

Using Eq. (5.1.8), the scattering phase matrix may be defined in terms of the transformation matrix as follows:

$$\frac{\mathbf{P}}{4\pi} = \frac{1}{\sigma_s k^2} \mathbf{F}. \tag{5.1.9}$$

The Stokes parameters can then be expressed in the form

$$\begin{bmatrix} I \\ Q \\ U \\ V \end{bmatrix} = \Omega_{\text{eff}} \frac{\mathbf{P}}{4\pi} \begin{bmatrix} I_0 \\ Q_0 \\ U_0 \\ V_0 \end{bmatrix}, \tag{5.1.10}$$

where $\Omega_{\text{eff}} = \sigma_s / R^2$, denoting the effective solid angle associated with scattering.

If no assumption is made on the shape and position of the scatterer, the scattering phase matrix consists of 16 independent elements:

$$\mathbf{P} = \begin{bmatrix} P_{11} & P_{12} & P_{13} & P_{14} \\ P_{21} & P_{22} & P_{23} & P_{24} \\ P_{31} & P_{32} & P_{33} & P_{34} \\ P_{41} & P_{42} & P_{43} & P_{44} \end{bmatrix}. \tag{5.1.11}$$

For an incident unpolarized light beam, $I = \Omega_{\text{eff}} I_o P / 4\pi$, where $P = P_{11}$ is the conventional phase function.

If the scatterers are randomly oriented in space such that every scatterer has a plane of symmetry, the law of reciprocity may be applied (Perrin, 1942; van de Hulst, 1957). We may reverse the directions of the incident and scattered polarized beams, the final results being equal. It follows that the amplitude functions (S_3, S_4) in Eq. (5.1.5c) must be equivalent to $(-S_4, -S_3)$. Consequently, the following six relationships between the phase matrix elements are valid: $P_{12} = P_{21}, P_{13} = -P_{31}, P_{14} = P_{41}, P_{23} = -P_{32}, P_{24} = P_{42}$ and $P_{34} = -P_{43}$. Next, consider an incident light beam described by negative ellipticity and orientation angles, $-\beta$ and $-\chi$. From Eqs. (5.1.4a)–(5.1.4d), the Stokes parameters are $(I_0, Q_0, -U_0, -V_0)$. The scattered beam from an isotropic medium composed of randomly oriented scatterers must have the same form for the Stokes parameters: $(I, Q, -U, -V)$. The (I, Q) components by definition are invariant to the change in the incident Stokes parameters from (U, V) to $(-U, -V)$. Using Eq. (5.1.10), four relationships can be derived from which we must have $P_{13} = P_{14} = P_{23} = P_{24} = 0$, and $P_{31} = P_{32} = P_{41} = P_{42} = 0$. It follows that the scattering phase matrix reduces to six independent elements in the form

$$\mathbf{P} = \begin{bmatrix} P_{11} & P_{12} & 0 & 0 \\ P_{12} & P_{22} & 0 & 0 \\ 0 & 0 & P_{33} & P_{34} \\ 0 & 0 & -P_{34} & P_{44} \end{bmatrix}. \tag{5.1.12}$$

For spherical scatterers, $S_3 = S_4 = 0$. Thus, $P_{22} = P_{11}$ and $P_{44} = P_{33}$. Consequently, there are only four independent scattering phase matrix elements.

5.1.3   Light scattering and absorption by water droplets

Scattering and absorption of electromagnetic waves by spherical water droplets can be exactly solved by the Mie theory (Mie, 1908), which is a complete, formal theory of the interaction of a plane wave with a dielectric sphere. The Mie theory begins with Maxwell's equations, from which the vector wave equation in spherical coordinates can be derived. Using the separation of variables, the solution of the electric and magnetic field vectors of the incident wave may be expressed in terms of a number of mathematical functions. For the scattered field at a very large distance from the sphere, the scattering and extinction cross sections and the phase function may be derived in terms of an infinite series containing the associated Legendre polynomials and spherical Bessel functions.

The Mie theory is applicable to a single homogeneous sphere. In order to apply this theory to clouds, the characteristics of cloud particle size distribution must be known. Because cloud particles are sufficiently far from each other and because the distance between them is much greater than the incident wavelengths in the solar and terrestrial ir spectra, the independent (or incoherent) scattering concept may be applied: Scattering by one particle in terms of the electric field may be treated independently of other particles. It is in this context that we apply the Mie theory to cloud droplets.

For a sample of spherical water droplets in which every droplet has a plane of symmetry and is randomly located in space, the scattering phase matrix reduces to four independent elements, in the form

$$\mathbf{P} = \begin{bmatrix} P_{11} & P_{12} & 0 & 0 \\ P_{12} & P_{11} & 0 & 0 \\ 0 & 0 & P_{33} & P_{34} \\ 0 & 0 & -P_{34} & P_{33} \end{bmatrix}. \tag{5.1.13}$$

After applying the Mie theory to water droplets having a size distribution function $n(r)$, the four phase matrix elements are defined in the forms

$$\frac{P_{11}(\Theta)}{4\pi} = \frac{1}{2k^2\beta_s} \int_{r_1}^{r_2} [i_2(\Theta, r) + i_1(\Theta, r)]n(r)\,dr, \qquad (5.1.14a)$$

$$\frac{P_{12}(\Theta)}{4\pi} = \frac{1}{2k^2\beta_s} \int_{r_1}^{r_2} [i_2(\Theta, r) - i_1(\Theta, r)]n(r)\,dr, \qquad (5.1.14b)$$

$$\frac{P_{33}(\Theta)}{4\pi} = \frac{1}{2k^2\beta_s} \int_{r_1}^{r_2} [i_4(\Theta, r) + i_3(\Theta, r)]n(r)\,dr, \qquad (5.1.14c)$$

$$\frac{P_{34}(\Theta)}{4\pi} = \frac{i}{2k^2\beta_s} \int_{r_1}^{r_2} [i_4(\Theta, r) - i_3(\Theta, r)]n(r)\,dr, \qquad (5.1.14d)$$

where r_1 and r_2 are the lower and upper limits of the droplet radius r, and the scattering coefficient is

$$\beta_s = \int_{r_1}^{r_2} \sigma_s(r)n(r)\,dr, \qquad (5.1.15)$$

with the scattering cross section defined by

$$\sigma_s(r) = \frac{\pi}{k^2} \int_0^\pi [i_2(\Theta, r) + i_1(\Theta, r)] \sin\Theta\,d\Theta. \qquad (5.1.16)$$

Note that a more general definition of the scattering cross section has been given in Eq. (5.1.8). In the above equations, the intensity functions in the Mie theory are related to the amplitude functions in the forms

$$i_1 = |S_1(\Theta)|^2, \qquad i_2 = |S_2(\Theta)|^2,$$
$$i_3 = S_2 S_1^*, \qquad i_4 = S_1 S_2^*. \qquad (5.1.17)$$

The amplitude functions are associated with the scattered electric vector in the far field:

$$S_1(\Theta) = \sum_{n=1}^{\infty} \frac{2n+1}{n(n+1)} [a_n \pi_n(\cos\Theta) + b_n \tau_n(\cos\Theta)], \qquad (5.1.18a)$$

$$S_2(\Theta) = \sum_{n=1}^{\infty} \frac{2n+1}{n(n+1)} [b_n \pi_n(\cos\Theta) + a_n \tau_n(\cos\Theta)], \qquad (5.1.18b)$$

where

$$\pi_n(\cos\Theta) = \frac{1}{\sin\Theta} P_n^1(\cos\Theta), \qquad (5.1.18c)$$

$$\tau_n(\cos\Theta) = \frac{d}{d\Theta} P_n^1(\cos\Theta), \qquad (5.1.18d)$$

and P_n^1 is an associated Legendre polynomial. The coefficients a_n and b_n can be determined from the boundary conditions at the surface such that the tangential components of the electric and magnetic vectors are continuous across the spherical surface, and are given by

$$a_n = \frac{\Psi'_n(y)\Psi_n(x) - m\Psi_n(y)\Psi'_n(x)}{\Psi'_n(y)\xi_n(x) - m\Psi_n(y)\xi'_n(x)}, \tag{5.1.19a}$$

$$b_n = \frac{m\Psi'_n(y)\Psi_n(x) - \Psi_n(y)\Psi'_n(x)}{m\Psi'_n(y)\xi_n(x) - \Psi_n(y)\xi'_n(x)}, \tag{5.1.19b}$$

where $x = kr$ and $y = mx$. The functions Ψ_n and ξ_n are related to the Bessel function $J_{n+1/2}$, and to the half-integral-order Hankel function of the second kind $H^{(2)}_{n+1/2}$, in the forms

$$\Psi_n(x) = (\pi x/2)^{1/2} J_{n+1/2}(x), \tag{5.1.20a}$$

$$\xi_n(x) = (\pi x/2)^{1/2} H^{(2)}_{n+1/2}(x). \tag{5.1.20b}$$

The Bessel and Hankel functions have zeros that increase in number with the size of the argument. For this reason, S_1 and S_2 can change rapidly with very small variations of x.

The removal of energy due to the presence of a particle in the direction of the incident beam (forward direction) is caused by both scattering to other directions and absorption within the particle. This removal is represented by the extinction cross section as if an area of the object had been covered up. In the context of the Mie theory, we have

$$\sigma_e = \frac{4\pi}{k^2} \mathrm{Re}[S(0)] = \frac{2\pi}{k^2} \sum_{n=1}^{\infty} (2n+1)\,\mathrm{Re}(a_n + b_n), \tag{5.1.21}$$

where $S(0) = S_1(0) = S_2(0)$ and Re denotes the real part. If a particle of any shape is much larger than the incident wavelength, the total energy is removed by scattering and absorption, giving an effective cross section area equal to the geometric area A. In addition, according to Babinet's principle, diffraction takes place through a hole in this area, giving a cross section area also equal to A. The total removal of incident energy is therefore twice the geometric area (i.e., $\sigma_e = 2A$). This is referred to as the *optical theorem for extinction*. In terms of the Mie theory for very large size parameters, $S(0) \to k^2 A/2\pi$, we also have $\sigma_e \to 2A$. For a given droplet size distribution, the extinction coefficient is defined as

$$\beta_e = \int_{r_1}^{r_2} \sigma_e(r)\, n(r)\, dr. \tag{5.1.22}$$

From Eq. (4.3.1) for the definition of the droplet number density, we may define the averaged extinction and scattering cross sections, $\bar{\sigma}_e = \beta_e/N$ and $\bar{\sigma}_s = \beta_s/N$.

A detailed discussion on the numerical techniques for the computation of the Mie solution has been provided by Deirmendjian (1969). Useful numerical results have also been presented in terms of tables for a number of aerosol, cloud, and rain models using various wavelengths. Moreover, with the more efficient formulations and vector structures of supercomputers, large numbers of computations of scattering functions in the Mie solution may be efficiently carried out even for very large size parameters (Wiscombe, 1980). It appears that the computational problems involving scattering and absorption by spherical particles have been completely resolved.

Below, we present some representative results for phase function, linear polarization, averaged extinction cross section, and single-scattering albedo for water clouds, as discussed in Subsection 4.3.1. Using the droplet size distribution of a typical fair weather cumulus, the phase functions and linear polarization patterns for the 0.5, 1.6, 3.7, and 10 μm wavelengths are presented in Fig. 5.2. At 0.5 μm, the feature shown at a scattering angle of $\sim$140° is the well-known primary rainbow that arises from rays undergoing one internal reflection in the drops. This cloudbow reduces in strength at 1.6 μm because of higher absorption and the smaller size parameters involved. At 0.5 μm, the minor maximum located at $\sim$130° is the secondary rainbow, which is produced by rays undergoing two internal reflections. The maximum that occurs at $\sim$180° for 0.5 μm is called the *glory*, which is often observed on mountain tops and from aircraft because of the required backscattering geometry. This maximum appears to be produced by the movement of surface waves around spherical droplets. The glory feature moves toward smaller scattering angles at 1.6 μm. At 3.7 and 10 μm, rainbow and glory features vanish due to the significant absorption by cloud droplets. Also noted is the reduction of the Fraunhofer diffraction peak when the size parameter is decreased. The polarization patterns for 0.5 μm contain many strong imprints of scattered light in the directions of the rainbow and glory angles. The positions of the primary rainbows for 1.6 and 3.7 μm shift to larger scattering angles due to the effect of the refractive index and the size parameter. For the 10 μm wavelength, scattering is primarily associated with external reflection, which produces a maximum positive polarization at the 90° scattering angle. The phase functions and linear polarization patterns for stratus and nimbostratus are substantially similar to those for cumulus.

Figure 5.3 shows the averaged extinction cross section, single-scattering albedo, and asymmetry factor for cumulus clouds as functions of wavelength. The averaged extinction cross section is normalized with respect to its value at 0.5 μm. In the visible wavelengths, the averaged size parameter is sufficiently large so that the extinction efficiency is approximately equal to 2, based on the optical theorem described previously. Extinction efficiency varies with the size parameter and the refractive index. The single-scattering albedo generally resembles the maximum and minimum patterns in the imaginary refractive index for water, as displayed in Fig. 5.1. Three large minima occur at $\sim$3, 6, and 10 μm. For wavelengths longer than 10 μm, the size parameter effect becomes more important. The asymmetry factor remains about the same, with values of about 0.82–0.86, for wavelengths up

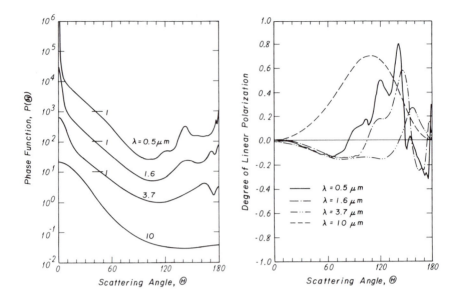

Fig. 5.2   Phase function and degree of linear polarization for incident wavelengths of 0.5, 1.6, 3.7, and 10 μm involving the droplet size distribution of cumulus clouds. For the phase function, the vertical scale applies to the lowest curve, while the upper curves are displayed upward by a factor of 10. The phase functions and polarization patterns for stratus and nimbostratus are similar.

to $\sim$10 μm, with the exceptions of $\sim$3 and 6 μm; at these wavelengths, diffraction predominates and produces strong forward scattering. Due to the reduction of the size parameter, the asymmetry factor decreases drastically for wavelengths longer than 10 μm. The extinction cross sections for various cloud types at 0.5 μm are shown in Table 5.1. Since absorption at 0.5 μm may be neglected, the scattering cross section is about equal to the extinction cross section. The extinction cross section depends on the droplet size spectrum. Cb has the largest value, while Sc has the smallest value, associated with the droplet size distributions displayed in Fig. 4.10.

5.1.4   *Light scattering and absorption by ice crystals*

As noted in Section 4.3.2, cirrus clouds are largely composed of nonspherical bullet rosettes, columns, and plates. Unlike the scattering of light by spherical water droplets, which is governed by the Mie solution, the light scattering and absorption properties of these hexagonal ice crystals are extremely difficult to determine. Below, we present approaches that may be followed to evaluate the single-scattering properties of ice crystals.

5.1.4.1   Geometric optics approach

Because of the lack of a proper definition of the coordinate systems for hexagonal

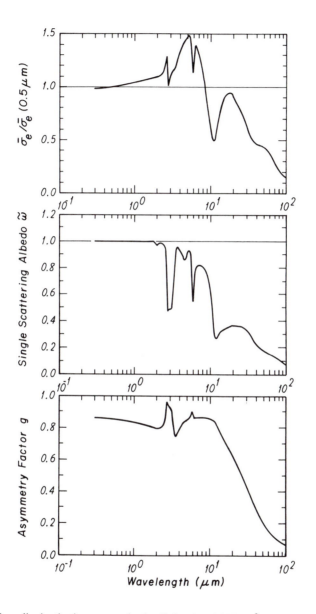

Fig. 5.3 Normalized extinction cross section $[\bar{\sigma}_e(0.5\ \mu m) = 165.9\ \mu m^2]$, single-scattering albedo, and asymmetry factor for the droplet size distribution of cumulus clouds. The spectral wavelengths used in the calculations using the notations $\lambda_i(\Delta\lambda_{ij})\lambda_j$, are: 0.3 (0.1) 1 (0.2) 2.6 (0.05) 3.5 (0.1) 8 (0.2) 10 (0.5) 20 (1) 30 (2) 50 (10) 100 μm. For stratus clouds, the results are similar.

Table 5.1 Extinction cross sections for various cloud types at 0.5 μm

Cloud type	Cu (fair weather)	Cu (congestus)	Cb	Sc	Ns	St
$\bar{\sigma}_e(\mu m^2)$	165.9	468.3	481.6	120.4	410.1	473.1

ice crystals, it is very difficult, if not impossible, to solve the scattering problem involving ice crystals by means of the wave equation approach. The alternative method would be to follow the geometric optics approach, which has been used to identify the positions of various optical phenomena produced by water droplets and ice crystals, such as rainbows and halos.

The laws of geometric optics may be applied to the scattering of light by an ice crystal under the condition that its size is much larger than the incident wavelength. In this case, a light beam may be thought of as consisting of a bundle of separate rays that hits the ice crystal. The width of the light beam is much larger than the wavelength and yet small compared with the crystal's size. Each ray that hits the crystal will undergo reflection and refraction and will pursue its own specific path along a straight line. The rays that emerge from various directions will have different amplitudes and phases.

According to Snell's law, the angles corresponding to incident and refracted waves on a smooth surface are related to the wave velocities in two media in the form

$$\frac{\sin \theta_i}{\sin \theta_t} = \frac{v_1}{v_2} = m, \tag{5.1.23}$$

where θ_i and θ_t denote the angles of the incident and refracted rays, respectively, with respect to the normal to the surface, v_1 and v_2 are the velocities in the two media, and m is the refractive index for the second medium with respect to the first. The energies that are reflected and refracted from the surface are governed by Fresnel formulas, which are results of continuity requirements for the tangential components of the electric and magnetic vectors at the interface (Born and Wolf, 1975). For the two components of the electric vectors that are perpendicular and parallel to a reference plane, the conventional Fresnel reflection coefficients, $R_{1,2}$, can be modified to account for absorption as follows:

$$|R_1|^2 = \frac{(\cos \theta_i - u)^2 + v^2}{(\cos \theta_i + u)^2 + v^2}, \tag{5.1.24}$$

$$|R_2|^2 = \frac{[(m_r^2 - m_i^2)\cos \theta_i - u]^2 + (2m_r m_i \cos \theta_i - v)^2}{[(m_r^2 - m_i^2)\cos \theta_i + u]^2 + (2m_r m_i \cos \theta_i + v)^2}, \tag{5.1.25}$$

where

$$u^2 = \frac{1}{2}\left\{ m_r^2 - m_i^2 - \sin^2 \theta_i + \left[(m_r^2 - m_i^2 - \sin^2 \theta_i)^2 + 4m_r^2 m_i^2 \right]^{1/2} \right\}, \tag{5.1.26a}$$

$$v^2 = \frac{1}{2} \left\{ -(m_r^2 - m_i^2 - \sin^2 \theta_i) + \left[(m_r^2 - m_i^2 - \sin^2 \theta_i)^2 + 4m_r^2 m_i^2 \right]^{1/2} \right\}.$$

(5.1.26b)

Based on the energy conservation principle, the refracted (transmitted) part of the energy is given by

$$|T_{1,2}|^2 = 1 - |R_{1,2}|^2.$$

(5.1.27)

Thus the reflected and refracted energies are proportional to $|R|^2$ and $|T|^2$, respectively.

To apply the geometric ray-tracing program to a hexagonal ice crystal, we must define the geometry of the orientation of the hexagon with respect to the incident electric vector of a geometric ray. There are eight faces into which light rays may enter. The two components of the electric vectors that are parallel and perpendicular to the scattering plane are required in ray tracing. Rays that undergo reflection and refraction can be traced on the basis of the geometry that has been defined. Electric fields reflected and refracted on a given surface can be computed using Eqs. (5.1.24)–(5.1.27). The distances between the points of entry and departure, as well as the phase shifts of the electric field due to reflection and refraction, can be evaluated. The electrical field vector of all incident rays that undergo external reflection, two refractions, and internal reflections may be obtained by summing the outgoing electric field vectors that have the same direction in space.

In the limits of geometric optics, half of the incident energy is associated with the diffracted rays. Based on the Fraunhofer diffraction theory for the far field, the wave disturbance of a light beam at an arbitrary point P may be expressed by (Born and Wolf, 1975)

$$u_p = -\frac{iu_0}{\lambda R} \iint_{B'} e^{ikR} \, dx' \, dy',$$

(5.1.28a)

where u_0 represents the disturbance in the original wave at point O on the plane wave front with wavelength λ, and R is the distance between point P and point $O'(x', y')$ on the aperture with an area B', as shown in Fig. 5.4(a). The eight apexes $B_i'(i = 1$–$8)$ denote the projections of the eight vertices $B_i(i = 1$–$8)$ of a hexagonal crystal on the plane perpendicular to an oblique incident light ray. Using the scattering and azimuthal angles Θ and Φ defined in this figure, we find

$$u_p(\Theta, \Phi) = -\frac{iu_0}{\lambda R} \iint_{B'} \exp\left[-ik(x' \cos \Phi + y' \sin \Phi) \sin \Theta \right] dx' \, dy'.$$

(5.1.28b)

Integration can be carried out analytically once the eight vertices in terms of the major axis L and minor axis $2r$ are given. If the hexagonal ice crystals are

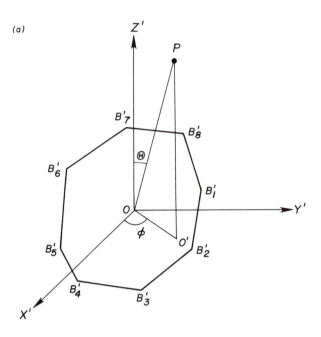

Fig. 5.4(a)   Geometry for Fraunhofer diffraction at an arbitrary point P. The projections of the eight vertices of a hexagonal crystal on the plane normal to an oblique incident ray are denoted by $B'_i (i = 1\text{–}8)$. Θ and Φ are the scattering and azimuthal angles of the diffracted light beam.

randomly oriented, u_p is only a function of the scattering angle. Unknown coefficients in Eq. (5.1.28b) may be determined by a proper normalization procedure for phase function.

The total electric field vector is obtained by summing the results from reflection and refraction, and diffraction. A detailed geometric ray-tracing procedure for computing the scattering phase matrix and scattering cross section for arbitrarily oriented hexagonal crystals has been presented in Cai and Liou (1982). A complete scattering diagram can be derived as a function of the scattering angle, the complementary angle α between the incident ray and the c axis defined in Fig. 5.4(b), and the rotational angle β about this axis. The c axis is the axis perpendicular to the basal faces of the hexagon. Figure 5.5 is a schematic representation of the components of the phase function for randomly oriented hexagonal ice crystals (see the following for further discussion on random orientation). Peaks at the $22°$ and $46°$ scattering angles are halo features produced by rays that undergo two refractions through the $60°$ and $90°$ prism angles. The diffraction pattern is primarily confined within scattering angles less than $\sim 10°$. In addition, a large energy component is produced by rays that undergo two refractions through parallel planes. This component is called the δ-*function transmission*, which occurs only at the $0°$ scattering angle, and is important for both absorption and nonabsorption

(b)

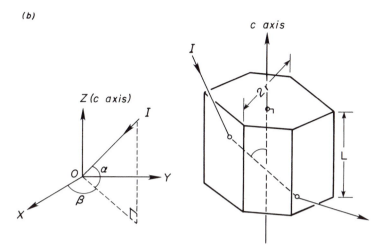

FIG. 5.4(b)   Geometry of a light ray in an ice crystal: $\frac{\pi}{2} - \alpha$ is the angle between the incident ray and the c axis defined in the diagram, β is the angle of rotation about this axis, I denotes the incident ray, L is the length of the ice crystal, and $2r$ is the width.

cases. Normalization of the phase function must properly take this component into account.

Let **G** represent 4×4 scattering phase matrix for a single crystal. For an ensemble of ice crystals with the same sizes randomly oriented in space, integration of the phase matrix over all possible orientations must be carried out so that

$$\mathbf{P}(\Theta) = \frac{6}{\pi} \int_0^{\pi/6} \int_0^{\pi/2} \mathbf{G}(\alpha, \beta) \cos \alpha \, d\alpha \, d\beta. \qquad (5.1.29)$$

There are only six independent elements, as shown in Eq. (5.1.12). Because of the symmetry of a hexagon, integration of the angle β is from 0 to $\pi/6$. The scattering cross-section for randomly oriented ice crystals is given by

$$\hat{\sigma}_s = \frac{6}{\pi} \int_0^{\pi/6} \int_0^{\pi/2} \sigma_s(\alpha, \beta) \cos \alpha \, d\alpha \, d\beta, \qquad (5.1.30)$$

where the scattering cross section for a single crystal, $\sigma_s(\alpha, \beta)$, can be computed by integration of the phase matrix element G_{11} over a 4π solid angle.

According to the optical theorem, the extinction cross section of a single particle is twice the geometric cross section area in the limits of geometric optics. The geometric cross-section area for an arbitrarily oriented hexagon is

$$\sigma(\alpha, \beta) = \frac{3\sqrt{3}}{2} r^2 \sin \alpha + 2rL \cos \alpha \cos \left(\frac{\pi}{6} - \beta \right), \qquad (5.1.31)$$

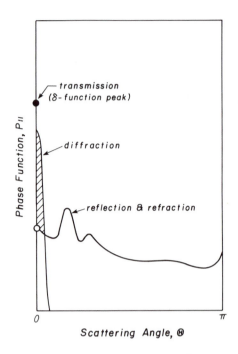

FIG. 5.5   A schematic representation of the components of the phase function P_{11} for randomly oriented hexagonal ice crystals.

where L is the length of the hexagon and r is the radius. Thus, the extinction cross section for a sample of randomly oriented ice crystals of the same size may be expressed in the form

$$\hat{\sigma}_e = \frac{6}{\pi} \int_0^{\pi/6} \int_0^{\pi/2} 2\sigma(\alpha, \beta) \cos \alpha \, d\alpha \, d\beta = \frac{3r^2}{2} \left[\sqrt{3} + 4(\frac{L}{2r}) \right] = \frac{S}{2}, \quad (5.1.32)$$

where S is the surface area of a hexagonal cylinder (Vouk, 1948).

Computations of the angular scattering patterns for hexagonal crystals based on the geometric ray-tracing method were first reported by Jacobowitz (1971), assuming infinitely long hexagonal columns. Wendling et al. (1979) and Coleman and Liou (1981) have undertaken a more comprehensive ray-tracing analysis to compute the phase functions for finite hexagonal columns and plates. Cai and Liou (1982), Takano and Jayaweera (1985), and Muinonen et al. (1989) have developed scattering models for arbitrarily oriented hexagonal columns and plates. Figure 5.6 illustrates the phase functions for randomly oriented ice columns (120 μm) and plates (20 μm) for an incident wavelength of 0.7 μm (left diagram). Also shown for comparison is the measured phase function derived from a number of laboratory scattering experiments for small plates with sizes of ~5 μm that were generated in a cold chamber illuminated by a 0.6328 μm laser beam, as reported by Sassen and

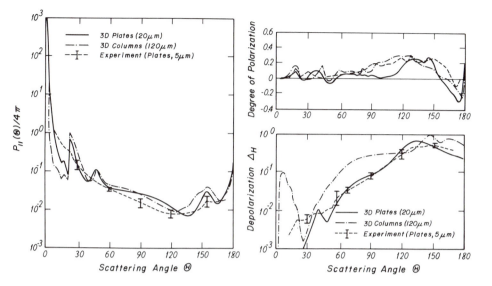

FIG. 5.6 Comparison of the computed and measured scattering phase functions (left), degree of linear polarization (upper right), and depolarization ratio (lower right) for randomly oriented columns and plates. The modal dimension of the plates observed in the scattering experiments is $\sim 5 \ \mu m$ (after Cai and Liou, 1982).

Liou (1979). The vertical bars depict the standard deviation of the measured data. The scattering patterns for hexagonal columns and plates are similar, with the 22° and 46° halos as well as a broader peak at $\sim$150° scattering angle. Experimental data show small maxima at $\sim$22° and 150° scattering angles, in general agreement with theoretical results. On the right of this figure are the computed and measured degree of linear polarization and depolarization ratios for columns and plates using horizontally polarized light. The theoretical results for 20 μm size plates closely match experimental data for plates having a modal diameter of $\sim$5 μm.

Takano and Liou (1989a) have developed a geometric ray-tracing program that accounts for the ice crystal size distribution as well as the possibility of horizontal orientation. New endeavors include a proper summation of contributions due to geometric reflection and refraction and Fraunhofer diffraction, and incorporation of the birefringent properties of ice. The computed phase function and polarization patterns from this program have been cross checked with available laboratory measured data (Dugin and Mirumyants, 1976; Nikiforova et al., 1977; Volkovitsky et al., 1980). The computed positions of optical features have also been used to identify numerous arcs and halos that occur in the atmosphere (Greenler, 1980).

As discussed in Subsection 4.3.2, the aspect ratio $L/2r$ of ice crystals may be related to the crystal length L, based on ice crystal observations. If the ice crystal size distribution is denoted by $n(L)$, then the phase matrix for a sample of ice crystals of different sizes may be obtained from

$$\mathbf{P}(\Theta) = \frac{\int_{L_1}^{L_2} \mathbf{P}(\Theta, L)\hat{\sigma}_s(L)n(L)\,dL}{\int_{L_1}^{L_2} \hat{\sigma}_s(L)n(L)\,dL},$$ (5.1.33)

where L_1 and L_2 are the lower and upper limits of crystal length. The scattering and extinction cross sections for a sample of ice crystals of different sizes are then given by

$$\sigma_{s,e} = \frac{1}{N}\int_{L_1}^{L_2} \hat{\sigma}_{s,e}(L)n(L)\,dL,$$ (5.1.34)

where N is the total number of ice crystals.

Table 5.2 lists the extinction coefficient, single-scattering albedo, and asymmetry factor for the ice crystal size distributions shown in Fig. 4.12 and for six wavelengths in the solar spectrum. Cirrus uncinus contains a significant number of large ice crystals in the second peak of the size distribution. For this reason, the extinction coefficient for this cloud is much larger than it is for the three other cloud types. The complex refractive indices of ice at each wavelength are averaged values over the wavelength band listed in the table, weighted by the solar irradiance. The extinction coefficients in the limit of geometric optics for a given size distribution are the same, regardless of the wavelength. The optical depth, $\tau = \beta_e \Delta z$, corresponding to each size distribution, can be obtained if the cloud thickness Δz is given. There is significant absorption at the $3.0\,\mu m$ wavelength with single-scattering albedos of about 0.53 that are almost independent of the ice crystal size distribution. Considerable absorption is seen at the 1.6 and $2.2\,\mu m$ wavelengths for cirrus uncinus consisting of large ice crystal sizes. At a given wavelength, the asymmetry factor for Cs, Ci (warm), and Ci (cold) are about the same since the phase functions for these clouds are very similar. For a given size distribution, the asymmetry factor increases toward the longer wavelengths, where absorption is increased. A typical phase function for an ice crystal cloud (Cs) at $0.5\,\mu m$ is tabulated in Table 5.3.

Based on the computational results for the scattering and absorption properties of hexagonal ice crystals and equivalent ice spheres, an equivalent sphere with the same area or volume as a hexagonal ice crystal is inadequate to reproduce single-scattering properties of hexagonal ice crystals. In general, equivalent ice spheres generate larger asymmetry factors and smaller single-scattering albedos than hexagonal ice particles. This is especially evident for wavelengths at which moderate absorption occurs. The geometric ray-tracing approach for the calculation of the scattering and absorption properties of ice crystals should be valid for solar wavelengths from 0.2 to $\sim 3.5\,\mu m$, in view of the observed ice crystal sizes in cirrus clouds (~ 20–$2000\,\mu m$). However, for thermal ir wavelengths (e.g., $10\,\mu m$) this approach may not be appropriate for small ice crystals. Also, ice crystals such

Table 5.2 Extinction coefficient $\beta_e (km^{-1})^a$, single-scattering albedo $\tilde{\omega}$, and asymmetry factor g, for four cirrus models and six solar wavelengths; the real and imaginary refractive indices are averaged values for the spectral band with limits λ_1 and λ_2.

$\lambda(\mu m)$ (λ_1, λ_2)	m_r	m_i		Cs	Ci uncinus	Ci (warm)	Ci (cold)
0.55	1.311	3.110×10^{-9}	β_e	0.3865	2.6058	0.6525	0.1662
(0.2, 0.7)			$\tilde{\omega}$	0.9999	0.9999	0.9999	0.9999
			g	0.7824	0.8404	0.7889	0.7724
1.0	1.302	1.931×10^{-6}	β_e	0.3865	2.6058	0.6525	0.1662
(0.7, 1.3)			$\tilde{\omega}$	0.9995	0.9981	0.9994	0.9997
			g	0.7905	0.8448	0.7945	0.7780
1.6	1.290	2.128×10^{-4}	β_e	0.3865	2.6058	0.6525	0.1662
(1.3, 1.9)			$\tilde{\omega}$	0.9658	0.9004	0.9628	0.9810
			g	0.8100	0.8778	0.8129	0.7925
2.2	1.263	7.997×10^{-4}	β_e	0.3865	2.6058	0.6525	0.1662
(1.9, 2.5)			$\tilde{\omega}$	0.9185	0.7996	0.9154	0.9528
			g	0.8436	0.9116	0.8440	0.8220
3.0	1.242	1.424×10^{-1}	β_e	0.3865	2.6058	0.6525	0.1662
(2.5, 3.5)			$\tilde{\omega}$	0.5321	0.5309	0.5327	0.5338
			g	0.9653	0.9728	0.9634	0.9583
3.7	1.401	7.178×10^{-3}	β_e	0.3865	2.6058	0.6525	0.1662
(3.5, 4.0)			$\tilde{\omega}$	0.7127	0.5889	0.7269	0.7912
			g	0.8583	0.9356	0.8442	0.8063

[a] Note that β_e is independent of wavelength in the limits of geometric optics but is related to ice crystal number density. The number densities for Cs, Ci, Ci (warm), and Ci (cold) are 0.187, 0.213, 0.442, and 0.176 cm^{-3}, respectively.

as bullet rosettes, hollow columns, and irregular shapes cannot be treated exactly by the theoretical method for light scattering. Innovative approaches are required. Although the scattering patterns computed from the geometric ray-tracing program match some results derived from laboratory measurements, well-controlled laboratory scattering and ice crystal physics experiments are needed to produce reliable scattering and absorption data for independent verification of theory.

5.1.4.2 Light scattering by nonspherical particles: exact theoretical approach

Approximations for hexagonal ice crystals may be made using circular cylinders and plates. Since these particle shapes can be defined by an appropriate coordinate system, an exact solution for the scattered field may be obtained from the wave equation. Rayleigh (1918) has derived the exact solution for the scattering of a homogeneous dielectric infinite circular cylinder for normal incidence, that is, the light beam is perpendicular to the axis of the cylinder. The case involving arbitrarily oblique incidence has been solved by Wait (1955). Subsequent numerical investigations concerning scattering by infinite circular cylinders have been carried out by van de Hulst (1957), Greenberg et al. (1967), Kerker (1969), and Liou

Table 5.3. Phase function for the cirrostratus cloud model at $\lambda = 0.5\,\mu m$

Θ	P_{11}	Θ	P_{11}	Θ	P_{11}	Θ	P_{11}
0.00	1.083×10^5	32	1.13×10^0	82	3.18×10^{-1}	132	1.04×10^{-1}
0.10	6.037×10^4	33	9.85×10^{-1}	83	3.14×10^{-1}	133	1.02×10^{-1}
0.20	3.231×10^4	34	8.54×10^{-1}	84	3.11×10^{-1}	134	1.02×10^{-1}
0.30	1.809×10^4	35	7.24×10^{-1}	85	3.11×10^{-1}	135	9.96×10^{-2}
0.40	9.985×10^3	36	6.26×10^{-1}	86	3.08×10^{-1}	136	1.02×10^{-1}
0.50	5.477×10^3	37	5.55×10^{-1}	87	3.03×10^{-1}	137	1.12×10^{-1}
0.60	3.210×10^3	38	5.55×10^{-1}	88	2.98×10^{-1}	138	1.21×10^{-1}
0.70	2.106×10^3	39	4.45×10^{-1}	89	2.95×10^{-1}	139	1.27×10^{-1}
0.80	1.502×10^3	40	4.12×10^{-1}	90	2.91×10^{-1}	140	1.35×10^{-1}
0.90	1.095×10^3	41	4.03×10^{-1}	91	2.89×10^{-1}	141	1.44×10^{-1}
1.00	7.875×10^2	42	3.74×10^{-1}	92	2.87×10^{-1}	142	1.54×10^{-1}
1.10	5.550×10^2	43	3.57×10^{-1}	93	2.85×10^{-1}	143	1.66×10^{-1}
1.20	3.885×10^2	44	4.09×10^{-1}	94	2.83×10^{-1}	144	1.78×10^{-1}
1.30	2.758×10^2	45	5.44×10^{-1}	95	2.82×10^{-1}	145	1.91×10^{-1}
1.40	2.027×10^2	46	7.44×10^{-1}	96	2.81×10^{-1}	146	2.04×10^{-1}
1.50	1.563×10^2	47	8.88×10^{-1}	97	2.78×10^{-1}	147	2.13×10^{-1}
1.60	1.267×10^2	48	8.78×10^{-1}	98	2.75×10^{-1}	148	2.25×10^{-1}
1.70	1.069×10^2	49	8.10×10^{-1}	99	2.72×10^{-1}	149	2.43×10^{-1}
1.80	9.266×10^1	50	7.15×10^{-1}	100	2.68×10^{-1}	150	2.63×10^{-1}
1.90	8.151×10^1	51	6.56×10^{-1}	101	2.62×10^{-1}	151	2.87×10^{-1}
2.00	7.210×10^1	52	5.91×10^{-1}	102	2.56×10^{-1}	152	3.04×10^{-1}
3.00	2.223×10^1	53	5.09×10^{-1}	103	2.53×10^{-1}	153	3.08×10^{-1}
4.00	9.666×10^0	54	4.52×10^{-1}	104	2.48×10^{-1}	154	3.09×10^{-1}
5.00	5.198×10^0	55	4.16×10^{-1}	105	2.42×10^{-1}	155	3.07×10^{-1}
6.00	3.208×10^0	56	3.89×10^{-1}	106	2.35×10^{-1}	156	2.99×10^{-1}
7.00	2.182×10^0	57	3.72×10^{-1}	107	2.27×10^{-1}	157	2.80×10^{-1}
8.00	1.598×10^0	58	3.58×10^{-1}	108	2.21×10^{-1}	158	2.63×10^{-1}
9.00	1.236×10^0	59	3.47×10^{-1}	109	2.16×10^{-1}	159	2.52×10^{-1}
10.00	1.013×10^0	60	3.46×10^{-1}	110	2.11×10^{-1}	160	2.36×10^{-1}
11.00	8.296×10^{-1}	61	3.44×10^{-1}	111	2.05×10^{-1}	161	2.13×10^{-1}
12.00	7.494×10^{-1}	62	3.43×10^{-1}	112	1.99×10^{-1}	162	1.95×10^{-1}
13.00	6.753×10^{-1}	63	3.44×10^{-1}	113	1.96×10^{-1}	163	1.77×10^{-1}
14.00	6.220×10^{-1}	64	3.43×10^{-1}	114	1.89×10^{-1}	164	1.66×10^{-1}
15.00	5.681×10^{-1}	65	3.40×10^{-1}	115	1.79×10^{-1}	165	1.57×10^{-1}
16.00	5.248×10^{-1}	66	3.38×10^{-1}	116	1.64×10^{-1}	166	1.58×10^{-1}
17.00	4.883×10^{-1}	67	3.37×10^{-1}	117	1.42×10^{-1}	167	1.69×10^{-1}
18.00	4.598×10^{-1}	68	3.36×10^{-1}	118	1.27×10^{-1}	168	1.90×10^{-1}
19.00	4.409×10^{-1}	69	3.36×10^{-1}	119	1.18×10^{-1}	169	2.35×10^{-1}
20.00	4.227×10^{-1}	70	3.36×10^{-1}	120	1.11×10^{-1}	170	2.71×10^{-1}
21.00	1.935×10^0	71	3.35×10^{-1}	121	1.04×10^{-1}	171	2.84×10^{-1}
22.00	5.333×10^0	72	3.34×10^{-1}	122	9.93×10^{-2}	172	2.88×10^{-1}
23.00	6.137×10^0	73	3.33×10^{-1}	123	9.86×10^{-2}	173	2.63×10^{-1}
24.00	6.043×10^0	74	3.32×10^{-1}	124	9.79×10^{-2}	174	2.67×10^{-1}
25.00	4.660×10^0	75	3.31×10^{-1}	125	9.70×10^{-2}	175	3.04×10^{-1}
26.00	3.665×10^0	76	3.29×10^{-1}	126	9.67×10^{-2}	176	3.81×10^{-1}
27.00	2.955×10^0	77	3.27×10^{-1}	127	9.68×10^{-2}	177	5.76×10^{-1}
28.00	2.404×10^0	78	3.25×10^{-1}	128	9.76×10^{-2}	178	7.98×10^{-1}
29.00	1.982×10^0	79	3.24×10^{-1}	129	9.97×10^{-2}	179	1.01×10^0
30.00	1.638×10^0	80	3.22×10^{-1}	130	1.01×10^{-1}	180	1.18×10^0
31.00	1.342×10^0	81	3.21×10^{-1}	131	1.03×10^{-1}		

(1972). An exact solution for a finite circular cylinder cannot be obtained because the electric field must be periodic along the axis of the cylinder.

An exact scattering solution, similar to the Mie theory, may be derived for particles with spheroidal shapes. Prolate and oblate spheroids may be used to approximate cylinders and plates, respectively. The spheroidal coordinates are obtained by the rotation of an ellipse about an axis of symmetry. The spheroidal coordinate systems may be defined by the following: η, the angular coordinate; ξ, the radial distance; and ϕ, the azimuthal angle. Shown in Fig. 5.7 are the prolate spheroidal coordinates, which can be related to the Cartesian coordinates by the following transformation:

$$x = \ell(1 - \eta^2)^{1/2}(\xi^2 - 1)^{1/2} \cos\phi, \tag{5.1.35a}$$

$$y = \ell(1 - \eta^2)^{1/2}(\xi^2 - 1)^{1/2} \sin\phi, \tag{5.1.35b}$$

$$z = \ell\eta\xi, \tag{5.1.35c}$$

where ℓ is the semifocal distance, $-1 \leq \eta \leq 1$, $1 \leq \xi < \infty$, and $0 \leq \phi \leq 2\pi$. For the oblate system, the ξ term in x and y should be replaced by $(\xi^2 + 1)^{1/2}$ with $0 \leq \xi < \infty$. The size and shape of an ellipse are defined by the semifocal distance and the eccentricity e. For the prolate and oblate systems, $e = 1/\xi_0$ and $1/(\xi_0^2 + 1)^{1/2}$, respectively, where ξ_0 is the value of ξ at the surface. If $\ell = 0$, the spheroidal coordinates reduce to the spherical coordinates. In the far field, $\xi \to \infty$, $\eta \to \cos\theta$, and $\ell\xi \to r$.

If ψ satisfies the scalar wave equation

$$\nabla^2\psi + k^2 m^2 \psi = 0, \tag{5.1.35d}$$

vectors associated with electric and magnetic fields can be expressed in the spheroidal coordinate system such that they satisfy the vector wave equations (replacing ψ with electric or magnetic vectors). The solution of the scalar wave equation can be expressed in terms of scalar spheroidal wave functions.

Consider an ensemble of spheroids of the same size that are randomly oriented in space. In this case, the scattering phase matrix is given by Eq. (5.1.11) and may be obtained from

$$\frac{\mathbf{P}(\Theta)}{4\pi} = \frac{1}{k^2\bar{\sigma}_s} \int_0^{2\pi}\int_0^{\pi} \mathbf{Z}(\Theta, \Phi; \zeta, \chi)\sin\zeta \, d\xi \, d\chi, \tag{5.1.36}$$

where

$$\mathbf{Z}(\Theta, \Phi; \zeta, \chi) = \mathbf{L}(\pi - \gamma)\mathbf{F}(\theta, \phi)\mathbf{L}(-\chi). \tag{5.1.37}$$

The transformation matrix $\mathbf{F}$ is defined in Eq. (5.1.5), but it consists of six independent elements that are equivalent to those in Eq. (5.1.12). The transformation matrix elements are directly related to the solutions of the wave equation. In the

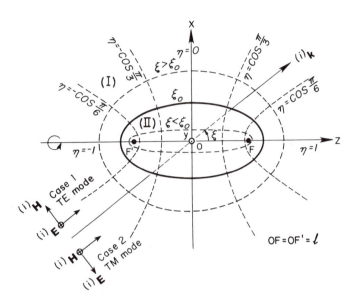

FIG. 5.7   Coordinate system (η, ξ, ϕ) for the scattering of a prolate spheroid with a semifocal distance ℓ. The z axis is chosen as the axis of revolution. The incident plane contains the incident direction and the x axis. The x axis is in the incident plane. For the TM mode, **E** is in the incident plane, while for the TE mode **H** is in the incident plane. The incident angle ζ is the angle in the incident plane between the incident direction and the z axis (after Asano and Yamamoto, 1975).

far field, these elements are functions of polar (θ) and azimuthal (ϕ) angles with respect to the symmetric axis of the spheroid (Fig. 5.7). Applying the rotational matrices **L** transforms **F** to the scattering plane. These matrices are defined by

$$
\mathbf{L}(\pi - y) = \mathbf{L}(-y) = \begin{bmatrix} 1 & 0 & 0 & 0 \\ 0 & \cos 2y & -\sin 2y & 0 \\ 0 & \sin 2y & \cos 2y & 0 \\ 0 & 0 & 0 & 1 \end{bmatrix},
\qquad (5.1.38)
$$

where $y = \gamma$ or χ. From the spherical geometry, (θ, ϕ) and γ are related to the scattering angle Θ, the azimuthal angle Φ, and the angles defining the orientation of the spheroid (ζ, χ) in the form

$$
\cos \theta = \cos \Theta \, \cos \zeta + \sin \Theta \, \sin \zeta \, \cos(\chi - \Phi), \qquad (5.1.38a)
$$

$$
\cos \phi = \frac{\cos \Theta \, \sin \zeta - \sin \Theta \, \cos \zeta \, \cos(\chi - \Phi)}{\pm \sin \theta}, \qquad (5.1.38b)
$$

$$
\cos \gamma = \frac{\cos \zeta \, \sin \Theta - \sin \zeta \, \cos \Theta \, \cos(\chi - \Phi)}{\pm \sin \theta}, \qquad (5.1.38c)
$$

where $+ \sin \theta$ is for $0 < (\chi - \Phi) < \pi$, and $- \sin \theta$ is for $\pi < (\chi - \Phi) < 2\pi$. The averaged scattering or extinction cross sections for randomly oriented spheroids may be computed by performing the integration over the orientation angle ζ as follows:

$$\bar{\sigma}_{s,e} = \frac{1}{2} \int_0^{\pi/2} \sigma_{s,e}(\zeta) \sin \zeta \, d\zeta. \tag{5.1.39}$$

The amplitude functions defined in Eq. (5.1.5c) have been derived by Asano and Yamamoto (1975) for spheroids and are given by

$$S_1(\theta, \phi) = \sum_{m=0}^{\infty} \sum_{n=m}^{\infty} [\alpha_{1,mn} \, \sigma_{mn}(\theta) + \beta_{1,mn} \, \chi_{mn}(\theta)] \cos m\phi, \tag{5.1.40a}$$

$$S_3(\theta, \phi) = \sum_{m=0}^{\infty} \sum_{n=m}^{\infty} [\alpha_{1,mn} \, \chi_{mn}(\theta) + \beta_{1,mn} \, \sigma_{mn}(\theta)] \sin m\phi, \tag{5.1.40b}$$

$$S_2(\theta, \phi) = \sum_{m=0}^{\infty} \sum_{n=m}^{\infty} [\alpha_{2,mn} \, \sigma_{mn}(\theta) + \beta_{2,mn} \, \chi_{mn}(\theta)] \cos m\phi, \tag{5.1.40c}$$

$$S_4(\theta, \phi) = \sum_{m=0}^{\infty} \sum_{n=m}^{\infty} [\alpha_{2,mn} \, \chi_{mn}(\theta) + \beta_{2,mn} \, \sigma_{mn}(\theta)] \sin m\phi. \tag{5.1.40d}$$

In Eqs. (5.1.40a – d), we have the following definitions:

$$\sigma_{mn}(\theta) = \frac{m S_{mn}(\cos \theta)}{\sin \theta}, \tag{5.1.41a}$$

$$\chi_{mn}(\theta) = \frac{d}{d\theta} S_{mn}(\cos \theta), \tag{5.1.41b}$$

where S_{mn} are the spheroidal angular functions. The terms $\alpha_{1,2}$ and $\beta_{1,2}$ are coefficients that can be determined from the four boundary conditions for the electric and magnetic fields at the surface of a spheroid. These are functions of the orientational angles (ζ, χ), incident wavelength, refractive index, and aspect ratio. The scattering cross section is dependent on ζ only (see Fig. 5.7) and can be evaluated from Eq. (5.1.8). The extinction cross section for a spheroid may be obtained from the extinction theorem [see Eq. (5.1.21)] and is given by

$$\sigma_e(\zeta) = \frac{4\pi}{k^2} R_e[S_1(\zeta, 0) + S_2(\zeta, 0)], \tag{5.1.42}$$

Figure 5.8 illustrates a number of scattering phase matrix elements as functions of the scattering angle for randomly oriented prolate spheroids. The size parameter $(2\pi a/\lambda)$, aspect ratio (semimajor axis a divided by the semiminor axis b), and refractive index used in the computations are 15, 5, and 1.33, respectively (Asano and Sato, 1980). Results for the area-equivalent spheres are also shown for comparison purposes. For the phase function, the rainbow features produced by spheres are absent in the case of spheroids. Spheroids have smaller backscattering than spheres. There are also significant differences in the linear polarization pattern. Large polarization is shown for prolate spheroids, which show large positive

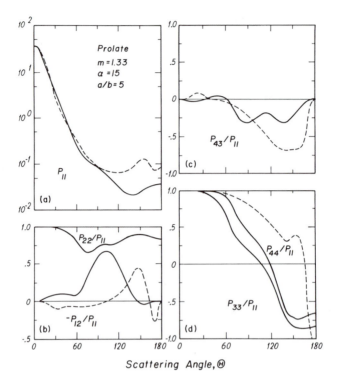

FIG. 5.8 Angular distribution of the normalized scattering phase matrix elements for randomly oriented prolate spheroids with $m = 1.33$, $\alpha = 15$, and $a/b = 5$: (a) phase function P_{11}, (b) P_{22}/P_{11} and the degree of linear polarization $-P_{12}/P_{11}$, (c) P_{43}/P_{11}, and (d) P_{33}/P_{11} and P_{44}/P_{11}. The scattering matrix elements for area-equivalent spheres are shown by dotted lines (after Asano and Sato, 1980).

polarization at ~90° scattering angle. Values of P_{43} and P_{44} also show noticeable differences between spheroids and spheres. In the case of spheres, $P_{11} = P_{22}$ and $P_{33} = P_{44}$. It is not computationally feasible to obtain the scattering and polarization properties of nonspherical particles by replacing them with area- (or volume-) equivalent spheres. The use of equivalent spheres would generally result in overestimation of the backscattering energy and, at the same time, an underestimation of the linear polarization of the scattered light beam.

The scattering problem involving nonspherical particles may also be solved by the integral equation method. In this approach, a solution is sought for the scattering field in terms of the incident field and the physical characteristics of the scattering object. According to the basic principle in electric and magnetic fields, the scattered field may be expressed by a set of surface currents located coincident with the surface of a scatterer. These surface currents may be determined by the internal fields, which may subsequently be expressed in terms of the incident field.

Consider a scattering volume bounded by a closed smooth surface and let $\mathbf{E}^i(\mathbf{r})$ denote the incident electric field at any point $\mathbf{r}$, with respect to an origin that is suitably located inside the scatterer. The scattered electric field may then be determined by

$$\mathbf{E}^s(\mathbf{r}) = \nabla \times \int_S [\mathbf{n} \times \mathbf{E}^t(\mathbf{r}')]g(k\mathbf{R})\,dS'$$
$$- \nabla \times \nabla \times \int_S \frac{1}{i\omega\epsilon_0}[\mathbf{n} \times \mathbf{H}^t(\mathbf{r}')]g(k\mathbf{R})\,dS', \qquad (5.1.43)$$

where $\mathbf{n} \times \mathbf{E}^t$ and $\mathbf{n} \times \mathbf{H}^t$ represent the equivalent magnetic and electric surface current densities, respectively; $\mathbf{E}^t$ and $\mathbf{H}^t$ are the total electric and magnetic fields external to the scatterer; $\mathbf{n}$ is the unit vector normal to the surface; ϵ_0 is the permittivity; $\mathbf{R} = \mathbf{r} - \mathbf{r}'$; ω is the circular frequency $(= kc)$; k is the wave number $(= 2\pi/\lambda)$; c is the velocity of light; and the free space Green's function is defined by

$$g(k\mathbf{R}) = e^{ik\mathbf{R}}/(4\pi\mathbf{R}). \qquad (5.1.44)$$

The terms in the integrals of Eq. (5.1.43) are expanded in terms of the transverse dyadic Green function, which may be expressed by vector spherical harmonics. By equating the total electric field to zero everywhere inside the scatterer, we have $-\mathbf{E}^i(\mathbf{r}) = \mathbf{E}^s(\mathbf{r})$. It follows that Eq. (5.1.43) may be used to evaluate the total electric field in terms of the incident electric field. This can be done by expanding all the fields in terms of vector spherical harmonics and applying the orthogonality properties of these functions. To determine the unknown coefficients in the expansion, we use the boundary conditions at the surface of the scatterer that require the continuity of the tangential components of the electric and magnetic fields. Subsequently, the scattered electric field may be computed from Eq. (5.1.43). The preceding method is referred to as the *extended boundary condition method* and has been used to compute the scattering and absorption properties of spheroids (Barber and Yeh, 1975; Iskander et al., 1983). The upper limit of the size parameter to which this method is applicable is ~20. However, there are computational difficulties with this method in the case of highly elongated or flattened particles, and a significant computational effort is required for concave particles.

In principle, the integral equation method may be employed to solve the scattering problem involving particles of any size and shape, since the use of a defined coordinate system is not required. However, the practical application of this method to the determination of scattering fields involving irregular particles, such as ice crystals, has yet to be worked out. A general survey of the methodologies that may be used to solve the scattering problem involving small irregular particles has been given by Bohren and Huffman (1983).

5.2   Radiative properties of clouds

5.2.1   *Intensity and polarization of sunlight reflected by clouds*

Intensity and polarization of sunlight reflected by clouds are useful information for the identification of the cloud optical depth and cloud microphysical properties such as thermodynamic phase and mean particle size. In this section, we present the information content of sunlight reflected by clouds. Radiative transfer methodologies presented in Chapter 3 can be used to determine the radiative properties of clouds under plane-parallel conditions.

5.2.1.1   Water clouds

Figure 5.9 shows the intensity (normalized by $F_\odot/\pi$) and linear polarization of solar radiation ($\mu_0 F_\odot = 1$) at 0.5 μm reflected by a cumulus cloud with a droplet size distribution depicted in Fig. 4.10. The adding method for radiative transfer (Section 3.2) was used in the calculations. The patterns are presented as functions of the emergent angle θ when the sun is overhead ($\mu_0 = 1$) for a number of optical depths ranging from 0.25 to 128. Note that $\theta = 0°$ corresponds to the scattering angle of 180°. The rainbow ($\sim40°$) and glory ($\sim0°$) features contained in the backscattering directions are largely lost as the optical depth increases. For optically thick clouds ($\tau > 8$), the scattered light becomes fairly isotropic. However, the linear polarization patterns retain the rainbow and glory features that are typical in the case of spherical cloud droplets, even though the polarization is largely reduced with increasing optical depth.

Measurements of the reflected intensity and linear polarization of sunlight from clouds have been made by Coffeen (1979) using the infrared polarimeter aboard the NASA Convair 990. The polarimeter had a 1.5° field of view and contained five band passes centered at 1.24, 1.60, 2.22, 3.08, and 3.38 μm. The near-ir wavelengths were selected to minimize the contribution of Rayleigh scattering. Observations of a few dozen cloud systems were carried out over the Caribbean and western Atlantic. The aircraft altitude was usually at about 12 km with the clouds located at a variety of altitudes. The measurements were performed on the solar principal plane, $\phi - \phi_0 = 0°/180°$. Figure 5.10 shows the results from a layer of maritime altostratus clouds for a wavelength of 2.22 μm. The polarimeter was not calibrated to yield the absolute intensity. Interpretations were made from radiative transfer calculations using the results computed from the droplet size distribution cited below. The curves and observations in the bottom part of the figure are for the degree of linear polarization (%), while those at the top are for the relative intensity. An optical depth of 8 gives the closest agreement with the observed patterns.

The theoretical curves are computed using the size distribution given by $n(r) = \text{const} \times r^{(1-3b)/b} \exp(-r/ab)$, where a and b represent a mean effective

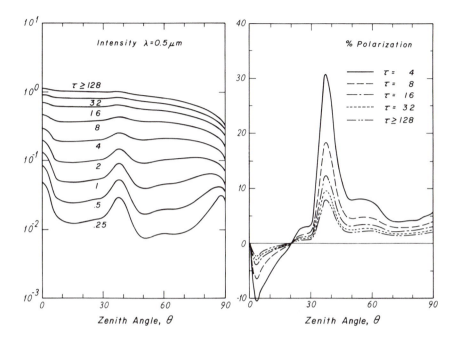

FIG. 5.9   Intensity and percentage of linear polarization of sunlight at 0.5 μm reflected by a cumulus cloud with the sun overhead ($\theta_0 = 0°$) as a function of the zenith angle. Results are shown for several optical thicknesses.

radius and an effective dispersion, respectively. In the calculations, b is set at 1/40 and three values of a (6, 14, and 28 μm) are used. The results from use of a mean effective radius of 14 μm fit the observed polarization values most closely. The observed polarization pattern consists of a maximum of 16% at about the 40° phase angle (180° minus the scattering angle), corresponding to the rainbow feature, and a maximum of about 4% at the 20° phase angle, corresponding to the supernumerary rainbow feature. Negative polarization (i.e., when the direction of polarization is predominantly parallel to the scattering plane) is shown in angles between the rainbow features and angles larger than 60°. The preceding presentation suffices to demonstrate the information content of the cloud optical depth and mean droplet size in the measured intensity and polarization patterns of sunlight reflected by water clouds.

5.2.1.2   Ice clouds

Scattering and polarization patterns for hexagonal ice crystals differ significantly from those for spherical water droplets (see Subsections 5.1.3 and 5.1.4). The reflected and transmitted (diffuse) intensities for randomly oriented ice columns with aspect ratios of 125 μm/50 μm as functions of the zenith angle for an overhead sun are shown in Fig. 5.11 (Takano and Liou, 1989b). These intensities as well as all

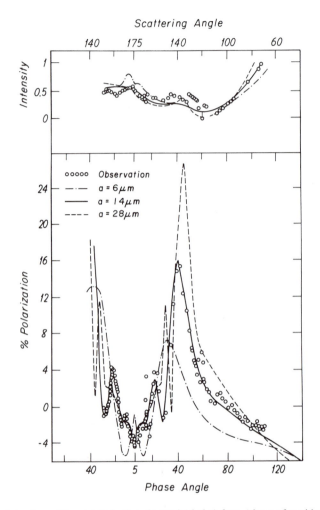

FIG. 5.10   Intensity and linear polarization observed (circles) from a layer of maritime altostratus clouds using a wavelength of 2.22 μm. Theoretical results use a droplet size distribution in the form: $r^{(1-3b)/b} \exp(-r/ab)$, where a and b are the effective mean radius and the variance of the size distribution, respectively. The optical depth used is 8, and b is fixed at $1/40$ (after Coffeen, 1979).

the intensity quantities presented hereafter are normalized with respect to $F_\odot/\pi$. For comparison, the results for area-equivalent ice spheres are also displayed. The reflected intensity increases with increasing optical depth. Significant differences between the reflected intensities for ice columns and spheres are seen. Ice spheres produce a peak intensity at $\theta = 45°$, associated with a combination of primary and secondary rainbow features due to single scattering, but ice columns have larger reflected intensity in other zenith angle regions. In the transmitted intensity pattern, the 22° and 46° halo features produced by ice columns are very distinct for small optical depths. They vanish when the optical depth is greater than about 16.

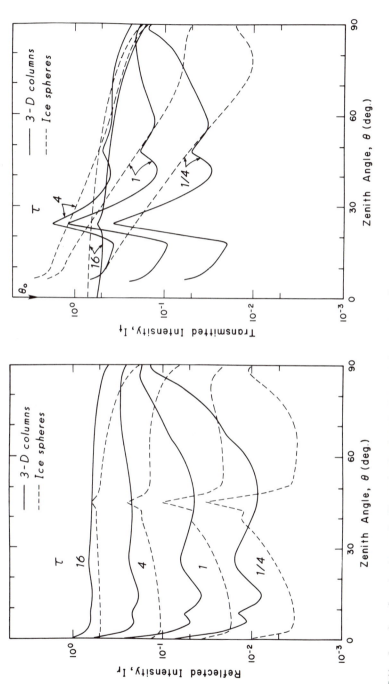

Fig. 5.11 Intensity reflected and transmitted by randomly oriented columns ($L/2r = 125\ \mu m/50\ \mu m$) and area-equivalent spheres with an overhead sun ($\theta_0 = 0°$) at $\lambda = 0.5\ \mu m$ (after Takano and Liou, 1989b).

Figure 5.12 shows the linear polarization of sunlight reflected by ice columns, ice plates, and ice spheres in the solar principal plane ($\phi - \phi_0 = 0°/180°$) when the position of the sun is at $\theta_0 = 50°$. For ice spheres, a maximum polarization of about 80% is shown at the $\sim 45°$ phase angle for an optical depth of 1. This maximum polarization is associated with the rainbow features produced by spherical particles and is absent for ice crystal clouds. Ice plates and columns both show negative polarization maximum at phase angles close to $0°$. In general, ice plates produce greater polarization than ice columns. Polarization decreases significantly for an optical depth of 16. However, its characteristic pattern does not vary with increasing optical depth. In particular, the neutral points (zero polarization) at phase angles of $\sim 18°$ for columns and $\sim 23°$ for plates remain nearly constant, regardless of the optical depth. These neutral points are also independent of the incident solar angle. It appears that the detection of neutral points may provide a means for the identification of particle shape.

Measurements of the linear polarization of sunlight reflected by cirrus clouds have also been reported by Coffeen (1979), as cited previously. Results are shown in Fig. 5.13. The observed cirrus is about 5 km thick. The computed polarization pattern used an optical depth of 64 and a solar zenith angle of $70°$ in order to match the observations (Takano and Liou, 1989b). The negative values around the $140°$ and $160°$ phase angles result from the outer and inner halos, respectively. Since there is no sharp peak due to the rainbow feature around the $\sim 45°$ phase angle in the observed polarization, the cloud particles must be nonspherical. The computed polarization for ice plates fits the observed values quite well. It appears that the cloud particles near the cloud top must be randomly oriented platelike ice crystals. There are deviations between the computed and observed polarization around the $0°$ phase angle (backscattering) because scattered light is partially blocked by the aircraft and because the cloud may contain irregular ice crystals, which would reduce polarization in backscattering directions. Concurrent cloud physics measurements are needed in order to develop remote sensing techniques for the detection of ice crystal clouds using the principle of multiple scattering and polarization.

5.2.2 *Near-infrared spectral characteristics of clouds*

As shown in Fig. 5.1, there are significant variations of the absorption properties of ice and water in the near-ir solar spectrum. In Subsection 3.8.2, we have introduced the absorption bands in the solar spectrum. In the near-ir region, absorption of solar radiation is primarily due to water vapor. Absorption of water vapor is strong in the band centers and falls off sharply away from these centers. Absorption by ice and water, on the other hand, is more consistent across each band, being less intense in the water vapor band centers and more intense away from the centers. Moreover, the maximum absorption due to ice and water does not coincide exactly with

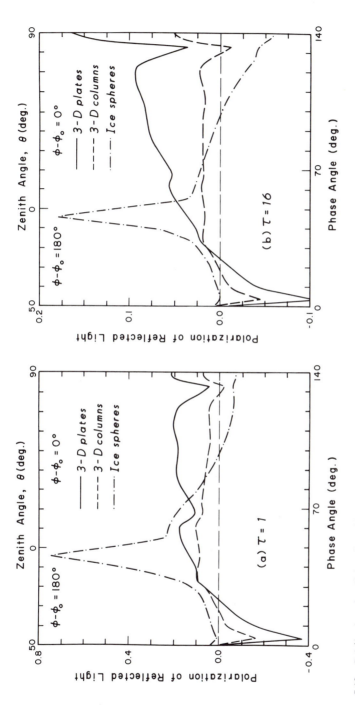

FIG. 5.12 Polarization of sunlight reflected by randomly oriented plates ($L/2r = 0.1$), columns ($L/2r = 2.5$), and area-equivalent spheres as a function of the phase angle in the solar principal plane ($\phi - \phi_0 = 0°/180°$) at $\lambda = 0.5~\mu m$. The solar zenith angle is $50°$. The optical depths considered are (a) 1 and (b) 16 (after Takano and Liou, 1989b).

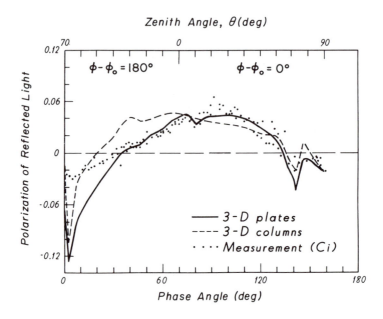

FIG. 5.13 Comparison of polarization of sunlight reflected by cirrus at $\lambda = 2.22\,\mu$m between measurements reported by Coffeen (1979) and the present computation. The ice crystal models used are 3D plates ($L/2r = 8\,\mu$m$/80\,\mu$m) and 3D columns ($L/2r = 100\,\mu$m$/40\,\mu$m) with an optical depth of 64 (after Takano and Liou, 1989b).

the water vapor band centers. There are regions in the near-ir where water vapor absorption dominates, whereas in other regions ice and water absorption is more important. Reflection of sunlight from clouds will depend on the optical depth and some measure of the particle size within the cloud. Variation of the reflectance in the near-ir region could provide a means for the influence of some physical properties of clouds. This possibility has been suggested by Hansen and Pollack (1970), who have interpreted spectral near-ir reflectances of clouds obtained from airborne measurements by Blau et al. (1966). Pollack et al. (1978) have illustrated that some characteristics of the clouds of Venus may be inferred from the observed spectral data in the near-ir region. Suggestions of the inference of the optical depth and mean radius from spectral reflectance measurements have also been made by Twomey and Seton (1980).

Measurements of the near-ir reflectance of ice crystal clouds that were generated in the laboratory have been reported by Zander (1966). Figure 5.14 shows the spectral reflectance for dense ice crystal clouds with temperatures ranging from $-28°$ to $-44°$C. These clouds are largely composed of small hexagonal plates and prisms less than $10\,\mu$m. The scattering angle of the observation was $150°$. The curve is an average of five runs indicated by the vertical bars. Minima at 1.5, 2, and 2.8 μm and maxima at 1.4, 1.8, and 2.3 μm of the reflectance patterns generally follow the behavior of the imaginary refractive index displayed in Fig. 5.1. Because of

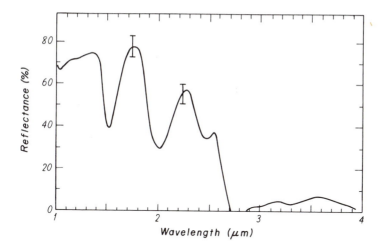

Fig. 5.14   Near-ir reflectance of ice clouds generated in the laboratory (after Zander, 1966).

the large absorption of ice for wavelengths longer than ∼3 μm, reflectance values are very small. Reflection of solar radiation by clouds in the near-ir region depends on both the scattering and absorption of cloud particles as well as on absorption due to water vapor within the clouds. The spectral reflectance in the near-ir is heavily dependent on the thermodynamic phase (ice or water), the particle size distribution and the optical depth.

The reflectance of clouds composed of water droplets in the near-ir has been determined from satellite measurements by Rozenberg et al. (1974) and from airborne radiometers by Twomey and Cocks (1982). Both attempted to interpret the observed data from theoretical calculations but without much success. In particular, Twomey and Cocks have performed radiative transfer calculations using the observed cloud droplet size distribution. The calculated spectral reflectances were markedly lower than the values attained from measured data. In order to match the observed reflectances, significant increases in the theoretical absorption coefficients were required. The disparities between measured and calculated spectral reflectances must be resolved before the inference of cloud microphysical properties may be made. King et al. (1990) have determined the spectral single-scattering albedo using a multi-wavelength scanning radiometer that measures the angular distribution of scattered radiation deep within stratus cloud layers at discrete wavelengths between 0.5 and 2.3 μm. In the diffusion domain, the ratio of the scattered intensities at nadir and zenith may be expressed in terms of the similarity parameter a, defined in Eq. (3.3.4e). The measured intensity ratio may then be used to evaluate the single-scattering albedo via the similarity parameter. The spectral single-scattering albedos determined from measurements are persistently smaller than calculated values, revealing more absorption within stratus clouds that could not be predicted from the present theory. (The anomalous absorption within clouds is discussed in Subsection 5.2.4.1).

As noted in Subsection 5.1.1 and Fig. 5.1, the imaginary refractive indices of ice and water have very similar patterns, except in the 1.6 μm region, where ice absorbs much more strongly than water. Phase functions for hexagonal ice crystals and spherical water droplets also exhibit profound differences, especially in the regions of the 22° and 46° halos for ice, as seen in Fig. 5.6, and the $\sim$138° rainbow for water, as seen in Fig. 5.2. In the backscattering directions, the rainbow feature will not be present for ice clouds as illustrated in Fig. 5.6. The specific optical properties of ice and water at the 1.6 μm wavelength and the backscattering characteristics of hexagonal ice crystals and water droplets could provide a means for differentiating between the two, using bidirectional reflectance measurements. As an illustration, Fig. 5.15 shows the reflected intensity ratios for a number of optical depths in the case of an overhead sun using the particle size distribution of cirrostratus (see Fig. 4.12) in the radiative transfer calculations. The adding method introduced in Section 3.2 was used. The ratios were obtained by dividing the reflected intensity at 1.6 μm by that at 0.5 μm. Water clouds have much larger values for all optical depths because of less absorption at 1.6 μm. Also evident is the significant rainbow feature for small optical depths at a zenith angle of $\sim$45°. Clearly, reflectance measurements at 1.6 μm from satellites would provide information about the cloud thermodynamic phase. When this wavelength is coupled with other solar wavelengths, it appears feasible that some physical properties of clouds such as the optical depth and mean particle size may be inferred from satellite measurements.

5.2.3   *Infrared radiative transfer in cloudy atmospheres*

In the thermal infrared, scattering in addition to emission and absorption takes place within clouds. The basic ir radiative transfer equation for gaseous absorption and emission, introduced in Section 2.1, must be modified to account for scattering processes.

Consider a plane-parallel cloud layer, and let the scattering coefficient for cloud particles be β_s and the absorption coefficient for cloud particles plus water vapor within the cloud be β_a. According to Kirchhoff's law, absorption is coupled with emission so that LTE is maintained. The source function in this case is the Planck function, B_ν. Let the source function associated with scattering be J_ν. The radiative transfer equation may then be written

$$\mu\frac{dI_\nu}{dz} = -\beta_a(I_\nu - B_\nu) - \beta_s(I_\nu - J_\nu)$$
$$= -\beta_e(I_\nu - S_\nu), \tag{5.2.1}$$

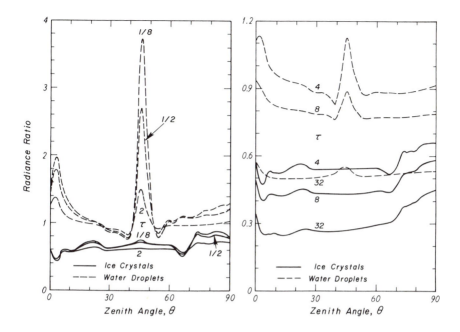

FIG. 5.15   Reflected intensity ratios for a number of optical depths in the case of overhead sun using the particle size distribution of cirrostratus. The ratios were obtained by dividing the reflected intensity at 1.6 μm by that at 0.5 μm.

where the extinction coefficients $\beta_e = \beta_s + \beta_a$ and the source function involving absorption and scattering processes is

$$S_\nu = (\beta_a B_\nu + \beta_s J_\nu)/\beta_e. \tag{5.2.2a}$$

It is an average of the two separate source functions, weighted by their respective absorption and scattering coefficients. The extinction coefficient is the inverse of the mean free path of a photon before scattering or absorption. Using the definition of the single-scattering albedo, $\tilde{\omega}_\nu = \beta_s/\beta_e$, upon which a free path will end with a scattering event, the source function may be expressed by

$$S_\nu = (1 - \tilde{\omega}_\nu)B_\nu + \tilde{\omega}_\nu J_\nu. \tag{5.2.2b}$$

The source function for scattering is associated with multiple scattering processes. In the thermal infrared, it suffices to take the azimuth-independent component:

$$J_\nu = \frac{1}{2} \int_{-1}^{1} P(\mu, \mu') I_\nu(\tau, \mu') \, d\mu', \tag{5.2.3}$$

where the azimuth-independent phase function, $P(\mu, \mu')$, has been defined in Eq. (3.1.15).

If the cloud as a whole is a blackbody, it would behave just like the earth's surface. In this case, radiation from below and above the cloud would not be able to penetrate the cloud. The emitted radiance at the cloud top or bottom is given by the Planck function. Most clouds that are composed of water droplets are black clouds, while clouds that are composed of ice crystals are generally non-black.

Measurements of the thermal ir radiance from space have been made by orbiting meteorological satellites, as illustrated in Fig. 2.1. In the late 1970s three Fourier spectrometers, in the spectral range from 400 to 1600 cm^{-1}, were launched on board METEOR satellites. Figure 5.16 shows the ir emission spectra measured in clear and cirrus cloudy atmospheres with a spectral resolution of 5 cm^{-1} (Spänkuch and Döhler, 1985). The presence of cirrus clouds significantly reduces the upwelling radiance in the entire spectral region, except in the center of the CO_2 15 μm band. In order to interpret these spectral radiances, radiative transfer calculations involving clouds are required. In the following we introduce the discrete-ordinates and adding methods for thermal ir radiative transfer.

5.2.3.1   Discrete-ordinates method

Based on the preceding discussion, the basic ir radiative transfer equation that accounts for absorption, emission, and scattering processes may be written in the form

$$\mu \frac{dI_\nu(\tau,\mu)}{d\tau} = I_\nu(\tau,\mu) - \frac{\tilde{\omega}_\nu}{2} \int_{-1}^{1} P(\mu,\mu')I_\nu(\tau,\mu')\,d\mu'$$
$$- (1-\tilde{\omega}_\nu)B_\nu(T_c), \qquad (5.2.4)$$

where T_c is the cloud temperature. The phase function can be expanded in the form of the Legendre polynomials, and the integral can be replaced by summation in a manner described in Subsection 3.2.1. Thus we have

$$\mu_i \frac{dI_\nu(\tau,\mu_i)}{d\tau} = I_\nu(\tau,\mu_i) - \frac{\tilde{\omega}_\nu}{2} \sum_{\ell=0}^{N} \tilde{\omega}_\ell P_\ell(\mu_i) \sum_j a_j P_\ell(\mu_j)I_\nu(\tau,\mu_j)$$
$$- (1-\tilde{\omega}_\nu)B_\nu(T_c). \qquad (5.2.5)$$

A particular solution for this set of differential equations is the Planck function. Thus the complete solution may be written in the form

$$I_\nu(\tau,\mu_i) = \sum_j L_j \phi_j(\mu_i)e^{-k_j\tau} + B_\nu(T_c), \qquad (5.2.6)$$

where k_j and ϕ_j are the eigenvalues and eigenvectors defined in Eq. (3.2.7). Consider now a cloud layer with emission and absorption atmospheres above and below. Let the optical depth at the cloud top and bottom be denoted as τ_t and τ_b,

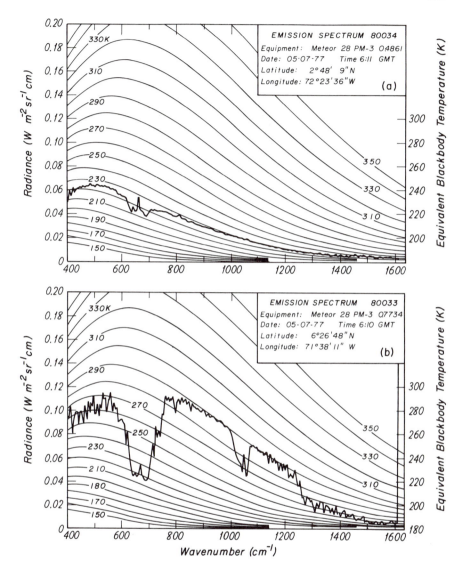

FIG. 5.16 Observed radiance spectra for (a) cirrus and (b) cloudless conditions obtained from METEOR satellites (after Spänkuch and Döhler, 1985).

respectively. As illustrated in Section 2.1, the solutions for the downward and upward radiances at the cloud top and bottom are given, respectively, by

$$I_\nu^-(\tau_t, -\mu_i) = \int_{\tau_t}^{\tau_\infty} B_\nu(T(\tau)) \exp\left(\frac{-(\tau - \tau_t)}{\mu_i}\right) \frac{d\tau}{\mu_i}, \tag{5.2.7}$$

$$I_\nu^+(\tau_b, \mu_i) = B_\nu(T_s)e^{-\tau_b/\mu_i} + \int_0^{\tau_b} B_\nu(T(\tau)) \exp\left(\frac{-(\tau_b - \tau)}{\mu_i}\right) \frac{d\tau}{\mu_i}, \tag{5.2.8}$$

where T_s is the surface temperature. The difference of two optical depths is defined as follows:

$$\tau_2 - \tau_1 = \int_{z_1}^{z_2} k_\nu(z)\rho_a(z)\,dz, \qquad (5.2.9)$$

where k_ν is the absorption coefficient and ρ_a is the density of absorbing gases. Having the boundary conditions given, the constant of proportionality L_j can be obtained. Upward radiances at the cloud top, $I_\nu(\tau_t, \mu_i)$, may then be evaluated via Eq. (5.2.6). Subsequently, attenuation of these radiances to TOA may be carried out using an expression similar to Eq. (5.2.8), except that $B_\nu(T_s)$ is replaced by $I_\nu(\tau_t, \mu_i)$ and the optical depth involved is from τ_t to τ_∞.

Because of large gaseous absorption in various absorption bands, the most important spectral wave numbers at which clouds interact most significantly with the thermal radiation field are in the window regions at 3.7 and 10 μm. In Subsection 2.2.5.2, we have presented the absorption coefficients of water vapor for window wavelengths. Finally, if the treatment of nonisothermal properties of clouds is necessary, the procedure outlined in Section 3.7.1 for radiative transfer in nonhomogeneous atmospheres may be followed with a modification by replacing the direct solar radiation term with the Planck function.

5.2.3.2 Adding method

The adding principle for radiative transfer presented in Subsection 3.2.2 may be used to evaluate the ir radiative transfer in cloud layers. If a thin layer, such that $\Delta\tau$ is very small, is considered, then the upward and downward emission from a gray body in the emergent direction μ may be expressed by

$$J^+(\mu) = J^-(\mu) \cong (1 - \tilde{\omega}_\nu) B_\nu(T) \Delta\tau / \mu. \qquad (5.2.10)$$

On the basis of tracing the light beam successively, as illustrated in Fig. 5.17, the upward (J_u) and downward (J_d) source terms at the interface of the two layers denoted by a and b are given by

$$
\begin{aligned}
J_d &= [1 + R_a^* R_b + (R_a^* R_b)^2 + \ldots] J_a^- \\
&\quad + [1 + R_a^* R_b + (R_a^* R_b)^2 + \ldots] R_a^* J_b^+ \\
&= (1 + S)(J_a^- + R_a^* J_b^+), \qquad (5.2.11a) \\
J_u &= [1 + R_b R_a^* + (R_b R_a^*)^2 + \ldots] J_b^+ \\
&\quad + [1 + R_b R_a^* + (R_b R_a^*)^2 + \ldots] R_b J_a^- \\
&= (1 + S^*)(J_b^+ + R_b J_a^-), \qquad (5.2.11b)
\end{aligned}
$$

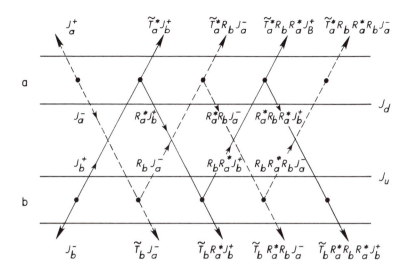

FIG. 5.17   Geometric configuration for the adding method involving thermal emission. The dashed and solid lines represent radiation emitted from layers a and b, respectively. All the notations are defined in the text.

where the superscript $*$ denotes quantities associated with radiation from below, and the multiple reflection terms are defined by

$$S = R_a^* R_b (1 - R_a^* R_b)^{-1}, \tag{5.2.12a}$$

$$S^* = R_b R_a^* (1 - R_b R_a^*)^{-1}. \tag{5.2.12b}$$

Using the upward and downward source terms at the interface, the upward and downward source terms at the top and bottom of the combined layer are given by

$$J_{ab}^+ = J_a^+ + \tilde{T}_a^* J_u, \tag{5.2.13a}$$

$$J_{ab}^- = J_b^- + \tilde{T}_b J_d, \tag{5.2.13b}$$

where the total transmission $\tilde{T}$ is the sum of the diffuse plus the direct components:

$$\tilde{T}_{a,b} = T_{a,b} + e^{-\tau_{a,b}/\mu}. \tag{5.2.14}$$

For a homogeneous layer, the reflection function and diffuse transmission function are the same regardless of radiation from below or above. That is, $R_{a,b}^* = R_{a,b}$ and $T_{a,b}^* = T_{a,b}$. It follows that the multiple reflection terms $S^* = S$ in Eqs. (5.2.12a) and (5.2.12b). The reflection and diffuse transmission functions can be computed from the phase function in the manner described in Eqs. (3.2.33) and (3.2.34), except that μ_0 is changed to μ', viz.,

$$R(\mu, \mu') = \frac{\tilde{\omega}\Delta\tau}{4\mu\,\mu'} P(\mu, -\mu'), \tag{5.2.15a}$$

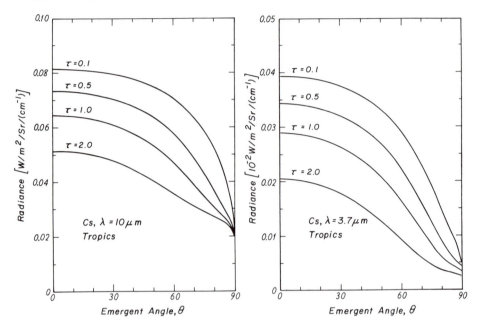

Fig. 5.18   10 and 3.7 μm radiances at TOA as a function of the emergent angle for a number of optical depths. The atmospheric temperature and water vapor profiles used in the calculations are for a tropical condition containing a cirrostratus cloud with a base height of 10 km.

$$T(\mu, \mu') = \frac{\tilde{\omega}\Delta\tau}{4\mu\,\mu'}P(-\mu, -\mu').\tag{5.2.15b}$$

Finally, surface emission can be considered as a separate layer, with emission only pointing upward.

Using the adding method for radiative transfer in a cirrostratus cloud with a base height of 10 km and including the contribution due to water vapor absorption, radiances at TOA have been computed for the two window wavelengths of 10 and 3.7 μm. The atmospheric temperature and water vapor profiles used are for tropical conditions. Figure 5.18 illustrates the emergent radiances for a number of optical depths ranging from 0.1 to 2. Significant variations in the radiance pattern with respect to optical depth for both wavelengths are clearly demonstrated. This suggests that the information content of cloud optical depth could be extracted from radiance observations in the window regions from space. In the 3.7 μm window region, radiance measurements during daytime will contain both the emitted ir and reflected solar components. For this reason, the development of remote sounding of cloud parameters using the ir radiative transfer principle must include reliable removal of the solar "contamination" contribution.

5.2.3.3   Broadband infrared radiative transfer involving black clouds

The energy exchange between clouds and the atmosphere depends on atmospheric temperature stratification. Consider a single black cloud layer and follow the isothermal broadband emissivity approach described in Subsection 2.7.2. Using this approach, the upward and downward fluxes can be expressed in terms of the temperature to the fourth power. Let the cloud top and base heights be denoted as z_t and z_b, respectively, and let their corresponding temperatures be T_t and T_b. To simplify the flux expressions, we may define the kernel function in the height coordinate in the form

$$K(|z - z'|) = -\frac{d}{dz'} \epsilon^f \left(|z - z'|, T(z') \right). \tag{5.2.16}$$

Analogous to Eq. (2.7.14a), the upward and downward fluxes at a given height above the cloud top are given by

$$F^+(z) = \sigma T_t^4 [1 - \epsilon^f (z - z_t, T_t)] + \int_{z_t}^z \sigma T^4(z') K(|z - z'|) \, dz',$$
$$\tag{5.2.17a}$$

$$F^-(z) = \int_{z_\infty}^z \sigma T^4(z') K(|z - z'|) \, dz'. \tag{5.2.17b}$$

Below the cloud, we have

$$F^+(z) = \sigma T_s^4 [1 - \epsilon^f (z, T_s)] + \int_0^z \sigma T^4(z') K(|z - z'|) \, dz', \tag{5.2.18a}$$

$$F^-(z) = \sigma T_b^4 [1 - \epsilon^f (z_b - z, T_b)] + \int_{z_b}^z \sigma T^4(z') K(|z - z'|) \, dz'. \tag{5.2.18b}$$

If the atmosphere contains several cloud layers, modification of the preceding equations may be made to obtain the atmospheric upward and downward fluxes. The net atmospheric flux is the difference between the upward and downward fluxes. The net flux within a black cloud may be estimated from the temperature difference using the Stefan–Boltzmann law. Consider a numerical model in which the cloud occupies several model layers with thicknesses denoted by Δz; the net flux at a given level within the cloud may be estimated from

$$F(z) = \sigma T^4(z - \Delta z) - \sigma T^4(z + \Delta z). \tag{5.2.19}$$

The fact that temperature generally decreases with height will ensure that strong cooling occurs at the cloud top. At the same time, because the net flux at the cloud base is normally less than it is in the cloud, heating will take place.

Finally, if a numerical model predicts partial cloudiness in the model grid area, the radiative heating rate within this area may be evaluated by a linear summation in the form

$$\left(\frac{\partial T}{\partial t} \right)^{pc} = \eta \left(\frac{\partial T}{\partial t} \right)^{ov} + (1 - \eta) \left(\frac{\partial T}{\partial t} \right)^{cl}, \tag{5.2.20}$$

where η denotes the fractional cloud cover, and the superscripts pc, ov, and cl stand for partly cloudy, overcast, and clear-sky conditions. The interactions between cloud side boundaries and the atmosphere are not accounted for in the equation.

5.2.4 *Observed and theoretical broadband radiative properties of clouds*

5.2.4.1 Broadband fluxes

1. Solar radiation. Broadband radiative fluxes can be obtained either from theoretical calculations using an appropriate radiative transfer method or from aircraft radiometric measurements. In the solar region, let the upward and downward fluxes at the cloud top and base be denoted by $[F^+(z_t), F^-(z_t)]$ and $[F^+(z_b), F^-(z_b)]$, respectively. Then the dimensionless reflectance (reflection or reflectivity) and transmittance (transmission or transmissivity) of solar radiation for a given solar zenith angle may be defined by

$$r = F^+(z_t)/F^-(z_t), \qquad (5.2.21a)$$

$$t = F^-(z_b)/F^-(z_t). \qquad (5.2.21b)$$

Note that $F^-(z_t)$ represents the available downward solar flux at the cloud top. Cloud absorptance (absorption) can be obtained from the divergence of the net flux at the cloud top and base in the form

$$A = \left\{ [F^-(z_t) - F^+(z_t)] - [F^-(z_b) - F^+(z_b)] \right\} / F^-(z_t). \qquad (5.2.22)$$

Figure 5.19 shows the absorption and reflection of solar radiation as functions of the cosine of the solar zenith angle for a number of cloud types: St, Cu, As, Ns, and Cb. The droplet size distributions for these cloud types have been displayed in Fig. 4.10. The climatological cloud bases and thicknesses of $(1.4, 0.1)$, $(1.7, 0.45)$, $(4.2, 0.6)$, $(1.4, 4.0)$, and $(1.7, 6.0)$ km are used, respectively, for the preceding cloud types. The standard atmospheric profile is used to obtain the saturation water vapor density in the clouds. Spectral radiative transfer calculations were carried out to obtain the flux distributions at the cloud top and base. Because of extremely large optical depths, Ns and Cb reflect $\sim$80–90% of the solar flux incident upon them; the differences in reflectance between these two cloud types are negligible. Absorptance within these clouds shows a value of $\sim$20% when the sun is overhead. Absorption of solar radiation by both water vapor and cloud particles in the near-ir is responsible for cloud absorption. Although the geometric thickness of cumulus is only $\sim$0.5 km, large reflectance values of $\sim$68–85% are obtained. About 9% of the solar flux incident on the cloud is absorbed when the sun is overhead. The reflectance and absorptance values for stratus with a geometric thickness of 0.1 km range from $\sim$45% to 72% and from $\sim$1% to 6%, respectively. Altostratus reflects $\sim$57–77% of the solar flux. The absorption of solar flux by As is between $\sim$8

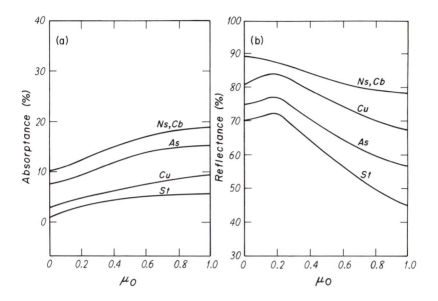

FIG. 5.19   (a) Solar absorptance and (b) solar reflectance for five cloud types as a function of the cosine of the solar zenith angle μ_0.

and 15%, which is larger than that by Cu and St due to its higher position in the atmosphere. In the case of As, the water vapor concentration above the cloud is relatively small, and a large portion of solar flux is available at the cloud top for absorption. This shows that the cloud position in the atmosphere is important in determining the absorption of solar flux in the cloud layer. Variations in cloud absorptance depend on the cloud location, thickness, and particle size distribution. Cloud reflectance is determined by the effective optical depth, which is a result of the combination of the cloud thickness, the extinction coefficient for the cloud particles, and the solar zenith angle.

Aircraft measurements of solar reflectance and absorptance for a number of cloud types in the tropics have been reported by Reynolds et al. (1975). For stratus and stratocumulus, measured reflectances and absorptances are ~37–42% and 12–36%, respectively. For developed cumulus, these values are ~66% and ~31%. Stephens et al. (1978) have obtained reflectances of ~50–75% and absorptances of ~4–20% for stratiform clouds over the east coast of Australia. Observed reflectances and absorptances for Arctic stratus clouds by Herman and Curry (1984) show values ranging from ~50 to 80% and ~1 to 23%. Foot (1988a) has measured reflectances and absorptances for stratiform clouds off the coast of the United Kingdom and derived values of ~80% and 10–15%. Reflectance and absorptance computed from theoretical programs for low and convective clouds using observed cloud parameters and solar zenith angles are generally higher and lower, respectively, than observed data. The largest cloud absorptance derived from theoretical calculations has an upper limit of ~20%, regardless of the thickness and particle size distribution employed (see Fig. 5.19a). However, the observed cloud absorptance could be larger than 30%.

The deviation of the theoretical value from observed data for the absorptance of water clouds has been noted by a number of researchers (Liou, 1976; Twomey, 1976; Stephens, 1978; Wiscombe et al., 1984; also see Fig. 5.24). There appear to be a number of possible reasons for "anomalous" absorption in cloud layers. First, suggestions have been made that the radiation that escapes from the sides of cumulus clouds may be observed by hemispheric radiometers as anomalous absorption (e.g., Welch et al., 1980). However, as pointed out previously, cloud absorptance greater than 20% has been measured in stratus and stratocumulus, in which the finiteness effects of clouds appear minimal. Second, the existence of very large cloud droplets has been explored as a possible cause for large cloud absorptance. For a given liquid water path, larger cloud droplets would reflect less solar flux than smaller droplets, because of their smaller cross-section areas. Hence, an addition of cloud droplets larger than 40–50 μm in radius would enhance cloud absorption. However, based on a detailed theoretical calculation, Wiscombe et al. pointed out that cloud absorptance can be increased by only ~2–4% with the presence of very large cloud droplets. Other likely candidates for anomalous cloud absorption are the purity of cloud particles and in-cloud aerosol particles: (a) Water droplets are generally formed on condensation nuclei that could be absorbing at solar wavelengths; (b) some of the aerosol particles within the cloud could be highly absorbing and not soluble. These particles may remain in the cloud if they are not activated. Through multiple-scattering processes, significant absorption within the cloud could be produced by (a) and (b). To perform theoretical computations, the refractive indices of a specific type(s) of aerosols in the solar spectrum must be known. Scattering and absorption programs must account for the nonhomogeneity of cloud particles and for radiation interactions between cloud and aerosol particles in cases where both are present. In-cloud water vapor absorption also appears to be a potential candidate for anomalous cloud absorption. Finally, the possibility that the spatial and temporal variations of clouds may affect the solar absorption in clouds has also been suggested (Stephens and Tsay, 1990). A special radiative transfer program is required to resolve the spatial nonhomogeneity of clouds.

Aircraft measurements of the broadband radiative properties and compositions of cirrus clouds over Socorro, New Mexico, have been reported by Paltridge and Platt (1981). Cirrus clouds less than 1 km thick with low concentration (~0.005 g m^{-3}) would have reflectance on the order of ~10% and absorptance of only a few percent. Observations reported by Foot (1988b) for extended cirrus over the vicinity of the United Kingdom illustrate the importance of the scattering and absorption properties of hexagonal ice crystals in the interpretation of measured radiative properties. Noticeable "anomalous" absorption within cirrus clouds has not been observed at this point.

2. Thermal infrared radiation. Infrared radiometric observations cannot separate radiant energy produced by emission and scattering by cloud particles. For this reason, the cloud ir radiative properties are usually represented by the apparent (or effective) emittance (emissivity) at the cloud top and base. The atmospheric and surface emission contributions must be removed in the determination of cloud emittance. If the upward flux reaching the cloud base from the atmosphere below the cloud and the surface is denoted by $F^+(z_b)$, we may define the apparent cloud top emittance in the form

$$\epsilon_t^f = [F^+(z_t) - F^+(z_b)]/[\sigma T_t^4 - F^+(z_b)]. \qquad (5.2.23a)$$

Likewise, if the downward flux reaching the cloud top from the atmosphere abovethe cloud is denoted as $F^-(z_t)$, the apparent cloud base emittance may be defined by

$$\epsilon_b^f = [F^-(z_b) - F^-(z_t)] / [\sigma T_b^4 - F^-(z_t)]. \qquad (5.2.23b)$$

The averaged cloud emittance may be taken as $\epsilon^f = (\epsilon_t^f + \epsilon_b^f)/2$. If the scattering effect may be neglected, the cloud ir transmittance (transmissivity) $t = 1 - \epsilon^f$. However, if scattering becomes relatively significant (e.g., with the presence of large ice particles), a non-negligible portion of the incident flux to the cloud base could be reflected back to the atmosphere below. Infrared radiometric observations cannot distinguish this reflection component, as pointed out previously. Aircraft observations of the apparent emittance have been carried out for non-black cirrus clouds. Measured results have been reported by Griffith et al. (1980) for the tropics, by Paltridge and Platt (1981) for New Mexico, and by Smith et al. (1990) for Green Bay, Wisconsin, during the First ISCCP Regional Experiment (FIRE), October–November, 1986. In the tropics, cirrus clouds having thicknesses greater than ~1 km are generally optically thick with apparent emittances close to 1. There are significant spatial variations in the cirrus ir emittance. Results presented by Paltridge and Platt are shown in Fig. 5.20, in which the ir emittance is expressed in terms of the ice water path. Individual cirrus emittance varies from 0.1 to 0.9 depending on the cloud thickness and the ice water content of the cloud. The solid curve is the best exponential fit to the data point with $k_c = 0.056$ (see Subsection 5.3.3 for further discussion).

5.2.4.2 Cloud heating rates

Radiative processes affect the dynamics and thermodynamics of the atmosphere through the generation of radiative heating and cooling rates, as well as net radiative fluxes at the earth's surface. Atmospheric radiative flux exchanges are largely

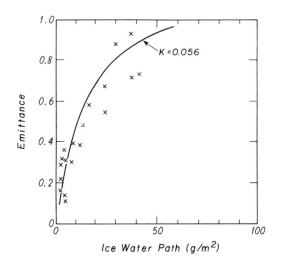

FIG. 5.20   Aircraft observations of broadband emittance from cirrus clouds as a function of ice water path. k is the absorption coefficient that is the best fit to observed data (after Paltridge and Platt, 1981, with modifications).

regulated by cloud fields. In Sections 2.1 and 3.1, we have defined the radiative heating rate in terms of the net flux divergence. The total radiative heating rate that is defined in the thermodynamic equation [Eq. (4.4.47) or (4.4.53)] may be expressed by

$$Q_R = \left(\frac{\partial T}{\partial t}\right)_s - \left(\frac{\partial T}{\partial t}\right)_{ir} = -\frac{1}{\rho C_p}\frac{\partial}{\partial z}(F_s - F_{ir}), \qquad (5.2.24)$$

where F_s and F_{ir} denote the net solar and ir fluxes, respectively. The net flux is the difference between upward and downward fluxes, which can be obtained from theoretical calculations and aircraft radiometric observations. In-cloud fluxes are needed to obtain cloud heating rates.

   1. Water clouds. Observations and computations of the heating rates in boundary-layer stratiform clouds have been reported by Stephens et al. (1978). Figure 5.21 shows an example of the results. The solid curve is a heating or cooling profile computed from a theoretical parameterization program. Observed averaged droplet size distributions, surface albedo, mean solar zenith angle, and temperature and moisture profiles were used in the calculation. The rms experimental errors are represented by the horizontal lines. The large deviations are indicative of the difficulties involved with aircraft flux measurements. The solar heating rates from calculations are consistently lower than those from observations, probably due to the presence of anomalous absorption, which has been discussed previously. The theoretical ir cooling rates are much larger than the measured values at the

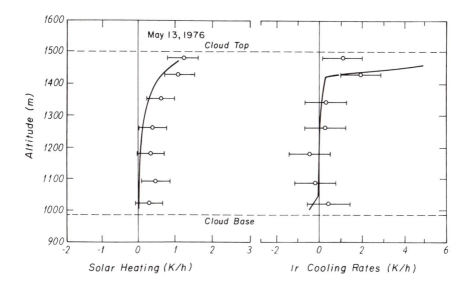

FIG. 5.21   Solar and ir profiles measured in the cloud layer. The rms experimental error is represented by the horizontal lines. The solid curve is the theoretically calculated profile (after Stephens et al., 1978).

cloud top. This discrepancy likely results from systematic errors in the ir radiometer outputs and is due, in part, to the inability to achieve accurate measurements of the structure of temperature and humidity discontinuity at the cloud top for use in theoretical calculations. Nevertheless, there appears to be a general agreement between theory and observations, especially concerning the shape of the heating and cooling profiles toward the cloud top.

2. Ice clouds. During the GARP (Global Atmospheric Research Program) Atlantic Tropical Experiment (GATE), aircraft observations of ir cooling rates within cirrus clouds have been made (Griffith et al. 1980). Figure 5.22 shows the averaged cooling rates derived from observed data, corresponding to a 1 km thickness within a thick cirrus cloud. Theoretical results computed from a band-by-band approach (Roewe and Liou, 1978) are shown for comparison. Using the observed IWC vertical profile, the theoretical value of $\sim 8 \mathrm{Kd}^{-1}$ at the cloud top layer is in close agreement with observations. Reliable measurements and interpretations of radiative fluxes involving high cirrus clouds are extremely difficult because data must be obtained from high-flying aircraft.

Figure 5.23 illustrates an example of heating rate profiles in and below a cirrus cloud in a tropical atmosphere with a base height of 16 km and a thickness of 1.5 km. The ice crystal size distribution for cirrostratus is used in the calculations. The IWC in this case is $\sim 3.2 \times 10^{-2} \mathrm{g\,m}^{-3}$. The ice crystals are assumed to be hexagonal plates and columns randomly oriented in space. The scattering and absorption programs for ice crystals are based on the geometric ray-tracing program for

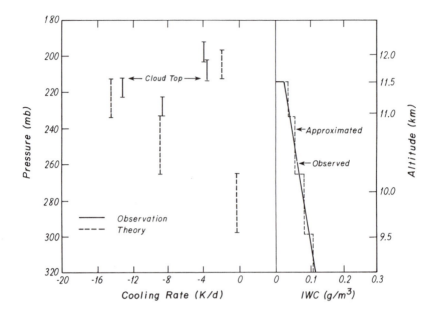

FIG. 5.22   Comparison of observed and calculated cooling rates within a thick cirrus. The observed values are taken from Griffith et al. (1980). Also illustrated are the observed and approximated vertical profiles of IWC.

size parameters larger than 30, and the scattering program for spheroids for size parameters smaller than 30. Four cosines of the solar zenith angle of 0 (nighttime condition), 0.167, 0.5, and 0.833, a solar constant of 1365 W m^{-2}, and a surface albedo of 0.1 were used in the calculations of solar and ir heating rates within and below the cloud. Except in the case when $\mu_0 = 0$, net heating is shown everywhere within the cloud with a maximum of 4–5 K h^{-1} at the cloud base. This heating is a result of the absorption by the very cold cirrus cloud of the ir fluxes emitted from the warmer underlying atmosphere and the surface. The effects of solar radiation on heating are primarily confined to the cloud top. Net heating within the cloud increases as the cosine of solar zenith angles increases. The presence of cirrus clouds suppresses ir cooling below the clouds. The dashed curves in Fig. 5.23 are net cooling rate profiles in a clear tropical atmosphere. Without clouds, the atmosphere shows cooling everywhere. In a cirrus cloudy condition when the cosine of the solar zenith angle is 0.833, net heating is shown above ~2 km. The modification of the net heating profile due to the presence of cirrus clouds in the tropics is important with respect to the maintenance of the Hadley circulation, as well as the time scale involved for the dissipation of tropical cirrus clouds, as discussed in Subsection 4.7.2.

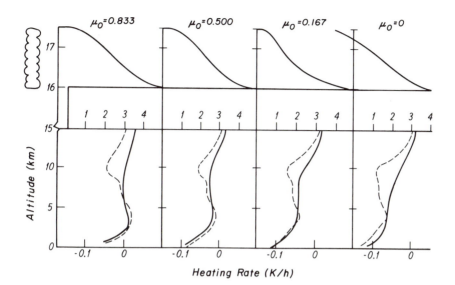

Fig. 5.23 Net heating rate (solar plus ir) profiles in a tropical atmosphere with and without the presence of a cirrostratus with a base height of 16 km and a thickness of 1.5 km. Four solar zenith angles are used. $\mu_0 = 0$ represents the nighttime condition. The upper scales are for heating rates within the cloud.

5.3 Parameterization of the radiative properties of clouds

5.3.1 *Solar radiative properties*

5.3.1.1 Water clouds

The characteristics of scattered light from clouds depend on the droplet size distribution, $n(r)$. The first parameter describing droplet size distribution should be some measure of the mean size. Since spherical droplets scatter an amount of light in proportion to their cross-sectional area, we may define a mean effective radius in the form

$$r_e = \int r\pi r^2 n(r)\,dr \bigg/ \int \pi r^2 n(r)\,dr. \qquad (5.3.1)$$

Mean effective radius differs from simple mean radius in that the droplet cross section is included as a weight factor. We wish to relate this mean parameter to the liquid water path (LWP) and optical depth.

From the definition of LWC for spherical droplets, LWP or vertically integrated LWC is given by

$$\mathrm{LWP} = \Delta z \frac{4\pi}{3} \rho_\ell \int r^3 n(r)\,dr. \qquad (5.3.2)$$

The optical depth for a given droplet size distribution is defined by

$$\tau = \Delta z \int \sigma_e n(r)\, dr = \Delta z \int Q_e \pi r^2 n(r)\, dr, \qquad (5.3.3)$$

where the extinction cross section $\sigma_e = Q_e \pi r^2$, and Q_e is the efficiency factor, which is a function of the droplet radius, wavelength and refractive index. For solar visible wavelengths, the efficiency factor $Q_e \cong 2$ for cloud droplets. Thus the mean effective radius can be related to LWP and τ in the form

$$r_e \cong \frac{3}{2}\mathrm{LWP}/\tau. \qquad (5.3.4)$$

The single-scattering co-albedo is defined by

$$1 - \tilde{\omega} = \sigma_a/\sigma_e, \qquad (5.3.5)$$

where σ_a denotes the absorption cross section, which is proportional to the product of the absorption coefficient, $k_i = 4\pi m_i/\lambda$, and the volume when absorption is small. From Eq. (5.3.1), the total droplet volume is given by $V = 4r_e G/3$, where $G = \int \pi r^2 n(r) dr$, representing the total cross sectional area. It follows that $\sigma_a = k_i 4 r_e G/3$. If an efficiency factor of 2 is used for extinction, then $\sigma_e \cong 2G$. Thus we have

$$1 - \tilde{\omega} \cong \frac{2}{3} k_i r_e. \qquad (5.3.6)$$

The other important factor is the asymmetry factor. Based on Mie scattering calculations, the asymmetry factor shows relatively little variation in the solar wavelengths, as illustrated in Fig. 5.3.

Parameterization of the solar radiative properties of water clouds has been carried out by Stephens (1978), Liou and Wittman (1979), and Slingo (1989). Slingo has developed a solar radiative transfer parameterization scheme based on the two-stream approximation described in Subsection 3.3.1. In this scheme, both LWP and r_e are used as the basic parameters for the computations of the spectral solar radiative properties of water clouds. Based on a number of droplet size distributions, the optical depth, single-scattering albedo, and asymmetry factor are expressed in terms of r_e in forms similar to those presented in Eqs. (5.3.4) and (5.3.6):

$$\tau_i = \mathrm{LWP}(a_i + b_i/r_e), \qquad (5.3.7a)$$
$$1 - \tilde{\omega}_i = c_i + d_i r_e, \qquad (5.3.7b)$$
$$g_i = e_i + f_i r_e, \qquad (5.3.7c)$$

where a_i–f_i are empirical coefficients. Four spectral bands in the solar region are sufficient to obtain accurate broadband solar radiative properties.

Broadband reflectance and absorptance are defined by

$$r = \sum_i r_i w_i, \qquad A = \sum_i A_i w_i, \tag{5.3.8}$$

where r_i and A_i denote the reflectance and absorptance for individual bands, and w_i is the fraction of solar fluxes in these bands. Figure 5.24 shows the broadband reflectance and absorptance of water clouds as functions of LWP and r_e for a solar zenith angle of 60°. Solar reflectance depends significantly on the mean effective radius. For an LWP of $100 \, \mathrm{g \, m^{-2}}$, the reflectance of a water cloud with a small r_e of 2 μm is a factor two larger than it is for one with an r_e of 16 μm. Absorption within clouds is less dependent on r_e for LWPs less than $\sim 100 \, \mathrm{g \, m^{-2}}$. For a given LWP, larger droplets reflect less but absorb more than smaller droplets, due to the fact that larger droplets have a smaller cross-sectional area and single-scattering albedo than small droplets.

5.3.1.2   Ice clouds

The LWC for a given ice crystal size distribution $n(L)$ is defined by

$$\mathrm{IWC} = \int V \rho_i n(L) \, dL, \tag{5.3.9}$$

where ρ_i denotes the density of ice, V is the volume of an individual ice crystal, and L is the major dimension. Consider a hexagonal ice crystal; the volume is given by

$$V = \frac{3\sqrt{3}}{8} L D^2. \tag{5.3.10}$$

The ice water path (IWP) for a cloud with a thickness of Δz is then

$$\mathrm{IWP} = \Delta z \, \mathrm{IWC}. \tag{5.3.11}$$

The optical depth for ice clouds is defined by

$$\tau = \Delta z \int \sigma_e n(L) \, dL. \tag{5.3.12}$$

For randomly oriented hexagonal ice crystals in the limits of geometric optics, the extinction cross section may be expressed by (Takano and Liou, 1989a)

$$\sigma_e = \frac{3}{2} D \left(\frac{\sqrt{3}}{4} D + L \right), \tag{5.3.13}$$

where $D(= 2r)$ is the minimum dimension (or width) of an ice crystal.

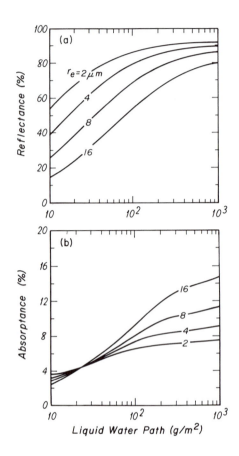

Fig. 5.24   Solar reflectance (a) and absorptance (b) of water clouds (percentage) at a solar zenith angle of 60° as functions of cloud liquid water path (LWP) and mean effective radius (r_e) (after Slingo, 1989, with modification).

Analogous to the mean effective radius, we may define a mean effective width (or size) corresponding to ice crystal size distribution in the form

$$D_e = \int D \cdot LD\, n(L)\, dL \bigg/ \int LD\, n(L)\, dL, \qquad (5.3.14)$$

where the numerator and denominator represent the volume- and area-weighted ice crystal size distributions, respectively. Using all the preceding equations, D_e, IWP, and optical depth may be related as follows:

$$\tau \simeq \text{IWP}(a + b/D_e), \qquad (5.3.15)$$

where a and b are certain coefficients. The optical depth is determined by two independent cloud physics variables, IWP and D_e, which are dependent on temperature (see Subsection 4.3.2).

A general parameterization of optical depth involving ice clouds may be written in the form

$$\tau_i = \left(\sum_{n=0}^{N} \frac{a_{n,i}}{D_e^n} \right) \text{IWP},$$

(5.3.15a)

where a_n are empirical coefficients and N is the total number of expansion coefficients.

The absorption cross section is approximately proportional to the product of the absorption coefficient k_i and volume for weak absorption. Thus, using the extinction cross section and volume for a hexagonal cylinder defined in Eqs. (5.3.10) and (5.3.13), the single-scattering co-albedo for randomly oriented hexagonal ice crystals may be parameterized in the form

$$1 - \tilde{\omega} = \frac{\sigma_a}{\sigma_e} \sim \frac{k_i \sqrt{3} L D^2}{D(\sqrt{3}D + 4L)}.$$

(5.3.16)

Since the maximum dimension L is related to the width D based on laboratory and aircraft observations, $1 - \tilde{\omega}$ must be proportional to the mean effective size D_e, defined in Eq. (5.3.14). Thus, we may express the single-scattering co-albedo in terms of the polynomials in D_e, in the form

$$1 - \tilde{\omega}_i = \sum_{n=0}^{N} b_{n,i} D_e^n,$$

(5.3.17)

where the subscript i denotes the index for the spectral band in the solar spectrum and b_n are empirical coefficients. The asymmetry factor may also be expressed in a likely expansion, viz.,

$$g_i = \sum_{n=0}^{N} c_{n,i} D_e^n,$$

(5.3.18)

where c_n again are empirical coefficients. To resolve the fluctuations of the refractive index of ice, it suffices to divide the solar spectrum into six bands, as denoted in Table 5.4. Using the light scattering program developed by Takano and Liou (1989a) and the modification of this program to account for size parameters less than ~ 30, spectral extinction cross sections, single-scattering albedos, and phase functions have been computed for the ice crystal size distributions presented in Subsection 4.3.2. Based on the spectral values computed from the light-scattering program, and the physical parameterization equations, empirical coefficients appearing in Eqs. (5.3.17), (5.3.18), and (5.3.15a) may be determined by numerical fittings. These coefficients are listed in Table 5.4, along with the fractional solar flux w_i in each band. Accuracy of the fittings is generally maintained within $\sim 1\%$.

The radiative properties of ice crystal clouds in the solar region can be computed by a radiative transfer method such as the δ–two-stream or δ–Eddington

Table 5.4 Empirical coefficients in the parameterization equations for the optical depth, single scattering co-albedo, and asymmetry factor involving ice crystal clouds.

Spectral Limits (μm)	$\tau(a)$ $1-\tilde{\omega}(b)$ $g(c)$	a_0 b_0 c_0	a_1 b_1 c_1	a_2 b_2 c_2	a_3 b_3 c_3	w_i
0.5 (0.2–0.7)	a	-6.6560×10^{-3}	3.6860×10^{0}	0	0	4.7330×10^{-1}
	b	1.0998×10^{-6}	-2.6101×10^{-8}	1.0896×10^{-9}	-4.7387×10^{-12}	
	c	7.6913×10^{-1}	-3.6524×10^{-4}	2.1697×10^{-7}	-1.1483×10^{-7}	
1.0 (0.7–1.3)	a	-6.6560×10^{-3}	3.6860×10^{0}	0	0	3.5138×10^{-1}
	b	2.0208×10^{-5}	9.6483×10^{-6}	8.3009×10^{-8}	-3.2217×10^{-10}	
	c	7.7002×10^{-1}	-2.9227×10^{-4}	2.1269×10^{-5}	-1.1460×10^{-7}	
1.6 (1.3–1.9)	a	-6.6560×10^{-3}	3.6860×10^{0}	0	0	1.1092×10^{-1}
	b	1.3590×10^{-4}	7.3453×10^{-4}	2.8281×10^{-6}	-1.8272×10^{-8}	
	c	7.7587×10^{-1}	2.4004×10^{-4}	1.6901×10^{-5}	-9.9085×10^{-8}	
2.2 (1.9–2.5)	a	-6.6560×10^{-3}	3.6860×10^{0}	0	0	3.6656×10^{-2}
	b	-1.6598×10^{-3}	2.0933×10^{-3}	-1.3977×10^{-6}	-1.8703×10^{-8}	
	c	7.9339×10^{-1}	9.2245×10^{-4}	8.4781×10^{-6}	-6.6995×10^{-8}	
3.0 (2.5–3.5)	a	-6.6560×10^{-3}	3.6860×10^{0}	0	0	2.3424×10^{-2}
	b	4.6180×10^{-1}	2.4471×10^{-4}	-2.7839×10^{-6}	1.0378×10^{-8}	
	c	9.3236×10^{-1}	9.9913×10^{-4}	-1.0901×10^{-5}	3.9323×10^{-8}	
3.7 (3.5–4.0)	a	-6.6560×10^{-3}	3.6860×10^{0}	0	0	4.3200×10^{-3}
	b	4.2362×10^{-2}	8.6425×10^{-3}	-7.5519×10^{-5}	2.4056×10^{-7}	
	c	6.9861×10^{-1}	5.5109×10^{-3}	-4.4816×10^{-5}	1.2736×10^{-7}	

approximation introduced in Section 3.3. If high accuracy is required, we may use the δ–four-stream approximation presented in Section 3.5. For the purpose of parameterization, we may use the analytic Henyey–Greenstein phase function defined in Eq. (3.4.1b) to obtain higher moments, viz., $\tilde{\omega}_\ell = (2\ell + 1)g^\ell$. To increase the accuracy of the radiation parameterization, the similarity principle for radiative transfer may be used to account for the fraction of scattered energy residing in the forward peak. In the context of the δ–four-stream approximation, the fractional energy $f = \tilde{\omega}_4/9$. The adjusted optical depth, single-scattering albedo, and phase function have been given in Eq. (3.4.18). The preceding parameterization for the solar radiative properties of ice clouds can be improved if more accurate single-scattering properties for irregular ice crystals have been determined from theory and laboratory experiments.

The broadband solar reflectance and absorptance [Eqs. (5.3.8)] for ice clouds as functions of D_e and IWP are shown in Fig. 5.25. A cosine of the solar zenith angle of 0.5 is used in the solar radiative transfer calculations. For a given D_e, solar reflectance increases with increasing IWP. But for a given IWP, a cloud with a smaller D_e reflects more solar radiation because of the larger effective optical depth. For example, for an IWP of $100\,\text{g m}^{-2}$, solar reflectances of about 70, 55, 40, and 30% are shown for D_e of 25, 50, 75, and $100\,\mu\text{m}$, respectively. For solar absorptance, small ice crystals also absorb more solar radiation for IWPs up to $75\,\text{g m}^{-2}$. However, absorption of solar radiation also depends on forward scattering by cloud particles. Larger particles would have stronger forward scattering and absorb more solar radiation when the IWP is larger than about $75\,\text{g m}^{-2}$.

5.3.2   Infrared emittance (emissivity) of clouds

To simplify the expression for cloud ir emittance, we may neglect the scattering contribution. This is a good approximation for small cloud particles ($< 10\,\mu\text{m}$) illuminated by long wavelengths ($\sim 10\,\mu\text{m}$). Based on the definition of transmittance [Eq. (2.3.3)], monochromatic cloud emittance may be expressed by

$$\epsilon_\nu = 1 - \exp(-k_\nu^c W), \qquad (5.3.22)$$

where k_ν^c is the mass absorption coefficient of cloud particles and W denotes either LWC or IWC. Furthermore, using Eq. (2.3.4), monochromatic flux emissivity is given by

$$\begin{aligned} \epsilon_\nu^f &= 1 - 2 \int_0^1 \exp(-k_\nu^c W/\mu)\mu\,d\mu \\ &= 1 - \exp(-k_\nu^c W/\bar{\mu}), \qquad (5.3.23) \end{aligned}$$

where $1/\bar{\mu}$ is the diffusivity factor. The cloud ir emittance approaches 1 when k_ν^c and/or W are very large.

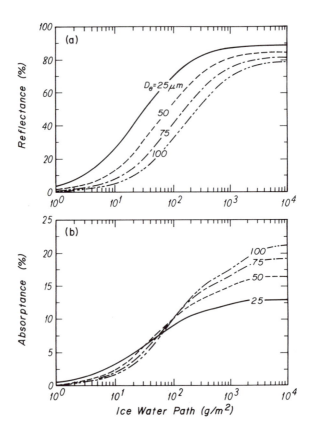

FIG. 5.25   Solar (a) reflectance and (b) absorptance for ice clouds as functions of the mean effective size (D_e) and ice water path (IWP) for $\mu_0 = 0.5$.

Because the variation in the mass absorption coefficient with wave number is relatively small due to the more continuous nature of water/ice absorption compared with molecular absorption, we may apply the gray approximation, viz.,

$$\epsilon \cong 1 - \exp(-k^c W), \tag{5.3.24a}$$

$$\epsilon^f \cong 1 - \exp(-k^c W/\bar{\mu}), \tag{5.3.24b}$$

where the wave-number-averaged mass absorption coefficient is defined by

$$k^c = \int_{\Delta\nu} k^c_\nu B_\nu(T)\,d\nu \bigg/ \int_{\Delta\nu} B_\nu(T)\,d\nu. \tag{5.3.25}$$

If k^c_ν is known through theoretical calculations for each infrared spectral region, k^c can be evaluated. Alternately, we may use measured spectral or broadband ir emittance values to obtain k^c.

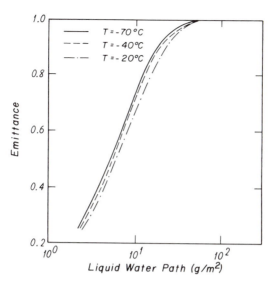

FIG. 5.26   ir emittance (emissivity) for water clouds as a function of liquid water path (*W* or LWP) (after Stephens, 1984).

The broadband emittance of water clouds has been computed by Stephens (1984) using the Mie theory and the representative droplet size distribution of a stratocumulus cloud type. Figure 5.26 shows the broadband emittance as a function of LWP for a number of cloud temperatures. The temperature dependence appears to be small. Emittance approaches 1 for LWPs greater than 50 g m^{-2}. The theoretical values for k^c range from 0.13 to 0.16.

Results of the broadband emittance of ice clouds have been presented in Liou (1986). Figure 5.27 illustrates cirrus broadband emittance as a function of the IWP derived by Liou and Wittman (1979), Kinne and Liou (1989), and exponential fits with $k^c = 0.1$ and 0.06. The computations used the ice crystal size distribution for cirrostratus described in Section 4.3. Multiple scattering by ice cylinders is accounted for in the calculations. The results of Liou and Wittman are limited to IWPs greater than 5 g m^{-2} and are not applicable to thin cirrus clouds. The theoretical results from Kinne and Liou closely match the gray approximation with a k^c value of 0.06. It appears that the effects of multiple scattering by Mie particles on the cloud emission are small.

Based on measured data from narrowband radiometers in the 10 μm region, k^c values of ~0.13 for stratus clouds and ~0.08 for cirrus clouds have been derived by Platt (1976) and Paltridge and Platt (1981), respectively. Broadband values of k^c have been determined from broadband hemispheric flux measurements. For stratus, it is ~0.08 (Stephens et al., 1978). For midlatitude cirrus, it is ~0.056, but it has values of 0.076–0.096 for tropical cirrus (Griffith et al., 1980). It is

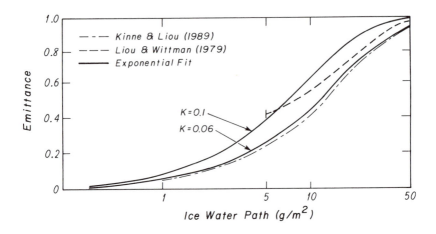

FIG. 5.27   Cirrus ir emittance (emissivity) as a function of ice water path (W or IWP) from various theoretical sources.

evident that k^c is a function of ice crystal size distribution. Parameterization of the broadband emittance for ice clouds, including the scattering contributions, may follow the procedure described in Subsection 5.3.1.2. The broadband emittance should be a function of both IWP and D_e.

5.4   Radiative transfer in finite clouds

From satellite cloud pictures, as well as our day-to-day experience, we see that a portion of the clouds and cloud systems that cover the earth are either finite in extent or occur in the form of cloud bands. This is especially evident in the tropics and the midlatitudes in the summertime. Figure 5.28 shows a Landsat image of a fair weather cumulus field over the western Atlantic Ocean off the coast of South Carolina for March 9, 1979 (Wielicki and Welch, 1986). A field of scattered cumulus with some evidence of open-celled convection patterns has been discussed in Subsection 4.1.2. Some mesoscale variations of cloud size within the field are also evident. The horizontal nonhomogeneity of the cloud system is clear from this illustration. One generally common feature is the presence of cumulus clouds whose horizontal dimensions are on the same order as their vertical dimensions. The classical radiative transfer theory that has been developed in Chapters 2 and 3 is applicable only for plane-parallel (i.e., horizontally homogeneous) media. The effect of the finiteness of clouds on reflection and absorption is the topic of this section.

   An estimate of solar radiation reflected and transmitted through finite cloud layers has been made using the Monte Carlo method, which is based on the tracing of photons within a finite geometry (Mckee and Cox, 1976; Busygin et al., 1973). Giovanelli (1959) has presented a procedure for the decomposition of the

Fig. 5.28   Landsat image of a fair weather cumulus cloud over the western Atlantic Ocean off the coast of South Carolina at 1510 GMT March 9, 1979. The white rectangles denote the regions used in the analysis (after Wielicki and Welch, 1986.)

three-dimensional transfer equation in Cartesian coordinates; this procedure is the basis on which a diffusion approximation may be developed. The diffusion approximation for radiative transfer in a three-dimensional space has frequently been used to evaluate the intensity distribution involving finite clouds (Davies, 1978; Liou and Ou, 1979; Ou and Liou, 1980). Radiative transfer in three-dimensional space has been a subject of extensive research in the field of neutron transport (Case and Zweifel, 1967). Crosbie and Linsenbardt (1978) have provided a comprehensive review of the earlier works on multidimensional radiative transfer. In the following, we introduce the diffusion approximation for radiative transfer in three-dimensional space and present some pertinent results regarding the effects of horizontal nonhomogeneity on the radiative properties of finite clouds.

5.4.1   *Diffusion approximation for three-dimensional radiative transfer*

Following the discussion presented in Sections 2.1 and 3.1, the basic steady-state solar radiative transfer equation governing diffuse intensity in three-dimensional space may be written

$$
\frac{1}{\beta_e}(\mathbf{\Omega} \cdot \nabla)I(\mathbf{s}, \mathbf{\Omega}) + I(\mathbf{s}, \mathbf{\Omega}) = \frac{\tilde{\omega}}{4\pi} \int_{4\pi} I(\mathbf{s}, \mathbf{\Omega})P(\mathbf{\Omega}, \mathbf{\Omega}')\, d\Omega'
$$
$$
+ \frac{\tilde{\omega}}{4\pi} P(\mathbf{\Omega}, \mathbf{\Omega_0})F_\odot e^{-\tau_s}, \qquad (5.4.1)
$$

where β_e denotes the extinction coefficient, which is assumed to be the same in all directions; Ω is a unit vector defining the direction of scattering through a position vector $\mathbf{s}$; and $\tau_s = \beta_e s(\Omega_0)$, the optical depth in the direction of the incident solar beam, where $s(\Omega_0)$ is the distance between the point of interest in the medium and the point of entrance of the incident beam in the direction of Ω_0. The phase function and diffuse intensity may be expressed in terms of the spherical harmonics expansion as follows:

$$P(\Omega, \Omega') = \sum_{\ell=0}^{N} \sum_{m=-1}^{\ell} \tilde{\omega}_\ell Y_\ell^m(\Omega) Y_\ell^{m*}(\Omega'), \qquad (5.4.2)$$

$$I(\mathbf{s}, \Omega) = \sum_{\ell=0}^{N} \sum_{m=-1}^{\ell} I_\ell^m(\mathbf{s}) Y_\ell^m(\Omega), \qquad (5.4.3)$$

where $\tilde{\omega}_\ell$ are certain coefficients, and N denotes the number of terms in the expansion of the spherical harmonics, which are defined by

$$Y_\ell^m(\theta, \phi) = (-1)^{(m+|m|)/2} \left(\frac{(\ell - |m|)!}{(\ell + |m|)!} \right) P_\ell^{|m|}(\cos\theta) e^{im\phi}, \qquad (5.4.4)$$

where P_ℓ^m is the associated Legendre polynomials discussed in Section 3.1, $|m|$ is the absolute value of m, and $i = \sqrt{-1}$. The complex conjugates of the spherical harmonics are given by

$$Y_\ell^{m*}(\theta, \phi) = Y_\ell^{-m}(\theta, \phi)/(-1)^m. \qquad (5.4.5)$$

The spherical harmonics are normalized such that

$$\frac{1}{4\pi} \int_0^{2\pi} \int_{-1}^{1} Y_\ell^m(\mu, \phi) Y_\alpha^{\beta*}(\mu, \phi) \, d\mu \, d\phi = \delta_\ell^\alpha \delta_m^\beta / (2\ell + 1), \qquad (5.4.6)$$

where δ_n^ℓ and δ_m^β are Kronecker delta functions.

In order to decompose Eq. (5.4.1) in accordance with spherical harmonics, we first insert Eqs. (5.4.2) and (5.4.3) into Eq. (5.4.1) to obtain

$$\frac{1}{\beta_e}(\Omega \cdot \nabla) \sum_{\ell=0}^{N} \sum_{m=-\ell}^{\ell} I_\ell^m(\mathbf{s}) Y_\ell^m(\Omega) = -\sum_{\ell=0}^{N} \sum_{m=-\ell}^{\ell} \gamma_\ell I_\ell^m(\mathbf{s}) Y_\ell^m(\Omega)$$

$$+ \frac{\tilde{\omega}}{4\pi} \sum_{\ell=0}^{N} \sum_{m=-\ell}^{\ell} \tilde{\omega}_\ell Y_\ell^m(\Omega) Y_\ell^{m*}(\Omega_0) F_\odot e^{-\tau_s}, \qquad (5.4.7)$$

where $\gamma_\ell = 1 - \tilde{\omega}\tilde{\omega}_\ell/(2\ell+1)$. Subsequently, we perform the following successive integrations:

$$\int_{4\pi} \text{Eq. (5.4.7)} \times Y_\alpha^{\beta*}(\Omega) \, d\Omega, \qquad \alpha = 0, 1, \ldots, N; \beta = -\alpha, \ldots, \alpha. \quad (5.4.8)$$

We find

$$
-\frac{1}{\beta_e} \sum_{\ell=0}^{N} \sum_{m=-\ell}^{\ell} \int_{4\pi} (\mathbf{\Omega} \cdot \nabla) Y_\ell^m(\mathbf{\Omega}) Y_\alpha^{\beta*}(\mathbf{\Omega}) I_\ell^m(\mathbf{s})\, d\Omega
$$

$$
= -\gamma_\alpha I_\alpha^\beta(\mathbf{s}) \frac{4\pi}{2\alpha + 1} + \frac{\tilde{\omega}\tilde{\omega}_\ell}{2\alpha + 1} Y_\alpha^{\beta*}(\mathbf{\Omega}_0) F_\odot e^{-\tau_s}. \tag{5.4.9}
$$

The left-hand side of this equation may be decomposed by using the recursion relationships in each coordinate system (Ou and Liou, 1982).

For application to the finite cloud problem, it suffices to make a first order approximation (i.e., $N=1$), the so-called *diffusion approximation*. Using the Cartesian coordinates and the definition of spherical harmonics and their recursion relationships, we obtain the following four partial differential equations:

$$
\frac{\partial I_1^0}{\partial z} + \frac{1}{\sqrt{2}} \left(\frac{\partial}{\partial x} - i \frac{\partial}{\partial y} \right) I_1^{-1} - \frac{1}{\sqrt{2}} \left(\frac{\partial}{\partial x} + i \frac{\partial}{\partial y} \right) I_1^1
$$

$$
= -\beta_e I_0^0 (1 - \tilde{\omega}) + \frac{\tilde{\omega}}{4\pi} F_\odot e^{-\tau_s}, \tag{5.4.10a}
$$

$$
\beta_e (1 - \tilde{\omega}g) I_1^{-1} = -\frac{1}{3\sqrt{2}} \left(\frac{\partial}{\partial x} + i \frac{\partial}{\partial y} \right) I_0^0
$$

$$
+ \frac{3}{\sqrt{2}} \frac{\tilde{\omega}g}{4\pi} (1 - \mu_0^2)^{1/2} (\cos\phi_0 + i\sin\phi_0) F_\odot e^{-\tau_s}, \tag{5.4.10b}
$$

$$
\beta_e (1 - \tilde{\omega}g) I_1^0 = -\frac{1}{3} \frac{\partial I_0^0}{\partial z} + \frac{3\tilde{\omega}g}{4\pi} \mu_0 F_\odot e^{-\tau_s}, \tag{5.4.10c}
$$

$$
\beta_e (1 - \tilde{\omega}g) I_1^1 = \frac{1}{3\sqrt{2}} \left(\frac{\partial}{\partial x} - i \frac{\partial}{\partial y} \right) I_0^0
$$

$$
- \frac{1}{\sqrt{2}} \frac{3\tilde{\omega}g}{4\pi} (1 - \mu_0^2)^{1/2} (\cos\phi_0 - i\sin\phi_0) F_\odot e^{-\tau_s}. \tag{5.4.10d}
$$

From the spherical harmonics expansion for diffuse intensity denoted in Eq. (5.4.3), we have

$$
I(x, y, z; \mathbf{\Omega}) = I_0^0 + I_1^{-1} Y_1^{-1}(\mathbf{\Omega}) + I_1^0 Y_1^0(\mathbf{\Omega}) + I_1^1 Y_1^1(\mathbf{\Omega}). \tag{5.4.11}
$$

If we substitute Eqs. (5.4.10b–d) into Eq. (5.4.11), we obtain

$$
I(x, y, z; \mathbf{\Omega}) = I_0^0 - \frac{3}{2h} \sum_{j=1}^{3} \frac{\partial I_0^0}{\partial x_j} \Omega_{xj} + \frac{9q}{2h} (\mathbf{\Omega} \cdot \mathbf{\Omega}_0) e^{-\tau_s}, \tag{5.4.12}
$$

where $x_1 = x$, $x_2 = y$, $x_3 = z$, $h = 3(1 - \tilde{\omega}g)/2$, $q = \tilde{\omega}gF_\odot/12\pi$, $\Omega_z = \mu$, $\Omega_x = (1 - \mu^2)^{1/2} \cos\phi$, and $\Omega_y = (1 - \mu^2)^{1/2} \sin\phi$.

Substituting Eqs. (5.4.10b–d) into Eq. (5.4.10a) yields the following nonhomogeneous diffusion equation:

$$\nabla^2 I_0^0 - k^2 \beta_e^2 I_0^0 = -\chi \beta_e^2 e^{-\tau_s}, \tag{5.4.13}$$

where the diffusion operator $\nabla^2 = \partial^2/\partial x^2 + \partial^2/\partial y^2 + \partial^2/\partial z^2$, and coefficients k and χ are given by

$$k^2 = 3(1 - \tilde{\omega})(1 - \tilde{\omega}g), \tag{5.4.14a}$$

$$\chi = 3\tilde{\omega} F_\odot (1 + g - \tilde{\omega}g)/4\pi. \tag{5.4.14b}$$

These values are exactly the same as those derived for the plane-parallel case illustrated in Eq. (3.3.11).

To seek a solution for Eq. (5.4.13), appropriate boundary conditions are required. We may assume that there is no inward diffuse intensity, $I_{in}(\Omega)$, on the boundary of the cloud; this condition is referred to as the *vacuum boundary condition*. Moreover, in order to satisfy the mathematical solutions that are derived from a finite expansion of intensity, we may set

$$\int_\Omega Y_\ell^m(\Omega) I_{in}(\Omega) \, d\Omega = 0, \qquad \ell = 1; m = -1, 0, 1, \tag{5.4.15a}$$

where the domain of integration is the inner half-plane. From the intensity expansion denoted in Eq. (5.4.3), we find

$$\sum_{\ell'=0}^{N} \sum_{m'=-\ell'}^{\ell'} \left(I_{\ell'}^{m'}(s) \int_\Omega Y_\ell^m(\Omega) Y_{\ell'}^{m'}(\Omega) \, d\Omega \right) = 0. \tag{5.4.15b}$$

To obtain the boundary conditions for a finite rectangle, let the six sides of this rectangle be $x = 0, a; y = 0, b;$ and $z = 0, c$ (see Fig. 5.29). The domains of integration are, respectively, $(0 < \theta < \pi; \ -\pi/2 < \phi < \pi/2)$, $(0 < \theta < \pi; \ \pi/2 < \phi < 3\pi/2)$, $(0 < \theta < \pi; \ 0 < \phi < \pi)$, $(0 < \theta < \pi; \ \pi < \phi < 2\pi)$, $(0 < \theta < \pi/2; \ 0 < \phi < 2\pi)$, and $(\pi/2 < \theta < \pi; \ 0 < \phi < 2\pi)$. After carrying out the integrations, we obtain the following six boundary conditions:

$$I_0^0 \pm \frac{2}{3} I_1^0 = 0, \qquad \text{at } z = 0(+), c(-), \tag{5.4.16a}$$

$$I_0^0 \pm \frac{\sqrt{2}}{3} i(I_1^{-1} + I_1^1) = 0, \qquad \text{at } y = 0(-), b(+), \tag{5.4.16b}$$

$$I_1^0 \pm \frac{\sqrt{2}}{3} (I_1^1 - I_1^{-1}) = 0, \qquad \text{at } x = 0(-), a(+). \tag{5.4.16c}$$

Using Eqs. (5.4.10a–d) for I_1^0, I_1^1 and I_1^{-1}, we find

$$\left(\frac{\partial I_0^0}{\partial x_j} \pm h\beta_e^2 I_0^0 \right)_{x_j = \binom{0}{a_j}} = \left(3q\beta_e^2 \Omega_{xj} e^{-\tau_s} \right)_{x_j = \binom{0}{a_j}}, \tag{5.4.17}$$

where $a_1 = a$, $a_2 = b$, $a_3 = c$, $\Omega_{x1} = (1 - \mu_0^2)^{1/2} \cos \phi_0$, $\Omega_{x2} = (1 - \mu_0^2)^{1/2} \sin \phi_0$, and $\Omega_{x3} = \mu_0$. With these boundary conditions, an analytic solution for I_0^0 in Eq. (5.4.13) may be derived if $\mu_0 = 1$ (i.e., if $\tau_s = \beta_e z$). This can be done by a separation of variables following the procedures for the solution of the nonhomogeneous partial differential equation (see, e.g., Berg and McGregor, 1966). The general form of the solution for I_0^0 is a double infinite series, in the form

$$I_0^0(x, y, z) = \sum_{n=1}^{\infty} \sum_{m=1}^{\infty} \left(c_+ e^{\zeta_{nm} z} + c_- e^{-\zeta_{nm} z} - t_{nm}/\zeta_{nm}^2 \right) U_n(x) V_m(y),$$

(5.4.18)

where the three-dimensional eigenvalue

$$\zeta_{nm}^2/\beta_e^2 = \lambda_n^2 + \lambda_m^2 + k^2,$$ (5.4.19a)

$$t_{nm} = \frac{k^2 \beta_e^2}{f_n g_m} S \int_0^b \int_0^a U_n(x) V_m(y) \, dx \, dy,$$ (5.4.19b)

$$f_n = \int_0^a U_n^2(x) \beta_e \, dx, \qquad g_m = \int_0^b V_m^2(y) \beta_e \, dy,$$ (5.4.19c)

$$U_n(x) = \cos(\lambda_n x) + \frac{h}{\lambda_n} \sin(\lambda_n x),$$ (5.4.19d)

$$V_m(y) = \cos(\lambda_m y) + \frac{h}{\lambda_m} \sin(\lambda_m y),$$ (5.4.19e)

and where λ_n and λ_m can be found by solving the equation

$$2 \cot(\lambda_\ell b) = \frac{\lambda_\ell}{h} - \frac{h}{\lambda_\ell}, \qquad \begin{cases} \ell = n \text{ or } m \\ \ell = 1, 2, \dots. \end{cases}$$ (5.4.19f)

The coefficients c_+ and c_- are constants to be determined from the boundary conditions given in Eqs. (5.4.16a–c). For solar radiation

$$S(\mu_0 = 1) = \chi e^{-\beta_e z}.$$ (5.4.19g)

However, for thermal ir radiation, S is the Planck function, $B_\nu(T)$.

5.4.2   Effects of finiteness on cloud radiative properties

The effects of the finiteness of clouds on the reflected intensity are illustrated by using the diffusion solution derived previously for two-dimensional space along the $x - z$ directions. The y direction is therefore infinite in this case. The sharp forward diffraction peak of the phase function for cloud particles can be incorporated in the calculations through the similarity relation discussed in Section 3.4. Figure 5.30 shows the reflected intensity normalized by $F_\odot/\pi$ at the cloud top for a number of vertical and horizontal depths. The case corresponds to conservative scattering,

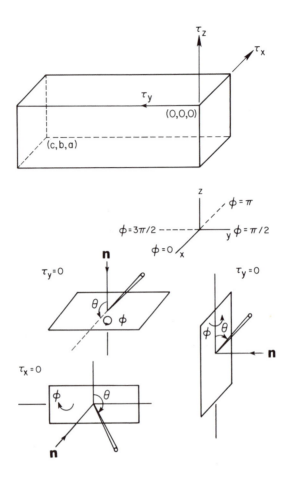

FIG. 5.29   Geometry of a finite cloud in the Cartesian coordinates with the domain of zenith and azimuthal angles for boundary conditions.

and an asymmetry factor of 0.85 is used. The results for a cloud with $\tau_z = 32$ and $\tau_x = 128$ compare well with those computed from the adding method for a plane-parallel cloud layer. When $\mu_0 = 1$ and $\phi - \phi_0 = 0$, the finite cloud reflects much less due to the leaking of radiation to the side directions. In the case of $\mu_0 = 0.5$, multi-dimensional radiative transfer calculations involve both ϕ and ϕ_0. For the case of $(\tau_z/\tau_x) = (32/128)$ and $\phi - \phi_0 = 0$, the reflected intensity patterns for various ϕ_0 differ significantly from the reflected intensity based on the plane-parallel assumption. This intensity decreases as the solar azimuthal angle increases due to the finite dimension in the x axis. At $\theta = 0$, the reflected intensity is independent of the azimuthal angle. The result at this angle computed from the diffusion approximation for a two-dimensional cloud is slightly greater than that computed from the adding method.

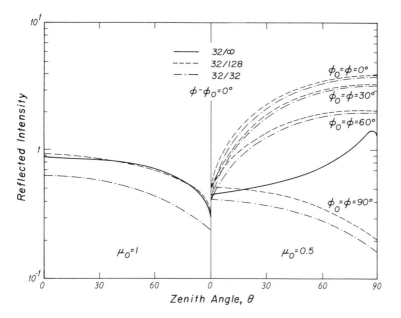

FIG. 5.30   Reflected intensity at the cloud top for a number of vertical and horizontal optical depths involving two-dimensional finite clouds. The single-scattering albedo and asymmetry factor used in the calculations are 1 and 0.85, respectively. The results for the plane-parallel cloud are computed from the adding method.

The Monte Carlo method has been used to compute the intensity distribution involving finite clouds. In this method, photons are followed until they leave the cloud. The distances between scattering events and changes in the scattering directions are simulated by random statistical probabilities. Figure 5.31 shows the reflected intensity as a function of the emergent angle for solar zenith angles of 1 and 0.866 (McKee and Cox, 1976). The histogram curves are derived by the Monte Carlo calculations in which the reflected intensities were averaged over the solid angle, that is, the ϕ_0 dependence presented in Fig. 5.4.3 was removed by the averaging procedure. As shown in this graph, the finite cloud reflects less solar radiation than its plane-parallel counterpart. The solid lines represent results computed from the diffusion approximation. These results appear to compare reasonably well with those simulated from the more exact Monte Carlo method, especially in the case for an overhead sun.

From the satellite image of a fair weather cumulus cloud field depicted in Fig. 5.28, it is clear that finite clouds are not usually isolated. According to Plank's (1969) observations of Florida cumuli, a cloud field consisting of an ensemble of cumuli accounts for a significant amount of the cloud cover over the globe (~20%). While this number may not be sufficiently accurate, the importance of the cumulus cloud field on the global radiation budget is evident. Wielicki and

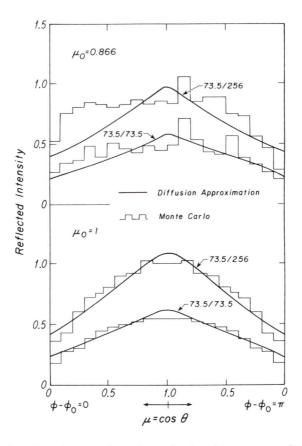

FIG. 5.31   Reflected intensity averaged over ϕ_0 as a function of the emergent angle. The histogram curves are results computed from the Monte Carlo method (McKee and Cox, 1976), and the smooth solid curves are from the diffusion approximation presented in the text.

Welch (1986) have analyzed Landsat Multispectral Scanner digital data for the mapping of cumulus cloud properties. Figure 5.32 shows the cloud size distribution for the western Atlantic cumulus scene presented in Fig. 5.28. The cloud size distribution is shown as the number of clouds per 1 km² surface area per 1 km cloud diameter range. Cloud size distribution is similar to the cloud droplet size distribution in which the normalization of droplets per unit volume is replaced by the normalization of clouds per unit surface area. The cloud amount varies according to the threshold method employed in the analysis. The cumulus cloud size appears to follow closely a power-law relationship with respect to the cloud diameter, except for very small sizes.

Using the Monte Carlo method, Aida (1977) has investigated the effects of cloud dimensions and orientations on the scattering pattern, assuming rectangularly shaped finite clouds. In particular, Aida showed that the clouds surrounding

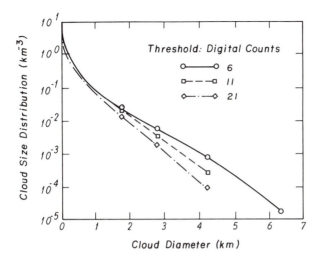

FIG. 5.32   Cloud size distribution for the western Atlantic cloud field. Values are given as the number of clouds per 1 km^2 surface area per 1 km cloud size class interval. Three curves denote results derived from different threshold methods (after Wielicki and Welch, 1986).

a cloud that is illuminated by the sun have significant interactions, if the cloud distances are within about five times their sizes. Weinman and Harshvardhan (1982) have also illustrated the importance of an array of horizontally finite clouds on solar reflection. The effects of an array of cuboidal clouds on the sensitivity of clouds on climate have been investigated by Harshvardhan (1982). While these illustrations are interesting, a realistic and quantitative assessment of the cloud horizontal nonhomogeneity with respect to the question of cloud albedo has yet to be developed.

5.5   Radiative transfer in an anisotropic medium

Cirrus clouds are composed of hexagonal ice crystals, whose single-scattering properties differ significantly from those computed for spherical particles. Figure 5.33(a) displays a number of optical phenomena in the presence of cirrus clouds at the South Pole (Greenler, 1980). Shown are 22° and 46° halos, sundogs, the parhelic circle, the circumzenithal arc of the 46° halo, and the upper tangent arc of the 22° halo. These optical features are produced by hexagonal ice plates and columns having a preferred orientation, that is, with their major axes oriented horizontally. Ice crystal habits in cirrus clouds have been presented in Fig. 4.11. With the exception of the 22° and 46° halos, which can be generated by randomly oriented ice crystals in three dimensional space, the majority of ice crystal optics are associated with the horizontal orientation of hexagonal columns and plates.

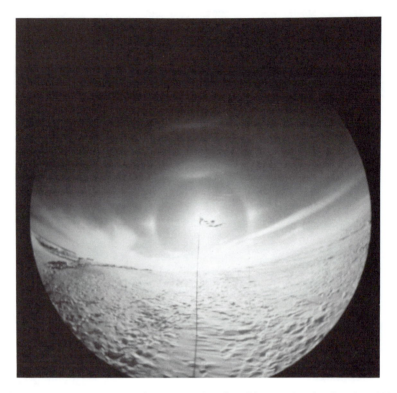

FIG. 5.33(a)   Halos and arcs produced by horizontally oriented ice crystals (after Greenler, 1980). Shown are $22°$ and $46°$ halos, sundogs, the parhelic circle, the circumzenithal arc of the $46°$ halo and the upper tangent arc of the $22°$ halo.

5.5.1 *Theory*

In the case of horizontally oriented ice crystals, the single-scattering parameters depend on the direction of the incident light beam. Thus, the conventional formulation for the multiple-scattering problem is no longer applicable. Liou (1980) has formulated the basic equation for the transfer of solar radiation in an optically anisotropic medium, in which the single-scattering properties vary with the incident angle of the light beam. Stephens (1980) and Asano (1983) have discussed the transfer of radiation through optically anisotropic ice clouds. The latter author used a hypothetical cloud model in which the scattering phase function was expressed in terms of the incident angle. Takano and Liou (1989b) have included realistic scattering parameters for horizontally oriented ice crystals in their discussion and analysis. The Stokes vector was also properly accounted for in the formulation.

Let the directions of incoming and outgoing light beams be denoted by (μ', ϕ') and (μ, ϕ), respectively, where μ is the cosine of the zenith angle and ϕ is the corresponding azimuthal angle. The scattering phase matrix $\mathbf{P}$ is a function of

(μ, ϕ, μ', ϕ') and cannot be defined by the scattering angle Θ alone, as in conventional radiative transfer. Moreover, the extinction (σ_e) and scattering (σ_s) cross sections vary with the direction of the incoming light beam (μ', ϕ').

Consider an anisotropic medium consisting of ice crystals randomly oriented in a horizontal plane. Because of the symmetry with respect to the azimuthal angle for the incoming light beam, the phase matrix and the extinction and scattering cross sections may be expressed by $\mathbf{P}(\mu, \phi, \mu', \phi')[= \mathbf{P}(\Theta, \Phi, \mu')], \sigma_e(\mu')$, and $\sigma_s(\mu')$, respectively, where Φ is the azimuthal angle associated with the scattering angle Θ. In this case, we may define the differential normal optical depth in the form

$$\frac{d\tilde{\tau}}{dz} = -\tilde{\sigma}_e N, \tag{5.5.1}$$

where the normal extinction cross section, $\tilde{\sigma}_e = \sigma_e(\mu' = 1), N$ is the number density of the particles, and z is the distance. Let the Stokes vector intensity $\mathbf{I} = (I, Q, U, V)$. The general equation governing the transfer of diffuse solar intensity may be written in the form

$$\mu \frac{d\mathbf{I}(\tilde{\tau}; \mu, \phi)}{d\tilde{\tau}} = \mathbf{I}(\tilde{\tau}; \mu, \phi)k(\mu) - \mathbf{J}(\tilde{\tau}; \mu, \phi), \tag{5.5.2}$$

where

$$k(\mu) = \sigma_e(\mu)/\tilde{\sigma}_e, \tag{5.5.3}$$

and the source function

$$\mathbf{J}(\tilde{\tau}; \mu, \phi) = \frac{1}{4\pi} \int_0^{2\pi} \int_{-1}^1 \tilde{\omega}(\mu')\mathbf{P}(\mu, \phi; \mu', \phi')\mathbf{I}(\tilde{\tau}; \mu', \phi')\, d\mu'\, d\phi'$$
$$+ \frac{1}{4\pi}\tilde{\omega}(-\mu_0)\mathbf{P}(\mu, \phi; -\mu_0, \phi_0)\mathbf{F}_\odot \exp\left[-k(-\mu_0)\tilde{\tau}/\mu_0\right]. \tag{5.5.4}$$

The first and second terms on the right-hand side represent contributions from, respectively, the multiple and single scattering of the direct solar intensity. The equivalent single-scattering albedo is defined as

$$\tilde{\omega}(\mu) = \sigma_s(\mu)/\tilde{\sigma}_e. \tag{5.5.5}$$

The general phase matrix, with respect to the local meridian plane, is given by (Liou, 1980)

$$\mathbf{P}(\mu, \mu'; \phi - \phi') = \mathbf{L}(\pi - i_2)\mathbf{P}(\Theta, \Phi, \mu')\mathbf{L}(-i_1), \tag{5.5.6}$$

where i_1 and i_2 denote the angles between the meridian planes for the incoming and outgoing light beams, respectively, and the plane of scattering, as illustrated

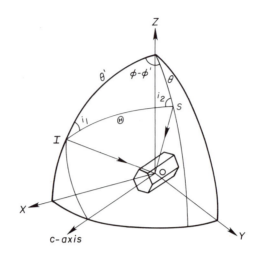

FIG. 5.33(b)   Scattering geometry involving horizontally oriented ice crystals. The terms IO and SO denote the incident (θ', ϕ') and scattered (θ, ϕ) directions, respectively. i_1 and i_2 are angles between the meridian planes for the incoming and scattered light beams, respectively. The scattering angle is denoted by Θ.

in Fig. 5.33(b). The transformation matrix for the Stokes vector is given by [see also Eq. (5.1.38)]

$$
\mathbf{L}(\chi) = \begin{bmatrix} 1 & 0 & 0 & 0 \\ 0 & \cos 2\chi & \sin 2\chi & 0 \\ 0 & -\sin 2\chi & \cos 2\chi & 0 \\ 0 & 0 & 0 & 1 \end{bmatrix},
\tag{5.5.7}
$$

where $\chi = -i_1$ or $\pi - i_2$. From spherical geometry, these angles are given by

$$
\cos i_1 = \frac{-\mu + \mu' \cos \Theta}{\pm(1 - \cos^2 \Theta)^{1/2}(1 - \mu^2)^{1/2}},
\tag{5.5.8}
$$

$$
\cos i_2 = \frac{-\mu' + \mu \cos \Theta}{\pm(1 - \cos^2 \Theta)^{1/2}(1 - \mu^2)^{1/2}}.
\tag{5.5.9}
$$

If $\mathbf{P}(\Theta, \Phi, \mu)$, $\sigma_e(\mu)$, and $\sigma_s(\mu)$ are known, then, in principle, Eq. (5.5.2) may be solved numerically.

We shall approach the multiple-scattering problem by means of the adding-method for radiative transfer described in Section 3.2.2. To proceed with the adding principle for radiative transfer in an anisotropic medium, we utilize the reflection and transmission matrices defined in Section 3.2.2 and consider an infinitesimal layer with a very small optical depth $\Delta\tilde{\tau}$, say 10^{-8}. Since the optical depth is so small, only single scattering takes place within the layer. From Eqs. (5.5.2) and

(5.5.4), the analytic solutions for the reflected and transmitted intensities undergoing single scattering may be derived. Subject to the condition that $\Delta\tilde{\tau} \to 0$, we find

$$\mathbf{T}(\mu, \mu_0; \phi - \phi_0) \cong \frac{\Delta\tilde{\tau}}{4\mu\mu_0}\tilde{\omega}(\mu_0)\mathbf{P}(\mu, \mu_0; \phi - \phi_0), \tag{5.5.10}$$

$$\mathbf{R}(\mu, \mu_o; \phi - \phi_0) \cong \frac{\Delta\tilde{\tau}}{4\mu\mu_0}\tilde{\omega}(\mu_0)\mathbf{P}(-\mu, \mu_0; \phi - \phi_0), \tag{5.5.11}$$

$$\mathbf{T}^*(\mu, \mu_0; \phi - \phi_0) \cong \frac{\Delta\tilde{\tau}}{4\mu\mu_0}\tilde{\omega}(\mu_0)\mathbf{P}(-\mu, -\mu_0; \phi - \phi_0), \tag{5.5.12}$$

$$\mathbf{R}^*(\mu, \mu_0; \phi - \phi_0) \cong \frac{\Delta\tilde{\tau}}{4\mu\mu_0}\tilde{\omega}(\mu_0)\mathbf{P}(\mu, -\mu_0; \phi - \phi_0). \tag{5.5.13}$$

where the phase matrix $\mathbf{P}$ has been defined in Eq. (5.5.6).

With modifications to account for the dependence of the optical properties on the incoming direction, the procedure for computing the reflection and transmission matrices for the composite layer may be described by the following equations:

$$\mathbf{Q}_1 = \mathbf{R}_a^* \mathbf{R}_b, \tag{5.5.14}$$

$$\mathbf{Q}_n = \mathbf{Q}_1 \mathbf{Q}_{n-1}, \tag{5.5.15}$$

$$\mathbf{S} = \sum_{n-1}^{M} \mathbf{Q}_n, \tag{5.5.16}$$

$$\mathbf{D} = \mathbf{T}_a + \mathbf{S}\exp[-k(\mu_0)\tilde{\tau}_a/\mu_0] + \mathbf{S}\mathbf{T}_a, \tag{5.5.17}$$

$$\mathbf{U} = \mathbf{R}_b\exp[-k(\mu_0)\tilde{\tau}_a/\mu_0] + \mathbf{R}_b\mathbf{D}, \tag{5.5.18}$$

$$\mathbf{R}_{a,b} = \mathbf{R}_a + \exp[-k(\mu)\tilde{\tau}_a/\mu]\mathbf{U} + \mathbf{T}_a^*\mathbf{U}, \tag{5.5.19}$$

$$\mathbf{T}_{a,b} = \exp[-k(\mu)\tilde{\tau}_b/\mu]\mathbf{D} + \mathbf{T}_b\exp[-k(\mu_0)\tilde{\tau}_a/\mu_0] + \mathbf{T}_b\mathbf{D}. \tag{5.5.20}$$

The exponential terms in the adding equations denote the direct transmission through layer a or b without scattering, where the anisotropic factor $k(\mu)$ is given in Eq. (5.5.3). The total transmission for the combined layer is the sum of the diffuse transmission $\mathbf{T}_{a,b}$ and the direct transmission, $\exp[-k(\mu_0)(\tilde{\tau}_a + \tilde{\tau}_b)/\mu_0]$, in the direction of the sun.

In order to compute the reflection and transmission matrices for the initial layer with a very small optical depth, via Eqs. (5.5.10)–(5.5.13), we need the phase matrix and single-scattering albedo, and the directions of the incoming and outgoing beams. The phase matrix elements must be expanded in terms of the incoming and outgoing directions denoted by μ, μ' and $\phi - \phi'$. For spherical particles, the phase matrix consists of four nonzero independent elements. These elements can be decoupled analytically in terms of functions associated with μ, μ' and $\phi - \phi'$ (Dave, 1970; Kattawar et al., 1973). However, for nonspherical particles, the decomposition of the phase matrix elements has yet to be worked out. For randomly oriented nonspherical particles that have a plane of symmetry, the phase matrix contains six independent elements. In this case, there are seven symmetrical relationships for these elements based on the reciprocity principle (Hovenier, 1969). From these relationships, the phase matrix elements can be expanded in terms of either cosine or sine Fourier components (Hansen, 1971). For randomly oriented ice plates or columns in a horizontal plane, there are 16 elements in the phase matrix. It can be proved that these elements obey a number of symmetrical relationships, based on which they can be expanded in terms of either cosine or sine Fourier components.

In view of the above discussion, the general phase matrix elements may be numerically expanded in the forms

$$\tilde{\omega}(\mu')P_{ij}(\mu,\mu';\phi-\phi') = P_{ij}^0(\mu,\mu')+2\sum_{m=1}^{N} P_{ij}^m(\mu,\mu')\times \begin{cases} \cos m(\phi-\phi'), & c \\ \sin m(\phi-\phi'), & s, \end{cases}$$
(5.5.21)

where

$$c, \quad ij = 11, 12, 21, 22, 33, 34, 43, 44$$

$$s, \quad ij = 13, 14, 23, 24, 31, 32, 41, 42$$

and $P_{ij}^m (m = 0, 1, ..., N)$ denotes the Fourier expansion coefficients. With this expansion, each term in the Fourier series may be treated independently in numerical computations. For spherical particles or randomly oriented nonspherical particle, Eq. (5.5.21) can be used except that $\tilde{\omega}(\mu') = 1$.

With respect to the normalization of the phase function P_{11}, the following procedures may be used. The phase function is normalized such that

$$\frac{1}{4\pi} \int_0^{2\pi} \int_{-1}^{1} P_{11}(\mu, \mu', \phi - \phi') \, d\mu \, d(\phi - \phi') = 1,$$
(5.5.22a)

where $d\mu \, d(\phi - \phi')$ denotes the differential solid angle. Using Eq. (5.5.21), we find

$$\frac{1}{2} \int_{-1}^{1} P_{11}^0(\mu, \mu') \, d\mu = \tilde{\omega}(\mu'),$$
(5.5.22b)

where $\tilde{\omega}(\mu')$ is defined in Eq. (5.5.5). In the case of randomly oriented nonspherical particles (or spherical particles), the single-scattering albedo $\tilde{\omega}$ is independent of μ' and is a constant. If there is no absorption, $\tilde{\omega} = 1$. In this case, Eq. (5.5.22b) can be derived from the expansion of the phase function in terms of the Legendre polynomial using the addition theorem for spherical harmonics (Liou, 1980). However, for randomly oriented ice crystals in a horizontal plane, $\tilde{\omega}$ is a function of the incident angle. Normalization of the phase function must be performed for each μ'.

The phase function P_{11} for ice crystals has a common sharp diffraction peak. In order to account for this peak in numerical integrations properly, thousands of

Fourier components are needed in the phase function expansion. To optimize the computational effort, the procedure proposed by Potter (1970) may be followed. In this procedure, the forward peak is truncated by extrapolating the phase function linearly from the scattering angles $10°$ to $0°$ in the logarithmic scale. Let the truncated phase function be P_{11}^t; then the truncated fraction of scattered intensity is given by

$$f = \int_{4\pi} (P_{11} - P_{11}^t)d\Omega/4\pi. \tag{5.5.23}$$

To use the truncated phase function in multiple-scattering computations while, at the same time, achieving the "equivalent" result, as in the case when the sharp diffraction peak was included in the computations, an adjustment must be made for the optical depth and single-scattering albedo in a manner described in Section 3.6.

5.5.2   *Effects of horizontal orientation of ice crystals on cloud radiative properties*

In order to compute the reflected and transmitted intensity of sunlight from oriented ice crystals, the phase function is required. The phase functions for horizontally oriented plates and columns have been computed by Takano and Liou (1989a) using the geometric ray-tracing technique. The phase function is a function of the incoming direction (μ, ϕ) and the outgoing direction (μ', ϕ'). The computations of the phase function for oriented two-dimensional (2D) plates (randomly oriented with their c axis vertical), Parry columns (the c axis and a pair of prisms facing horizontal), and 2D columns (the c axis horizontal with random rotational orientation about this c axis) lead to the interpretation of numerous halos and arcs observed in the atmosphere. Figure 5.34 illustrates the reflected and transmitted intensity patterns for 2D plates when the solar zenith angle, θ_0, is $75°$. The optical depth presented in this graph is a mean value averaged over the zenith angle θ. In the case of 2D plates, scattered sunlight is confined to four latitude belts, due to specific geometry. Based on the ray-tracing geometry for plates, the emergent zenith angle may be computed from the incoming solar zenith angle θ_0, in the forms

$$\theta^* = \begin{cases} \pi/2 - \sin^{-1}(m_r^2 - \sin^2\theta_0)^{1/2}, & \text{for } \theta_0 > \sin^{-1}(m_r^2 - 1)^{1/2} \simeq 58° \\ \sin^{-1}(m_r^2 - \cos^2\theta_0)^{1/2}, & \text{for } \theta_0 < \cos^{-1}(m_r^2 - 1)^{1/2} \simeq 32°. \end{cases}$$
$$\tag{5.5.24}$$

For $\theta_0 = 75°$, these latitude belts correspond to zenith angles of $\pm75°$ and $\pm27°$, with negative values representing mirror images. If the incident angle is $27°$, the four latitude belts are $\pm27°$ and $\pm75°$. Due to the symmetrical property of 2D plates with respect to incoming light beams, all multiply scattered light is also confined to the four latitude belts. The reflected and transmitted intensities for optical depths 1/4, 1, 4, and 16 are displayed as a function of the azimuthal

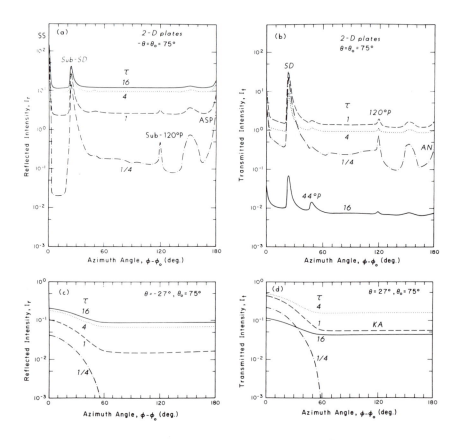

FIG. 5.34 Intensity of sunlight reflected and transmitted by 2D plates ($L/2r = 0.1$) as a function of the azimuth angle, $\phi - \phi_0$, for (a) $\theta = -\theta_0 = -75°$, (b) $\theta = \theta_0 = 75°$, (c) $(\theta, \theta_0) = (-27°, 75°)$ and (d) $(\theta, \theta_0) = (27°, 75°)$. The optical depth τ in 2D cases is a value averaged over all directions ($0 \le \mu \le 1$) (after Takano and Liou, 1989b).

angle, $\phi - \phi_0$. The reflected intensity increases with increasing optical depth. For the 75° emergent zenith angle, the subsun, subsundog, 120° subparhelion, and antisolar peak optical features are shown distinctly for an optical depth of 1/4, at which single scattering dominates. For an optical depth of 16, only the subsun and subparhelion are observed. For the transmitted intensity, the sundog is visible even for large cloud optical depths. In addition to these optical phenomena, the anthelion (AN) located at the 180° azimuthal angle is seen for the small optical depth of 1/4, due to double scattering, viz., the coupling of the subsun and antisolar peak. For an optical depth of 16, the 44° parhelion produced by double scattering (denoted as 44°P) is also observed. At the 27° zenith angle, the Kern's arc (KA) appears for $\tau \ge 1$ due to the effects of multiple scattering.

Figures 5.35 and 5.36 display the reflected and transmitted intensity patterns as a function of the zenith angle on the plane $\phi - \phi_0 = 0°$ for Parry and 2D columns,

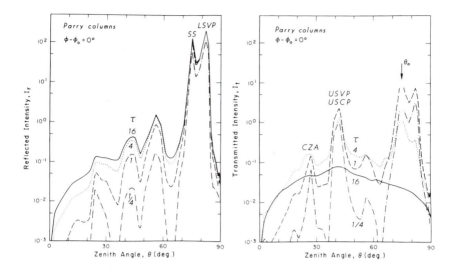

FIG. 5.35 Intensity of sunlight reflected and transmitted by Parry columns ($L/2r = 2.5$) in the solar principal plane ($\phi - \phi_0 = 0°$) at $\lambda = 0.5$ μm. The solar zenith angle is 75° (after Takano and Liou, 1989b).

respectively. The solar zenith angle used for the computation is 75°. These patterns for an optical depth of 1/4 are basically similar to those from single-scattering computations, except for the subpeak at $\theta = 82°$ in the transmitted intensity in Fig. 5.35. This peak is caused by the lower sunvex Parry arc of the subsun. For Parry columns, several optical features are observed for optical depths less than 4. These include the subsun (SS) and lower sunvex Parry arc (LSVP) in

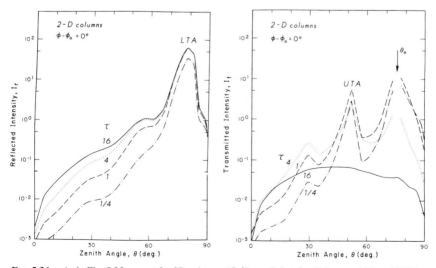

FIG. 5.36 As in Fig. 5.35, except for 2D columns ($L/2r = 2.5$) (after Takano and Liou, 1989b).

the reflected intensity (these are also visible for an optical depth of 16), and the circumzenith arc (CZA) and upper suncave (USCP) and sunvex (USVP) Parry arcs in the transmitted intensity. For 2D columns, the lower tangent arc (LTA) is noticeable in the reflected intensity for all optical depths. When the optical depth is large, the transmitted intensities have the lowest values in the zenith and nadir directions, as shown in Figs. 5.35 and 5.36. Except for these features, the reflected and transmitted intensities of 2D columns are similar to those of randomly oriented columns. Also, it is noted that the reflected and transmitted intensity distributions of Parry columns are similar to those of 2D columns for large optical depths ($\tau = 16$) because sufficient multiple scattering is present. Optical phenomena produced by 2D plates, Parry columns, and 2D columns and their causes are listed in Table 5.5. The causes of the frequently observed halos and arcs produced by horizontally oriented columns and plates are illustrated in Fig. 5.37, in which the rays that are required to generate these optics are displayed.

22° Halo, Upper & Lower Tangent Arcs,
Parry Arcs

22° Halo, Sundog

Sub-sundog

Parhelic Circle

Sub-parhelic Circle

120° Parhelion

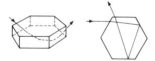

Subsun

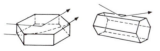

46° Halo, Circumzenithal Arc

46° Halo, Circumhorizontal Arc

46° Halo, Supralateral Arc

46° Halo, Infralateral Arc

Heliac Arc

Subhelic Arc

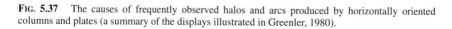

Wegener's & Hastings' Anthelic Arcs

FIG. 5.37 The causes of frequently observed halos and arcs produced by horizontally oriented columns and plates (a summary of the displays illustrated in Greenler, 1980).

Table 5.5 Optical phenomena caused by 2D plates, Parry columns, and 2D columns[a]

Abbreviation	Definition
	2D Plates
ASP	antisolar peak
CHA	circumhorizontal arc
CZA	circumzenithal arc
ER	external reflection
KA	Kern's arc
120° P	120° parhelion
120° Sub-P	120° subparhelion
PC	parhelic circle
SD	sundog (22° parhelion)
SS	subsun
Sub-CHA	subcircumhorizontal arc
Sub-CZA	subcircumzenithal arc
Sub-SD	sub-sundog (22° subparhelion)
	Parry Columns
ASA	antisolar arc
ASP	antisolar peak
CZA	circumzenithal arc
HA	heliac arc
HAA	Hastings' anthelic arc
LSCP	lower suncave Parry arc
LSVP	lower sunvex Parry arc
PC	parhelic circle
SHA	subhelic arc
SID	solar incident direction
SS	subsun
Sub-CZA	subcircumzenithal arc
Sub-PC	subparhelic circle
USCP	upper suncave Parry arc
SUVP	upper sunvex Parry arc
	2D Columns
ASA	antisolar arc
ASP	antisolar peak
DA	diffuse arc
ILA	infralateral arc
LTA	lower tangent arc
PC	parhelic circle
SHA	subhelic arc
SID	solar incident direction
SLA	supralateral arc
TAA	Tricker's anthelic arc
UTA	upper tangent arc
WAA	Wegener's anthelic arc

[a] After Takano and Liou (1989a).

REFERENCES

Aida, M., 1977: Scattering of solar radiation as a function of cloud dimensions and orientation. *J. Quant. Spectrosc. Radiat. Transfer*, **17**, 303–310.

Asano, S., 1983: Transfer of solar radiation in optically anisotropic ice clouds. *J. Meteor. Soc. Japan*, **61**, 402–413.

Asano, S., and G. Yamamoto, 1975: Light scattering by a spheroidal particle. *Appl. Opt.*, **14**, 29–49.

Asano, S., and M. Sato, 1980: Light scattering by randomly oriented spheroidal particles. *Appl. Opt.*, **19**, 962–974.

Barber, P., and C. Yeh, 1975: Scattering of electromagnetic waves by arbitrarily shaped dielectric bodies. *Appl. Opt.*, **14**, 2864–2872.

Berg, P. W., and J. L. McGregor, 1966: *Elementary Partial Differential Equations*. Holden-Day, Inc., San Francisco, 421 pp.

Bertie, J. E., H. J. Labbé, and E. Whalley, 1969: Absorptivity of ice I in the range of 4000–30 cm^{-1}. *J. Chem. Phys.*, **50**, 4501–4520.

Blau, H. H., R. P. Espinola, and E. C. Reifenstein, 1966: Near infrared scattering by sunlit terrestrial clouds. *Appl. Opt.*, **5**, 555–564.

Bohren, C. F., and D. R. Huffman, 1983: *Absorption and Scattering of Light by Small Particles*. Wiley, New York, 530 pp.

Born, M., and E. Wolf, 1975: *Principles of Optics*. Pergamon Press, Oxford, 808 pp.

Busygin, V. P., N. A. Yevstratov, and Ye. M. Feygel'son, 1973: Optical properties of cumulus clouds, and radiant fluxes for cumulus cloud cover. *Izv. Acad. Sci. USSR Atmos. Ocean. Phys.*, **9**, 1142–1151.

Cai, Q., and K. N. Liou, 1982: Polarized light scattering by hexagonal ice crystals: Theory. *Appl. Opt.*, **21**, 3569–3580.

Case, K., and P. F. Zweifel, 1967: *Linear Transport Theory*. Addison-Wesley, Reading, Mass., 342 pp.

Coffeen, D. L. 1979: Polarization and scattering characteristics in the atmosphere of Earth, Venus, and Jupiter. *J. Opt. Soc. Amer.*, **69**, 1051–1064.

Coleman, R. F., and K. N. Liou, 1981: Light scattering by hexagonal ice crystals. *J. Atmos. Sci.*, **38**, 1260–1271.

Crosbie, A. L., and T. L. Linsenbardt, 1978: Two-dimensional isotropic scattering in a semi-infinite medium. *J. Quant. Spectrosc. Radiat. Transfer*, **19**, 257–284.

Dave, J. V., 1970: Coefficients of the Legendre and Fourier series for the scattering functions of spherical particles. *Appl. Opt.*, **9**, 1888–1896.

Davies, R., 1978: The effect of finite geometry on the three-dimensional transfer of solar irradiance in clouds. *J. Atmos. Sci.*, **35**, 1712–1725.

Deirmendjian, D., 1969: *Electromagnetic Scattering on Spherical Polydispersions*. Elsevier, New York, 291 pp.

Dugin, V. P., and S. O. Mirumyants, 1976: The light scattering matrices of artificial crystalline clouds. *Izv. Acad. Sci. USSR Atmos. Ocean. Phys.*, **12**, 988–991.

Foot, J. S., 1988a: Some observations of the optical properties of clouds. I. Stratocumulus. *Quart. J. Roy. Meteor. Soc.*, **114**, 129–144.

Foot, J. S., 1988b: Some observations of the optical properties of clouds. II. Cirrus. *Quart. J. Roy. Meteor. Soc.*, **114**, 145–164.

Giovanelli, R. G., 1959: Radiative transfer in a non-uniform media. *Aust. J. Phys.*, **12**, 164–170.

Greenberg, J. M., A. C. Lind, R. T. Wang, and L. F. Libelo, 1967: Scattering by nonspherical system. In *Electromagnetic Scattering*, R. L. Rowell and F. S. Stein, Eds., Gordon and Breach, New York, pp. 3-54.

Greenler, R., 1980: *Rainbows, Halos, and Glories*. Cambridge University Press, Cambridge, 195 pp.

Griffith, K. T., S. K. Cox, and R. G. Knollenberg, 1980: Infrared radiative properties of tropical cirrus clouds inferred from aircraft measurements. *J. Atmos. Sci.*, **37**, 1077–1087.

Hale, G. M., and M. R. Querry, 1973: Optical constants of water in the 200- μm to 200-nm wavelength region. *Appl. Opt.*, **12**, 555–563.

Hansen, J. E., 1971: Multiple scattering of polarized light in panetary atmospheres. Parts I and II. *J. Atmos. Sci.*, **28**, 120–125, 1400–1426.

Hansen, J. E., and J. B. Pollack, 1970: Near-infrared light scattering by terrestrial clouds. *J. Atmos. Sci.*, **27**, 265–281.

Harshvardhan, 1982: The effect of brokeness on cloud-climate sensitivity. *J. Atmos. Sci.*, **39**, 1853–1861.

Herman, G. F., and J. A. Curry, 1984: Observational and theoretical studies of solar radiation in arctic stratus clouds. *J. Climate Appl. Meteor.*, **23**, 5–24.

Hobbs, P. V., 1974: *Ice Physics*, Oxford University Press, London, 835 pp.

Hovenier, J. W., 1969: Symmetry relationships for scattering of polarized light in a slab of randomly oriented particles. *J. Atmos. Sci.*, **26**, 488–499.

Irvine, W. M., and J. B. Pollack, 1968: Infrared optical properties of water and ice spheres. *Icarus*, **8**, 324–360.

Iskander, M.F., A. Lakhtakia, and C. Durney, 1983: A new procedure for improving the solution stability and extending the frequency range of the EBCM. *IEEE Trans. Antennas and Propagat.*, **AP-31**, 317–324.

Jacobowitz, H., 1971: A method for computing transfer of solar radiation through clouds of hexagonal ice crystals. *J. Quant. Spectrosc. Radiat. Transfer*, **11**, 691–695.

Kattawar, G. W., S. J. Hitzfelder, and J. Binstock, 1973: An explicit form of the Mie phase matrix for multiple scattering calculations in the I, Q, U, and V representation. *J. Atmos. Sci.*, **30**, 289–295.

Kerker, M., 1969: *The Scattering of Light and Other Electromagnetic Radiation*. Academic Press, New York, 666 pp.

King, M. D., L. F. Radke, and P. V. Hobbs, 1990: Determination of the spectral absorption of solar radiation by marine stratocumulus clouds from airborne measurements within clouds. *J. Atmos. Sci.*, **47**, 894–907.

Kinne, S., and K. N. Liou, 1989: The effects of the nonsphericity and size distribution of ice crystals on the radiative properties of cirrus clouds. *Atmos. Res.*, **24**, 273–284.

Liou, K. N., 1972: Light scattering by ice clouds in the visible and infrared: A theoretical study. *J. Atmos. Sci.*, **29**, 524–536.

Liou, K. N., 1976: On the absorption, reflection and transmission of solar radiation in cloudy atmospheres. *J. Atmos. Sci.*, **33**, 798–805.

Liou, K. N., 1980: *An Introduction to Atmospheric Radiation*. Academic Press, New York, 392 pp.

Liou, K. N., 1986: Influence of cirrus clouds on weather and climate processes: A global perspective. *Mon. Wea. Rev.*, **114**, 1167–1199.

Liou, K. N., and S. C. Ou, 1979: Infrared radiative transfer in finite cloud layers. *J. Atmos. Sci.*, **36**, 1985–1996.

Liou, K. N., and G. D. Wittman, 1979: Parameterization of the radiative properties of clouds. *J. Atmos. Sci.*, **36**, 1261–1273.

Lord Rayleigh, 1918: The dispersal of light by a dielectric cylinder. *Phil. Mag.*, **36**, 365–376.

Mckee, T., and S. K. Cox, 1976: Simulated radiance patterns for finite cubic clouds. *J. Atmos. Sci.*, **33**, 2014–2020.

Mie, G., 1908: Beiträge zur Optik trüber Medien, speziell kolloidaler Metallösungen. *Ann. Physik*, **25**, 377–445.

Muinonen, K., K. Lumme, J. Peltoniemi, and W. M. Irvine, 1989: Light scattering by randomly oriented crystals. *Appl. Opt.*, **28**, 3051–3060.

Nikiforova, N. K., L. N. Pavlova, A. G. Petrushin, V. P. Snykov, and O. A. Volkovitsky, 1977: Aerodynamic and optical properties of ice crystals. *J. Aerosol Sci.*, **8**, 243–250.

Ou, S. C., and K. N. Liou, 1980: Numerical experiments on the Helmholtz equation derived from the solar radiation transfer equation in three-dimensional space. *Appl. Math. Comp.*, **7**, 155–175.

Ou, S. C., and K. N. Liou, 1982: Generalization of the spherical harmonic method to radiative transfer in multi-dimensional space. *J. Quant. Spectrosc. Radiat. Transfer*, **28**, 271–288.

Paltridge, G. W., and C. M. R. Platt, 1981: Aircraft measurements of solar and infrared radiation and the microphysics of cirrus clouds. *Quart. J. Roy. Meteor. Soc.*, **107**, 367–380.

Perrin, F., 1942: Polarization of light scattered by isotropic opalescent media. *J. Chem. Phys.*, **10**, 415–527.

Plank, V. G., 1969: The size distribution of cumulus clouds in representative Florida populations. *J. Appl. Meteor.*, **8**, 46–67.

Platt, C. M. R., 1976: Infrared absorption and liquid water content in stratocumulus clouds. *Quart. J. Roy. Meteor. Soc*, **102**, 553–561.

Pollack, J. B., D. W. Strecker, F. C. Witteborn, E. F. Erickson, and B. J. Baldwin, 1978: Properties of the clouds of Venus, as inferred from airborne observations of its near-infrared reflective spectrum. *Icarus*, **23**, 28–45.

Potter, J. F., 1970: The delta function approximation in radiative transfer theory. *J. Atmos. Sci.*, **27**, 943–951.

Reynolds, D. W., T. H. Vonder Haar, and S. K. Cox, 1975: The effect of solar radiation absorption in the tropical troposphere. *J. Appl. Meteor.*, **14**, 433–443.

Roewe, D., and K. N. Liou, 1978: Influence of cirrus clouds on the infrared cooling rate in the troposphere and lower stratosphere. *J. Appl. Meteor.*, **17**, 92–106.

Rozenberg, G., M. Malkevich, V. Malkova, and V. Syachinov, 1974: Determination of the optical characteristics of clouds from measurements of reflected solar radiation by KOSMOS 320. *Izv. Acad. Sci. USSR Atmos. Ocean. Phys.*, **10**, 14–24.

Sassen, K., and K. N. Liou, 1979: Scattering of polarized laser light by water droplet, mixed phase, and ice crystal clouds: I. Angular scattering patterns. *J. Atmos. Sci.*, **36**, 838–851.

Schaaf, J. W., and D. Williams, 1973: Optical constants of ice in the infrared. J. *Opt. Soc. Amer.*, **63**, 726–732.

Seki, M., K. Kobayashi, and J. Nakahara, 1981: Optical spectra of hexagonal ice. *J. Phys. Soc. Japan*, **50**, 2643–2648.

Slingo, A., 1989: A GCM parameterization for the shortwave radiative properties of water clouds. *J. Atmos. Sci.*, **46**, 1419–1427.

Smith, W. L., Jr., P. F. Hein, and S. K. Cox, 1990: The 27–28 October 1986 FIRE IFO cirrus case study: In situ observations of radiation and dynamic propterties of a cirrus cloud layer. *Mon. Wea. Rev.*, **118**, 2389–2401.

Spänkuch, D., and W. Döhler, 1985: Radiative properties of cirrus clouds in the middle ir derived from Fourier spectrometer measurements from space. *Z. Meteor.*, **6**, 314–324.

Stephens, G. L., 1978: Radiation profiles in extended water clouds. II. Parameterization schemes. *J. Atmos. Sci.*, **35**, 2123–2132.

Stephens, G. L., 1980: Radiative transfer on a linear lattice: Application to anisotropic ice crystal clouds. *J. Atmos. Sci.*, **37**, 2095–2104.

Stephens, G. L., 1984: The parameterization of radiation for numerical weather prediction and climate models. *Mon. Wea. Rev.*, **112**, 826–867.

Stephens, G. L., G. W. Paltridge, and C. M. R. Platt, 1978: Radiation profiles in extended water clouds. III: Observations. *J. Atmos. Sci.*, **35**, 2133–2141.

Stephens, G. L., and S. C. Tsay, 1990: On the cloud absorption anomaly. *Quart. J. Roy. Meteor. Soc.*, **116**, 671–704.

Takano, Y., and K. Jayaweera, 1985: Scattering phase matrix for hexagonal ice crystals computed from ray optics. *Appl. Opt.*, **24**, 3254–3263.

Takano, Y., and K. N. Liou, 1989a: Solar radiative transfer in cirrus clouds. Part I: Single-scattering and optical properties of hexagonal ice crystals. *J. Atmos. Sci.*, **46**, 3–19.

Takano, Y., and K. N. Liou, 1989b: Solar radiative transfer in cirrus clouds. Part II: Theory and computation of multiple scattering in an anisotropic medium. *J. Atmos. Sci.*, **46**, 20–36.

Twomey, S., 1976: Computations of the absorption of solar radiation by clouds. *J. Atmos. Sci.*, **33**, 1087–1091.

Twomey, S., and T. Cocks, 1982: Spectral reflectance of clouds in the near-infrared: Comparison of measurements and calculations. *J. Meteor. Soc. Japan*, **60**, 583–592.

Twomey, S., and K. J. Seton, 1980: Inferences of gross microphysical properties of clouds from spectral reflectance measurements. *J. Atmos. Sci.*, **37**, 1065–1069.

van de Hulst, H. C., 1957: *Light Scattering by Small Particles*. Wiley, New York, 470 pp.

Volkovitsky, O. A., L. N. Pavlova, and A. G. Petrushin, 1980: Scattering of light by ice crystals. *Izv. Acad. Sci. USSR. Atmos. Ocean. Phys.*, **16**, 90–102.

Vouk, V., 1948: Projected area of convex bodies. *Nature*, **162**, 330–331.

Wait, J. R., 1955: Scattering of a plane wave from a circular dielectric cylinder at oblique incidence. *Can. J. Phys.*, **33**, 189–195.

Warren, S. G., 1984: Optical constants of ice from ultraviolet to the microwave. *Appl. Opt.*, **23**, 1206–1225.

Weinman, J. A., and Harshvardhan, 1982: Solar reflection from a regular array of horizontally finite clouds. *Appl. Opt.*, **21**, 2940–2944.

Welch, R. M., S. K. Cox, and J. M. Davis, 1980: *Solar Radiation and Clouds*. *Meteor. Monogr.*, No. 39, Amer. Meteor. Soc., Boston, 96 pp.

Wendling, P., R. Wendling, and H. K. Weickmann, 1979: Scattering of solar radiation by hexagonal ice crystals. *Appl. Opt.*, **18**, 2663–2671.

Wielicki, B. A., and R. M. Welch, 1986: Cumulus cloud properties derived using Landsat satellite data. *J. Climate Appl. Meteor.*, **25**, 261–276.

Wiscombe, W. J., 1980: Improved Mie scattering algorithms. *Appl. Opt.*, **19**, 1505–1509.

Wiscombe, W. J., R. M. Welch, and W. D. Hall, 1984: The effects of very large drops on cloud adsorption. Part I: Parcel models. *J. Atmos. Sci.*, **41**, 1336–1355.

Zander, R., 1966: Spectral scattering properties of ice clouds and hoarfrost. *J. Geophys. Res.*, **71**, 375–378.

6
ATMOSPHERES IN RADIATIVE AND THERMAL EQUILIBRIUM

In this chapter is discussed the subject of radiation and thermal equilibrium in the earth's atmosphere, with a specific emphasis on the cloud effect. Radiative equilibrium involves the balance between incoming solar radiation, which drives the general circulation of the atmosphere and the oceans, and outgoing thermal infrared (ir) radiation emitted from the earth–atmosphere system. Consideration of this equilibrium provides a fundamental step toward a physical understanding of the thermal structure of the atmosphere, as a first approximation. Radiative equilibrium has been extensively studied in the fields of stellar atmospheres, atmospheres of other planets, and the earth's primitive atmosphere. Indeed, some of the earlier pioneering research efforts on the subject of radiative equilibrium can be found in the discussion of temperature for stars (Chandrasekhar, 1939), in the discussion of the vertical structure of planetary atmospheres (Chamberlain and Hunter, 1987), and in the discussion of the structure of the earth's stratosphere (Goody and Yung, 1989).

Although the earth–atmosphere system is in radiative equilibrium at the top of the atmosphere (TOA) on the mean annual basis, the system is evidently not in radiative equilibrium within the atmosphere and at the surface. This phenomenon has been briefly introduced in Section 1.4, using the globally averaged condition as a prototype. The temperature structure of the atmosphere is regulated not only by radiative processes, but also by the general circulation of the atmosphere and by processes involving cloud formation and precipitation. At the surface, eddy transports of sensible and latent heat fluxes are significant in modulating the energy and hydrological balances that determine surface temperatures.

Consider the global earth and atmosphere as a unit. Equilibrium of this unit, referred to as *thermal equilibrium*, can be achieved only through radiative and convective processes. The convective nature of the earth–atmosphere system is fundamental in weather and climate processes. This simple thermal equilibrium model is a valuable vehicle for examining the physical principles that govern the vertical temperature structure. We begin our discussion on the atmospheres in radiative and thermal equilibrium with the energy input from the sun: solar insolation.

6.1   Solar insolation

6.1.1   *Solar insolation and earth–sun geometry*

The solar flux incident on a horizontal unit area at TOA for a given point on the earth at a given time can be obtained from

$$F(t) = S \left(\frac{r_0}{r}\right)^2 \cos \theta_0, \qquad (6.1.1)$$

where S is the solar constant corresponding to the mean earth–sun distance r_0 (further discussion is given in the following), and r is the actual distance between the earth and sun at a given time t.

The solar zenith angle θ_0 is related to the latitude ϕ, the declination of the sun δ [the angle between the sun's direction and the plane of the earth's equator; see Fig. 6.1(a)], and the hour angle h. From spherical geometry we have

$$\cos \theta_0 = \sin \phi \sin \delta + \cos \phi \cos \delta \cos h. \qquad (6.1.2a)$$

The hour angle, which is the angle at which the earth must turn to bring the meridian of the given point directly under the sun, is given by $h = 2\pi t/\Delta t_\odot$, where $\Delta t_\odot = 86,400$ s.

At the poles, $\phi = \pm \pi/2$, so that $\cos \theta_0 = \sin \delta$ or $\theta_0 = \pi/2 - \delta$. The solar elevation angle $(\pi/2 - \theta_0)$ is equal to the declination of the sun. During the six months of daylight, the sun simply circles around the horizon and never rises more than about 23.5°. At solar noon at any latitude, the hour angle $h = 0$. Thus, $\cos \theta_0 = \cos(\phi - \delta)$ or $\theta_0 = \phi - \delta$. Except at the poles, $\theta_0 = \pi/2$ at sunrise or sunset. From Eq. (6.1.2a), the half-day (i.e., from sunrise to noon or from noon to sunset) H is defined by

$$\cos H = - \tan \phi \tan \delta. \qquad (6.1.2b)$$

If $\phi = 0$ (equator) or $\delta = 0$ (equinoxes), then $\cos H = 0$. The length of the solar day is 12 hours. The latitude of the polar night $(H = 0)$ is determined by $\phi = 90° - |\delta|$ in the winter hemisphere.

In order to calculate the instantaneous flux at a given latitude, the values for (r_0/r) and δ are required. The computation of these two values is rather involved. First, a number of geometric positions and angles with respect to the earth's orbit about the sun must be defined. The plane of the earth's orbit is called the *plane of the ecliptic*, as shown in Fig. 6.1(b). The position of the sun is located at one of the foci of an ellipse. Let a and b denote the semimajor and semiminor axes of the ellipse. The earth orbit eccentricity is defined as the ratio of the distance between

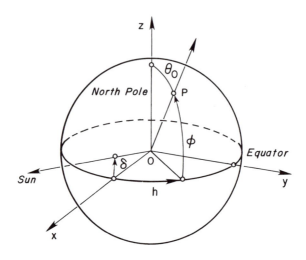

FIG. 6.1(a)   Definitions of the solar zenith angle (θ_0), latitude (ϕ), declination of the sun (δ), and the hour angle (h). P is a point along the local zenith.

the two foci to the major axis of the ellipse and is given by $e = (a^2 - b^2)^{1/2}/a$. The tilt of the earth's axis with respect to normal to the ecliptic plane is defined by the oblique angle ϵ. The longitude of the perihelion relative to the vernal equinox is defined by the angle ω. For a given time, the position of the earth is defined by the true anomaly ν in reference to the perihelion.

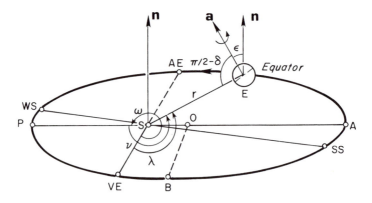

FIG. 6.1(b)   The earth–sun geometry. P denotes the perihelion, A the aphelion, AE the autumnal equinox, VE the vernal equinox, WS the winter solstice, SS the summer solstice; $\mathbf{n}$ is normal to the ecliptic plane; $\mathbf{a}$ is parallel to the earth's axis; δ is the declination of the sun, ϵ the oblique angle of the earth's axis, ω the longitude of the perihelion relative to the vernal equinox, ν the true anomaly of the earth at any given time, λ the true longitude of the earth, O the center of the ellipse, OA (or $OP = a$) the semimajor axis, $OB(= b)$ the semiminor axis, S the position of the sun, E the position of the earth, and $ES(= r)$ the distance between the earth and the sun.

From the geometry shown in Fig. 6.1(b), the declination of the sun can be expressed in terms of the oblique angle of the earth's axis, the longitude of the perihelion relative to the vernal equinox, and the true anomaly of the earth at a given time. Also, letting the true longitude of the earth, counted counterclockwise from the vernal equinox, be λ, from three-dimensional geometry we find

$$\sin \delta = \sin \epsilon \sin(\nu + \omega) = \sin \epsilon \sin \lambda. \tag{6.1.3}$$

Having defined the relevant geometric parameters, we shall now outline the procedure by which the solar flux at a given latitude can be computed. The planet's orbital ellipse can be described by Kepler's first law (the law of orbits), in which the distance is related to the true anomaly in the form

$$r = \frac{a(1 - e^2)}{1 + e \cos \nu}. \tag{6.1.4}$$

This equation can be derived from the geometry of an ellipse. To compute the solar flux over a certain time period, Kepler's second law (the law of areas) is required. This law is a statement of the conservation of angular momentum that the radius vector, drawn from the sun to the planet, sweeps out equal areas in equal times. Letting $\tilde{T}$ denote the tropical year (i.e., the time between successive arrivals of the sun at the vernal equinox, which is equal to 365.2422 mean solar days) and noting that the area of an ellipse is πab, we have

$$r^2 \frac{d\nu}{dt} = \frac{2\pi a^2}{\tilde{T}} (1 - e^2)^{1/2}. \tag{6.1.5}$$

The true anomaly of the earth's orbit at a given date is computed indirectly from the eccentric anomaly E, defined by

$$\cos E = \frac{a - r}{ae}. \tag{6.1.6}$$

From Kepler's equations, and the definition of the eccentric anomaly, we can prove that

$$\frac{(1 - e^2)^{1/2}}{1 + e \cos \nu} d\nu = dE. \tag{6.1.7a}$$

Based on Eq. (6.1.7a), an alternate equation can be derived for the calculation of the eccentric anomaly in the form

$$E - e \sin E = 2\pi t/\tilde{T} + (E_0 - e \sin E_0), \tag{6.1.7b}$$

where E_0 is the eccentric anomaly on March 21, at which $t = 0$ is chosen. The true anomaly is related to the eccentric anomaly by Lacaille's formula, which is the integration result of Eq. (6.1.7a), as follows:

$$\tan \frac{\nu}{2} = \left(\frac{1 + e}{1 - e} \right)^{1/2} \tan \frac{E}{2}. \tag{6.1.8}$$

The computation of the true anomaly and the declination of the sun requires the values of eccentricity e, the oblique angle ϵ, and the longitude of the perihelion ω. Based on celestial mechanics, the secular variations in these three parameters are associated with the perturbations that other principal planets exert on the earth's orbit. Milankovitch (1941) has developed mathematical expressions for the computation of solar insolation including orbital parameters. More recently, Berger (1978) developed the following trigonometric expansions for the efficient computation of these parameters:

$$\epsilon = \epsilon^* + \sum_i A_i \cos(f_i t + \delta_i), \qquad (6.1.9a)$$

$$e \sin \omega = \sum_i P_i \sin(\alpha_i t + \xi_i), \qquad (6.1.9b)$$

$$e \cos \omega = \sum_i P_i \cos(\alpha_i t + \xi_i), \qquad (6.1.9c)$$

$$e = e_0 + \sum_i E_i \cos(\lambda_i t + \phi_i). \qquad (6.1.9d)$$

In the computation, $t = 0$ refers to A.D. 1950 and t is negative for the past. The constants of integration, $\epsilon^* = 23°321$ and $e_0 = 0.0287$, are deduced from the initial conditions. The terms (A_i, P_i, E_i) are amplitudes, $(f_i, \alpha_i, \lambda_i)$ the mean rates, and $(\delta_i, \xi_i, \phi_i)$ the phases.

Figure 6.2 shows the three basic astronomical parameters for the last 200,000 years. The eccentricity varies from about 0.01 to 0.04 (the mean is $\sim$0.017), with a characteristic period of 100,000 years. The oblique angle varies from about 22° to 24.5° with a dominant period of about 41,000 years. The longitude of the perihelion has a periodicity of about 21,000 years.

In order to compute solar insolation, the mean distance between the earth and the sun must be defined. From Kepler's second law, we may define a mean distance based on the conservation of angular momentum such that

$$r_0^2 = \frac{1}{2\pi} \int_0^{2\pi} r^2 d\nu = a^2(1 - e^2)^{1/2}. \qquad (6.1.10)$$

The solar constant is defined in reference to this mean distance. Kepler's third law (the law of periods) states that

$$a^3/\tilde{T}^2 = k, \qquad (6.1.11)$$

where $\tilde{T}$ is the planet's period of revolution around the sun, and k has the same value for all planets. This law is a consequence of the balance between gravitation and centripetal forces governing a planet in orbit. The semimajor axis of the earth's orbit is invariant. Because the factor $(1 - e^2)^{1/2}$ is very close to 1, the mean

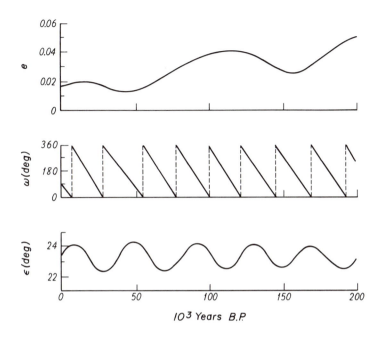

FIG. 6.2   The eccentricity e, the obliquity of the ecliptic ϵ, and the longitude of the perihelion ω of the earth as a function of year before the present (after Berger, 1978).

distance between the earth and the sun may be set as the invariant semimajor axis (i.e., $r_0 = a$).

The daily solar insolation is the integration of the instantaneous flux density from sunrise, $-t_0$, to sunset, t_0. Let H represent the half-day as defined in Eq. (6.1.2b). Since r and δ do not vary significantly over one day, from Eq. (6.1.1) the daily solar insolation may be expressed by

$$\bar{F} = \int_{-t_0}^{t_0} F(t)dt = S \left(\frac{a}{r}\right)^2 \int_{-H}^{H} (\sin\phi\sin\delta + \cos\phi\cos\delta\cos h)\frac{dh}{2\pi}$$

$$= \frac{S}{\pi}\left(\frac{a}{r}\right)^2 (H\sin\phi\sin\delta + \sin H\cos\phi\cos\delta). \tag{6.1.12}$$

The daily solar insolation, as a function of latitude and the time of year, can be computed with the aid of Eqs. (6.1.2b)–(6.1.12). The results are shown in Fig. 6.3. The distribution of solar insolation is independent of the longitude and is slightly asymmetric between the Northern and Southern Hemispheres. The sun is closest to the earth in January (winter in the Northern Hemisphere) so that the maximum solar insolation received in the Southern Hemisphere is greater than that received in the Northern Hemisphere. At the equinoxes, solar insolation is at a maximum at the equator and is zero at the poles. At the summer solstice of the Northern Hemisphere, the daily insolation reaches a maximum at the North Pole because of

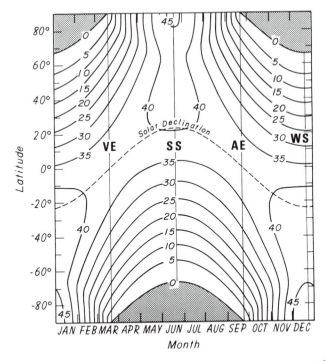

FIG. 6.3 Daily solar insolation as a function of latitude and day of year in units of $10^6\,\mathrm{Jm^{-2}d^{-1}}$ based on a solar constant of 1365 W m^{-2}. The hatched areas denote zero insolation. The positions of vernal equinox (VE), summer solstice (SS), autumnal equinox (AE), and winter solstice (WS) are indicated with solid vertical lines. Solar declination is indicated with a dashed line (courtesy of Eric A. Smith).

the 24-hour-long solar day. At the winter solstice, the sun does not rise above the horizon north of $\sim66.5°$, where the solar insolation is zero.

From Eq. (6.1.12), the solar insolation over a period of time (t_1, t_2) is

$$\bar{F} = \int_{t_1}^{t_2} \frac{S}{\pi} \left(\frac{a}{r}\right)^2 (H \sin\phi \sin\delta + \sin H \cos\phi \cos\delta)\, dt. \qquad (6.1.13)$$

Using Eqs. (6.1.5), (6.1.3) and (6.1.2b), we obtain

$$\bar{F} = \frac{S}{\pi} \frac{T \sin\phi \sin\epsilon}{2\pi(1 - e^2)^{1/2}} \int_{\lambda_1}^{\lambda_2} (H - \tan H) \sin\lambda\, d\lambda, \qquad (6.1.14)$$

where the half-day H from Eqs. (6.1.2b) and (6.1.3) is given by

$$\cos H = -\frac{\tan\phi \sin\epsilon \sin\lambda}{(1 - \sin^2\epsilon \sin^2\lambda)^{1/2}}. \qquad (6.1.14a)$$

In the domain of the true longitude of the earth, the year can be divided into astronomical spring $(0-\pi/2)$, summer $(\pi/2-\pi)$, autumn $(\pi-3\pi/2)$, and winter

$(3\pi/2\text{--}2\pi)$. The seasonal solar insolation can be evaluated using this division. If we replace λ with $\pi - \lambda$ in Eq. (6.1.14), the values are the same for spring and summer as well as for autumn and winter. Thus, it suffices to consider the summer half-year (spring plus summer) and winter half-year (autumn plus winter) for the calculation of seasonal solar insolation.

The annual solar insolation for a given latitude is given by

$$\bar{\bar{F}}_a = \frac{S\tilde{T}\tilde{s}(\phi, \epsilon)}{\pi(1 - e^2)^{1/2}}, \tag{6.1.15}$$

where

$$\tilde{s}(\phi, \epsilon) = \frac{\sin\phi \sin\epsilon}{2\pi} \int_0^{2\pi} (H - \tan H) \sin\lambda \, d\lambda. \tag{6.1.16}$$

The solar insolation for the summer and winter half-years can be expressed in terms of annual solar insolation. Let $\lambda' = \lambda - \pi$; it can be shown from Eq. (6.1.14a) that $H(\lambda) = \pi - H(\lambda')$ and $\tan H(\lambda) = -\tan H(\lambda')$. Consider the total solar insolation for the winter $(\pi, 2\pi)$ half-year and express this insolation in terms of that for the summer $(0, \pi)$ half-year by using Eq. (6.1.14). We find

$$\bar{\bar{F}}_{s,w} = \frac{S\tilde{T}}{2\pi(1 - e^2)^{1/2}}[\tilde{S}(\phi, \epsilon) \pm \sin\phi \sin\epsilon], \tag{6.1.17}$$

where the positive and negative signs are for the summer solar insolation $\bar{\bar{F}}_s$ and the winter solar insolation $\bar{\bar{F}}_w$, respectively. It follows that the difference between the amount of solar insolation for summer and winter periods is

$$\bar{\bar{F}}_s - \bar{\bar{F}}_w = \frac{S\tilde{T}}{\pi(1 - e^2)^{1/2}} \sin\phi \sin\epsilon. \tag{6.1.18}$$

This difference is caused by the change in the solar irradiances available at a given latitude and the difference in the lengths of the summer $(\tilde{T}_s)$ and winter $(\tilde{T}_w)$ half-years, which can be evaluated with the aid of Kepler's first and second laws. By neglecting the high-order terms in the eccentricity expansion, we have

$$\tilde{T}_{s,w} = \frac{\tilde{T}}{2}\left(1 \pm \frac{4}{\pi}e \sin\omega\right). \tag{6.1.19}$$

The annual global solar insolation can be evaluated by using the instantaneous solar insolation for the entire earth, which is given by $S(a/r)^2\pi a_e^2$. Distributing this energy over the surface area of the earth, $4\pi a_e^2$, the mean solar insolation for one day is given by $\Delta t_\odot(a/r)^2S/4$. We then perform an integration over a year and use Eq. (6.1.5) to obtain

$$\bar{\bar{F}}_{at} = \int_0^{2\pi} \frac{S\Delta t_\odot}{4}\left(\frac{a}{r}\right)^2 \frac{dt}{\Delta t_\odot} = \frac{S}{4}\tilde{T}(1 - e^2)^{-1/2}. \tag{6.1.20}$$

The annual global solar insolation is proportional to $(1-e^2)^{-1/2}$, but is independent of the declination of the sun δ and the true anomaly ν. Since e is small, the annual global insolation is approximately proportional to $(1 + e^2/2)$.

6.1.2   Solar constant

As is evident from Eq. (6.1.1), solar insolation is dependent on the solar constant, which represents the energy output from the sun. The solar constant as defined in Eq. (6.1.1) is a function of the temperature of the sun's photosphere. It has been determined from ground-based radiometers over more than 50 years. Based on thousands of observations at various locations around the world, a value of 1353 W m^{-2} has been proposed. The presence of aerosols in the atmosphere imposes limitations on the accuracy of determining solar constant from ground-based radiometer measurements; these limitations have been discussed by Reagan et al. (1986). To minimize atmospheric effects, measurements of solar irradiance have also been made in the upper troposphere, including measurements from balloons in the 27–35 km altitude range, jet aircraft at $\sim$12 km, and rocket aircraft at $\sim$82 km. From these measurements, a value of 1353 ($\pm$21) W m^{-2} was recognized as a standard solar constant (Thekaekara, 1976).

More recently, solar constant data have been derived from solar total irradiance measurements made by the self-calibrating pyrheliometers of the Earth Radiation Budget (ERB) mission on board Nimbus 7 (launched February, 1978), the Solar Maximum Mission (SMM) Active Cavity Radiometer Irradiance Monitor 1 (ACRIM1) (February, 1980), and the Earth Radiation Budget Experiment (ERBE) on board the NASA Earth Radiation Budget Satellite (ERBS, October, 1984), NOAA 9 (December, 1984), and NOAA 10 (July, 1986). A correction of the measured solar irradiance to the mean earth–sun distance, that is, half the sum of the perihelion and aphelion distances (of the earth from the sun), gives the solar constant. Hickey et al. (1988) have analyzed the solar irradiance measurements taken by the cavity sensor of the ERB on Nimbus 7 over a 9-year period, and have derived a value of about 1370 W m^{-2} for the solar constant. A value of about 1367 W m^{-2} has been determined by Willson and Hudson (1988) using the solar irradiance data gathered from SMM ACRIMI. Lee at al. (1988) have analyzed the solar irradiance measurements from ERBS, NOAA 9, and NOAA 10. There have been slight disparities in the measured solar irradiance data from these satellites. A mean value for the solar constant of 1365 W m^{-2} has been obtained. This value appears to be the best typical value for the solar constant at this time. The goal of the solar constant measurements is to obtain an accuracy of $\pm$3 W m^{-2}.

Figure 6.4 shows the solar constant values derived from ERBS solar monitor measurements over a three-year period. Also shown is a second-order polynomial fit to the data points. The most obvious features in the time series are the decreasing trend before the middle part of 1986 at a rate of about 0.02% per year and the

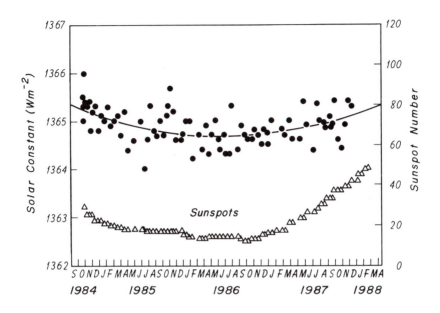

FIG. 6.4   The solar constant as a function of time. The solar constant values were derived from the ERBS solar monitor. The solid curve represents a second-order polynomial fit to the solar constant values, while the dotted curve represents the smoothed (12-month running mean) international sunspot number (after Lee et al., 1988).

increasing trend after that time at about the same rate. Measurements by NOAA 9 and NOAA 10 show similar trends. The most apparent phenomenon that could cause this variation in solar output is sunspots.

Sunspots are relatively dark regions on the sun's photosphere, ranging in size from barely visible to more than 150,000 km. The spots usually occur in pairs or in complex groups that trail a leader spot in the direction of the sun's rotation. Small sunspots persist for several days or a week, while the largest spots may last for several weeks, long enough for them to reappear during the course of the sun's 27-day rotation. Sunspots are entirely confined to the latitude zone between 40° and the equator; they never appear near the sun's poles. Sunspots are cooler regions with an average temperature of about 4000 K, which is 2000 K lower than the average temperature of the sun's photosphere. The number of sunspots that appear on the solar disk, averaged over a period of time, is highly variable. The number and position of sunspots have been recorded daily for more than 200 years. The average length of time between sunspot maxima is about 11 years, a period referred to as the *11-year sunspot cycle*. The origin and dynamics of sunspots are still uncertain; however, it appears that magnetic fields near the photosphere must play an important role. A magnetic field exerts a force on the charged gas particles that move in the field. This force may prevent the development of convection and the transport of hot interior matter upward to the photosphere, which may explain the fact that sunspots appear as relatively dark areas and are cooler than their surroundings. Clearly, solar irradiance must be affected by the activities of sunspots.

Solar irradiance is also perturbed by larger bright areas known as *faculae*, which surround sunspots. After sunspots develop, faculae increase and raise the solar irradiance. The solar irradiance is lowest at sunspot minimum, if the radiant contribution by faculae is also at minimum. Figure 6.4 also gives the smoothed international sunspot number as an indicator of solar activity. A sunspot minimum occurred around September, 1986. The decrease and increase of sunspots appear to correlate with variations in the solar constant values over this time period. Reliable observed data on the solar constant have been employed to validate solar models that may be used to characterize past solar variabilities, as well as to predict future changes in the solar constant that are critical to the earth's climate.

6.2   Global radiative equilibrium

6.2.1   *Monochromatic radiative equilibrium*

We shall first discuss monochromatic radiative equilibrium, under which the radiative energy is balanced for each wave number. Consider a nonscattering atmosphere that is in local thermodynamic equilibrium. The basic equation describing the transfer of monochromatic intensity for such an atmosphere has been presented in Eq. (2.1.7) and is duplicated here for the discussion of radiative equilibrium:

$$\mu \frac{dI_\nu(\tau, \mu)}{d\tau} = I_\nu(\tau, \mu) - B_\nu(T(\tau)). \tag{6.2.1}$$

The monochromatic optical depth is defined by

$$\tau = \int_z^{z_\infty} k_\nu(z') \rho_a(z') \, dz', \tag{6.2.2}$$

where k_ν denotes the absorption coefficient, ρ_a is the density of the absorbing gas, and z is the height.

The net flux at level τ (or z) is defined by

$$F_\nu(\tau) = 2\pi \int_{-1}^{1} I_\nu(\tau, \mu) \mu \, d\mu, \tag{6.2.3}$$

Applying an integration over μ from -1 to 1 to Eq. (6.2.1) leads to

$$\frac{dF_\nu}{d\tau} = 4\pi(J_\nu - B_\nu), \tag{6.2.4}$$

where J_ν is the mean intensity defined by

$$J_\nu(\tau) = \frac{1}{2} \int_{-1}^{1} I_\nu(\tau, \mu) \, d\mu. \tag{6.2.5}$$

Moreover, by multiplying Eq. (6.2.1) by μ and performing an integration over μ from -1 to 1, we obtain

$$2\pi \frac{d}{d\tau} \int_{-1}^{1} I_\nu(\tau, \mu)\mu^2 \, d\mu = F_\nu(\tau). \tag{6.2.6}$$

Following the procedure discussed in Subsection 3.3.1, we may use the two-stream approximation for the intensity transfer. Equations (6.2.5) and (6.2.6) may then be written

$$J_\nu \approx \frac{1}{2}(I_\nu^+ + I_\nu^-), \tag{6.2.7}$$

$$F_\nu \approx \frac{d}{d\tau} 2\pi \mu_1^2 (I_\nu^+ + I_\nu^-). \tag{6.2.8}$$

It follows that

$$F_\nu \approx 4\pi \mu_1^2 \frac{dJ_\nu}{d\tau}. \tag{6.2.9}$$

Further, by differentiating Eq. (6.2.4) with respect to τ and noting that $\mu_1^2 = 1/3$, we obtain

$$\frac{d^2 F_\nu}{d\tau^2} = 3F_\nu - 4\pi \frac{dB_\nu}{d\tau}. \tag{6.2.10}$$

The concept of radiative equilibrium requires that the net flux be constant with respect to height for each wave number so that the radiative energy is balanced. We first examine the case of monochromatic radiative equilibrium. Since F_ν is a constant, we have

$$\frac{dB_\nu}{d\tau} = \frac{3}{4\pi} F_\nu. \tag{6.2.11}$$

The boundary condition at TOA is

$$B_\nu(\tau = 0) = B_\nu(T_\infty), \tag{6.2.12}$$

where T_∞ is the temperature at $\tau = 0$. Thus, the solution for Eq. (6.2.11) is

$$B_\nu(\tau) = B_\nu(T_\infty) + \frac{3}{4\pi} F_\nu \tau. \tag{6.2.13}$$

Under the two-stream approximation, the net flux is given by

$$F_\nu \approx 2\pi \mu_1 (I_\nu^+ - I_\nu^-). \tag{6.2.14}$$

Also, under the condition of constant monochromatic fluxes, Eq. (6.2.4) may be expressed by

$$B_\nu = J_\nu = \frac{1}{2}(I_\nu^+ + I_\nu^-). \tag{6.2.15}$$

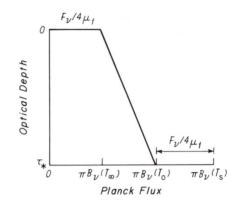

FIG. 6.5   Planck flux as a function of the optical depth under the monochromatic radiative equilibrium condition.

Using Eqs. (6.2.14) and (6.2.15), we obtain the solution for the upward and downward fluxes in the forms

$$2\mu_1\pi I_\nu^+(\tau) = 2\mu_1\pi B_\nu(\tau) + F_\nu/2, \tag{6.2.16}$$

$$2\mu_1\pi I_\nu^-(\tau) = 2\mu_1\pi B_\nu(\tau) - F_\nu/2. \tag{6.2.17}$$

At TOA ($\tau = 0$, and $T = T_\infty$), there is no downward radiation, that is, $I_\nu^-(0) = 0$, so that

$$0 = 2\mu_1\pi B_\nu(T_\infty) - F_\nu/2. \tag{6.2.18}$$

At the surface ($\tau = \tau_*$ and $T = T_s$) the upward flux emitted from the black surface is $I_\nu^+(\tau_*) = B_\nu(T_s)$. Let the air temperature immediately above the surface be denoted by T_0. Then

$$2\mu_1\pi B_\nu(T_s) = 2\mu_1\pi B_\nu(T_0) + F_\nu/2. \tag{6.2.19}$$

Equations (6.2.18) and (6.2.19) imply that in the special case of monochromatic radiative equilibrium, there is a discontinuity in temperature at the surface, with $T_s > T_0$. The air at TOA approaches a certain temperature, T_∞, under radiative equilibrium (Fig. 6.5). The atmospheric lapse rate in this case is $(T_0 - T_\infty)/\tau_*$.

Finally, substituting F_ν from Eq. (6.2.18) into Eq. (6.2.13), we obtain

$$\pi B_\nu(\tau) = \pi B_\nu(T_\infty)(1 + 3\mu_1\tau). \tag{6.2.20}$$

Also, from Eqs. (6.2.16) and (6.2.18) the outgoing flux at TOA is

$$\pi I_\nu^+(\tau = 0) = 2\pi B_\nu(T_\infty). \tag{6.2.21}$$

Hence, atmospheric thermal radiation is characterized by the thermal emission at $\tau = 1/3\mu_1 = 1/\sqrt{3}$; that is, at this optical depth, $\pi B_\nu(\tau) = \pi I_\nu^+(\tau = 0)$.

6.2.2 Gray atmosphere heated from the surface

To obtain the equilibrium temperature of planetary atmospheres, all wave numbers must be considered. We begin with Eq. (6.2.11) and use the definition for optical depth to obtain

$$-\frac{d\pi B_\nu}{\rho_a k_\nu\, dz} = \frac{3}{4} F_\nu. \tag{6.2.22}$$

In a gray atmosphere, we may define a mean absorption coefficient such that

$$\frac{1}{\bar{k}} = \frac{1}{\sigma T^4} \int_0^\infty \frac{\pi B_\nu}{k_\nu}\, d\nu, \tag{6.2.23}$$

or

$$\bar{k} = \frac{1}{F} \int_0^\infty k_\nu F_\nu\, d\nu, \tag{6.2.24}$$

where σT^4 is the result from the Stefan–Boltzmann law introduced in Eq. (2.0.2), and the total flux is defined by

$$F = \int_0^\infty F_\nu\, d\nu. \tag{6.2.25}$$

The $\bar{k}$ defined in Eq. (6.2.23) is referred to as the *Rosseland mean*, which is widely used in astrophysics, while the $\bar{k}$ defined in Eq. (6.2.24) is referred to as the *Chandrasekhar mean*. It suffices to point out that while both definitions will yield the same final result, the physical rationale for using Eq. (6.2.23) or (6.2.24) to define the mean absorptivity for a gray atmosphere depends on the problem being considered. For example, Eq. (6.2.23) may work well for stellar atmospheres, but there are difficulties in applying this equation to planetary atmospheres because the Planck radiance varies with height and $\bar{k}$ cannot be treated as a constant. On the other hand, since F_ν is not known, the evaluation of $\bar{k}$ from Eq. (6.2.24) requires iterations or approximations. In any event, for the purpose of the present discussion, we may assume that $\bar{k}$ is a known quantity, so that Eq. (6.2.22) becomes

$$\frac{d\sigma T^4}{\rho_a \bar{k}\, dz} = -\frac{3}{4} F. \tag{6.2.26}$$

Thus,

$$\sigma T^4(z) = \sigma T_\infty^4 + \frac{3}{4} F \bar{\tau}(z), \tag{6.2.27}$$

where

$$\bar{\tau} = \int_z^{z_\infty} \bar{k}\rho_a \, dz. \tag{6.2.28}$$

Following the same procedure as in the case of monochromatic radiative equilibrium, we may perform wave-number integrations on Eqs. (6.2.18), (6.2.19), and (6.2.21) to obtain

$$\sigma T_\infty^4 = F/4\mu_1, \tag{6.2.29}$$

$$\sigma T_s^4 = \sigma T_0^4 + F/4\mu_1, \tag{6.2.30}$$

$$\pi I^+(z_\infty) = 2\sigma T_\infty^4, \tag{6.2.31}$$

where

$$\pi I^+(z_\infty) = \int_0^\infty \pi I_\nu^+(z_\infty) \, d\nu.$$

It follows from Eqs. (6.2.27) and (6.2.29) that

$$\sigma T^4(z) = \sigma T_\infty^4 [1 + 3\mu_1 \bar{\tau}(z)]. \tag{6.2.32}$$

Based on Eq. (6.2.31), the outgoing net flux in the two-stream approximation may be defined by $F^+(z_\infty) = 2\mu_1\pi I^+(z_\infty)$. Under the constant flux assumption, we have $F^+(z_\infty) = F$. We define an equilibrium temperature T_e that is related to the constant net flux as follows:

$$\sigma T_e^4 = F^+(z_\infty) = F. \tag{6.2.33}$$

Thus the equilibrium temperature is related to the emitting temperature at TOA by

$$\sigma T_\infty^4 = \frac{\sqrt{3}}{4}\sigma T_e^4. \tag{6.2.34}$$

or, $T_\infty = 0.8112T_e$.

By virtue of Eqs. (6.2.27) and (6.2.29), we find

$$\sigma T^4(u) = \frac{F}{4}[\sqrt{3} + 3\bar{\tau}(u)], \tag{6.2.35}$$

where the path length u is defined by

$$u = \int_z^{z_\infty} \rho_a \, dz.$$

By utilizing the definition of the equilibrium temperature in Eq. (6.2.33), and assuming that $\bar{k}$ is independent of path length, we have

$$\sigma T^4(u) = \sigma T_e^4(\sqrt{3} + 3\bar{k}u)/4. \tag{6.2.36}$$

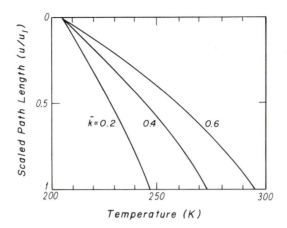

Fig. 6.6   Atmospheric temperature profiles as functions of the scaled path length.

The mean absorption coefficient for a gray atmosphere is related to the broadband flux emissivity in the form

$$\epsilon^f \approx 1 - e^{-\bar{k}u_1},$$ (6.2.37)

where u_1 is the total path length. The temperature profile as a function of path length, as defined in Eq. (6.2.36), depends on T_e and $\bar{k}$.

Let the equilibrium temperature of the earth be 255 K (see Table 6.1). The atmospheric temperature is then given by

$$T(u) = 255[(\sqrt{3} + 3\bar{k}u)/4]^{1/4}.$$ (6.2.38)

Consider a water vapor atmosphere with a total path length $u_1 \approx 2.86\,\mathrm{g\,cm^{-2}}$, which corresponds to the standard atmospheric condition, and let $\bar{k}$=0.2, 0.4, and 0.6 cm² g⁻¹. Atmospheric temperature profiles as functions of the scaled path length, u/u_1, are depicted in Fig. 6.6. Based on Eq. (6.2.37), the corresponding flux emissivities for the above $\bar{k}$ values are 0.44, 0.68, and 0.82, respectively. The surface air temperatures for these three cases are 245.7, 271.8, and 292.0 K.

6.2.3   *Global radiative equilibrium with the sun as the energy source*

In previous discussions of the global radiative equilibrium problem, we have assumed that the atmosphere is heated from below without accounting for an external radiation source. We shall now consider the sun as the external source and derive the global temperature profile for an atmosphere that is in radiative equilibrium. Assuming that the atmosphere is motionless, the local time rate of change of temperature for a given atmospheric layer is associated with the net flux divergence in the form

$$\rho C_p \left(\frac{\partial T}{\partial t}\right)_R = -\frac{\partial}{\partial z}(F_s - F_{ir}),$$ (6.2.39)

where ρ is the air density, C_p is the specific heat at constant pressure, and F_s and F_{ir} denote the net solar and thermal infrared fluxes. Under steady-state conditions (i.e., $t \to \infty$) $\partial T/\partial t = 0$. Also, at TOA we require that $F_s(z_\infty) = F_{ir}(z_\infty)$; that is, the absorbed solar flux must be balanced by the emitted thermal ir flux from the surface and the atmosphere. With this boundary condition, an integration from z to z_∞ over Eq. (6.2.39) yields

$$F_s(z) = F_{ir}(z). \tag{6.2.40}$$

This equation represents the steady-state global radiative equilibrium condition. Based on our previous analysis, described in Section 2.7, the broadband thermal ir flux may be expressed in terms of an integral form as follows:

$$
\begin{aligned}
F_{ir}(z) &= \int_0^{z_\infty} \sigma T^4(z') K(|z - z'|) \, dz' \\
&= \sigma T_s^4 [1 - \epsilon^f(z, T)] - \int_0^{z_\infty} \sigma T^4(z') \frac{d\epsilon^f(|z - z'|, T(z'))}{dz'},
\end{aligned}
\tag{6.2.41}
$$

where K is a general kernel function associated with the weighting function, and ϵ^f is the broadband flux emissivity.

We may define an equilibrium temperature for a planet in a manner described in the Introduction:

$$F_{ir}(z_\infty) = \sigma T_e^4. \tag{6.2.42}$$

This equation can also be derived from Eq. (6.2.41) by using the isothermal temperature T_e. The equilibrium temperature is therefore the equivalent mean temperature of the atmosphere and the surface, if the planet as a whole is considered as a blackbody. Let the solar constant be S and the global albedo of the planet be $\bar{r}$. Balancing the absorbed solar flux and emitted thermal ir flux yields

$$\frac{\pi a_e^2 S(1 - \bar{r})}{4\pi a_e^2} = \sigma T_e^4, \tag{6.2.43a}$$

where a_e denotes the radius of the planet. The left-hand side of the equation represents the total solar flux absorbed by the planet, with the cross-sectional area being πa_e^2. The absorbed solar flux spreads over the entire spherical area, $4\pi a_e^2$. The equilibrium temperature is then given by

$$T_e = [(1 - \bar{r})S/4\sigma]^{1/4}. \tag{6.2.43b}$$

The solar constant can be evaluated from the energy balance equation, if the emittance F of the sun is known, viz., $F4\pi a_s^2 = S4\pi r_0^2$, where a_s and r_0 denote

Table 6.1 Relative distance of the planets from the sun with respect to the earth ($\bar{d}$), global albedo ($\bar{r}$), and equilibrium temperature (T_e)

Planet	$\bar{d}$	$\bar{r}(\%)$	$T_e(K)$
Mercury	0.39	6	$\sim$441
Venus	0.72	78	$\sim$226
Earth	1.00	30	$\sim$255
Mars	1.52	17	$\sim$217
Jupiter	5.20	45	$\sim$106

the radius of the solar disk and the mean distance between the sun and the planet, respectively. The effective temperature of the sun's photosphere is $\sim$5800 K. Based on Eq. (6.2.42), $F = \sigma T_e^4$. Using this temperature, a mean radius (visible disk) of 6.96×10^5 km and a mean distance between the earth and the sun of 1.49598×10^8 km, the equilibrium temperatures of various planets with albedos, $\bar{r}$, have been calculated (see Table 6.1.).

The equilibrium temperature is extremely sensitive to the global albedo of the planet, which depends on the internal composition and surface characteristics; it is also sensitive to the solar constant, which is governed by the composition of the sun (see Subsection 6.1.2).

6.3 Radiative and thermal equilibrium in nongray atmospheres

6.3.1 *Radiative equilibrium*

In a nongray atmosphere the radiative equilibrium temperature is determined by the balance between solar and thermal ir heating rates. The solar and ir heating rate profiles produced by H_2O, CO_2, and O_3 have been illustrated in Figs. 2.17, 2.18, and 3.16. Radiative equilibrium calculations require information on the cosine of the solar zenith angle (μ_0), the length of the solar day ($\bar{t}$), and the solar constant (S), as well as atmospheric compositions and the earth's surface albedo ($\bar{r}_s$). The H_2O and O_3 concentrations for the standard atmosphere and other atmospheric conditions are displayed in Fig. 6.7. Background aerosol and molecular distributions have been presented in Fig. 3.11. The globally averaged cloud type, cloud top and base heights, and cloud cover are listed in Table 6.2. These values are derived from the zonally averaged cloud climatology data provided by London (1957) for the Northern Hemisphere and Sasamori et al. (1972) for the Southern Hemisphere. Also indicated in this table are the solar reflectance and transmittance, and ir emissivity (emittance) for various cloud types (Liou and Ou, 1983). For the mean annual condition, we use $\mu_0 = 0.5$ and $\bar{t} = 12$ h. The solar constant $S = 1365$ W m^{-2}, and the globally averaged surface albedo $\bar{r}_s \cong 0.159$. The CO_2 concentration is assumed to have a uniform mixing ratio of 330 ppm.

Table 6.2 Cloud Parameters and mean radiative properties for various cloud types

Cloud type	Base (km)	Top (km)	Cover	Solar reflectance[a]	Solar transmittance[a]	ir emissivity
Low	1.70	2.52	0.102	0.769	0.092	1
Middle	3.50	4.26	0.065	0.825	0.032	1
Ci	9.24	10.94	0.138	0.105	0.880	0.475
Ns	1.35	3.56	0.070	0.835	0.026	1
Cb	1.70	5.20	0.030	0.837	0.025	1
St	1.35	1.45	0.104	0.524	0.375	1
Clear	—	—	0.489	—	—	—

[a] For $\mu_0 = 0.5$.

Figure 6.8 displays the solar heating and ir cooling rates for H_2O, CO_2, and O_3, corresponding to the standard atmosphere. In the troposphere, ir cooling rates are primarily produced by the H_2O rotational band and continuum, while solar heating rates are produced by various combination and overtone bands in the near-ir region. In the stratosphere and lower mesosphere, solar heating and ir cooling rates are largely associated with O_3 and CO_2, respectively. Contributions to ir cooling by O_3 are also noticeable. In the troposphere, ir cooling rates outweigh

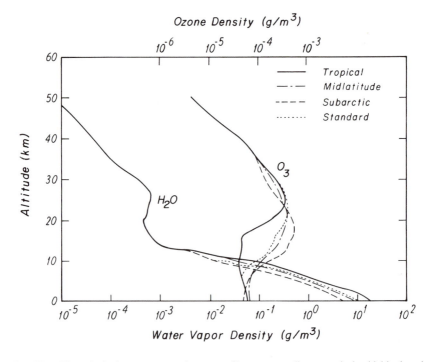

Fig. 6.7 Climatological water vapor and ozone profiles corresponding to tropical, midaltitude, subarctic, and standard conditions.

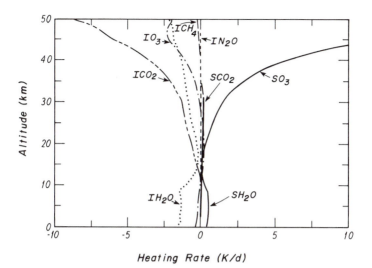

FIG. 6.8   Components of solar heating (solid lines) corresponding to $\mu_0 = 0.5$ and 12 h solar days, and IR cooling rates (dashed lines) for H_2O, CO_2, O_3, CH_4, and N_2O. The calculations use the standard atmospheric profiles presented in Fig. 6.7 and a line-by-line program, except for ozone heating rates.

solar heating rates by a factor of about two for the mean annual condition. Clouds have profound effects on the heating and cooling profiles. Some quantitative results will be illustrated in Subsection 6.5.4.

To obtain the vertical temperature profile under radiative equilibrium, we may carry out an iterative and time marching procedure as follows:

$$T^{(n+1)}(z) = T^{(n)}(z) + \left(\frac{\partial T}{\partial t}\right)_R^{(n)} \Delta t, \qquad (6.3.1)$$

where n is the time step of the integration and Δt is the time interval. A numerical differencing scheme and an initial guess of the temperature are needed. Radiative equilibrium is reached when the temperatures at the $(n + 1)$ and n time steps differ by a small preset value. Specifically, radiative equilibrium at the surface and at TOA must be satisfied.

The time marching method, which involves considerable computational effort, will give the evolution of the temperature as a function of time under the radiative equilibrium condition (Manabe and Strickler, 1964). However, if one is interested in the temperature profile in the equilibrium state, the simplification given in Eq. (6.2.40) may be used. The solar flux must be balanced by the thermal ir flux. Thus we write

$$F_s(z) = F_{ir}(z) = \int_0^{z_\infty} \sigma T^4(z') K(|z - z'|) \, dz'. \qquad (6.3.2)$$

The kernel function K, defined in Eq. (6.2.41), depends largely on atmospheric gaseous profiles, principally those for H_2O, CO_2 and O_3, as well as on the distribution of cloud fields. As a good approximation, the kernel function may be considered to be independent of temperature. Thus, once the solar net flux profile has been given, the temperature profile may be determined.

The atmosphere can be divided into finite layers, so that Eq. (6.3.2) may be expressed in a finite difference form:

$$F_s(z_i) = \sum_{j=1}^{N} \sigma T^4(z_j) K(|z_i - z_j|) \Delta z_j, \qquad i = 1.2, \ldots, N. \qquad (6.3.3)$$

In compact matrix form we write

$$\mathbf{F_s} = \mathbf{K} \cdot \sigma \mathbf{T}^4, \qquad (6.3.4)$$

where $\mathbf{F_s}$ and $\sigma \mathbf{T}^4$ are column vectors and $\mathbf{K}$ is an $N \times N$ matrix. Inverting the matrix leads to

$$\sigma \mathbf{T}^4 = \mathbf{K}^{-1} \mathbf{F_s}, \qquad (6.3.5)$$

where $\mathbf{K}^{-1}$ denotes the inverse of the kernel function matrix.

Figure 6.9 shows the atmospheric temperature profiles under radiative equilibrium with and without the cloud contribution. Without the contribution of clouds, the surface temperature under radiative equilibrium is $\sim$340 K and the temperature in the tropopause is $\sim$215 K. In cloudy conditions, temperature decreases significantly in the troposphere due to the reflection of solar flux by clouds. In particular, a $\sim$30 K reduction in surface temperature is seen. The temperature profile for average cloudiness, shown in Fig. 6.9, is obtained by accounting for clear and cloudy areas. This radiative equilibrium temperature has been presented in Fig. 1.7. It is evident that the radiative equilibrium temperature is much too warm near the surface and too cold in the tropopause. On the mean annual basis the earth–atmosphere system is in radiative equilibrium at TOA. However, it is clearly not in radiative equilibrium within the atmosphere and at the surface. In terms of the one-dimensional globally averaged condition, the only mechanism that can bring the system into thermodynamic equilibrium is the vertical transport of heat by means of eddies. The convective nature of the earth and the atmosphere is fundamental to weather and climate systems, as well as in the numerical modeling of these systems.

6.3.2 Radiative equilibrium and convective adjustment

We have introduced and outlined the convective adjustment scheme in Subsection 4.8.1. This scheme is a numerical procedure by which the computed lapse rate is set to a critical lapse rate whenever the computed value becomes supercritical. This process is performed for all supercritical layers without changing the mass or

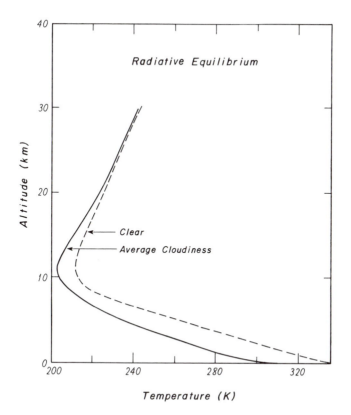

FIG. 6.9   Vertical distributions of radiative equilibrium temperatures with and without clouds.

total energy of the column. The critical lapse rate is usually taken to be 6.5 K km^{-1} for a standard atmosphere. This value is a result of the modification of the radiative equilibrium profile by free and forced convection and by vertical heat transports due to eddies. Radiative equilibrium profiles are unstable under conditions of moist and dry convection. The convective adjustment would allow for the transport of heat in the time evolution of numerical computations by adjusting the temperature between the vertical layers to the critical lapse rate. The critical lapse rate of 6.5 K km^{-1} is much lower than the value that would be critical for dry convection, whose lapse rate is $g/C_p (\cong 9.8$ K km$^{-1})$ for the earth's atmosphere. However, the moist adiabatic lapse rate from Eq. (4.9.25) is generally lower than 6.5 K km^{-1} over the large range of temperatures and humidities that is characteristic of the lower atmosphere.

Extensive numerical experiments on convective adjustment have been carried out by Manabe and Strickler (1964) using a fixed distribution of absolute humidity. Manabe and Wetherald (1967) have extended the convective adjustment experiment assuming a given distribution of relative humidity, which is a more realistic

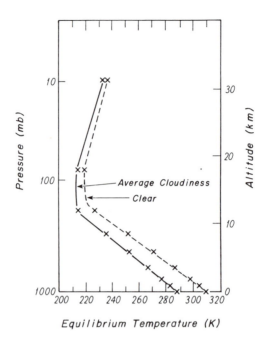

FIG. 6.10   Vertical distributions of radiative–convective equilibrium temperatures with and without clouds (after Manabe and Wetherald, 1967).

condition for the earth's atmosphere. Manabe and Wetherald performed the first sensitivity experiment of the radiative–convective model to examine radiative forcings such as a doubling of CO_2 (see Subsection 7.1.1 for further discussion). Figure 6.10 illustrates the effects of clouds on the vertical distribution of equilibrium temperature. The solid curve denotes the equilibrium temperature profile for an atmosphere with average cloudiness. This profile is close to the standard temperature profile (see Fig. 6.9). Temperatures without clouds are much higher than they are with clouds. The surface temperature differences are as much as 20 K. Temperature differences decrease with height.

The role of the parameterization of vertical convection on equilibrium temperature has been investigated by Lindzen et al. (1982) in connection with the climatic impact of doubling CO_2. Three models have been used in a one-dimensional radiative–convective model. These include the conventional 6.5 K km^{-1} lapse rate adjustment (model 1), the moist adiabatic lapse rate adjustment (model 2), and a cumulus convection parameterization (model 3). A comparison of the equilibrium temperatures computed from these three models is shown in Fig. 6.11. The conditions used in the calculations are typical, but without clouds. The incoming solar flux used is 353 W m^{-2}. Models 2 and 3 are both affected by humidity and have smaller lapse rates at lower levels than model 1. The surface temperature computed from model 1 is greater than that computed from models 2 and 3 by about

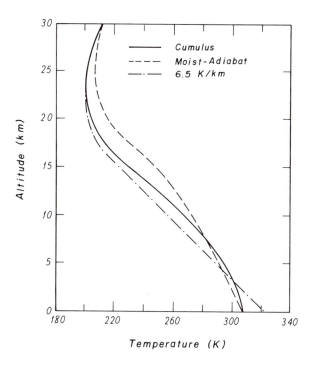

FIG. 6.11   Vertical distributions of equilibrium temperatures in a clear tropical atmosphere for the three convective models indicated in the graph (after Lindzen et al., 1982, with modifications).

10 K. The transport of latent heat through the convective adjustment has the effect of decreasing the sensitivity of radiative–convective models to external radiative forcings.

6.3.3   Radiative and turbulent equilibrium

Because the mean vertical velocity $\bar{w} \approx 0$ in the context of a one-dimensional model, it is appropriate to consider the coupling of radiative and turbulent transfers in determining the vertical temperature profile. In reference to Eqs. (4.6.20) and (4.6.21), the vertical eddy flux may be expressed in terms of the eddy thermal diffusion coefficient k_h in the form

$$F_v = -\rho C_p k_h \left(\frac{\partial T}{\partial z} + \gamma_d + \frac{L}{C_p} \frac{\partial q}{\partial z} \right), \qquad (6.3.6)$$

where we have set the notations $T = \bar{T}, q = \bar{q}$, and $\rho = \bar{\rho}$ for simplicity of presentation. We have also set the eddy diffusion coefficient for water vapor $k_e \approx k_h$.

Analyses of climatological temperature and humidity profiles compiled by Oort and Rasmusson (1971) reveal that the atmosphere near the ground is mostly stable. However, observed data provided by Budyko (1982) show evidence that there are upward eddy fluxes near the surface for almost all latitudes. This inconsistency has been found in the boundary layer heat transfer area by a number of researchers. Deardorff (1966) has proposed a factor γ_*, referred to as the *countergradient lapse rate*, to be subtracted from $(\partial T/\partial z + \gamma_d)(\approx \partial\Theta/\partial z)$. For a time scale on the order of hours and an area of $10 \times 10 \, \text{km}^2$, a value of $\sim 0.7 \, \text{K km}^{-1}$ for γ_* has been suggested. Deardorff (1972) has derived a theoretical expression relating γ_* to potential temperature and vertical velocity in the form

$$\gamma_* = g\overline{\Theta'^2}/\Theta_0\overline{w'^2}, \qquad (6.3.7)$$

where Θ_0 denotes the averaged potential temperature in the boundary layer. For longer time periods and larger spatial scales, the variability in the potential temperature will be much greater than in the vertical velocity. Consequently, γ_* should have a larger value than $0.7 \, \text{K km}^{-1}$. Deardorff has illustrated that γ_* should be increased by at least a factor of two over a sufficiently long time average. Saltzman and Ashe (1976) have proposed a value of $5 \, \text{K km}^{-1}$ for γ_* over the Northern Hemisphere and a time scale of a month in the climate modeling studies. It is clear from Eq. (6.3.7) that γ_* is subject to large variations, since $\overline{\Theta'^2}$ and $\overline{w'^2}$ depend on complicated atmospheric conditions associated with time and spatial scales. In an annual climate model developed for the Northern Hemisphere, Ou and Liou (1984) have shown that γ_* is on the order of $\sim 10 \, \text{K km}^{-1}$.

In light of the preceding discussion and considering the balance between radiative and turbulent fluxes, a radiative–turbulent model may be described by the following first-order, differential–integral equation in the form

$$-\rho C_p k_h \left(\frac{\partial T}{\partial z} + \gamma_c\right) + \int_0^\infty \sigma T^4(z')K(|z - z'|)\,dz' = F_s(z), \qquad (6.3.8)$$

where γ_c is similar to the critical lapse rate defined in the convective adjustment scheme, and the ir kernel function K has been introduced in Eq. (6.2.41). Equation (6.3.8) relates the solar and outgoing thermal ir fluxes to the turbulent flux in terms of eddy thermal diffusion processes. If the atmospheric lapse rate $(\partial T/\partial z)$ is greater than γ_c, there will be a positive vertical transport due to the unstable condition caused by excess solar flux. A negative transport takes place when $(\partial T/\partial z)$ is less than γ_c. When γ_c is equal to the atmospheric lapse rate, the atmospheric temperature field is defined by the balance between solar and thermal ir fluxes. A model involving the balance between turbulent and radiative fluxes in the atmosphere has been used by Gierasch and Goody (1968) in the study of the temperature structure of the Martian atmosphere.

To complete the one-dimensional radiative-turbulent model, it is necessary to discuss the manner in which the eddy thermal diffusion coefficient, k_h, is determined. In accordance with the analysis presented in Section 4.6, k_h in the constant flux layer ($z < 100\,\mathrm{m}$) may be expressed by

$$k_h \propto k^2 z^2 \left[\frac{g}{T} \left(\frac{\partial T}{\partial z} + \gamma_d \right) \right]^{1/2}. \tag{6.3.9}$$

In the troposphere between 100 m and the tropopause ($\sim$100 mb), the mixing length theory may be used to calculate k_h, viz.,

$$k_h \propto \overline{\ell'^2} \left| \frac{\partial \bar{u}}{\partial \bar{z}} \right|, \tag{6.3.10}$$

where the mixing length $\overline{\ell'^2}$ is $\sim$30 m and the mean wind shear may be approximated by a linear equation in terms of height in the form

$$\frac{\partial \bar{u}}{\partial z} = az + b, \tag{6.3.11}$$

where a and b are estimated from the known k_h values at the top of the constant flux layer and the tropopause. The latter k_h values for various atmospheric profiles may be computed from those presented by Reed and German (1965), who have provided the seasonal eddy thermal diffusion coefficients for the Northern Hemisphere for layers at 100, 50 and 30 mb.

Figure 6.12 shows the eddy thermal diffusion coefficient for tropical, midlatitude, subarctic, and standard atmospheric profiles based on the foregoing computational procedures. The computation of these coefficients makes use of the radiative equilibrium temperature profile under clear conditions. In the context of steady-state radiative equilibrium, a temperature profile could have been established, but it would be unstable because too much solar energy is stored at the surface. To achieve thermal equilibrium, vertical eddy transports must subsequently occur in which the eddy thermal diffusion process is responsible for stabilizing the thermodynamic system. Eddy thermal diffusivity is largest in the troposphere under the tropical condition. Except in the subarctic condition, k_h has values on the order of 10^6 cm^2 s^{-1} near the surface. In the stratosphere, k_h values are two to three orders of magnitude less than those in the troposphere. Since the subarctic region is generally stable, k_h values in the troposphere are relatively small.

In order to find a solution for the temperature field from Eq. (6.3.8), a radiative equilibrium temperature T_0, defined by the balance between solar and ir fluxes, may be evaluated from

$$\int_0^\infty \sigma T_0^4(z') K(|z - z'|)\, dz' = F_s(z). \tag{6.3.12}$$

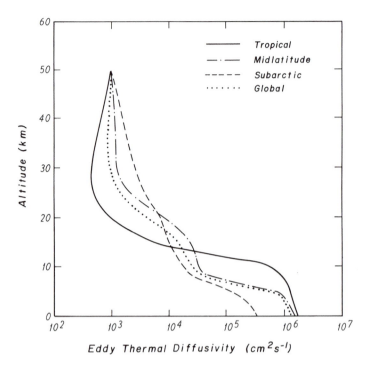

FIG. 6.12   Eddy thermal diffusion coefficients for topical, mid-latitude, subarctic, and global (standard) atmospheres (after Liou and Ou, 1983).

We wish to derive a perturbation equation from which the thermal equilibrium temperature can be computed numerically in reference to a perturbation parameter χ, defined by the difference between the Planck fluxes of radiative-turbulent and pure radiative temperatures. We define χ and a number of other parameters in the forms

$$\chi(z) = [\sigma T^4(z) - \sigma T_0^4(z)]/k', \qquad k' = c\rho C_p,$$
$$H(|z - z'|) = K(|z - z'|)/f(z), \qquad f(z) = k_h/c, \tag{6.3.13}$$

where c is a scaling constant, such that $f(z) = 1$ when $z = 0$. Using Eqs. (6.3.12) and (6.3.13), Eq. (6.3.8) may be rewritten in the form

$$1 + \frac{1}{\gamma_c}\frac{dT}{dz} = \int_0^\infty \chi(z')H(|z - z'|)\,dz'. \tag{6.3.14}$$

The perturbation parameter can be expressed by (see, e.g., Carrier, 1974)

$$\chi = \sum_{n=1}^N g_n \chi_n, \tag{6.3.15}$$

where g_n are arbitrary constants. In reference to Fig. 6.13, we wish to find an equilibrium temperature T from the radiative equilibrium temperature T_0, with

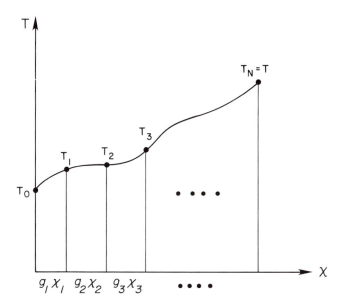

FIG. 6.13   Configuration of temperature perturbation with respect to a perturbation parameter χ defined in the figure, where T_0 is the radiative equilibrium temperature and T the equilibrium temperature to be obtained (after Liou and Ou, 1983).

respect to the components of the perturbation parameter χ_n. Using the Taylor series expansion, each adjacent temperature may be expressed by

$$
\begin{aligned}
T_m &= T_{m-1} + \sum_{n=1}^{N} \frac{1}{n!} \left(\frac{\partial^n T}{\partial \chi^n} \right)_{T=T_{m-1}} (g_m \chi_m)^n \\
&= T_{m-1} + \left(\frac{\partial T}{\partial \chi} \right)_{T=T_{m-1}} (g_m \chi_m) + g_m^2 R_m, \qquad m = 1, 2, \ldots, N,
\end{aligned}
$$

$$(6.3.16)$$

where Taylor series expansion terms higher than the first order are defined by

$$
g_m^2 R_m = T_m - T_{m-1} - \left(\frac{\partial T}{\partial \chi} \right)_{T=T_{m-1}} (g_m \chi_m). \qquad (6.3.17)
$$

Also, from the definition of χ in Eqs. (6.3.13) and (6.3.15), the perturbation temperature can be expressed by

$$
\sigma T_m^4 = \sigma T_0^4 + k' \sum_{n=1}^{m} g_n \chi_n. \qquad (6.3.18)
$$

From Eq. (6.3.16), summing all the perturbation temperatures leads to

$$T = T_N = T_0 + \sum_{m=1}^{N} \left[\left(\frac{\partial T}{\partial \chi} \right)_{T=T_{m-1}} (g_m \chi_m) + g_m^2 R_m \right]. \qquad (6.3.19)$$

On substituting Eq. (6.3.19) into Eq. (6.3.14), we find

$$1 + \frac{1}{\gamma_c} \left\{ \frac{dT_0}{dz} + \sum_{m=1}^{N} \frac{d}{dz} \left[\left(\frac{\partial T}{\partial \chi} \right)_{T=T_{m-1}} (g_m \chi_m) + g_m^2 \frac{dR_m}{dz} \right] \right\}$$

$$= \sum_{n=1}^{N} g_n \int_0^\infty \chi_n(z') H(|z - z'|) \, dz'. \qquad (6.3.20)$$

From Eq. (6.3.18), the derivative of the perturbation temperature with respect to χ, defined in Eq. (6.3.15), is

$$\left(\frac{\partial T}{\partial \chi} \right)_{m-1} = k'(4\sigma T_{m-1}^3)^{-1}. \qquad (6.3.21)$$

From the standpoint of numerical computations, we may set $g_{m+1} = g_m$ and $g_1 = 1$. Upon collecting terms with the same order in g_1 and using Eqs. (6.3.17) and (6.3.21), we obtain the perturbation equations in the forms

$$1 + \frac{1}{\gamma_c} \left[\frac{dT_0}{dz} + \frac{d}{dz} \left(\frac{k'}{4\sigma T_0^3} \chi_1 \right) \right] = \int_0^\infty \chi_1(z') H(|z - z'|) \, dz', \qquad (6.3.22)$$

$$\frac{d}{dz} \left(T_{m-1} - T_{m-2} - \frac{k'}{4\sigma T_{m-2}^3} \chi_{m-1} \right) + \frac{d}{dz} \left(\frac{k'}{4\sigma T_m^3} \chi_m \right)$$

$$= \gamma_c \int_0^\infty \chi_m(z') H(|z - z'|) \, dz',$$

$$m = 2, 3, \ldots, N. \qquad (6.3.23)$$

Since $T_0(z)$ has been determined previously from Eq. (6.3.12), Eqs. (6.3.22) and (6.3.23), which represent the same type of equations, can be successively used to obtain χ_1, T_1, χ_m and T_m (i.e., $\sigma T_m^4 = \sigma T_{m-1}^4 + k' \chi_m$) by a similar numerical procedure. Using the finite-difference formats for these equations, the unknown variable χ_m, as a function of z, can also be evaluated using a standard matrix inversion method. The convergence that uses the present perturbation scheme has been shown to be extremely efficient (Liou and Ou, 1983). Normally, three to four computations are sufficient to converge the temperature solution to within about 0.1%.

Using the atmospheric and cloud data presented in Subsection 6.3.1, equilibrium temperatures for clear, high, middle and low cloud conditions are shown in

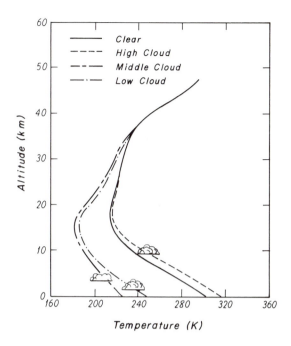

FIG. 6.14   Effects of high, middle, and low clouds on equilibrium temperature in standard atmospheric conditions (after Liou and Ou, 1983).

Fig. 6.14. Cloud covers for the three cloud types are 100%. Equilibrium temperatures computed from an atmosphere containing high clouds are greater than those from a clear atmosphere. Surface temperature differs by about 15 K. Warmer temperatures in the troposphere are due to the downward ir emissions of high clouds. Both temperature profiles are warmer than the standard temperature profile (see Fig. 1.7). For middle and low clouds, temperatures below about 30 km are much lower than those in a clear atmosphere because of the reflection of solar radiation. Middle clouds produce more cooling than low clouds because the former have greater solar reflectance resulting from the higher cloud droplet concentration used in calculating solar radiative properties.

6.4   Earth radiation budget as viewed from space

The earth radiation budget (ERB) at TOA has been derived from satellite observations since the beginning of the meteorological satellite era. Four factors have influenced the evolution of ERB instrumentation: (1) spacecraft limitation of power, data storage, mode of stabilization, and altitude control; (2) viewing angles of nonscanning and scanning medium- and high-resolution radiometers; (3) spectral band-pass requirements; and (4) on-board calibration. The first-generation ERB instruments included black and white hemispheric sensors using thermistor detectors to measure the sensor temperature on board the first U.S. meteorological satellite, Explorer 6, launched February 17, 1959 (Suomi, 1957). This first generation also included a five-channel scanning radiometer on board the TIROS satellite series. Each of the channels had a narrow instantaneous field of view, and the measurements could be interpreted as radiances rather than flux densities for the hemispheric sensors.

In the second generation of satellite missions during the 1960s and 1970s, polar, sun-synchronous orbits became possible due to powerful rockets, and provided the opportunity for daily global coverage of the earth. The duration of spacecraft measurements extended to several years. Flat nonscanning radiometers were installed on several research and ESSA operational satellites and, later, medium and wide field-of-view radiometers were also deployed on the NOAA satellites. Medium- and high-resolution infrared radiometers were used on Nimbus satellites for the detection of shortwave (solar) and longwave (ir) radiation. Nimbus 2 and 3 contained five-channel medium-resolution scanning radiometers and provided the first observations of the ERB for the entire globe. The aperture of the radiometers was about 2.5°, resulting in a spatial resolution of about 50 km near nadir and about 110 km at a nadir angle of 40°. Spectral band passes for these radiometers were 0.3–4 μm for shortwave and 5–50 μm for longwave. The computation of outgoing flux densities from measured radiances required conversion of measured filtered radiances to radiances covering the entire solar and thermal ir spectra, integration over all angles of measurements using bidirectional models, and estimation of the average flux density over a 24-hour period. NOAA polar-orbiting satellites performed scanning measurements in the visible and ir window regions. The transformation of these narrow spectral interval data to broadband estimates of flux densities required several assumptions and models. The narrowband scanning radiometers had a spatial resolution of about 4 km at nadir. Valuable data sets have been constructed from the narrowband measurements of ERB components.

The third generation of satellite observational systems led to the development of ERB instruments that measured direct solar irradiance, reflected shortwave radiation, and emitted longwave radiation. Nimbus 6 and Nimbus 7 satellites contained wide field-of-view and scanning radiometers that provided valuable observations of the ERB. The scanning measurements observed the directional and bidirectional reflecting and emitting properties of the earth–atmosphere system varying in both time and space, and were important in developing directional models for the conversion of radiances to flux densities. The longest record of solar constant measurements was made available by Nimbus 7 observations. The sun-synchronous, polar-orbiting satellites observed each location at a fixed local time. As a consequence, the observations were insufficient to provide a more detailed quantitative estimate of the temporal and spatial sampling errors. Studies of the ERB using data from the geostationary satellites were especially useful, because they provided a regular sample of the atmospheric diurnal cycle. This enabled a wide range of spatial and temporal radiation variations to be investigated. Observations from GOES and METEOSAT satellites have been used in numerous studies of the ERB. Radiometers on board geostationary satellites were confined to narrow bands and had spatial resolutions that varied from about 0.5 to 10 km at nadir. The processing of observed data from geosynchronous altitudes required assumptions similar to those used to interpret the measurements from NOAA polar-orbiting satellites.

In order to provide comprehensive data sets for studying the diurnal and annual cycles of the ERB as well as the role of clouds in the ERB, the fourth generation of satellite observation systems, referred to as the Earth Radiation Budget Experiment (ERBE), was launched in the 1980s. The experiment consisted of scanning and nonscanning radiometers on three satellites. The NASA Earth Radiation Budget satellite (ERBS) performed a $57°$ inclination orbital precession around the earth once every 2 months. The other two satellites were NOAA 9 and NOAA 10 operational meteorological satellites. An excellent review of the characteristics of radiometric equipment and the required data-processing techniques has been given by House et al. (1986).

6.4.1 *Radiation budget of the earth–atmosphere system*

The ERB is usually presented in terms of the emitted longwave flux, F_{ir}, referred to as outgoing longwave radiation (OLR); the planetary albedo (or, simply, albedo) r, defined as the ratio of the reflected solar flux to the incident flux at TOA; and the net radiative flux, defined by

$$F_N = (1 - r)Q - F_{ir}, \qquad (6.4.1)$$

where the incoming solar flux, $Q = S/4$, and S is the solar constant. The first term on the right-hand side of Eq. (6.4.1) represents the absorbed solar flux within the earth–atmosphere system, F_s. The globally and annually averaged albedo and OLR have been derived by a number of researchers from various data sources. The best estimate for the global albedo is $\sim$30%. For the global OLR, it is $\sim$229 W m^{-2} (Jacobowitz et al., 1984).

Figure 6.15 shows the annual variations of albedo, OLR, and net radiative flux (Stephens et al., 1981). The ERB was derived from a composite of 48 monthly mean radiation budget maps. The global net flux albedo and OLR display a distinct annual cycle. The monthly global albedo reaches a maximum of about 32% near the December solstice. Variations are in part produced by hemispheric differences in land–ocean, cloud, and snow–ice distributions. The annual variation in OLR shows a maximum close to June. This variation results from the greater surface area occupied by dry continental areas in the Northern Hemisphere at this time

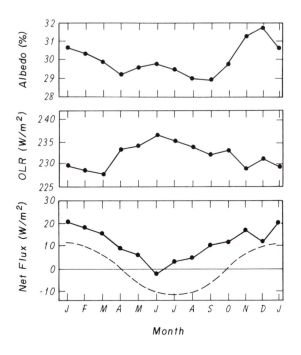

FIG. 6.15   Annual variation of the global albedo, emitted longwave flux and net flux. Also shown in the dashed line is the monthly deviation from the annual mean insolation using a solar constant of 1376 W m^{-2} (after Stephens et al., 1981).

of year. The global mean surface temperature is highest in July ($\sim$16.7°C) and lowest in January ($\sim$13.1°C). The surface temperature variation that regulates OLR is dominated by the large annual cycle over the continents of the Northern Hemisphere. Over moist areas, variations in clouds can also modulate annual variations in OLR.

Also shown in Fig. 6.15 is the monthly deviation from the annual mean solar insolation due to the earth's orbital parameters. To calculate this deviation, a solar constant of 1376 W m^{-2} was used. The net flux pattern at TOA resembles the insolation deviation pattern, which has a difference of about 22 W m^{-2} between January and July. The annual variation in absorbed solar flux is smaller than the annual variation in solar insolation by about a factor of two because the albedo increases as insolation increases. The annual variation in OLR is on the same order of magnitude as absorbed solar flux but is 180° out of phase. The net flux pattern is the difference between the absorbed solar flux and OLR.

The latitudinal distribution of annual and seasonal net fluxes at TOA is displayed in Fig. 6.16 (Stephens et al., 1981). The annual net flux pattern is approximately symmetric between the Northern and Southern Hemispheres, with a maximum occurring at the equator. This maximum is a result of minimum OLR

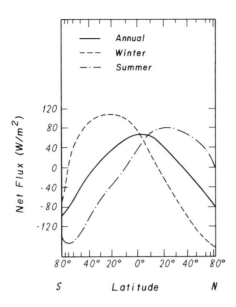

FIG. 6.16   Latitudinal distribution of the annual, winter (December, January, and February), and summer (June, July, and August) net fluxes (after Stephens et al., 1981, with modifications).

due to towering cumulus clouds associated with the intertropical convergence zone (ITCZ) and the large absorbed solar flux in the equatorial regions. Similar results have been presented in Fig. 1.8. In the polar regions, large negative net fluxes are due to the high albedo of snow and ice. The net flux pattern is also produced in part by the variation of the mean solar zenith angle with latitude. In the winter season, the net flux pattern has a maximum at about 30° S, while in the summer season, the maximum net flux is located at about 15–20° N.

Global distributions of OLR, albedo, and net flux averaged for the mean annual condition have been derived by Stephens et al. (1981) and Smith and Smith (1987). The latter authors used a 5-year (1979–83) dataset obtained from the Nimbus 7 ERB experiment. The albedos reveal a distinct land-ocean contrast equatorward of 30° N and 30° S, as well as high convective clouds, in particular Asian monsoons, where high albedo and low emission are evident (Fig. 6.17). Poleward of 30° N and 30° S the radiation budgets are relatively uniform zonally, especially in the Southern Hemisphere. At low latitudes, regions of relative gain and loss of energy are evident for a given zone. Distinct variations are shown in the net fluxes in the tropical and subtropical zones where the deserts of Africa and Arabia appear as negative or small positive anomalies. The convective regions near Asia show large positive anomalies. In general, the albedos are negatively correlated with OLR, principally due to the presence of clouds. The exception is over desert regions, where cloud covers are at minima and the surfaces are relatively bright and warm. A net radiative gain is evident throughout almost the entire zone between ~40° N and 40° S, and is flanked by radiation sinks that generally deepen toward the poles. It is clear that the radiation budgets of the earth and the atmosphere are largely regulated by clouds and temperature fields.

ERB components vary with the local time at a given location. Absorbed solar flux and albedo vary over the daylight hours as a result of the dependence of scattering and absorption processes on the solar zenith angle and atmospheric conditions, principally the variation of cloud types and amounts with local time. OLR varies throughout the day because of diurnal variations in clouds, humidity, trace gases, and temperature fields. Radiometrically gathered data by geosynchronous satellites provide excellent temporal and spatial resolution for the study of regional diurnal variations in cloudiness and ERB components. Using GOES-East observations of visible albedo and window radiance, Minnis and Harrison (1984) have investigated the diurnal variation of cloud cover and OLR during the daylight hours over the southeast Pacific stratus region and over the Amazon Basin. Mean hourly cloud amount, albedo, and OLR components are shown in Fig. 6.18 In the stratus region, cloud amount decreases from about 90% at 0600 local time to about 50% in the early afternoon and remains constant until sunset. The OLR from clear and cloudy areas remains fairly constant throughout the day, suggesting that temperature fields remain largely unchanged. Over the Amazon Basin, cloud amount varies by about 10%, with highest values occurring near sunrise and sunset. Cloudy and total albedos vary by more than 30% in response to the cloud amount variations. Clear-sky albedo is largely affected by the solar zenith angle. There are large diurnal variations in OLR patterns, although cloud amount varies only by about 10%. This appears to be a result of diurnal variation in cloud top height. Maximum OLR occurs in the morning when cloud tops are relatively low and warm. Cloud tops rise after noon, and OLR decreases. In clear sky, OLR varies slightly and is controlled by the uniform surface temperature, which is regulated by surface evaporation. The above description is sufficient to illustrate the significance of diurnal variation of ERB with respect to variations in cloud structure and humidity.

ERB data inferred from narrowband radiometers on board NOAA satellites over more than a decade have been used to investigate intraseasonal and interannual variabilities, which are largest in the tropics and closely associated with changes in atmospheric circulation and sea surface temperature. The 30–50 day oscillation and El Niño/Southern Oscillation (ENSO) are the two most important phenomena ever found in the low-frequency motion of the tropical atmosphere. Anomalies in OLR can be used to characterize local diabatic heating in the tropics. When OLR is anomalously low, deep convective clouds are present. Latent heat and radiative flux exchanges associated with convective clouds lead to a net heating of the atmosphere. This heating drives large-scale circulations, which produce the moisture flux necessary to sustain cumulus clouds.

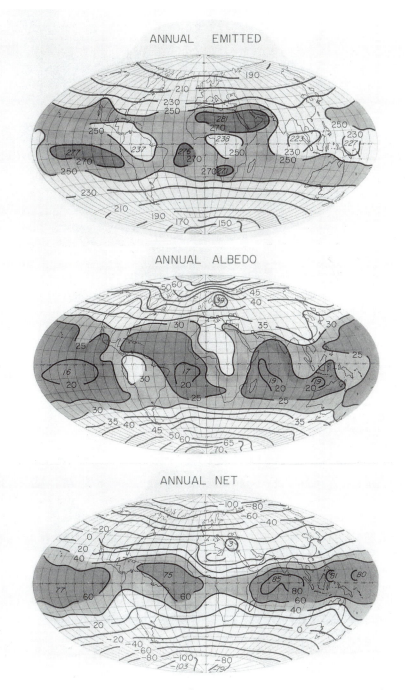

Fɪɢ. 6.17 Mean annual infrared flux (W m⁻²), planetary albedo (%), and net flux (W m⁻²) (after Stephens et al., 1981).

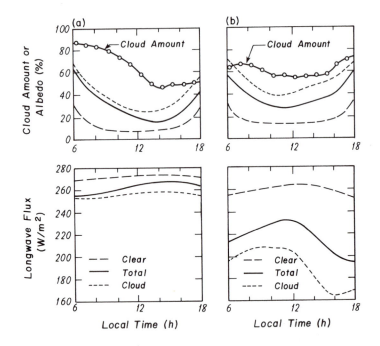

FIG. 6.18   Mean hourly, daytime and regional cloud amounts, albedo and OLR over (a) the southeast Pacific stratus region (21.4° S, 86.3° W) and (b) the Amazon Basin (10.1° S, 55.1° W) (after Minnis and Harrison, 1984).

Figure 6.19 shows the time-longitude cross section of 5-day mean OLR anomalies between 10° S and the equator over the Indian and Pacific Oceans before and during the 1982–1983 ENSO (Lau and Chan, 1985). A number of 30–50 day waves originated from the western Indian Ocean and progressed eastward toward the central Pacific during the period from July, 1981 to June, 1982. An abrupt change in the appearance of convection took place in June and July, 1982. Enhanced convection began to form over eastern Indonesia and New Guinea and propagated slowly eastward reaching 120° W by May, 1983. From August, 1982, to May, 1983, the positive and negative anomaly pattern (dipole anomaly) was very pronounced. The maximum anomaly occurred in February and March with a value of about 60 W m^{-2} over the central Pacific, revealing a significant increase in deep convection over this region. The 30–50 day waves were superimposed on the dipole anomaly and continued to propagate eastward from the Indian Ocean but were suppressed over the Indonesian region. The foregoing illustration gives an excellent example of the usefulness of ERB data in the study of disturbances in the tropical atmosphere.

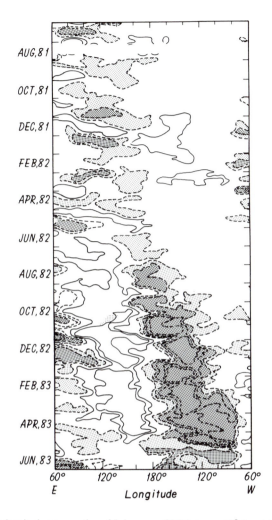

Fig. 6.19 Time-longitude cross section of 5-day mean OLR between $10°$ S and the equator over the Indian and Pacific Oceans before and during the 1982–83 ENSO. Contour interval is $10 \ \mathrm{W \ m^{-2}}$ and shaded areas denote negative anomalies (after Lau and Chan, 1985).

6.4.2 Extracting cloud effects from earth radiation budget data

Clouds regularly occupy at least 50% of the sky on a global scale, as described in Sections 4.1 and 4.2. Clouds are the most important regulators of ERB. The presence of clouds greatly increases the solar flux that is reflected back to space. This effect, known as the *solar albedo effect*, reduces the solar flux available to the earth–atmosphere system and results in a cooling of the earth–atmosphere system. On the other hand, clouds reduce the thermal radiation emitted to space by absorbing the ir flux from the earth's surface and the atmosphere below the cloud layer, and by emitting thermal radiation at normally colder cloud top temperatures.

This effect, known as the *ir greenhouse effect*, increases the radiation budget, which, in general, results in a warming of the earth–atmosphere system. The net radiation available to the earth–atmosphere system, and thus the differential heating of the system, is strongly regulated by the horizontal extent, vertical position, thermodynamic phase, liquid/ice content, and particle size distribution of clouds.

Many attempts have been made to extract cloud effects from the ERB data inferred from satellite data. In reference to Eq. (6.4.1) and considering the specific effect of cloud cover η, partial differentiation leads to

$$\frac{\partial F_N}{\partial \eta} = -Q\frac{\partial r}{\partial \eta} - \frac{\partial F_{ir}}{\partial \eta} = \frac{\partial F_s}{\partial \eta} - \frac{\partial F_{ir}}{\partial \eta}. \qquad (6.4.2a)$$

Or we may write

$$\frac{\partial F_N}{\partial \eta} = \frac{\partial F_N}{\partial F_{ir}}\frac{\partial F_{ir}}{\partial \eta} = -\frac{\partial F_{ir}}{\partial \eta}\left(Q\frac{\partial r}{\partial F_{ir}} + 1\right). \qquad (6.4.2b)$$

The first and second terms on the right-hand side of these equations represent the effect of cloud cover on absorbed solar and emitted ir fluxes. To estimate these terms, the method of regression on the observed data at a given grid point may be used. Since albedo increases with increasing cloud cover, $\partial r/\partial \eta$ is always positive. In contrast, ir flux decreases with increasing cloud cover so that $\partial F_{ir}/\partial \eta$ is negative. It follows that if $\partial F_N/\partial \eta$ is positive, the ir greenhouse effect is greater than the solar albedo effect. The reverse is true if $\partial F_N/\partial \eta$ is negative.

Hartmann and Short (1980) have used daily observations of the ERB from NOAA scanning radiometer data for the periods June–August, 1975, and December, 1975–February, 1976, to estimate the relative magnitudes of cloud effects on solar and ir radiation. Ohring et al. (1981) have estimated the sensitivity parameter from ERB data obtained from the NOAA scanning radiometer data set. Monthly mean values of OLR and albedo for a period of 45 months (June, 1974–February, 1978) and for a 2.5° × 2.5° grid were used in the analysis. The net and ir flux sensitivities to changes in cloud cover were determined from a knowledge of either the clear-sky and surface albedos or the planetary albedos. Koenig et al. (1987) have estimated cloud cover sensitivity in the ERB using climatology data obtained from the analysis of the 3DNEPH cloud data base and ERB data derived from Nimbus 7 broadband radiation budget measurements. The importance of oceanic stratus clouds on the solar albedo effect has been specifically noted. However, because of the uncertainty in estimating high cloud covers, the significance of cirrus clouds on the solar albedo and ir greenhouse effects has not been comprehensively examined from ERB data. All of the preceding studies, based on ERB data, have demonstrated that the effects of cloud cover are substantially larger on solar radiation than they are on ir radiation. Thus, the result of an increase in cloud cover would be a reduction in net radiation.

The simplest way of extracting the effects of clouds from ERB data is to identify the ERB in clear sky. Consider a region that is partially covered by clouds. This
region consists of an overcast region (cloudy) with a fractional area of coverage
of η and a clear-sky region with a fractional area of coverage of $1 - \eta$. Let F be
the observed OLR, absorbed solar flux or net flux over the region. Then we may
write

$$F = (1 - \eta)F^{cl} + \eta F^{ov}, \tag{6.4.3a}$$

where F^{cl} and F^{ov} are the clear-sky and cloudy fluxes, respectively. The effects
of clouds on F may be evaluated from

$$C = F^{cl} - F = \eta(F^{cl} - F^{ov}). \tag{6.4.3b}$$

The term C is referred to as *cloud radiative forcing* (e.g., see Hartmann et al.,
1986). A separation of solar and IR effects from Eq. (6.4.3a) leads to

$$C_{ir} = F^{cl}_{ir} - F_{ir}, \tag{6.4.4a}$$
$$C_s = Q(r^{cl} - r), \tag{6.4.4b}$$

where F^{cl}_{ir} and r^{cl} are clear-sky OLR and albedo, respectively.

Clouds are almost always more reflective than the oceans and land, except
when snow is present. Thus, when clouds are present, more solar flux is reflected
back to space than with clear sky. Cloud solar forcing C_s, which is the difference
between the clear-sky and cloudy-sky reflected solar fluxes, gives a quantitative
estimate of cloud effects for solar radiation. On the other hand, less thermal ir flux
is emitted to space from a cloudy region than from clear sky. Cloud ir forcing C_{ir},
which is the difference between the clear-sky and cloudy-sky emitted ir fluxes, is
a measure of cloud effects for thermal ir radiation.

Estimates of the global distributions of cloud radiative forcing have been
obtained from ERBE. As described previously, ERBE includes three satellites in
different orbits: ERBS, NOAA 9, and NOAA 10. Harrison et al. (1990) have used
the data gathered by the scanning radiometers on board ERBS and NOAA 9 to
evaluate cloud radiative forcing. Table 6.3 shows a summary of the seasonal cloud
radiative forcing values for global and hemispheric scales. The seasonal effect of
cloud radiative forcing is generally small. The global annual C_s and C_{ir} are -48
and $31\,\mathrm{Wm^{-2}}$, respectively, resulting in net cloud radiative forcing of $-17\,\mathrm{Wm^{-2}}$.
April has the smallest net cloud radiative forcing value, whereas January has the
largest. Variation of this forcing between hemispheres is significant during winter
and summer months. C_s and C_{ir} nearly cancel each other in the winter hemisphere,
while in the summer hemisphere, the magnitude of C_s is twice that of C_{ir}. On a
global scale for all seasons the albedo effect from clouds is more dominant than
the greenhouse effect.

Table 6.3 Global cloud radiative forcing in units of Wm^{-2} estimated from ERBE data.[a]

Date	Geographical extent	C_{ir}	C_s	C
April, 1985	N.H.	31.2	−49.8	−18.6
	S.H.	31.5	−40.4	−8.9
	Global	31.3	−45.1	−13.8
July, 1985	N.H.	33.8	−66.3	−32.5
	S.H.	26.4	−27.2	−0.8
	Global	30.1	−46.7	−16.6
October, 1985	N.H.	34.1	−40.1	−6.0
	S.H.	30.2	−60.2	−30.0
	Global	32.2	−50.1	−17.9
January, 1986	N.H.	26.6	−26.7	−0.1
	S.H.	34.7	−76.8	−42.1
	Global	30.6	−51.7	−21.1
Annual	N.H.	31.4	−45.7	−14.3
	S.H.	30.7	−51.1	−20.4
	Global	31.1	−48.4	−17.3

[a] After Harrison et al. (1990).

Cloud radiative forcing has significant regional characteristics, and interpretation of the results must be handled with care. For a given region, cancellation of cloud solar and ir forcing, which are defined from ERB components at TOA, does not imply a negligible role of clouds in the regional climate. The vertical gradients of radiative heating and cooling produced by clouds, which are critical in weather and climate processes, are not accounted for in the preceding discussion of cloud radiative forcing (see Subsection 7.4.2 for further comments).

6.5   Energy budget of the atmosphere and the surface

In the previous section dealing with radiation budgets, we noted that, in the mean, there is a radiation excess in the tropical region and a radiation deficit in the middle and high latitudes. Significant temperature and water vapor gradients occur in the atmosphere near the surface. Thus, there must be poleward as well as upward energy transports in order to produce an overall energy balance. Energy exchanges at TOA are solely due to radiative processes. However, within the earth–atmosphere system, these exchanges involve a number of mechanisms of which radiative energy transfer is only one component of the total energy budget.

6.5.1   *Development of horizontal and vertical heat transport equations*

In this subsection, expressions for the horizontal and vertical transports of sensible and latent heat fluxes are developed. In reference to Eq. (4.4.44) for moist static energy, we separate the time-averaged variables into a mean and a deviation component along the longitudinal direction (zonal average), and define

$$[\bar{\chi}] = \int_0^{2\pi} \bar{\chi}(t, z, \lambda, \phi) \frac{d\lambda}{\Delta\lambda}, \tag{6.5.1}$$

where χ is any atmospheric variable. Thus, $\bar{\chi} = [\bar{\chi}] + \bar{\chi}^*$, where the superscript * denotes the deviation from the longitudinal average, ϕ is the latitude, λ is the longitude, and $[\bar{\chi}^*] = 0$. Carrying out the zonal average and using the spherical coordinates, Eq. (4.4.44) for moist static energy becomes

$$\frac{\partial}{\partial t}\rho[\bar{E}] + \frac{\partial F_v}{\partial z} + \frac{1}{a_e \cos\phi}\frac{\partial}{\partial\phi}(F_h \cos\phi) = -\frac{\partial F}{\partial z}, \tag{6.5.2}$$

where a_e is the radius of the earth, $F = [\bar{F}]$ is the mean net radiative flux, and the horizontal and vertical fluxes are defined, respectively, in the forms

$$F_h = \rho\left([\bar{v}][\bar{E}] + [\bar{v}^*\bar{E}^*] + [\overline{v'E'}]\right), \tag{6.5.3a}$$

$$F_v = \rho\left([\bar{w}][\bar{E}] + [\bar{w}^*\bar{E}^*] + [\overline{w'E'}]\right). \tag{6.5.3b}$$

The terms $[\bar{v}][\bar{E}]$, $[\bar{v}^*\bar{E}^*]$, and $[\overline{v'E'}]$ represent the transports of moist static energy by mean motions, stationary eddies, and transient eddies in the meridional direction, respectively. Likewise, $[\bar{w}][\bar{E}]$, $[\bar{w}^*\bar{E}^*]$, and $[\overline{w'E'}]$ represent their counterparts in the vertical direction. Stationary eddies in the vertical, $[\bar{w}^*\bar{E}^*]$, are usually very small and can be disregarded for practical applications to atmospheric problems. Large-scale horizontal flux is composed of a mean and an eddy component. The mean meridional circulation transports the mean moist static energy. The eddy term consists of transports by transient and stationary eddies.

Moist static energy can be decomposed into dry static energy, $s = C_pT + gz$, and latent heat, Lq; that is, $E = s + Lq$. The net flux of sensible heat by atmospheric currents, that is, the rate of diabatic heating or cooling of air, may be expressed in terms of the rate of change of the enthalpy C_pT and geopotential $\Phi = gz$, through the first law of thermodynamics and hydrostatic equilibrium. By definition, dry static energy represents the sum of internal and potential energies. Separating the dry static energy (simply referred to here as sensible heat) and the latent heat components, the poleward transports of energy can be written

$$F^T = \rho\left([\overline{v's'}] + [\bar{v}^*\bar{s}^*] + [\bar{v}][\bar{s}]\right), \tag{6.5.4a}$$

$$LF^q = \rho L\left([\overline{v'q'}] + [\bar{v}^*\bar{q}^*] + [\bar{v}][\bar{q}]\right). \tag{6.5.4b}$$

The first terms on the right-hand side of these equations are the transient eddies representing the covariance (approximate correlation coefficient) between local variations in time of wind and temperature and of wind and specific humidity. If the time interval is on the order of a month or longer, these terms denote the part of the mean meridional heat and moisture transport that is largely due to traveling cyclones and anticyclones. The second and third terms are, respectively, the stationary eddy and the meridional circulation transports, which involve the covariance and product of the time-averaged quantities and can, in principle, be predicted explicitly by a mean-motion model. However, the transient eddy transport must be parameterized in terms of the mean quantities.

To obtain the vertically integrated values for the atmosphere, we perform an integration of Eq. (6.5.2) from the surface ($z = 0$) to TOA ($z = z_\infty$). At TOA, the vertical flux $F_v(z_\infty) = 0$ for all latitudes. Thus,

$$S_A = \frac{\partial}{\partial t} \int_0^{z_\infty} \rho[\bar{E}] \, dz$$
$$= [F_v(0) + F(0)] - F_a - F(z_\infty), \qquad (6.5.5)$$

where the first term represents the net surface flux; the second term, the divergence of atmospheric transport of heat; and the third the TOA net radiative flux; S_A denotes the zonal mean rate of energy storage per unit area, $F_v(0)$ is the surface vertical flux, $F(0)$ and $F(z_\infty)$ are the net radiative flux at the surface and TOA, respectively, and the divergence of the atmospheric transport of moist static energy is defined as

$$F_a = \frac{1}{a_e \cos \phi} \frac{\partial}{\partial \phi} \int_0^{z_\infty} F_h \cos \phi \, dz. \qquad (6.5.6)$$

This term can be decomposed into sensible and latent heat components as follows:

$$F_a^T = \frac{1}{a_e \cos \phi} \frac{\partial}{\partial \phi} \int_0^{z_\infty} F^T(z) \, dz, \qquad (6.5.7a)$$

$$LF_a^q = \frac{1}{a_e \cos \phi} \frac{\partial}{\partial \phi} \int_0^{z_\infty} LF^q(z) \, dz. \qquad (6.5.7b)$$

It follows that $F_a = F_a^T + LF_a^q$. Equation (6.5.5) is the basic energy balance equation for an atmospheric column in which the total heat storage rate is balanced by the surface input minus the sum of the divergence of the poleward transport of heat and the net radiative flux at TOA.

At the surface, the energy balance is governed by the rate of energy storage, the net surface flux and the divergence of the transport of sensible heat below the earth's surface. The heat capacity of the continental surface is generally very small, as in the cases of snow and ice. Thus, on a large scale, the principal heat exchange must occur between the atmosphere and the oceans. Analogous to Eq. (6.5.5) for atmospheric storage, the zonal mean rate of energy storage per unit area for the oceans may be expressed by

$$S_O = \frac{\partial}{\partial t} \int_{z_B}^0 \rho c_o T \, dz$$
$$= -[F_v(0) + F(0)] - F_o, \qquad (6.5.8)$$

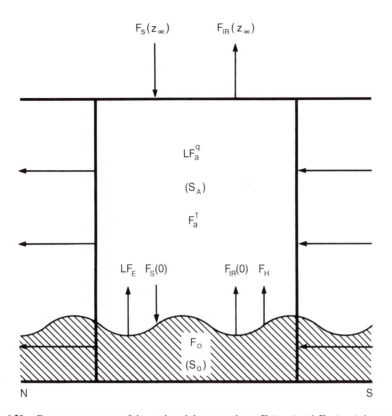

FIG. 6.20   Energy components of the earth and the atmosphere. $F_s(z_\infty)$ and $F_{ir}(z_\infty)$ denote the absorbed solar flux and the emitted ir flux at TOA, $F_S(0)$ and $F_{ir}(0)$ are the absorbed solar flux and the emitted ir flux at the surface, LF_a^q and F_a^T are the divergence of meridional transports of latent and sensible heat, F_o is the divergence of oceanic transports of sensible heat, LF_E and F_H are the vertical fluxes of latent and sensible heat, the sum of which is denoted by $F_v(0)$, and S_A and S_0 are the zonal mean rate of energy storage per unit area for the atmosphere and the oceans, respectively.

where the first term represents the net surface flux; the second, the divergence of oceanic transport of heat; z_B denotes the bottom of the ocean layer; c_o is the heat capacity of the oceans; and F_o is the divergence of the oceanic transport of sensible heat.

A schematic representation of the preceding vertical and horizontal transports and radiative components is illustrated in Fig. 6.20. For a given atmospheric column, the heating mechanisms are: (1) absorption of solar flux, (2) condensation of water vapor via the horizontal water vapor flux transport, (3) increase of sensible heat via the horizontal sensible heat flux transport in the column, and (4) increase of sensible heat via the horizontal oceanic transport of sensible heat in the column. On the other hand, the cooling mechanisms are: (1) emission of thermal ir flux to space, (2) evaporation of water via the horizontal water vapor flux transport, (3) decrease of sensible heat via the transport of horizontal sensible heat flux out of

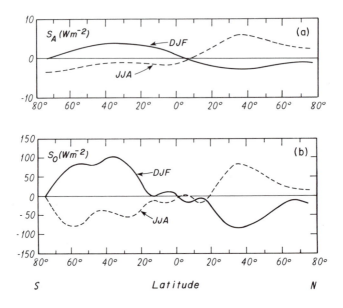

FIG. 6.21   Zonal mean rate of energy storage per unit area (a) in the atmosphere (from Oort and Péixoto, 1983) and (b) in the oceans (after Levitus, 1982).

the column, and (4) decrease of sensible heat via the horizontal oceanic transport of sensible heat out of the column.

6.5.2   *Atmospheric and oceanic transports of energy*

Based on surface reports and rawinsonde soundings from May, 1958, to April, 1963, Oort and Rasmusson (1971) have generated extensive atmospheric general circulation statistics for the Northern Hemisphere. Oort (1983) has expanded these statistics to include the Southern Hemisphere and extended the period of data analysis to cover 1958 to 1973. The data sets consist of zonal and meridional wind fields, geopotential height, temperature, humidity, and flux fields for the transport of sensible and latent heat. Oort and Péixoto (1983) and Ou et al. (1989) have presented various energy budget components for seasonal and annual means.

The zonal mean rate of energy storage per unit area in the atmosphere in units of W m^{-2} [Eq. (6.5.5)] is shown in Fig. 6.21(a). The annual mean is zero. For the December, January, February (DJF) and June, July, August (JJA) seasons, the heat storage in the atmosphere is less than 10 W m^{-2}. In the transitional seasons, March, April, May (MAM) and September, October, November (SON), larger storage values of about 20 W m^{-2} occur at high latitudes. In general, the atmosphere has a limited heat capacity. A seasonal variation of the zonal mean rate of energy storage per unit area for the oceans [Eq. (6.5.8)] has been estimated by Levitus (1982) and

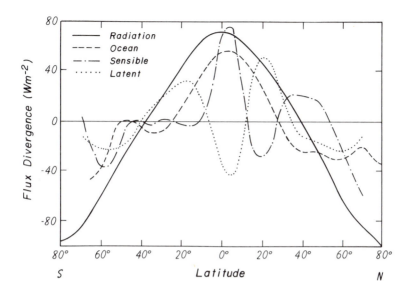

FIG. 6.22   Divergence of atmospheric transports of sensible and latent heat, and oceanic transports of sensible heat. Also shown is the net radiative flux density at TOA. (Data from Oort and Péixoto, 1983.)

is shown in Fig. 6.21(b). The results of the rate of heat storage are computed based on global monthly mean temperature analyses in the first 275 m of the world ocean. A variation of $100\,\mathrm{W\,m^{-2}}$ is shown in the middle latitudes of each hemisphere. A large difference in heat storage is evident near 60° S. However, near 60° N oceanic storage is negligible since, for the most part, there are no oceans at this latitude.

The divergence of the vertically integrated atmospheric transport of sensible and latent heat [Eq. (6.5.6)] is shown in Fig. 6.22. Positive values represent divergence of heat, while negative values indicate convergence of heat. The atmospheric latent heat component has a minimum between about 10° N and 10° S because of excess precipitation in the tropical convective zone. Minima between about 40° and 60° in both hemispheres occur as a result of surplus precipitation produced by cyclonic storm activity. On the other hand, maxima patterns are evident in the subtropics between about 20° and 30° in both hemispheres, where evaporation rates are highest. Evaporation in the northern subtropics is less pronounced due to the presence of large desert areas. The atmospheric sensible heat component shows a large maximum in the tropics associated with the Hadley circulation. Small maxima at 40° N and S are related to the transport of heat by eddies and mean meridional circulation. On an annual basis about $20\,\mathrm{W\,m^{-2}}$ of sensible and latent heat is lost between 40° S and 40° N. About 50 to 70 $\mathrm{W\,m^{-2}}$ is gained poleward of 60°. In the middle and high latitudes, the transport of sensible and latent heat is largely due to transient eddies. The mean circulation contribution to heat transport is most pronounced in the tropics but also has some contribution in middle and high latitudes.

The transport of sensible heat by oceanic currents plays a very important role in the annual variation of the earth's heat balance. For annual mean conditions, the divergence of the oceanic transport of sensible heat may be estimated from Eq. (6.5.5) as a residual, if the atmospheric heat storage, the atmospheric transport of sensible and latent heat, and the net radiative flux at TOA are given (Oort and Péixoto, 1983). About 40 W m^{-2} of heat is transported by ocean currents out of the tropics, and about 25 W m^{-2} of heat is transported by ocean currents into latitudes poleward of $40°$, as shown in Fig. 6.22. Since there are no oceans south of about $70° \text{ S}$, the energy convergence of heat in the oceans should disappear there. The results of annual oceanic transport in the Antarctic areas based on the residual method require improvements. Finally, the sum of the divergence of atmospheric and oceanic transports of heat must be balanced by the net radiative flux at TOA on the annual basis.

To obtain the total horizontal flux (energy/time) for the entire atmosphere, we may perform an integration of the horizontal flux [Eq. (6.5.3a)] over the latitudinal cross section to obtain

$$f_a = 2\pi a \cos \phi \int_0^{z_\infty} F_h \, dz. \tag{6.5.9}$$

A separation of sensible and latent heat components can be made using Eqs. (6.5.4a) and (6.5.4b). A similar operation can be used to compute the oceanic transport. Figure 6.23 shows the meridional profiles of the poleward transports of sensible and latent heat in the atmosphere, the poleward transport of sensible heat in the oceans, and the net radiative flux (energy/time) at TOA. The oceanic transports dominate in low latitudes with maximum poleward transports of about $3 \times 10^{15} \text{ W}$ near $25° \text{ N}$ and of about $-3.5 \times 10^{15} \text{ W}$ near $20° \text{ S}$. It should be noted that the uncertainty in the estimates of oceanic transports based on the residual method is quite large, especially in the Southern Hemisphere, where the radiosonde network is inadequate to measure atmospheric transports reliably. Latent heat is transported both toward the equator and toward the poles from about $20° \text{ N}$ and S, where the evaporation maxima are located. The transports of sensible heat in the atmosphere show double maxima at about $10°$ and $50°$ in both hemispheres with values of about $2 \times 10^{15} \text{ W}$. The atmospheric transports of sensible and latent heat largely cancel each other out in lower latitudes between $20° \text{ N}$ and $20° \text{ S}$. As a result, the total atmospheric transport of energy is more important in middle and high latitudes.

As pointed out previously, the transport of latent heat by the atmosphere is associated with the hydrological cycle of the earth–atmosphere system. Consider an atmosphere column. The water storage in this column, S_{WA}, must be the net result

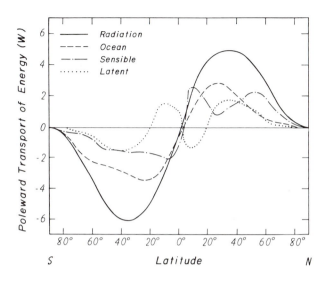

FIG. 6.23   Meridional profiles of the poleward transports of sensible and latent heat in the atmosphere, and of sensible heat in the oceans. Also shown is the net radiative flux. All the terms are in units of 10^{15} W. (Data used in this figure are from Oort and Péixoto, 1983).

of precipitation P, evaporation E, and divergence (or convergence) of moisture, denoted in Eq. (6.5.7b), as follows:

$$S_{WA} = E - P - \Delta F_a^q. \qquad (6.5.10a)$$

The atmosphere can contain only relatively small amounts of water at any of its phases. Thus, the storage term is usually much smaller than the other components. On the annual basis, $S_{WA} \cong 0$, so that

$$\Delta F_a^q = E - P. \qquad (6.5.10b)$$

At the surface, the water balance is related to runoff in the form

$$S_{WS} = P - E - \Delta F_r, \qquad (6.5.11a)$$

where S_{WS} is the water storage at the surface and ΔF_r is the water runoff. Again, on the annual basis, $S_{WS} \cong 0$. In this case, we have

$$P = E + \Delta F_r. \qquad (6.5.11b)$$

It follows that annually, we must have $\Delta F_r = -\Delta F_a^q$. The annual runoff from any region must be balanced by an influx of moisture into the atmosphere column above that region. Over the entire globe, the horizontal redistribution of surface water always equals zero, because areas of runoff are exactly balanced by areas of inflow. Thus, precipitation must be equal to evaporation.

Meridional profiles of precipitation, evaporation, and runoff have been estimated by Baumgartner and Reichel (1975) using surface observations, which were divided into two types: ocean and land. Figure 6.24 shows the annual values of the preceding three components. For global latitude zones, precipitation maxima occur at ~5° N (~2000 mm) associated with the ITCZ, and at ~45–50° S and 45–55° N associated with cyclonic storm activities. Minima are located at subtropical highs (20–30° N, 25–30° S). The global evaporation pattern shows a minimum at 0–5° N and maxima at 10–15° N and S, but this pattern is relatively uniform. For runoff, maximum surpluses occur at 5–10° N as well as at 50–55° N and S. Maximum deficits appear at 20–25° N and S. The annual runoff pattern is similar to the divergence of atmospheric transport of latent heat, except for a sign change, as pointed out previously.

Some specific features of the hydrological cycle are noted below. Over the oceans, especially over the Pacific, the precipitation maximum at ~5° N is produced by the pronounced convergence zone. Maxima at ~45–50° N and S are associated with storm tracks. The evaporation pattern shows two maxima at subtropical highs produced by the absorption of solar radiation by the oceans. The runoff surplus is greatest in the tropical zone at ~5° N. Maxima of runoff deficits are evident at subtropical high regions. Over land the precipitation maxima are associated with convective tropical rain (~5° S) and orographic rain on the west coasts of South America and New Zealand produced by westerlies between 45 and 50° S. The evaporation maximum at ~5° S is caused by high precipitation and higher temperature over land. Low evaporation over Antarctica is the result of glaciation and elevation. The runoff pattern is similar to the precipitation pattern and is positive due to the relatively small amount of evaporation over land.

6.5.3 Surface energy budget

The net surface flux defined in Eq. (6.5.5) consists of the vertical flux and the net radiative flux, which can be divided into solar and ir components. A large fraction of the solar flux at TOA is transmitted through the clear atmosphere, as is evident in Fig. 3.1. In cloudy conditions the solar flux available at the surface is largely dependent on the solar zenith angle and cloud optical depth, which is a function of cloud liquid/ice water content and particle size distribution, as discussed in Subsection 5.3.1. Let the solar flux reaching the surface be $F_s^-(0)$. Then the absorbed portion of the flux is given by

$$F_s(0) = F_s^-(0)(1 - r_s), \qquad (6.5.12)$$

where r_s is the surface albedo.

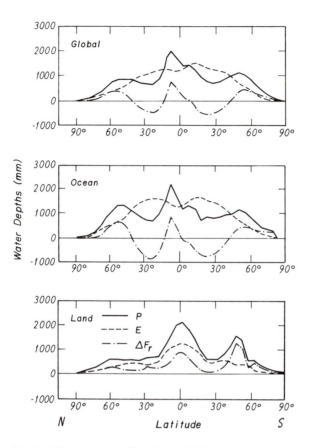

Fig. 6.24 Precipitation (P), evaporation (E) and runoff (ΔF_r) in units of mm water depths over land, oceans, and the globe (after Baumgartner and Reichel, 1975).

The net ir flux at the surface is the difference between the upward ir flux emitted by the surface and the downward ir flux from the atmosphere reaching the surface, $F_{ir}^-(0)$. If the surface emissivity is denoted as ϵ_s, then we have

$$F_{ir}(0) = \epsilon_s \sigma T_s^4 - \varepsilon_s F_{ir}^-(0), \qquad (6.5.13)$$

where T_s is the surface temperature.

The vertical transport of energy from the surface to the atmosphere consists of the flux of sensible heat from the surface F_H, and the evaporative water loss from the surface F_E. Thus, we write

$$F_v(0) = L F_E + F_H, \qquad (6.5.14)$$

where L is the coefficient for the latent heat of evaporation (or sublimation).

Table 6.4 Surface albedos averaged over the solar spectrum[a]

Water surfaces			Soil, dry clay or gray	20–35
	0° latitude	6	Soil, dry light sand	25–45
Winter	30° latitude	9	Concrete, dry	17–27
	60° latitude	21	Road, black top	5–10
	0° latitude	6	**Natural surfaces**	
Summer	30° latitude	6	Desert	25–30
	60° latitude	7	Savanna, dry season	25–30
Bare areas and soils			Savanna, wet season	15–20
Snow, fresh fallen		75–95	Chaparral	15–20
Snow, several days old		40–70	Meadows, green	10–20
Ice, sea		30–40	Forest, deciduous	10–20
Sand dune, dry		35–45	Forest, coniferous	5–15
Sand dune, wet		20–30	Tundra	15–20
Soil, dark		5–15	Crops	15–25
Soil, moist gray		10–20		

[a] Data from Sellers (1965).

On the basis of the energy conservation principle, the net flux of energy absorbed by the surface must equal the rate at which the surface is storing energy S_s, so that

$$S_s = F_s^-(1 - r_s) - \epsilon_s \sigma T_s^4 + \epsilon_s F_{ir}^-(0) - LF_E - F_H. \qquad (6.5.15)$$

Over the oceans, we have $S_s = S_O + F_o$, from Eq. (6.5.8). Over land and on a microscopic scale, S_s may be expressed in terms of the rate of heat conduction into the underlying surface. Equation (6.5.15) represents the basic surface energy equation. Local variations such as the melting of snow and ice have been disregarded. The assumption that $S_s = 0$ has been frequently used in weather prediction and climate models because of its simplicity. This assumption is approximately correct for land surfaces, averaged over 24 hours, but would lead to large errors in determining surface temperature over a diurnal cycle. If $S_s = 0$, surface temperature can be directly computed from Eq. (6.5.15), provided that all the radiative and turbulent fluxes are known.

The surface albedo is dependent on the type of surface, as well as being a function of the solar zenith angle and wavelength. Table 6.4 lists the surface albedos averaged over the entire solar spectrum for water and various land surfaces. For water surfaces, the albedo ranges from 6 to 9%, except for cases involving the low solar angle that is associated with the high latitudes of the winter hemisphere. A significant albedo variation, ranging from 10 to 40%, is evident over land surfaces. The albedos of snow and ice are greater than about 40%. Variation of the albedo with solar wavelength is large. For example, for dry desert soil, the visible albedo ($\lambda < 0.7 \, \mu m$) is 26%; however, in the near-ir ($\lambda > 0.7 \, \mu m$), the albedo is larger by a factor of two. The importance of the albedo over land surfaces in climate modeling has been articulated by Dickinson (1983).

The thermal ir emissivities from water and land surfaces are normally between 90 and 95%. It is usually assumed that the earth's surfaces are approximately black in infrared radiative transfer calculations. Exceptions include snow and some sand surfaces whose emissivities are wavelength dependent and could be less than 90%.

The transport of sensible heat and water vapor from surfaces is governed by turbulent motions. Section 4.6 introduced the means by which the vertical transport of fluxes by eddies may be expressed in terms of mean quantities. In the context of surface flux we have

$$F_H = -\rho_0 C_p K_h \left(\frac{\partial T}{\partial z}\right)_{z \to 0}, \tag{6.5.16a}$$

$$LF_E = -\rho_0 L K_e \left(\frac{\partial q}{\partial z}\right)_{z \to 0}, \tag{6.5.16b}$$

where ρ_0 is the density of air at the surface. For practical applications, equations for these fluxes can be obtained by integrating Eqs. (6.5.16a) and (6.5.16b) from the surface to a given height, say 2 m, above the surface. This procedure leads to aerodynamic transfer formulas for sensible and latent heat fluxes:

$$F_H \cong \rho_0 C_p C_{DH} |\mathbf{v_0}|(T_s - T_a), \tag{6.5.17a}$$

$$LF_E \cong \rho_0 L C_{DE} |\mathbf{v_0}|(q_s - q_a), \tag{6.5.17b}$$

where C_{DH} and C_{DE} are the drag coefficients for heat and moisture, respectively, $|\mathbf{v_0}|$ is the wind velocity at a reference height, q_s is the specific humidity of air immediately adjacent to the surface, and T_a and q_a are the temperature and specific humidity at a reference height. Generally, we may set $C_{DH} \cong C_{DE} = C_D$. The drag coefficient depends on the Richardson number [see Eq. (4.6.27)] and on the reference height used. Solar and ir fluxes reaching the surface for a given atmospheric condition can be computed using the methodologies presented in Chapters 2 and 3.

Latitudinal distributions of net solar and ir fluxes and the total turbulent flux at the earth's surface are displayed in Figs. 6.25(a)–(c). For solar and ir fluxes, seasonal values are also shown. The computations of radiative fluxes use a radiative transfer scheme and climatological profiles for clouds, specific humidity, temperature, and ozone (Ou et al., 1989). In Fig. 6.25(a), the annual net surface solar flux decreases poleward. There is a slight hemispheric asymmetry. Compared with the annual absorbed solar flux (see Fig. 1.8), the solar flux absorbed in the atmosphere is nearly independent of latitude. The seasonal variation is largely controlled by solar insolation and the solar zenith angle. In Fig. 6.25(b), latitudinal distributions of net surface ir flux show double maxima at about 35° in both hemispheres; these

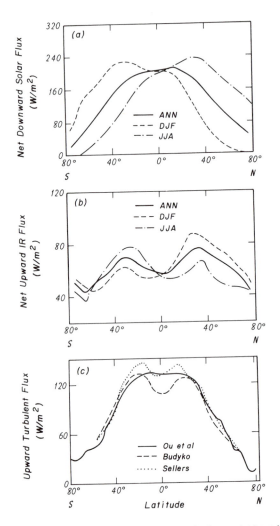

FIG. 6.25 Latitudinal distributions of (a) net solar flux, (b) ir flux, and (c) total turbulent flux at the earth's surface. (a) and (b) include seasonal values; (c) includes estimates from Ou et al. (1989), Budyko (1982), and Sellers (1965).

maxima are associated with subtropical highs. The minimum pattern in the tropics is due to the large downward ir fluxes produced by the extensive cloud cover in the ITCZ. This pattern is similar to the dip that occurs in the emitted ir flux at TOA. The seasonal variation is largely governed by surface temperature.

Estimates of the vertical turbulent fluxes of sensible and latent heat using available climatological data have been carried out by Budyko (1982), Sellers (1965), and, more recently, Ou et al. (1989). Latitudinal distributions of the surface turbulent flux are displayed in Fig. 6.25(c). The surface turbulent flux decreases from about 100–150 W m^{-2} in the tropics to near zero at the poles. There are large uncertainties in the estimate of vertical flux in the tropics. A large fraction of the vertical transport takes place through cumulus convection, which occurs in only about 3–4% of the total tropical area. For this reason, it is very difficult to quantify vertical transport based on data obtained from limited observational points. On the annual basis, the sum of the turbulent flux and the net ir flux at the surface is approximately equal to the solar flux absorbed by the surface.

6.5.4   Atmospheric radiative heating

Radiative processes directly affect the dynamics and thermodynamics of the atmosphere through the generation of radiative heating and cooling rates as well as the net radiative flux available at the surface. Some typical solar heating and ir cooling profiles produced by radiatively active gases in clear atmospheres have been shown in Fig. 6.8. Observed radiative heating profiles associated with stratus and cirrus have been presented in Figs. 5.22 and 5.21. The radiative flux exchange is largely controlled by the cloud field. Consider the radiative heating/cooling rates in the context of the basic thermodynamic equation, expressed in the pressure coordinate in the form

$$\frac{\partial T}{\partial t} + \frac{u}{a_e \sin \theta} \frac{\partial T}{\partial \lambda} + \frac{v}{a_e} \frac{\partial T}{\partial \theta} + \omega \left(\frac{\partial T}{\partial p} - \frac{RT}{C_p p} \right) = Q + Q_R + F^T + g \frac{\partial \tau^T}{\partial p},$$
(6.5.18)

where $\theta = \pi/2 - \phi$, the co-latitude, Q is latent heat, Q_R is radiative heating, F^T is the horizontal eddy flux divergence for temperature, τ^T is the vertical eddy sensible heat flux, and other notations are conventional.

Net radiative heating, $Q_R = Q_s + Q_{ir}$, is the result of solar and ir net radiative heating. Radiative heating rates have been defined in Eqs. (2.1.17), (3.1.24), and (6.2.39). As shown in Eq. (5.2.20), the heating rate for an atmospheric column containing a fraction of cloud cover η may be written in the form

$$Q_{s,ir} = \eta Q_{s,ir}^{ov} + (1 - \eta) Q_{s,ir}^{cl},$$
(6.5.19)

where the superscripts ov and cl denote overcast and clear-sky conditions.

ir cooling profiles for clear and cloudy atmospheres containing high, middle and low clouds and using standard atmospheric temperature and humidity profiles are shown in Fig. 6.26(a). The positions of high, middle and low clouds are located between 200–450 mb, 450–735 mb, and 735–950 mb, respectively, in an 18-layer model. The absorption gases considered are H_2O, CO_2 and O_3. The radiative transfer calculations use an adding/doubling method, described in Subsection 3.2.2, and the cloud particle size distributions employed for the high, middle, and low clouds are for Cs, As, and St, respectively (Section 4.3). The positions of the clouds are shown in the diagram, and the heating rates for 100% cloud cover are averages over the model layer. In a clear atmosphere, maximum cooling rates on the order

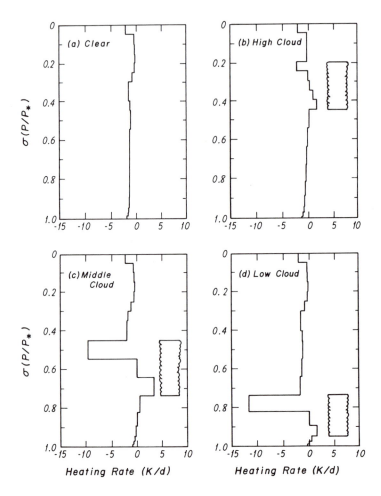

FIG. 6.26   ir cooling rate profiles computed from an adding/ doubling radiative transfer scheme based on the 1962 U.S. Standard Atmosphere for different cases: (a) clear, (b) high cloud, (c) middle cloud, and (d) low cloud. The calculations are performed for an 18-layer vertical model. The vertical bars represent the average heating rates for model layers.

of $\sim 2\,K\,d^{-1}$ are seen near the surface. Cooling rates decrease up to a height of $\sim 12\,km$ and then increase as a function of height due to carbon dioxide and ozone absorption and emission. When a model high cloud is present, a maximum cooling rate of $\sim 4\,K\,d^{-1}$ is shown at the cloud top. There is slight heating at the cloud base. In the case of middle cloud, strong cooling ($\sim 11\,K\,d^{-1}$) at the cloud top and significant heating ($\sim 4\,K\,d^{-1}$) at the cloud base are shown. Heating just below the cloud is also evident. A similar cooling/heating pattern can be seen in the low cloud case. It is clear that the presence of clouds significantly alters the vertical cooling profile in a clear atmosphere. The principal feature is strong cooling at the cloud top and significant warming at the cloud base. This feature is largely

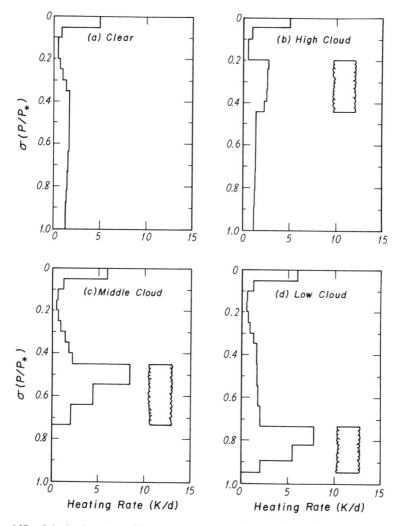

FIG. 6.27 Solar heating rate profiles computed from an adding/doubling radiative transfer scheme based on the 1962 U.S. Standard Atmosphere for different cases: (a) clear, (b) high cloud, (c) middle cloud, and (d) low cloud. The vertical structure is the same as that in Fig. 6.26. The solar constant is 1365 W m^{-2}, the surface albedo is 0.2, and the cosine of the zenith angle is 0.8.

produced by the temperature gradient between cloud and surface. The degree of cooling/heating is a function of the cloud position and the optical depth. Cooling rates for a partly cloudy region can be obtained from Eq. (6.5.19).

The solar counterparts are shown in Fig. 6.27(b). The solar constant used is 1365 W m^{-2}, the surface albedo is 0.2, and the cosine of the solar zenith angle μ_0 is 0.8, which represents the maximum averaged value. In general, solar heating rates should be less than those presented in this figure. Solar heating rates in a clear atmosphere are normally less than $\sim$1.5 K d^{-1}. High cloud produces a heating rate

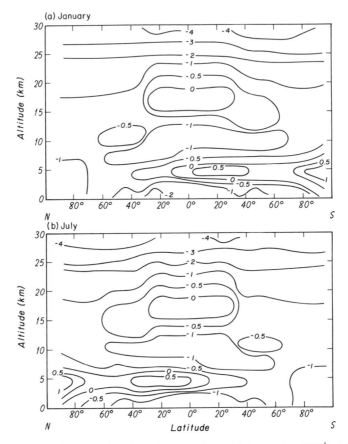

FIG. 6.28   Latitudinal cross section of zonally averaged net radiative heating (Kd^{-1}) of the atmosphere for (a) January and (b) July.

of $\sim 2\,\mathrm{K\,d^{-1}}$ at the cloud top. Middle and low clouds generate heating rates of about $\sim 8\,\mathrm{K\,d^{-1}}$ at the cloud tops. At night, there would be no solar heating.

In a clear atmosphere, cooling outweighs heating essentially everywhere in the atmosphere. In various cloudy atmospheres, it is also the general tendency that maximum cooling rates at cloud tops are greater than the maximum solar heating rates. Therefore, the effect of clouds on the vertical heating profile is to produce cooling at the cloud top and warming at the cloud base.

Figures 6.28(a) and (b) show the latitudinal cross section of the zonally averaged net radiative heating of the atmosphere for January and July. The results are obtained from a radiative transfer program using climatological data for temperature, humidity, ozone and cloud profiles (Freeman and Liou, 1979). ir cooling dominates solar heating almost everywhere in the atmosphere. In the upper stratosphere above about 25 km, intense cooling on the order of 4 to 4.5 K d^{-1} is due to ir cooling produced by CO_2 and O_3, which overshadows the solar heating caused by O_3. The cooling in the troposphere is due primarily to water vapor, and thus has a maximum in the tropics. The presence of clouds tends to moderate cooling in the lower levels of the atmosphere. As shown in Figs. 6.26 and 6.27, the presence of clouds reduces ir cooling below their bases and increases ir cooling and solar heating above their tops. Cloud effects vary with latitude and season as cloud distribution varies. The weak net heating pattern, extending from the summer pole into the tropical latitude of the winter hemisphere at a level of about 5 km, is due to the absorption of solar flux by H_2O augmented by solar geometry. Heating by clouds also contributes significantly to this feature. Below the cloud bases, the small amount of solar heating is supplemented by reduced ir cooling. The maximum heating is found in this region near the summertime pole, where the length of the period of solar heating offsets the low solar zenith angle. ir cooling in this region is relatively small because of low temperatures and cloud effects. In both hemispheres, cooling maxima of ~ 2 K d^{-1} are found in the surface layer of the winter tropics. This is due to the maximum H_2O concentration near the surface layer and, in part, to a relative minimum of cloudiness in the winter tropics as compared with the summer tropics. It is apparent that cooling produced by ir flux divergences generally outweighs solar heating at every latitude for both seasons.

REFERENCES

Baumgartner, A., and E. Reichel, 1975: *The World Water Balance*. Elsevier, Amsterdam, 179 pp.

Berger, A. L., 1978: Long-term variations of daily insolation and quaternary climatic changes. *J. Atmos. Sci.*, **35**, 2362–2367.

Budyko, M. I., 1982: *The Earth's Climate: Past and Future*. Academic Press, New York, 307 pp.

Carrier, G. F., 1974: Perturbation methods. In *Handbook of Applied Mathematics*, C. E. Pearson, Ed., Van Nostrand Reinhold, pp. 761–828.

Chamberlain, J. W., and D. M. Hunter, 1987: *Theory of Planetary Atmospheres: An Introduction to their Physics and Chemistry*. Academic Press, New York, 481 pp.

Chandrasekhar, S., 1939: *An Introduction to the Study of Stellar Structure*. University of Chicago Press, Chicago, 509 pp.

Deardorff, J. W., 1966: The counter-gradient heat flux in the lower atmosphere and in the laboratory. *J. Atmos. Sci.*, **23**, 503–506.

Deardorff, J. W., 1972: Theoretical expression for the counter-gradient vertical heat flux. *J. Atmos. Sci.*, **77**, 5900–5904.

Dickinson, R. E., 1983: Land surface processes and climate–surface albedos and energy balance. *Adv. Geophys.*, **25**, 305–353.

Freeman, K. P., and K. N. Liou, 1979: Climatic effects of cirrus clouds. *Adv. Geophys.*, **21**, 231–287.

Gierasch, P., and R. M. Goody, 1968: A study of thermal and dynamical structure of the Martian lower atmosphere. *Planet. Space Sci.*, **16**, 615–646.

Goody, R. M., and Y. L. Yung, 1989: *Atmospheric Radiation: Theoretical Bases.* Oxford University Press, New York, 519 pp.

Harrison, E. F., P. Minnis, B. R. Barkstrom, V. Ramanathan, R. D. Cess, and G. G. Gibson, 1990: Seasonal variation of cloud radiative forcing derived from the Earth Radiation Budget Experiment. *J. Geophys. Res.*, **95**, 18,687–18,703.

Hartmann, D. L., and D. A. Short, 1980: On the use of earth radiation budget statistics for studies of clouds and climate. *J. Atmos. Sci.*, **37**, 1233–1250.

Hartmann, D. L., V. Ramanathan, A. Berroir, and G. E. Hunt, 1986: Earth radiation budget data and climate research. *Rev. Geophys.*, **24**, 439–468.

Hickey, J. R., B. M. Alton, H. L. Kyle, and D. Hoyt, 1988: Total solar irradiance measurements by ERB/Nimbus-7. A review of nine years. *Space Sci. Rev.*, **48**, 321–342.

House, F. B., A. Gruber, G. E. Hunt, and A. T. Mecherikunnel, 1986: History of satellite missions and measurements of the earth radiation budget (1957–1984). *Rev. Geophys.*, **24**, 357–377.

Jacobowitz, H., R. J. Tighe, and Nimbus 7 ERB Experiment Team, 1984: The earth radiation budget derived from the Nimbus 7 ERB experiment. *J. Geophys. Res.*, **89**, 4997–5010.

Koenig, G., K. N. Liou, and M. Griffin, 1987: An investigation of cloud/radiation interactions using three-dimensional nephanalysis and earth radiation budget data bases. *J. Geophys. Res.*, **92**, 5540–5554.

Lau, K. M., and P. H. Chan, 1985: Aspects of the 40–50 day oscillation during the northern winter as inferred from outgoing longwave radiation. *Mon. Wea. Rev.*, **113**, 1889–1909.

Lee, R., B. Barkstrom, E. Harrison, M. Gibson, S. Natarajan, W. Edmonds, A. Mecherikunnel, and H. Kyle, 1988: Earth Radiation Budget Satellite extraterrestrial solar constant measurements: 1986–1987 increasing trend. *Adv. Space Res.*, **8**, 11–13.

Levitus, S., 1982: *Climatological Atlas of the World Ocean.* NOAA Prof. Pap. 13 (DOC C 55.25:13), 173 pp.

Lindzen, R. S., A. Y. Hou, and B. F. Farrell, 1982: The role of convective model choice in calculating the climate impact of doubling CO_2. *J. Atmos. Sci.*, **39**, 1189–1205.

Liou, K. N., and S. C. Ou, 1983: Theory of equilibrium temperatures in radiative–turbulent atmospheres. *J. Atmos. Sci.*, **40**, 214–229.

London, J., 1957: *A study of the Atmospheric Heat Balance.* Final Report, Contract AF19 (122)–166, Dept. of Meteor. and Oceanogr., New York Univ. (ASTIA 117227, Air Force Geophysics Laboratory, Hanscom AFB), 99 pp.

Manabe, S., and R. Strickler, 1964: Thermal equilibrium of the atmosphere with a convective adjustment. *J. Atmos. Sci.*, **21**, 361–385.

Manabe, S., and R. T. Wetherald, 1967: Thermal equilibrium of the atmosphere with a given distribution of relative humidity. *J. Atmos. Sci.*, **24**, 241–259.

Milankovitch, M., 1941: Canon of Insolation and the Ice Age Problem. *Königlich Serbische Akademie*, Belgrade, 484 pp. (English translation by the Israel Program for Scientific Translation, published by the U.S. Department of Commerce and the National Science Foundation).

Minnis, P., and E. F. Harrison, 1984: Diurnal variability of regional cloud and clear-sky radiative parameters derived from GOES data. III. November 1978 radiative parameters. *J. Climate Appl. Meteor.*, **23**, 1032–1051.

Ohring, G., P. F. Clapp, T. R. Heddinghaus, and A. F. Krueger, 1981: The quasi-global distribution of the sensitivity of the earth-atmosphere radiation budget to clouds. *J. Atmos. Sci.*, **38**, 2539–2541.

Oort, A. H., 1983: *Global Atmospheric Circulation Statistics, 1958–1973*. NOAA Prof. Pap. 14 (DOC C 55.25:14), 180 pp.

Oort, A. H., and J. P. Péixoto, 1983: Global angular momentum and energy balance requirements from observations. *Adv. Geophys.*, **25**, 355–490.

Oort, A. H., and E. M. Rasmusson, 1971: *Atmospheric Circulation Statistics*. NOAA Prof. Pap. 5 (DOC C 55.25:5), 323 pp.

Ou, S. C., and K. N. Liou, 1984: A two-dimensional radiative–turbulent climate model: I. Sensitivity to cirrus radiative properties. *J. Atmos. Sci.*, **41**, 2289–2309.

Ou, S. C., K. N. Liou, and W. J. Liou, 1989: The seasonal cycle of the global zonally averaged energy balance. *Theor. Appl. Climatol.*, **40**, 9–23.

Reagan, J., L. Thomason, and B. Herman, 1986: Assessment of atmospheric limitations on the determination of the solar spectral constant from ground-based spectroradiometer measurements. *IEEE Tras. Geosci. and Remote Sensing*, **24**, 258–266.

Reed, R. J., and K. E. German, 1965: A contribution to the problem of stratospheric diffusion by large-scale mixing. *Mon. Wea. Rev.*, **93**, 313–321.

Saltzman, B., and S. Ashe, 1976: The variance of surface temperature due to diurnal and cyclone-scale forcing. *Tellus*, **28**, 307–322.

Sasamori, T., J. London, and D.V. Hoyt, 1972: Radiation budget of the Southern Hemisphere. In *Meteorology of the Southern Hemisphere*, C. W. Newton, Ed., *Meteor. Monogr.*, No. 13, pp. 9–24.

Sellers, W. D., 1965: *Physical Climatology*. University of Chicago Press, Chicago, 272 pp.

Smith, E. A., and M. R. Smith, 1987: Interannual variability of the tropical radiation balance and the role of extended cloud systems. *J. Atmos. Sci.*, **44**, 3210–3234.

Stephens, G. L., G. G. Campbell, and T. H. Vonder Haar, 1981: Earth radiation budgets. *J. Geophys. Res.*, **86**, 9739–9760.

Suomi, V. E., 1957: The radiation balance of the earth from a satellite. *Amer. Int. Geophys. Year*, **6**, 331–340.

Thekaekara, M. P., 1976: Solar irradiance: Total and spectral and its possible variations. *Appl. Opt.* **15**, 915–920.

Willson, R. C., and H. S. Hudson, 1988: Solar luminosity variation in solar cycle 21. *Nature*, **322**, 810–812.

7

THE ROLE OF RADIATION AND CLOUD PROCESSES IN ATMOSPHERIC MODELS

This chapter presents the role of radiation and cloud processes in various types of atmospheric models that are primarily designed to study the effects of external radiative forcings on the earth's climate and on global climate change. Atmospheric models are required for the physical understanding of atmospheric dynamic and thermodynamic processes and for the prediction of future states of the atmosphere in terms of the fundamental variables governing its behavior. A model, as such, is constructed on the basis of certain aspects of physical principles and is designed for application to a specific atmospheric process. The distinction between atmospheric models and the real atmosphere should be delineated. Because of the complexity of the real atmosphere with respect to temporal and spatial scales, a hierarchy of models with increasing sophistication has been developed to study many processes that occur in the atmosphere.

One-dimensional radiative–convective models (variation only in the vertical) in conjunction with the determination of the atmospheric thermal equilibrium have been introduced in Section 6.3. Because of their simplicity, these models have been widely utilized to investigate temperature perturbations due to external radiative forcings. One-dimensional energy-balance models (variations only in latitude) have been primarily used to examine the impact of solar insolation variations on climate and, in particular, the question of the advance and retreat of polar ice in the geological time scale. Two-dimensional models include variations both in the vertical and in latitude. Two specific examples, the extended drought in Africa and high contrail cirrus produced by human activities, are discussed in association with these types of models.

Finally, we present some aspects of radiative and cloud processes in the context of general circulation models. The degree of temperature perturbation due to external radiative forcings is subject to model approximations, assumptions, and postulations and will not be elaborated in this discussion. However, focus will be made on the interactions and feedbacks involving radiation and cloud physics in dynamic and thermodynamic systems.

7.1   Radiation and cloud effects in one-dimensional radiative–convective models

7.1.1   *The greenhouse effects of carbon dioxide and feedbacks*

One of the major concerns in climate studies over the past two decades has been the impact on surface temperature of the steady increase in atmospheric carbon dioxide content produced by the rapid burning of fossil fuels. Since the beginning of the Industrial Revolution more than a century ago, carbon has been removed from the earth in the form of coal, petroleum, and natural gas. In burning processes, carbon dioxide is formed through the oxidation reaction, $C+O_2 \rightarrow CO_2$. Of the CO_2 that has been produced, about half is believed to remain in the atmosphere, while the other half is dissolved in the oceans or absorbed by the earth's biomass, primarily the forests.

Carbon dioxide is a colorless gas. Below $-78.5°$ C, CO_2 assumes a solid form, known as dry ice. As discussed in Section 3.3, CO_2 is virtually transparent to solar radiation. However, it is a strong absorber in the $15\,\mu$m band ($\sim$12–18 μm) of the thermal ir spectrum, as described in various parts of Section 2.2. The $15\,\mu$m band consists of the ν_2 fundamental, the combination bands, the hot bands, and the P, Q, and R branches of the rotational transitions. An increase in atmospheric CO_2 content can cause the additional trapping of the outgoing thermal ir radiation emitted from the surface and lower atmosphere and can enhance the greenhouse effect. To what degree and extent the anticipated CO_2 concentration increase will influence atmospheric and surface temperatures is a question that must be determined by atmospheric models.

Variations in CO_2 over different geographical areas are relatively small because the variability in the sources and sinks of CO_2 is small at the earth's surface. Atmospheric CO_2 is removed from and released into the atmosphere through a number of natural processes. A considerable amount of CO_2 dissolves annually into the oceans and returns to the atmosphere by a reverse process. A significant component of atmospheric CO_2 consumption appears to be photosynthesis, via the reaction $CO_2 + H_2O + h\tilde{\nu} \rightleftharpoons CH_2O + O_2$. The sedimentary layers of the earth's crust, the lithosphere, contain a considerable amount of carbon from which CO_2 is formed through oxidation processes. These CO_2 cycles appear to introduce little change in the total amount of CO_2 in the atmosphere. However, anthropogenic increases in CO_2 generate important climatic responses through oceanic processes.

The amount of carbon dioxide in the atmosphere has risen from an estimated 280–290 parts per million by volume (ppmv) in 1900 to the present 350 ppmv (1990). The mean monthly concentration of atmospheric CO_2 at Mauna Loa, Hawaii, since 1958 has been displayed in Fig. 1.2. The yearly oscillation is produced by the annual cycle of photosynthesis and respiration of plants in the Northern Hemisphere. A steady increase in the atmospheric CO_2 concentration is evi-

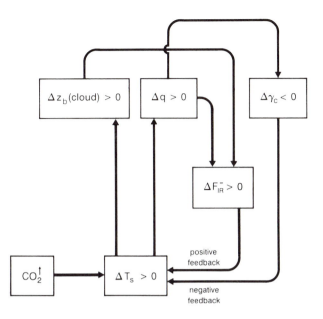

FIG. 7.1   Water vapor feedbacks in the CO_2 climate problem. z_b denotes the cloud base, q is the specific humidity, F_{ir} is the downward ir flux, T_s is the surface temperature, and γ_c is the temperature lapse rate. Positive and negative feedbacks are also identified.

dent. The CO_2 concentration has also been continuously monitored at the South Pole. The records show that there is a lag of a couple of years between the Northern Hemisphere, where most of the carbon dioxide is released, and the Southern Hemisphere, because of the slow exchange of air between hemispheres. Based on the model calculations for the carbon cycle, which include the takeup of anthropogenic CO_2 by the oceans and biomass, and an assumed quasiexponential increase in the rate of fossil fuel burning, notably coal, in the next half-century or more, it is estimated that the CO_2 concentration could reach 380–390 ppmv by A.D. 2000 and may double the amount of $\sim$300 ppmv by the middle of the next century (Bacastow and Keeling, 1973).

Extensive numerical experiments using a one-dimensional radiative–convective model have been performed to investigate the effects of CO_2 doubling on climatic temperature increases, beginning with the pioneering work of Manabe and Wetherald (1967). Various feedbacks involving moisture that are associated with CO_2 doubling have been investigated by Hansen et al. (1981) using a one-dimensional model. First, using a fixed relative humidity, a lapse rate of 6.5 K km^{-1}, and a prescribed cloud altitude, the equilibrium surface temperature T_s increases. This is because rising temperatures in the atmosphere and at the surface increase the water vapor concentration in the atmosphere, which in turn traps more thermal ir radiation and, to a lesser degree, absorbs more solar radiation. The feedback due to increasing water vapor through higher temperatures is positive. Next, a moist adiabatic lapse rate is used instead of a fixed lapse rate. This causes the equilibrium surface temperature to be less sensitive to radiative perturbations, since the lapse rate decreases as more moisture is added to the atmosphere. Subsequently, the temperature differences between the top of the convective region and the surface are reduced, and ΔT_s decreases. Thus the feedback due to the use of a moist adiabatic lapse rate is negative. The temperature change, ΔT_s, is sensitive to the assumption of fixed versus varying cloud altitude. The assumption that clouds move to a higher altitude as temperature increases results in a greater ΔT_s than for the case in which cloud altitude is fixed. Clouds move aloft, enhance the trapping of thermal ir emission from the surface by water vapor, and exert a positive feedback. However, hypothetical black clouds are used, and the cloud radiative properties have not been properly accounted for. The feedbacks due to water vapor in the CO_2 climate problem are summarized in Fig. 7.1.

The exchange of latent and sensible heat and radiative fluxes between the oceans and the atmosphere through the planetary boundary layer provides an important source of ocean–atmospheric interaction. Hydrological cycle feedbacks to the surface warming caused by the doubling of CO_2 are important processes in the climate problem. In the context of the one-dimensional radiative–convective model, the surface warming due to doubled CO_2 may be divided into three processes. The first two processes involve direct surface and atmospheric heating due to the greenhouse effect of doubled CO_2. The third process is related to the interactions among the ocean surface temperature, the hydrological cycle, and the tropospheric convective adjustment. Surface warming due to the first two processes, illustrated in Fig. 7.2, enhances the water vapor evaporation into the troposphere, which indirectly amplifies the surface warming via the latent heat release within the troposphere and increases the tropospheric absolute humidity that in turn increases downward ir emission to the surface. According to the one-dimensional model calculation, the third process is most significant in the feedback process (Ramanathan, 1981). It is clear that hydrological cycle feedbacks play an essential role in the surface warming due to a doubling of CO_2. Overall, the one-dimensional models show a sensitivity of surface temperature increase of about 2–3° C in the doubling of CO_2 concentration experiments.

7.1.2   *Radiation–chemistry–climate interactions and feedbacks*

Atmospheric chemical processes involving possible changes in ozone and other trace gases such as methane (CH_4), chlorofluorocarbons (CFCs = $CFCl_3$, CF_2Cl_2, etc.), and nitrous oxide (N_2O), as well as the associated effects of ultraviolet (uv) fluxes on climate, have been subjects of great concern in recent years. In particular, there has been continuous concern with the question of whether the ozone layer

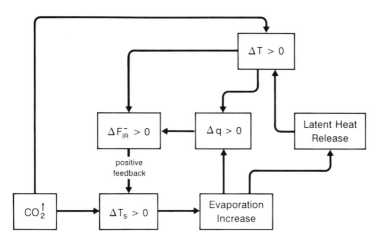

FIG. 7.2   The role of evaporation and condensation in the CO_2 climate problem. The most important feedback is via the hydrological cycle involving surface evaporation, condensation of water vapor, and the addition of moisture to the atmosphere. Effects of clouds and precipitation are not included.

in the upper atmosphere is actually thinning globally because of the increased use of CFCs.

A decrease of Antarctic ozone during the austral spring, just after the end of the Antarctic polar night, has recently been discovered. Data from the British Antarctic Survey collected at the Halley Bay station indicate a 40% total ozone decrease during the month of October between 1977 and 1984. This decrease has been confirmed by data derived from the Total Ozone Mapping Spectrometer (TOMS) and the Solar Backscatter Ultraviolet (SBUV) Spectrometer on board Nimbus 7. The largest percentage decrease corresponds roughly to the polar vortex, covering approximately the entire area of the Antarctic continent. It has been suggested that the so-called *ozone hole* is the precursor of a major decrease in the ozone layer worldwide caused by the addition of CFCs to the atmosphere. It has also been noted that the ozone hole appears to have a dynamic origin and is related to the rather special meteorological conditions prevalent over Antarctica in October of each year. The cold polar vortex traps the air parcels for weeks, during which time polar stratospheric clouds are formed. These clouds may facilitate certain chemical reactions that favor destruction of ozone. Comprehensive observations were carried out in 1987 by the Airborne Antarctic Ozone Experiment, which employed some 150 scientists to investigate the question of ozone reduction and to seek physical explanations for the ozone hole. Two theories have been proposed. One theory assumes that the addition of CFCs is a cause of the ozone depletion. The other emphasizes the natural shift in the air movements that transport ozone-rich air into the polar stratosphere during the austral spring. Based on the available data the growing suspicion that CFCs contribute importantly to the ozone hole appears to gain some weight (Stolarski, 1988).

Ozone is produced primarily in the middle and upper stratosphere by three-body collisions involving molecular oxygen, O_2, and its atomic form, O, and a third body via $O_2 + O + M \rightarrow O_3 + M$. The atomic oxygen is produced by photolysis of O_2 due to the absorption of solar fluxes with wavelengths shorter than 2423 Å; $O_2 + h\tilde{\nu} \rightarrow O + O$. The destruction of O_3 is mainly due to the catalytic effects of various free radical species, including nitrogen oxides, chlorine oxides, and hydrogen oxides. The excited atomic oxygen in the ^1D state, produced by the photodissociation of ozone from solar wavelengths shorter than 3100 Å, via $O_3 + h\tilde{\nu} \rightarrow O_2 + O(^1D)$, is essential for the production of these radicals.

Ozone absorbs solar radiation in the uv Hartley and Huggins bands and in the visible Chappuis band, as discussed in Section 3.8. As shown in that section, the major heating at altitudes above $\sim$45 km is due to absorption in the Hartley band, while the heating below $\sim$30 km is due to absorption in the Chappuis band. In the infrared, O_3 exhibits a number of vibrational–rotational bands. The 9.6 μm band is most important because of its location in the window region (see Subsection 2.2.5).

On an annual and global basis, O_3 contributes $\sim$20% of the total downward ir flux from the stratosphere to the troposphere. A reduction in stratospheric O_3 could lower the temperature in the region in which it occurs and, at the same time, produce two competing effects on tropospheric and surface temperatures. First, more solar uv and visible fluxes could reach the troposphere and surface, leading to a warming effect. On the other hand, the greenhouse effect, due to the trapping of ir fluxes by O_3, could be reduced and, in turn, cooling could result. The competition of these two effects determines whether the troposphere is warmed or cooled due to the perturbation caused by O_3.

Based on one-dimensional radiative–convective models, a reduction in O_3 concentrations would lead to a decrease in atmospheric and surface temperatures. Cooling of a few tenths of a degree could result if the stratospheric O_3 concentration were reduced by 20–30% (Ramanathan and Dickinson, 1979). Moreover, temperature perturbations due to other radiative forcings, such as a doubling of CO_2, may lead to changes in the reaction rates involving ozone. This in turn affects the ozone concentration. Increases in the CO_2 concentration reduce stratospheric temperatures, resulting in an increase in the total ozone due to a slowing down of the ozone loss reaction rates. Because of the uncertainty involved in chemical reaction rates, this feedback is very difficult to quantify. The assessment at this point, using one-dimensional models, indicates that the surface temperature perturbation associated with a CO_2-induced O_3 change would be small.

As described in Subsection 2.2.5, methane exhibits an absorption band at $\sim$1400 cm^{-1} and is a greenhouse gas. The current tropospheric concentration of methane is about 1.7 ppmv. There are some indications that atmospheric CH_4 appears to be biological in origin. It has been suggested that a possible, and the most likely, cause of CH_4 increase is the increase in biogenic emissions associated with a growing human population. Production from rice paddies also appears to be a major source of CH_4. There is also a possibility that termites are a significant source of atmospheric CH_4. Overall, changes in the CH_4 budget are clearly related to increasing human and corresponding animal populations and changes in land use. Based on one-dimensional radiative–convective models, a doubling of CH_4 from 1.7 to 3.4 ppmv would lead to an increase in the surface temperature by ~0.2–$0.4°$ C (Wang et al., 1986). Moreover, changes in the CH_4 concentration may affect global O_3 distributions through reactions with hydroxyl radicals and other trace gases. An increase in CH_4 may lead to an increase in O_3 via the net reaction, $CH_4 + 4O_2 \rightarrow CH_2O + H_2O + 2O_3$, if enough nitric oxide is present (Crutzen, 1983).

Nitrous oxide exhibits complex ir absorption bands that overlap with the CH_4 bands. The mixing ratio of N_2O is ~0.3 ppmv. Measurements indicate a possible global increase in N_2O concentrations at a rate of $\sim0.2\%$ per year. This increase is attributed to the increase in fossil fuel combustion and fertilizer denitrification. Based on a one–dimensional radiative–convective model, the surface temperature is estimated to increase by ~0.3 to $0.4°$ C due to a doubling of N_2O from 0.3 to 0.6 ppmv (Wang et al., 1986). Dissociation of N_2O by excited oxygen atoms, $O(^1D)$, is the major source of nitrogen oxides ($NO_x = NO, NO_2$). NO_x is important in determining the distribution of both tropospheric and stratospheric O_3. At ~25 km, the net effect of NO_x additions to the stratosphere will be to lower O_3 concentrations. However, below ~25 km in the stratosphere NO_x protects ozone from destruction (Crutzen, 1983). The effect of the feedback between tropospheric NO_x and O_3 on climatic temperature perturbations is a subject requiring quantification.

Although carbon monoxide does not exhibit a significant ir absorption band, it is a climatically important gas because of its chemical reactions involving O_3 and CO_2 in the troposphere. The surface mixing ratio of CO is ~0.12 ppmv. The anthropogenic component of CO production is produced primarily by transportation, industrial fossil fuel combustion processes, deforestation, biomass burning, and modification of CH_4 sources. With sufficient NO_x, O_3 and CO_2 are produced via the net reaction, $CO + 2O_2 \rightarrow CO_2 + O_3$. On the other hand, a competing reaction is the destruction of O_3 through the net reaction, $CO + O_3 \rightarrow CO_2 + O_2$. In both cases, CO_2 is produced (Crutzen, 1983).

As discussed in Subsection 2.2.5.6, chlorofluorocarbons are important greenhouse gases. A direct greenhouse effect due to increased CFCs has been investigated by several researchers (Ramanathan et al., 1987). However, because of the uncertainties in the absorption coefficients, the degree of temperature perturbation differs significantly among models. The indirect effects of CFCs on the greenhouse effect are associated with the catalytic destruction of ozone by chlorine released from the photodissociation of CFCs, principally CF_2Cl_2 and $CFCl_3$. Such ozone destruction is produced by the reactions: $Cl + O_3 \rightarrow ClO + O_2$ and

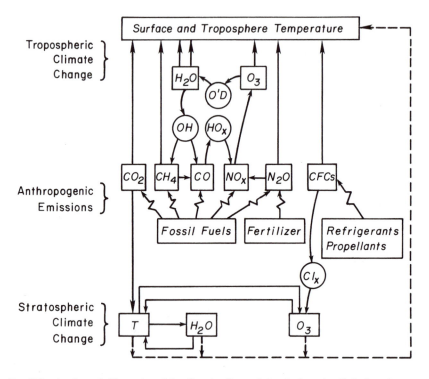

FIG. 7.3   A schematic illustration of the climatic effects of chemically and radiatively active gases (after Ramanathan et al., 1987).

$ClO + O \rightarrow Cl + O_2$. The reduction of ozone in the stratosphere will have significant impact on stratospheric temperatures and may also be related to surface warming. The ozone-climate effect resulting from the addition of chlorines to the atmosphere is a subject requiring further study. Figure 7.3 shows a schematic illustration of the climatic effects of chemically and radiatively active trace gases.

7.1.3   *Aerosols, radiation, and climate*

Aerosol particles range in size from $\sim 10^{-3}$ to $\sim 20\,\mu$m. Volcanic dust, smoke from forest fires, wind-blown dust, particles from sea spray, and the small particles produced by the chemical reactions of natural gases are natural aerosols. Primary man-made aerosols include particles directly emitted during combustion processes and particles formed from the gases emitted during combustion. Aerosol concentrations vary with locality, and generally decrease rapidly with height in the troposphere. The largest concentration occurs in the urban and desert areas. Some aerosols are effective condensation and ice nuclei upon which cloud particles may form. For the hygroscopic type, the sizes of the aerosol particles are affected by relative humidity. As discussed in Subsection 5.2.4.1, cloud absorption is significantly affected by soluble and insoluble aerosols within the cloud.

Aerosol particles in the atmosphere can affect the radiation balance of the earth–atmosphere system by reflecting sunlight back into space, by absorbing sunlight, and by absorbing and emitting ir radiation. A cloud of aerosol particles generally tends to warm the atmospheric layer it occupies. However, it may either warm or cool the underlying layers and the surface, depending on whether the particles absorb ir radiation more efficiently than they reflect and/or absorb radiation from the sun. The degree to which aerosols will cool the surface by blocking sunlight (albedo effect) or warm the surface by enhancing the trapping of ir radiation (greenhouse effect) depends on the size of the particles. If the average diameter of the particles is less than a typical ir wavelength on the order of $10\,\mu$m, the ir optical depth will be less than the visible optical depth. Accordingly, even fine aerosols that weakly absorb sunlight should have a significant cooling effect on the lower atmosphere and the surface. This is also true for larger particles consisting of highly absorbing materials such as soot. The relative albedo versus greenhouse effect also depends on the thickness and density of the aerosol. Because the solar flux decreases exponentially with height when a large amount of aerosols is present, the dominant climatic effect tends to be strong surface cooling.

Volcanic particles are normally less than $10\,\mu$m and have weak absorption properties at visible wavelengths. Hence, their presence would cause cooling in the lower atmosphere and at the surface. The largest volcanic clouds may disturb the earth's radiation balance enough to cause anomalous weather. Much more significant climatic disturbances could result from the huge clouds of dust that would be thrown into the atmosphere by the impact of an asteroid or a comet with a diameter of several kilometers or more. The radiative effects of aerosols on the temperature of a planet depend not only on their optical depth, visible absorptance, and average size, but also on the variation of these properties with time. The response of surface temperature anomalies would be a function of the length of time that the aerosols could be sustained in the atmosphere.

The possible response of the global mean temperature field to an increase in aerosol concentrations is far more difficult to estimate than changes in CO_2 concentration and the solar constant. This greater difficulty has three sources. First, the optical properties of various aerosol species in terms of the real and imaginary parts of the refractive index, with respect to wavelength, have not been quantified completely. Second, unlike the increase of CO_2, which appears to be uniform over the globe, increases in atmospheric aerosols are likely to be regional. Third, there are questions concerning particle shape and the particle size distribution as a function of altitude.

Despite all these uncertainties, one-dimensional radiative–convective models have been used to clarify our understanding of how the perturbation caused by a specific aerosol might influence the radiation field of the earth–atmosphere system and its temperature structure. Temperature perturbation due to an increase in stratospheric aerosols has been investigated based on one-dimensional radiative–convective models with specific emphasis on the increase of volcanic dust caused by the explosive eruption of Mount Agung in Indonesia in 1963 (Hansen et al., 1978). A slight surface cooling and a substantial stratospheric warming are shown in model results. After the Agung eruption, average tropospheric temperatures decreased by a few tenths of a degree within a time scale on the order of about 1 year. The temperature changes computed from the model appear to agree well with those observed. The overall effect of the added aerosols on the bulk of the atmosphere and the surface is cooling. This is because sulfuric acid used in the one-dimensional model is highly reflective to solar radiation and, therefore, tends to decrease the amount of solar flux absorbed by the earth–atmosphere system.

The potential global atmospheric and climatic consequences of nuclear war have also been investigated using the one-dimensional radiative–convective models that were developed to study the effects of volcanic eruption (Turco et al., 1983). Significant hemispherical attenuation of solar radiation and subfreezing land temperatures may result due to the fine dust particles produced by high-yield nuclear surface bursts and the smoke from city and forest fires ignited by airbursts. Subsequent studies based on general circulation models appear to confirm the so-called *nuclear winter* theory.

As pointed out previously, the impact of atmospheric aerosols on climate depends on their optical properties, which, in turn, are a function of aerosol size distribution and the spectral refractive indices. Charlock and Sellers (1980) have carried out a comprehensive study of the effects of aerosols on temperature perturbation using a one-dimensional radiative–convective model. Aerosol optical depth and single-scattering albedo in the visible are used as the two basic parameters in the perturbation studies. The former parameter is an indicator of the attenuation power of aerosols, while the latter represents the relative strength of scattering and absorption by aerosols. Figure 7.4 illustrates how the perturbed surface temperature computed from the one-dimensional model responds to visible optical depth and the single-scattering albedo. For aerosols with weak absorption, surface temperature decreases as the optical depth increases because of the domination of backscattering. For aerosols with strong absorption, however, warming could occur as the optical depth increases. The dashed line in Fig. 7.4 represents changes in surface temperature using the aerosol model proposed by Toon and Pollack (1976). A general survey of the problem area involving the climatic impact of aerosols has been provided by Lenoble (1984).

Perhaps the most significant role that aerosols play in climate is through their interactions with clouds. In Subsection 5.2.4, "anomalous" absorption in cloud layers has been discussed and linked to aerosols. Furthermore, some of the aerosols are effective cloud condensation nuclei (CCN), which can affect the droplet size dis-

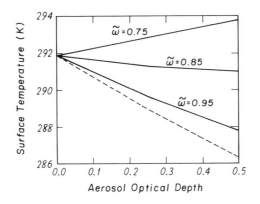

FIG. 7.4   Surface temperature change as a function of aerosol optical depth for various single-scattering albedos $\tilde{\omega}$ (after Charlock and Sellers, 1980).

tribution and, hence, the radiative properties of clouds and precipitation processes (see the next subsection for further discussion).

7.1.4   *The role of clouds in the carbon dioxide climate problem*

The importance of clouds in climate and climatic perturbations has been discussed in various sections of Chapters 4 and 6. In particular, Section 6.3 highlights cloud effects in its discussion of radiative and thermal equilibrium.

Numerical experiments using one-dimensional radiative–convective models have been carried out to aid in the understanding of the role of clouds in climate. Nonblack high cloud often produces a warming effect in the troposphere and low stratosphere due to the combined effects of the transmission of solar fluxes and emission of ir fluxes from the cloud. For middle and low clouds, a significant cooling in the atmosphere and at the surface is produced due to significant reflection by the clouds. The degree of the warming and cooling that results depends on the radiative properties of the clouds and their positions, as used in the model calculations. Since optically thin high clouds are largely composed of ice crystals and optically thick middle and low clouds primarily contain water droplets, it appears appropriate to suggest that ice clouds are greenhouse elements and that the presence of water clouds would exert a significant solar albedo effect. The effects of high, middle and low clouds on the equilibrium temperature have been shown in Fig. 6.14. This hypothetical example illustrates that clouds exert two competing effects on the radiation field of the earth–atmosphere system. On the one hand, clouds reflect a significant portion of the incoming solar flux, and on the other, clouds trap the outgoing thermal ir fluxes emitted from the atmosphere below the clouds and from the surface. The competition between the solar albedo

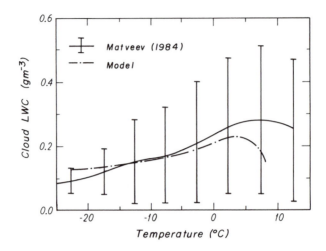

FIG. 7.5   Observed and computed LWC as a function of temperature. The observed data are based on statistical averages of the aircraft observations presented in Matveev (1984), with the bars denoting standard deviations.

and ir greenhouse effects determines whether the surface will undergo cooling or warming.

The first issue concerning the role of clouds in CO_2 greenhouse perturbation is the possible variation in cloud position and cover. If a formation of high clouds were to rise higher in the atmosphere, there would be a positive feedback because of the enhanced downward ir flux. A positive feedback would also be evident if high cloud cover increased due to greenhouse perturbations. The reverse would be true if middle and low cloud covers increased as a result of greenhouse warming.

Some evidence exists, based on cloud physics observations, that increasing temperature leads to an increase in cloud liquid water content (LWC) (Feigelson, 1981; Matveev, 1984). Shown in Fig. 7.5 is cloud LWC as a function of temperature as presented by Matveev. These values are derived from the aircraft data gathered between 1957 and 1968 at various locations in the Soviet Union. The vertical bars in the figure denote deviations from the mean. The observed mean LWCs increase with temperature but show a slight decrease with temperatures higher than about 5°C, due to the fact that higher temperatures correspond to the cloud base region, where a discontinuity in LWC occurs. Also shown in Fig. 7.5 is computed LWC from a one-dimensional cloud model (Liou and Ou, 1989). As a result of greenhouse warming, due to the anticipated increase in carbon dioxide, clouds would reflect more solar radiation because of a greater LWC, leading to a negative feedback (Charlock, 1982; Somerville and Remer, 1984). Using an interactive cloud model, which included the formation of cloud cover and liquid water, but without accounting for precipitation, Liou et al. (1985) also found that, overall, clouds exert a negative feedback to the temperature increases produced by CO_2 warming.

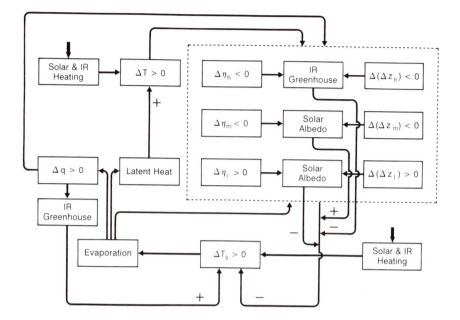

FIG. 7.6   Flow chart depicting interactions and feedbacks due to positive radiative forcing. The plus and minus signs represent positive and negative feedbacks (after Liou et al., 1985).

Figure 7.6 is a flow diagram illustrating various possible interactions and feedbacks due to positive radiative forcings, such as a doubling of CO_2 and a 2% increase in the solar constant. The increase in surface temperature caused by positive radiative forcing would lead to an increase in surface evaporation. This, in turn, would generate additional water vapor fluxes from the surface to the atmosphere. The subsequent latent heat release would be a positive feedback for atmospheric heating. Additional water vapor would also enhance the ir greenhouse effect by producing larger downward ir fluxes, leading to a further increase in surface temperature. This process is also a positive feedback. At the same time, more clouds may form in the low level, while the probability of the formation of middle and high clouds is reduced. The solar albedo effect would prevail in the case involving an increase in low cloud cover. The reduction of high cloud cover would also decrease the ir greenhouse effect. This, in effect, would enhance the solar albedo effect. Although a decrease in middle cloud cover would reduce the solar albedo effect, the net result due to all clouds, as illustrated in Fig. 7.6, is a negative feedback.

Cloud cover and LWC are related to precipitation. In Subsection 4.5.2, we show that precipitation is produced by the autoconversion of cloud droplets in which the time constant is proportional to the mean droplet radius to the fourth power. Perturbations to cloud microphysics, particularly the droplet size distribution, may affect the atmospheric hydrological cycle and, hence, the climate by inducing changes in cloud LWC and precipitation. Variations in cloud LWC would affect the planetary radiation balance, and variations in precipitation would affect the washout of CCN, producing a return feedback loop to cloud microphysics. Liou and Ou (1989) have investigated the potential link between cloud microphysical processes and climate. The cloud program includes temperature-dependent parameterization equations for condensation, evaporation, and precipitation, as presented in Section 4.5. Using the doubling of CO_2 as the radiative forcing, the importance of the cloud particle radius in relation to precipitation in climate has been investigated. A schematic illustration of the interactions and feedbacks involving cloud particle size, LWC, precipitation, and radiation is depicted in Fig. 7.7. Two theories have been proposed. First, if the perturbed mean radii are less than the climatological mean value, precipitation decreases, leading to increases in LWC. Thus the solar albedo effect outweighs the ir greenhouse effect, and the temperature perturbation due to a positive radiative forcing is stabilized. Second, if the perturbed mean radii are larger than the climatological mean value, precipitation increases, leading to decreases in the cloud LWC. Thus, the ir greenhouse effect outweighs the solar albedo effect, and the temperature perturbation due to a positive radiative forcing is amplified.

The perturbations of climatological particle mean radii due to temperature feedbacks have the following two possibilities. (1) Theory and experiments indicate that the cloud droplet concentration is approximately proportional to CCN, and that the size of the droplets decreases as the CCN in water clouds increase. Over the oceans the major source of CCN is non-sea-salt sulfate, which is produced from the emission of dimethylsulfide (DMS) by marine organisms. The highest rate of DMS emission to the atmosphere is associated with the warmest, most saline, and most intensely illuminated regions of the oceans (Charlson et al., 1987). An increase in the surface temperature could cause an increase in DMS emission and, hence, in CCN. Over land, numerous observations have verified that pollution increases the number of CCNs (Hobbs et al., 1974; Braham, 1974). The increase in CCN over land leads to more small cloud droplets per unit volume (Twomey et al., 1984). Such an increase is caused by anthropogenic sources and is not directly related to greenhouse temperature perturbations. The preceding discussions illustrate the possibility of the existence of small cloud droplets compared to the climatological mean in the global atmosphere. The additional CCN over the oceans as a result of greenhouse warming and over land as a result of pollution, provide a restoring mechanism (i.e., negative feedback) to the potential runaway greenhouse effect and stabilize the temperature perturbation. (2) The competing mechanism for the potential existence of larger cloud droplets compared to the climatological mean in the atmosphere is as follows: In the model simulation, due to greenhouse warming, precipitation increases. Precipitation is considered to be the primary mechanism for the removal of atmospheric aerosols, including CCN. Because of the increase

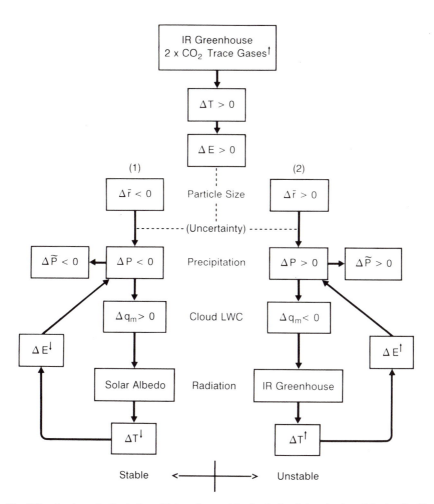

FIG. 7.7 A schematic illustration of interactions and feedbacks involving cloud particle size, liquid water content (LWC), precipitation, and radiation. T is the surface temperature, E is the surface evaporation, $\bar{r}$ is the mean radius, P is the precipitation generation rate, $\tilde{P}$ is the precipitation flux, and q_m is LWC. The Δ denotes the change due to climatic perturbations; the arrows indicate that the perturbation parameters (ΔT or ΔE) are increased or decreased. It is postulated that there are uncertainties in the cloud particle size in the atmosphere. If the perturbed mean radii are increased because of greenhouse warming, then the precipitation generation rate is enhanced, resulting in a decrease in LWC. In this case, the ir greenhouse effect outweighs the solar albedo effect, leading to a positive feedback and surface temperature increase. The reverse is true if the perturbed mean radii are decreased because of greenhouse warming (after Liou and Ou, 1989).

in precipitation, the number of CCN may be reduced. Larger cloud droplets could be formed due to a low concentration of efficient CCN. This process provides a means for positive feedback involving precipitation, as well as temperature. The potential positive feedback associated with particle size in cloud–climate feedback problems has been noted by Liou and Ou (1989) and Albrecht (1989).

The preceding discussion points out the complexity of the role of clouds in climate and climatic perturbations due to external radiative forcings. The results derived from one-dimensional models shed some light on the possibility of interactions and feedbacks of the cloud height, cloud cover, cloud LWC, and cloud particle size and phase on temperature increases due to greenhouse warming. However, one-dimensional models do not take into account horizontal variations in cloud parameters. In view of the fact that the distribution of clouds is global in nature, as shown in Section 4.1, the role of clouds in climate and global change must ultimately be resolved with models that can simulate both vertical and horizontal cloud structure. (See subsection 7.4.4 for further discussion).

7.2   The role of radiation in one-dimensional energy balance climate models

Subsection 6.5.1 developed expressions for vertical and horizontal transports of energy. In reference to Eqs. (6.5.5) and (6.5.8) for the zonal mean rates of energy storage per unit area in the atmosphere and the oceans, we may relate energy storage to the local time rate of change of surface temperature, as a first approximation, in the form

$$S_A + S_O = c(\phi) \frac{\partial T}{\partial t}, \tag{7.2.1}$$

where the surface temperature is denoted by T in this section and c is referred to as the *thermal inertia coefficient*. On eliminating the net surface flux from Eqs. (6.5.5) and (6.5.8) and letting the total atmospheric and oceanic transports of energy be

$$R = F_a + F_o, \tag{7.2.2}$$

we have

$$-c(x) \frac{\partial T(t, x)}{\partial t} = -[F_s(t, x) - F_{ir}(t, x)] + R(t, x), \tag{7.2.3}$$

where $x = \sin \phi$. This is the basic equation for the one-dimensional energy balance climate model. To seek a solution for surface temperature, it is necessary to relate both the infrared and horizontal fluxes to surface temperature. We must also express the absorbed solar flux as a function of surface albedo.

We may approach the determination of equilibrium surface temperature using the steady-state condition to obtain

$$F_s(x) = F_{ir}(x) + R(x). \tag{7.2.4}$$

Table 7.1 Empirical coefficients for Eq. (7.2.5)[a]

	Northern Hemisphere	**Southern Hemisphere**
a_1 (W m^{-2})	257	262
a_2 (W m^{-2})	91	81
b_1 (W m^{-2}°C^{-1})	1.63	1.64
b_2 (W m^{-2}°C^{-1})	0.11	0.09

[a] After Cess (1976).

In an infinitesimal latitude belt under equilibrium, the absorbed solar flux must be balanced by the emitted infrared flux and the net horizontal flux divergence, including sensible and latent heat transports by atmospheric motions and sensible heat transports by ocean currents (see Fig. 6.20). In the following subsection we discuss the manner in which these three variables may be parameterized in relation to surface properties.

7.2.1   Radiation and transport parameterizations

The outgoing longwave radiation fluxes (OLR) at the top of the atmosphere may be related to surface temperature and cloud cover. Using monthly mean atmospheric temperature and humidity profiles, and cloud cover observed at 260 stations, Budyko (1969) has calculated the OLR and found that it correlates well with surface temperature and cloud cover η, in the linear form

$$F_{ir}(x) = a_1 + b_1 T(x) - [a_2 + b_2 T(x)]\eta, \qquad (7.2.5)$$

where a_1, b_1, a_2, and b_2 are empirical constants based on statistical fittings. Cess (1976) has checked the validity of this empirical equation by using the climatological records of zonal surface temperature, cloud cover, and satellite-observed OLR. Cess found that this equation provided an excellent fit to the dataset used in the study. Table 7.1 lists the empirical coefficients in Eq. (7.2.5) for both the Northern and Southern Hemispheres. Remarkably, these empirical constants exhibit only minor differences from one hemisphere to the other. If cloud cover is taken to be constant (~ 0.5), Eq. (7.2.4) then becomes

$$F_{ir}(x) = a + bT(x), \qquad (7.2.6)$$

where $a = a_1 - a_2\eta$, $b = b_1 - b_2\eta$, and T is in °C. The approximation for the linear relation between OLR and surface temperature may be argued from the fact that the temperature profiles have more or less the same shape at all latitudes, and that the OLR, which depend on temperatures at all levels, may be expressed as a function of the surface temperature. Based on Eq. (7.2.6), it is clear that the variation in OLR with latitude depends only on b. This equation is the most straightforward relation for climate perturbation studies using one-dimensional models.

Next, the absorbed solar flux of the earth–atmosphere system is related to insolation and planetary albedo r in the form

$$F_s(x) = Qs(x)[1 - r(x)], \tag{7.2.7}$$

where $Q = S/4$, S is the solar constant, and $s(x)$ is the normalized mean annual distribution of insolation at each latitude defined by

$$\int_0^1 s(x)\,dx = 1.$$

Table 7.2 lists the normalized annual mean insolation as a function of latitude. A simple parameterization for the planetary albedo may be developed by accounting for ice-covered and ice-free surfaces. The mean annual isotherm of about $-10°$ C represents the boundary of permanent ice cover for the Northern Hemisphere. Poleward of this isotherm, the surface may be assumed to be covered by ice, whereas equatorward it is ice-free. The planetary albedos for ice-covered and ice-free cases may be assigned the following values:

$$r(x, x_s) = \begin{cases} r_1, & x > x_s \\ r_2, & x = x_s \\ r_3, & x < x_s, \end{cases} \tag{7.2.8}$$

where $x_s = \sin \phi_s$ and ϕ_s is the ice-line latitude. For the present climate conditions, $x_s = 0.95, r_1 = 0.62, r_2 = 0.5$, and $r_3 = 0.32$ (Budyko, 1969). In this manner, the planetary albedo can be varied according to surface temperature.

An alternate form for the albedo parameterization has also been suggested, in which the latitude dependence is accounted for by (Sellers, 1969)

$$r(x) = \begin{cases} d(x) + eT(x), & T < T_r \\ d(x) - 2.55, & T \geq T_r, \end{cases} \tag{7.2.9}$$

where T_r is set at 283.16 K, $e = -0.009 \text{ K}^{-1}$, and $d(x)$ is the value that would match the observed latitudinal albedos. Values for $d(x)$ are listed in Table 7.3. Because the albedos of clouds and surfaces are highly dependent on the position of the sun, the planetary albedos of ice-free regions would be high at high latitudes and low at low latitudes. For this reason, albedo changes due to surface changes from ice-free to ice-covered conditions would be large at low latitudes but small at high latitudes.

To determine the horizontal heat flux, a simple empirical equation has been derived by comparing the observed mean latitudinal values of $R(x)$ with the differences between the mean annual surface temperature at a given latitude and the mean global surface temperature $\bar{T}$ in the following form (Budyko, 1969):

$$R(x) = c[T(x) - \bar{T}], \tag{7.2.10}$$

Table 7.2 Normalized annual mean insolation[a]

ϕ (deg)	x	$s(x)$
0	0.000	1.224
5	0.087	1.219
10	0.174	1.214
15	0.259	1.189
20	0.342	1.160
25	0.423	1.120
30	0.500	1.075
35	0.574	1.021
40	0.643	0.961
45	0.707	0.892
50	0.766	0.834
55	0.819	0.770
60	0.866	0.694
65	0.906	0.624
70	0.940	0.565
75	0.966	0.531
80	0.985	0.510
85	0.996	0.500
90	1.000	0.496

[a] After Chylek and Coakley (1975).

Table 7.3 Albedo coefficient d in Eq. (7.2.9).[a]

$\Delta\phi(°N)$	d	$\Delta\phi(°S)$	d
80–90	2.924	80–90	2.900
70–80	2.927	70–80	2.992
60–70	2.878	60–70	2.989
50–60	2.891	50–60	2.937
40–50	2.908	40–50	2.922
30–40	2.870	30–40	2.865
20–30	2.826	20–30	2.815
10–20	2.809	10–20	2.798
0–10	2.808	0–10	2.801

[a] After Sellers (1969).

where c is an empirical constant based on statistical fittings. Here, the horizontal heat transport is expressed in terms of a linear function of the surface temperature.

The horizontal transports of sensible and latent heat fluxes by atmospheric motions and the sensible heat flux by ocean currents have been described in Subsection 6.5.1. The horizontal fluxes may be expressed as functions of a linear function of surface temperature differences. This characterization is especially valid for sensible heat fluxes. Therefore, the flux divergence may be related to a second-order equation in ΔT. The second-order equation in ΔT for the horizontal heat transport is, in essence, the diffusion approximation. A thermal diffusion form $-\nabla(K'\nabla T)$ may be used for the horizontal heat transport, where K' denotes an exchange coefficient that is to be determined by fitting model results with present climate values. All the transport processes are parameterized within a single coefficient. This is similar to an eddy diffusion approach to dispersion by macroturbulence in the entire geo-fluid system. Using the spherical coordinates for the Laplace operator and noting that the surface temperature varies only with latitude ϕ, we find

$$-\nabla(K'\nabla T) = \frac{-1}{a_e^2 \sin\theta} \frac{d}{d\theta} K' \left(\sin\theta \frac{dT}{d\theta} \right) = -\frac{d}{dx} K(1-x^2) \frac{dT(x)}{dx}, \quad (7.2.11)$$

where the co-latitude $\theta = \pi/2 - \phi$, $K = K'/a_e^2$, and a_e is the earth's radius. The diffusion approximation for the transport of sensible and latent heat fluxes may work well for the largest temporal and spatial scales, but it is not valid for smaller scales (Lorenz, 1979).

7.2.2 Theory of the ice-covered earth: solar constant perturbation

Paleoclimatic studies have shown that the ice cover over the oceans varied significantly in the past. It is believed that the ice cover over the oceans repeatedly advanced to the middle latitudes during the Quaternary period, which corresponds to the periods when large-scale continental glaciation developed. Variations in the ice-cover position in the Arctic during the last millennium have also been noted from paleogeographical data and historical sources. Arctic polar ice represents an important component in the earth's climate. Perturbations in the solar constant (see Subsection 6.1.2) have been linked to Quaternary glaciation. The Budyko–Sellers-type model (Budyko, 1969; Sellers, 1969) has been widely used to investigate the ice-cover variation based on perturbations in the solar constant. In the following we introduce this model and discuss the critical factor that can amplify radiative perturbations, namely, the ice–albedo feedback.

On the basis of the parameterizations described in Eqs. (7.2.5), (7.2.6), and (7.2.9), and on substituting these equations into Eq. (7.2.3), we obtain

$$Qs(x)[1 - r(x)] = [a + bT(x)] + c[T(x) - \bar{T}]. \quad (7.2.12)$$

Integration over x from 0 to 1 gives

$$Q(1 - \bar{r}) = a + b\bar{T}, \quad (7.2.13)$$

where the global surface temperature and global albedo, assuming hemispheric symmetry, are defined, respectively, by

$$\bar{T} = \int_0^1 T(x)\,dx, \qquad \bar{r} = \int_0^1 s(x)r(x)\,dx. \quad (7.2.14)$$

From Eq. (7.2.12), the solution for the surface temperature is given by

$$T(x) = \frac{Qs(x)[1 - r(x)] - a + c\bar{T}}{c + b},\tag{7.2.15}$$

where $\bar{T}$ is defined in Eq. (7.2.14). Once $s(x)$ and $r(x)$ are known, the surface temperature can be evaluated as a function of latitude.

To study the model sensitivity to the perturbation of the solar constant, a linear analysis of the global surface temperature, solar constant, and global albedo may be carried out by setting $\bar{T} = \bar{T}_0 + \Delta\bar{T}, Q = Q_0 + \Delta Q$, and $\bar{r} = \bar{r}_0 + \Delta\bar{r}$, where $\bar{T}_0, Q_0$, and $\bar{r}_0$ denote the mean values for the present climate condition. It follows that

$$\begin{aligned}
T(x) = \frac{1}{c + b} &\left[Q_0 s(x)[1 - r(x)]\left(1 + \frac{\Delta Q}{Q_0}\right) - a + c\bar{T}_0 \right.\\
&\left. + \frac{cQ_0}{b}\left(\frac{\Delta Q}{Q_0}(1 - \bar{r}_0 - \Delta\bar{r}) - \Delta\bar{r}\right) \right].
\end{aligned}\tag{7.2.16}$$

In this equation, we must have an expression for $\Delta\bar{r}$ in order to close the model. From Eq. (7.2.8), the difference between the planetary albedos for ice-covered and ice-free areas is 0.3. $\Delta\bar{r}$ must be proportional to this value, the additional ice-covered (or ice-free) area, and the mean solar flux in the zone of change in the ice-covered area. Thus, $\Delta\bar{r}$ may be parameterized in the form

$$\Delta\bar{r} = 0.3\,\ell\,(x, \bar{x}_s)\,s^*(x_s),\tag{7.2.17}$$

where $\bar{x}_s$ is the sine of the mean ice-line position for the present climate conditions; ℓ is the ratio of change in the ice-covered area to the total area of the Northern Hemisphere, which is equal to $(\bar{x}_s - x_s)$; and s^* denotes the mean insolation in the zone of change in the ice-covered area with respect to the mean insolation for the entire hemisphere, that is,

$$s^*(x_s) = \int_{x_s}^{\bar{x}_s} s(x)\,dx.$$

Equation (7.2.17) is critical in the sensitivity study, because it allows for variation in the ice albedo.

Using Eq. (7.2.16), calculations can be performed to investigate the position of the ice line for different values of $\Delta Q/Q_0$. In these calculations, the existing mean value for the solar constant and the global albedo are taken to be 1353 W m^{-2} and 0.33, respectively. The ice-line temperature, $T(x_s)$, is taken to be $-10°$ C based on climatological data. The step function defined in Eq. (7.2.8) is used to determine the planetary albedo. With a 1% change in the incoming solar flux, the global surface temperature is reduced by about $5°$ C. Further, a 1.5% decrease in the

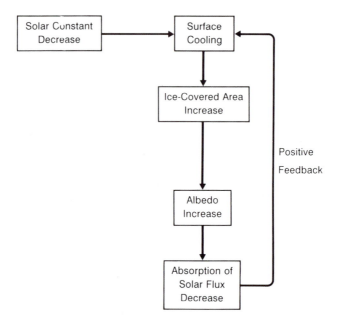

FIG. 7.8   An illustration of the ice–albedo feedback due to the radiative perturbations of the solar constant.

incoming solar flux decreases the global surface temperature by about $9°$ C. The response to these decreases in temperature is a southward advance of glaciation by $\sim 18°$ of latitude. This advance approximately corresponds to that of Quaternary glaciation. When $\Delta Q/Q_0$ is reduced by about 1.6%, the ice line reaches about $50°$ N. At this latitude, the global surface temperature decreases to several tens of degrees below zero. As a result, the ice sheet continues to advance southward to the equator with no further reduction in the solar constant required. This process describes Budyko's theory of the runaway ice-covered earth, which is based on the ice–albedo feedback illustrated in Fig. 7.8. It should be noted that if a constant global mean albedo (i.e., $\Delta \bar{r} = 0$) is assumed, a 1% decrease in $\Delta Q/Q_0$ lowers the global surface temperature by only ~ 1.2–$1.5°$C.

The preceding model and sensitivity analyses are based on the primitive assumption that the large-scale transport of sensible and latent heat fluxes may be parameterized by the difference between the surface temperature and its global mean. It is important to find out whether a different assumption concerning horizontal heat transports would alter the basic argument of the runaway ice-covered earth theory. A more physically based parameterization for large-scale transports would utilize the diffusion approximation described in Eq. (7.2.11). From Eq. (7.2.3) we have

$$\frac{d}{dx}K(1-x^2)\frac{dT(x)}{dx} = a + bT(x) - Qs(x)\left[1 - r(x, x_s)\right]. \tag{7.2.18}$$

This is a second-order differential equation and, when K is independent of x, analytical solutions may be derived subject to the boundary conditions imposed. For a mean annual model with symmetrical hemispheres, the boundary conditions must require that there be no heat flux transports at the poles or across the equator, so that $\nabla T(x) = 0$ when $x = 1$ and 0. In the one-dimensional case, we must have

$$-K(1 - x^2)^{1/2} \frac{dT(x)}{dx}\bigg|_{x=0,1} = 0. \tag{7.2.19}$$

Because the Legendre polynomials $P_n(x)$ are eigenfunctions of the diffusion operator, we have

$$-\frac{d}{dx}(1 - x^2) \frac{dP_n(x)}{dx} = n(n + 1)\, P_n(x). \tag{7.2.20}$$

Moreover, $P_n(x)$ values also satisfy the boundary conditions denoted in Eq. (7.2.19) when n is even. Thus we may expand the surface temperature in the form

$$T(x) = \sum_{n=\text{even}} T_n\, P_n(x). \tag{7.2.21}$$

We note the following orthogonal property of the Legendre polynomials for an even index:

$$\int_0^1 P_n(x)P_k(x)\, dx = \begin{cases} 1/(2n + 1), & k = n, n = \text{even} \\ 0, & \text{otherwise.} \end{cases} \tag{7.2.22}$$

Also, we define

$$H_n(x_s) = (2n + 1) \int_0^1 s(x)[1 - r(x, x_s)]P_n(x)\, dx. \tag{7.2.23}$$

This term can be evaluated from the known values for $s(x)$ and $r(x, x_s)$. It follows that the solution for T_n is given by

$$T_n = \frac{QH_n}{n(n + 1)K + b} - \frac{a}{b}\delta_{n,0}, \qquad n = 0, 2, \ldots, \tag{7.2.24}$$

where $\delta_{n,0} = 1$ when $n = 0$. Otherwise, $\delta_{n,0} = 0$. For the mean global mode, we have

$$T_0 = \frac{Q(1 - \bar{r}) - a}{b}, \tag{7.2.25}$$

which is the same as the global surface temperature given in Eq. (7.2.13). Finally, the diffusion transport coefficient K must be determined. This can be done empirically by varying K in Eq. (7.2.23) until the present climate condition is satisfied. Substituting Eq. (7.2.24) into Eq. (7.2.21) yields

$$T(x) = \sum_{n=\text{even}} \left(\frac{QH_n(x_s)}{n(n + 1)K + b} - \frac{a}{b}\delta_{n,0} \right) P_n(x). \tag{7.2.26}$$

Diffusivity can be determined by setting $x = \bar{x}_s$ and $T(\bar{x}_s) = -10°$ C. From Eq. (7.2.26), the solution for the solar constant as a function of the ice-line position may be expressed by

$$Q(x_s) = [a + bT(x_s)] \left(b \sum_{n=\text{even}} \frac{H_n(x_s)P_n(x_s)}{n(n + 1)K + b} \right)^{-1}. \tag{7.2.27}$$

The results derived from the preceding diffusion model show the multiple-branch nature of the solution. A southward advance of glaciation due to a decrease in the solar constant by about 8 to 9% is indicated (North, 1975). After the ice line reaches $\sim$45–50° N, its southward advance continues even though the incoming solar flux increases. This result is basically in agreement with Budyko's theory, although the decrease in the solar constant required to cause an ice-covered earth is substantially greater. Diffusion models, with nonlinear coefficients such that K is proportional to dT/dx, produce virtually the same results and conclusions (Held and Suarez, 1974; North et al., 1981).

Lindzen and Farrell (1977) have argued that the simple climate models described here are not in reasonable agreement with the nearly isothermal surface temperatures observed within 30° latitude of the equator, and that substantial reductions in the solar constant are necessary for an ice-covered earth. It is important to introduce transports by Hadley cell circulation into simple climate models. Lindzen and Farrell assumed that there is a heat flux that drops to zero in latitudes higher than about $\sim$25° N. The term $Qs(x)[1 - r(x, x_s)]$ in the Budyko model is replaced by its average over the region south of this latitude. Using this adjustment, the reduction in the solar constant must be on the order of about 15–20% before glaciation will advance continuously southward.

7.2.3   Orbital theory of climate change: solar insolation perturbation

The cause of fluctuations in the ice sheets of the Pleistocene period has been a topic of scientific debate and speculation. The Pleistocene period, about 600,000 years ago, is the first epoch of the Quaternary period in the Cenozoic Era, characterized by the spreading and recession of continental ice sheets and by the appearance of the modern human. A number of external factors have been hypothesized as the major causes of the earth's climatic variations during this period, including variations in the output of the sun, seasonal and latitudinal distributions of incoming solar radiation due to the earth's orbital changes, the volcanic dust content of the atmosphere, and the distribution of carbon dioxide between the atmosphere and the oceans. The orbital theory of climate change proposed by Milankovitch (1941) has gained increasing scientific support in recent years. In Subsection 6.1.1, we have detailed the variation in solar insolation as a function of the orbital parameters: eccentricity, oblique angle and the longitude of the perihelion (precession).

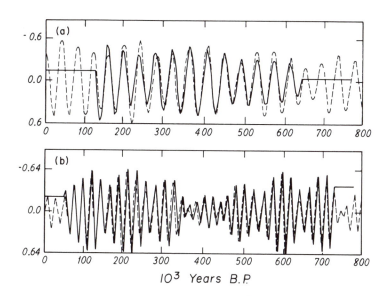

FIG. 7.9   Variations in obliquity, precession, and the corresponding frequency components of $\delta^{18}O$ over the past 780,000 years. Solid lines are filtered versions of the $\delta^{18}O$ record. Dashed lines are phase-shifted versions of (a) obliquity and (b) precession curves. All curves have been transformed to have zero means with arbitrary scales (after Imbrie et al., 1984).

Hays et al. (1976) have reconstructed the climatic record for the earth up to ~500,000 years ago, based on measurements of the oxygen isotopic composition of planktonic foraminifera from deep-sea sediment cores taken in the southern Indian Ocean. Summer sea-surface temperatures at the core site have been estimated based on statistical analyses of radiolarian assemblages. Spectral analyses of paleoclimatic time series indicate significant peaks in the frequencies at which the earth's orbital parameters are known to vary. These observations are of fundamental importance to our understanding of climatic sensitivity. If the climatic response to perturbations in the orbital parameters is indeed observable in paleoclimatic time series, we have a unique record of responses to known changes in external parameters with which to test theories of climatic sensitivity. Hays et al. showed that the principal periods of climatic variation (~100,000, 42,000, and 23,000 years) correspond to periods of orbital changes in eccentricity (~105,000 years), obliquity (~41,000) and precession (~23,000 and 19,000 years).

It has been generally accepted that down-core variations in $\delta^{18}O$ reflect changes in oceanic isotopic composition, and that these changes are caused primarily by the waxing and waning of the great Pleistocene ice sheets. Measurements of the ratio of ^{18}O and ^{16}O have been reported with respect to an international standard as $\delta^{18}O$ in parts per thousand. Imbrie et al. (1984) have used observations of $\delta^{18}O$ in five deep-sea cores to develop a geological time scale for the past 780,000 years and to evaluate the orbital theory of the Pleistocene ice ages. Figure 7.9

shows the variations in obliquity and precession, and the corresponding frequency components of $\delta^{18}O$ over the past 780,000 years. The orbital and isotopic signals are strikingly coherent in the 41,000 and 22,000 year components. Imbrie et al. concluded that variations in the geometry of the earth's orbit around the sun are the prime cause of the succession of the late Pleistocene ice ages.

Encouraged by the evidence of climate variations extracted from observations of $\delta^{18}O$, a number of scientists have employed one-dimensional energy balance models in an attempt to explain glacial cycles in terms of orbital changes. Based on Eq. (6.1.20) for mean annual global insolation, the fractional changes are approximately given by $\Delta e^2/2$. Since the eccentricity is less than about 0.07, the resulting changes in the solar constant are less than $\sim$0.2%. A 0.2% change in the solar constant corresponds to a change of about 0.4° C in the global mean surface temperature, based on a one-dimensional energy balance model (North et al., 1981). This change in surface temperature is an order of magnitude smaller than the changes reported by Hays et al.

It has been speculated that the glacial maximum of 18,000 years ago was due to a variation in the declination of the sun from about 22.2° to its present value of 23.45°. This change resulted in less insolation at the poles and more at the equator. Based on the annual energy balance model, this obliquity change causes only a 1–2° equatorward shift in the ice line, which is much smaller than the results presented by the CLIMAP Project Members (1976), who suggested an ice-line shift of about 15°.

The small responses of the annual energy balance model to orbital parameters have led to the suggestion that seasonal variations must be incorporated in the model to produce proper orbital solar forcings. Suarez and Held (1979) have developed a seasonal energy balance model to investigate orbital forcings. While there is a qualitative similarity between this model's paleoclimatic record and the actual record for the past 150,000 years, a significant discrepancy exists between the model's present climate and the actual record. Moreover, the response of the surface temperature to orbital parameters in the seasonal energy balance model is fairly linear and does not explain the strong surface temperature variations derived from the spectral analyses of the paleoclimatic record over the last 100,000 years, as reported by Hays et al. It would seem that, for the energy balance model to be successful in reproducing the paleoclimatic record, physical factors such as oceanic heat transports, variations in cloudiness, and atmospheric transports of sensible and latent heat must be accounted for more comprehensively.

7.3 Radiation and cloud effects in two-dimensional models

7.3.1 Introduction

In one-dimensional radiative–convective models, all the atmospheric dynamic processes are lumped together in one convective parameterization. For this reason, horizontal transports of heat and momentum by atmospheric motions cannot be accounted for in these models. In one-dimensional energy balance models, on the other hand, the vertical variations in the atmospheric parameters are absent, and the dynamic effects are treated via a large-scale diffusion approximation. These models are useful in the investigation of the sensitivity of external radiative forcings to the temperature field and the associated feedbacks, as illustrated in the preceding two sections. However, if some aspects of atmospheric processes are important and necessary in conjunction with the physical understanding of dynamic and thermodynamic interactions and feedbacks, a more sophisticated model must be used. A two-dimensional model may explicitly include many atmospheric processes and, at the same time, has the advantage of computational economy, as compared with a three-dimensional model.

Based on the dynamic and thermodynamic principles presented in Section 4.4, consider the following prototype equations in the pressure–latitude domain for a zonally averaged two-dimensional model: Zonal momentum

$$\frac{\partial \bar{u}}{\partial t} + \frac{\partial(\bar{u}\bar{v}_*)}{\partial y} + \frac{\partial(\bar{u}\bar{\omega})}{\partial p} = f_*\bar{v} + \text{eddy terms.} \tag{7.3.1}$$

Meridional momentum

$$\frac{\partial \bar{v}}{\partial t} + \frac{\partial(\bar{v}\bar{v}_*)}{\partial y} + \frac{\partial(\bar{v}\bar{\omega})}{\partial p} = -f_*\bar{u} - g\frac{\partial \bar{z}}{a_e \partial \phi} + \text{eddy terms.} \tag{7.3.2}$$

Thermodynamics

$$\frac{\partial \bar{T}}{\partial t} + \frac{\partial(\bar{T}\bar{v}_*)}{\partial y} + \frac{\partial(\bar{T}\bar{\omega})}{\partial p} = \frac{R\bar{T}\bar{\omega}}{C_p p} + \frac{Q + Q_R}{C_p} + \text{eddy terms.} \tag{7.3.3}$$

Continuity

$$\frac{\partial \bar{v}_*}{\partial y} + \frac{\partial \bar{\omega}}{\partial p} = 0. \tag{7.3.4}$$

Hydrostatic balance

$$g\frac{\partial \bar{z}}{\partial p} + \frac{R\bar{T}}{p} = 0. \tag{7.3.5}$$

Moisture

$$\frac{\partial \bar{q}}{\partial t} + \frac{\partial(\bar{q}\bar{v}_*)}{\partial y} + \frac{\partial(\bar{q}\bar{\omega})}{\partial p} = -\frac{Q}{L} + \text{eddy terms.} \tag{7.3.6}$$

where $y = a_e \sin\phi, \bar{v}_* = \bar{v}\cos\phi, f_* = f + \bar{u}\tan\phi/a_e, \omega = dp/dt, a_e$ is the earth's radius, f is the Coriolis parameter, and other notations have been defined in Section 4.4. As shown in Section 4.6, the eddy terms for momentum, temperature, and moisture consist of vertical and horizontal components. The transport of sensible and latent heat by transient and stationary eddies has been discussed in Subsection 6.5.1. On the basis of Eq. (7.3.1), divergence of the horizontal eddy momentum flux, $\overline{u'v'}$, is important in terms of the transfer of zonal momentum and with respect to the driving of a meridional circulation that redistributes sensible and latent heat through Eqs. (7.3.3) and (7.3.6). In middle to high latitudes, the transport of heat by horizontal eddies, $\overline{v'T'}$, is significant. This transport is a prime factor in determining the latitudinal variation of atmospheric temperature. A good approximation for the meridional momentum in midlatitudes is the geostrophic balance in which the horizontal pressure gradient force is balanced by the Coriolis torque. To find solutions for the mean circulation, temperature, and moisture fields, parameterizations of the vertical and horizontal eddy transports are required.

In weather prediction and climate models, the vertical eddy terms are usually expressed in terms of mean quantities based on the mixing length theory. In this theory, the eddy transfer coefficients for momentum and heat are determined by empirical methods. Section 4.6 has provided an introduction to this subject. The parameterization of large-scale horizontal eddies that transport momentum and heat is a difficult subject in dynamic meteorology. It has been argued that since baroclinic waves are driven by the meridional temperature gradient, it would seem appropriate to assume that the transport of these waves by eddies may be expressed in terms of this gradient, viz.,

$$\overline{v'T'} = -K_T \frac{\partial \bar{T}}{\partial y}, \qquad (7.3.7)$$

$$\overline{u'v'} = -K_M \frac{\partial \bar{T}}{\partial y}, \qquad (7.3.8)$$

where K_T and K_M are eddy transfer coefficients for sensible heat and momentum, respectively. These coefficients may be determined from observations. A more rigorous approach to deriving parameterizations for eddy transports has been based on models of baroclinic waves (Stone, 1974). However, the forms of these parameterizations are similar to those presented in Eqs. (7.3.7) and (7.3.8). Hoskins (1983) has provided a discussion and review of the parameterization of large-scale eddy transports.

In addition to eddy transport parameterization, a proper definition is required for the diabatic heating produced by condensation and radiative transfer through the thermodynamic equations. This has been comprehensively discussed in Chapters 2, 3, and 4. Finally, the solutions for wind, temperature, and humidity fields require that initial and boundary conditions be imposed on the partial differential equations along with appropriate numerical methods.

Two-dimensional climate models may be generally grouped into two types. First are the so-called statistical-dynamic momentum models, as reviewed in Saltzman (1983). These models emphasize the effects of atmospheric perturbations on circulation patterns. Usually the radiative and cloud processes described in

Eq. (7.3.3) are highly simplified. The other kind of two-dimensional models is based on the thermodynamic energy balance principle via Eq. (7.3.3) in which the mean circulation of the atmosphere is prescribed. Li et al. (1982) have used a multilayer energy balance model to investigate the sensitivity and feedbacks to changes in the solar constant and to the doubling of CO_2 concentrations. Ou and Liou (1984) have developed a two-dimensional radiative–turbulent model to investigate the perturbation and feedbacks to changes in cirrus cloud emissivity and position. Li et al. (1987) have incorporated an ocean model in a two-dimensional energy balance model that allows for seasonal variations in order to study long-term transient climate response to increases in CO_2 concentrations. In the following subsections, two specific examples of radiation and cloud effects in two-dimensional models are described.

7.3.2   *Radiation effects on the dynamics of deserts*

During the 1970s and 1980s, drought and famine spread throughout the Sahel region at the edge of the Sahara. The human suffering resulting from the extended drought in the Sahel bears agonizing witness to the disastrous potential of climate variability. In 1972–1973, annual rainfall was about 50% of the long-term mean. Figure 7.10 shows the annual rainfall departures for four sub-Saharan zones between about 5 and 20° N from 1900 to 1980 (Nicholson, 1983). These departures were calculated for each station based on a statistical method: $\Delta R_j = \sum_i (r_{ij} - \bar{r}_i)/(\sigma_i I_i)$, where r_{ij} represents the annual total for station i and year j, $\bar{r}_i$ is the annual mean value averaged over about 50 years, σ_i is the standard deviation, and I_i is the number of stations available for the year. From this record, it is seen that the years 1950–1960 were wetter than normal, whereas the years 1970–1980 were drier than normal. According to the historical rainfall record, there have been several drought periods, followed by wet periods in the Sahel region. During the period 1820–1920 three rainfall anomalies occurred, two dry periods and one wet period, each lasting about 25 years. In the years with low rainfall, the temperature was higher at the surface (Tanaka et al., 1975). The significance of the African drought has also been described by Lamb (1982).

It has been suggested that desertification through overgrazing or excessive cultivation of marginal lands played an important role in the African drought. Charney (1975) has proposed a theory linking the creation and perpetuation of deserts and droughts to the reflection of solar radiation by the surface. The feedback relationship between sinking motion and the surface albedo was studied using a two-dimensional model. Desert areas reflect more solar radiation to space and form a radiative heat sink. Because the land surface stores little heat, the air loses heat, descends, and compresses adiabatically. This sinking motion suppresses rainfall in desert areas. Precipitation produced from the model in the high-albedo experiment is substantially less than that in the low-albedo experiment. Similar sensitivity

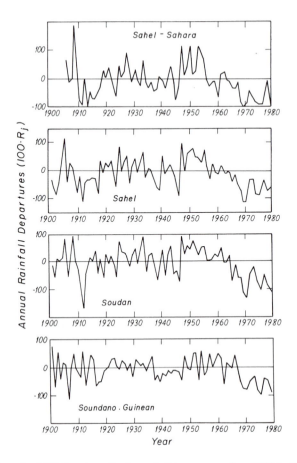

Fig. 7.10 Annual rainfall departures for four sub-Saharan zones from 1900 to 1980 in terms of $100 \, \Delta R_j$ defined in the text. These departures are approximately equivalent to a regionally averaged percentage of standard departures (after Nicholson, 1983).

results have also been obtained by Elsaesser et al. (1976) using a zonally averaged two-dimensional model. Using the time of series of precipitation data and satellite derived radiation budget over North Africa from the period 1979–1983, Smith and Sohn (1990) have shown that rainfall anomalies during that period appear to be correlated to net radiation perturbations, which are driven by the surface processes.

In Charney's original theory, temperature decreases after the surface albedo is increased. This conclusion differs from the observation that the surface temperature should be warmer in the desert than over the vegetated surface. Evapotranspiration due to vegetation must be accounted for in the model in order for the results to match the observed temperature variation. Less evaporation would lead to high surface temperatures over the desert areas. The effects of soil moisture and vegetation cover on precipitation, temperature, and wind patterns are more complicated than originally postulated by Charney, as he pointed out in a later paper (Charney et al., 1977). Soil moisture content affects atmospheric conditions by influencing not only the surface albedo of bare soil but also evaporation, and hence the energy balance, at the surface. The role of surface hydrology and energy budgets involving vegetation, moisture, and albedo can be extremely significant in the evolution of climate.

The physical processes that take place over the vegetated land include the following: First, precipitation incident on the surface is, in part, intercepted by the vegetation foliage. The intercepted precipitation either falls to the ground along with the nonintercepted precipitation or is re-evaporated by the vegetation. Part of the water incident on the ground infiltrates the soil, while the rest travels along the surface and enters streams. The water in the soil passes downward and may travel below the active surface layer of soil that is trapped by roots into groundwater reservoirs and may eventually supply stream flow. Second, plants extract water from the soil through their roots and move this water to the atmosphere by transpiration through their leaves. The amount of evapotranspiration that takes place is determined by the energy budget of a plant canopy. Third, shading of the ground by the canopy would alter the energy balance at the ground level and would, in turn, affect ground evaporation and snow melting. Dickinson (1984) has developed a comprehensive parameterization scheme that couples a vegetation canopy model and a two-layer moisture model for use in climate models.

The climate pattern in Africa is fairly zonally symmetric. Thus, a zonally averaged two-dimensional model is well suited for the study of the African climate. Xue et al. (1990) have developed a two-dimensional model that combines the atmosphere and the soil and vegetation layers to investigate the biogeophysical feedback on the African drought. The types of vegetation and the albedos at each latitude in Africa are varied according to climatological data. The albedo for soil is a function of soil moisture. The area between 17.5 and 32.5° N is designated as desert in the model, and the soil type is taken to be sandy. The ocean areas are set at north of 37.5° N. In the control runs, the general features of the July African zonal patterns, including temperature, humidity, circulation and precipitation, are simulated. To test the feedback of surface parameters to climate, the desert is extended to 10° N to ensure computational significance. As the desert expands, cloud cover and precipitation are reduced. At the same time, the temperature increases and the westerly winds become stronger. In contrast, when the desert is replaced by a vegetation layer, the opposite occurs. These results are in general agreement with the observed features of Africa's dry and wet years.

The physical processes by which the surface influences the atmosphere are schematically displayed in Fig. 7.11. When the desert expands, the surface albedo increases. This leads to a decrease in solar fluxes, which is somewhat compensated for by a smaller cloud amount in the desert area. As a result of less net radiative flux, the sum of latent and sensible heat also decreases. Because the ground has a limited heat capacity, the sum of the change in latent and sensible heat is close to the change in net radiation. Sensible heat becomes greater because the ground temperature increases due to desertification, while latent heat reduces to ensure that the sum of the two has the same sign as the net radiative fluxes. The reduction in the water source results in less evaporation. Also, the drag coefficient in the desert area decreases, which leads to a further reduction in the horizontal convergence of water vapor. The reduction in evaporation and the horizontal convergence of water vapor, coupled with the increase in atmospheric temperature, results in less relative humidity, and hence, less precipitation and cloud cover. While the oceans may also play an important role in the African drought, it seems reasonable to suggest that desertification in the Sahel area through surface radiative perturbations could be one of the major factors linked to the African drought.

7.3.3   Climatic effects of contrail cirrus

Although evidence of changes in global average cloudiness does not exist at present, there have been suggestions that localized cloudiness has increased. Machta and Carpenter (1971) have reported secular increases in the amount of high cloud cover in the absence of low or middle clouds at a number of midlatitude stations in the United States between 1948 and 1970. It has been suggested that there may be a link between this increase in cloudiness and the expansion of jet aircraft flights in the upper troposphere and lower stratosphere (*Study of Man's Impact on Climate*, 1971).

The exhaust plumes of jet aircraft, which consist primarily of water vapor, carbon dioxide, and some hydrocarbons, produce so-called *contrail cirrus*. The water vapor within jet plumes may undergo homogeneous and/or heterogeneous nucleation processes, upon which ice particles form and grow. Contrails may persist only a short time if the ambient air is very dry. In humid conditions, however, they may persist for minutes and several hours and spread into linear formations a few kilometers in width and tens of kilometers in length. Furthermore, contrails tend to cluster in groups.

A detailed analysis of cloud observations, sunshine measurements, surface temperatures, and air traffic data over the northern mid-west United States has been carried out by Changnon (1981) for the period 1901–1977. There appears to be a general downward trend in the annual number of clear days, implying that the frequency of cloud formation has been increasing. In particular, high cloudiness increased over north-central Illinois and Indiana during the period 1951–1976, which approximately corresponds to the period of rapid expansion of air traffic. Moreover, the number of months with moderated temperature, defined as the difference between the monthly average maximum temperature and minimum temperature that is less than normal difference, has been increasing. This trend is consistent with the increase in cloudiness.

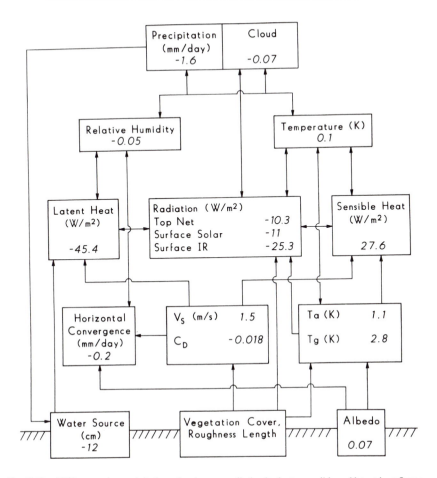

FIG. 7.11   Differences in precipitation, cloud cover, radiation budgets, sensible and latent heat fluxes, temperature, and surface parameters between the desert expanded and control runs. Interactions and feedbacks of the surface processes to the generation of moisture and radiation fields are illustrated (after Xue et al., 1990).

High cloudiness for Salt Lake City, Utah, has been analyzed by Liou et al. (1990) for the period 1949–1982 using cloud data obtained from surface observations. The monthly average high cloudiness was computed using only those cases in which the sum of the low and middle cloud amount was 0.5 tenths or less. The annual high cloudiness was then computed from the monthly average values. Figure 7.12(a) shows the mean high-cloud amounts with no low or middle intervening cloudiness for the period 1949–1982. Based on statistical analyses, the high cloud record may be considered as two separate populations corresponding to the period from 1949 to 1964 and the period from 1965 to 1982. The mean high-cloud amounts for these two time periods are 11.8% and 19.6%, respectively. Domestic jet fuel consumption from 1956 to 1982 is also shown in Fig. 7.12(a). A significant

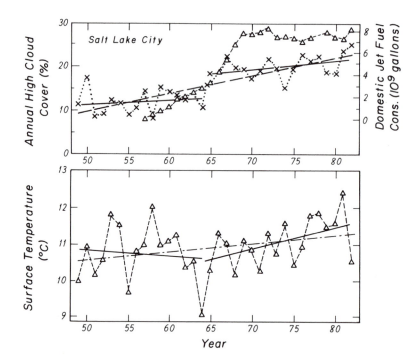

FIG. 7.12   (a) Mean annual high cloud cover in Salt Lake City from 1948 to 1982 (×) and domestic jet
fuel consumption (Δ). The two solid lines are the statistical fitting curves for high cloud cover during
the periods 1948–1964 and 1965–1982. The statistical fitting curve for the entire period is denoted by
the dashed line. (b) Mean annual surface temperature in Salt Lake City from 1948 to 1982. The two
solid lines are the statistical fitting curves for surface temperature during the periods 1948–1964 and
1965–1982. The statistical fitting curve for the entire period is denoted by the dashed line (after Liou
et al., 1990).

increase in consumption in the mid-1960s is evident. However, very little increase
is shown for the period 1969–1982. The sharp anomalous increase in annual high
cloudiness appears to coincide with the significant increase in domestic jet fuel
consumption in the mid-1960s. The sharp increase appears to suggest a possible
causative relationship between increased jet aircraft traffic and the increase in annual high-cloud amount since 1965. During the 1970–1982 period, the frequency
of occurrence of high-cloud observations decreased, suggesting that the increase
in mean annual high-cloud amount was due primarily to an increase in daily cloud
amount. Thus, it appears probable that the increase in high-cloud amount could be
linked to the increased jet aircraft traffic.

Annual surface temperature values for Salt Lake City for the period 1949–
1982 have also been calculated using the surface temperature data published by
the National Climate Center [see Fig. 7.12(b)]. The temperature data are divided
into two time periods that coincide with the time periods denoted above for high
cloud amount. The slope for the linear fit of temperature for the first time period

is negative, but the coefficient of determination is fairly small. However, for the second time period, the slope is positive, and the coefficient of determination is approximately five times greater than that of the first time period. An increase in regional surface temperatures could be caused by a number of factors, ranging from natural climatic variabilities to human influence on climate. Nevertheless, the correlation between the mean high-cloud amount and mean surface temperature for the period 1965–1982 appears to suggest that the increase in surface temperature could be related to the increase in high-cloud amount.

In Subsection 7.1.4, we point out that nonblack high cirrus clouds often produce a warming effect at the surface and in the low troposphere due to the downward emission of ir fluxes from the cloud. The effects of high clouds on equilibrium temperature have also been presented in Fig. 6.14. It is clear that the degree of warming is dependent on the radiative properties and positions of high clouds and on the feedbacks due to thermodynamic processes involving cloud formation. The composition of cirrus clouds has been discussed in Subsection 4.3.2, and some aspects of the cirrus radiative properties have been described in Subsection 5.2.4. However, the composition of contrail and invisible cirrus clouds has not been determined from in situ aircraft observations.

The heating effects of cirrus clouds on surface temperatures have been investigated by Cox (1971) and determined to be positive or negative depending on the cirrus cloud emissivity. Manabe (1975) has discussed the effect of contrails on the surface temperature and points out the importance of cirrus blackness on the temperature sensitivity experiment. Freeman and Liou (1979) have investigated the effect of increased contrail cirrus cover in midlatitudes on the radiative budget of the earth–atmosphere system. Extensive one-dimensional numerical experiments have been carried out by Liou and Gebhart (1982) to study the effects of cirrus clouds on equilibrium temperatures. Their results show that high cirrus clouds above about 8 km would produce a warming effect at the surface. The degree of warming is a function of the cirrus cloud optical depth (or emissivity). Ou and Liou (1984) have constructed a two-dimensional climate model based on the energy balance approach to study the effect of the radiative properties of cirrus clouds on global temperature perturbations. The importance of cirrus emissivity and position has been illustrated by the results derived from this two-dimensional energy balance model. The research efforts pertaining to cirrus clouds and climate have been comprehensively reviewed by Liou (1986).

The climatic effects of contrail cirrus have been specifically studied using a two-dimensional cloud–climate model, in view of the fact that the increase in contrail cirrus is primarily confined to midlatitudes (Liou et al., 1990). This model is a combination of a two-dimensional energy balance climate model and an interactive cloud formation model that generates cloud covers and LWCs based on the thermodynamic principles. The climate and cloud models are interactive through the radiation program. The broadband radiative properties of cirrus clouds are pre-

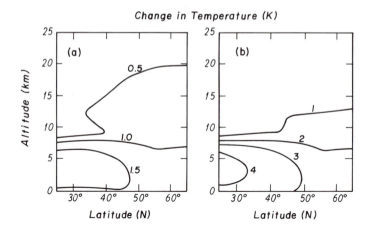

FIG. 7.13   Changes in atmospheric temperatures due to increases in high-cloud cover of (a) 5% and (b) 10% (after Liou et al., 1990).

scribed as functions of the vertically integrated IWC. Middle and low clouds are treated as blackbodies in the infrared. The cloud–climate model is first adjusted to the present climate state in terms of the temperature field, climatological cloud cover and radiation budgets at TOA. Humidity and surface albedo feedbacks are included in the model. Because of the complexity of the dynamic transport feed-back when the cloud model is interactive with the climate model, this feedback is not considered in the perturbation experiment.

The effects of contrail cirrus cover on cloud formation and temperature fields are investigated by increasing the cloud cover between 20 and 70° N, roughly corresponding to the region with the most jet aircraft traffic. A 5% increase in high-cloud cover leads to a substantial further amplification in high-cloud cover increase (∼15%) at 20–40° N, caused by the increase in specific humidity. Low and middle clouds also increase slightly because of the additional moisture supply (solar albedo effect). Overall, enhanced downward thermal ir emission from additional high clouds causes a temperature increase in the troposphere of the lower latitudes (ir greenhouse effect), as shown in Fig. 7.13. A 10% increase in high-cloud cover in the perturbation experiment shows a temperature increase of more than a factor of two in the troposphere. In the case of a 5% increase in high-cloud cover, surface temperature increases by about 1 K, but varies with latitude. When the increase in high cloud is doubled, a surface temperature increase of about 2.5 K is obtained in the experiment. In summary, all perturbation experiments involving an increase in high-cloud cover indicate increases in atmospheric and surface temperatures, caused by a positive greenhouse feedback from high-cloud cover and specific humidity.

7.4   The role of radiation and cloud processes in general circulation models

Radiative processes directly influence the dynamics and thermodynamics of the atmosphere through the generation of radiative heating/cooling rates, as well as net radiative fluxes available at the surface. Section 1.5 illustrates how the Hadley circulations in the tropics and arctic regions are maintained by differential radiative heating. Radiative flux exchanges are largely controlled by cloud fields. Through phase changes and the subsequent latent heat release, clouds also directly affect temperatures and wind fields. In this section, we shall confine our discussion to the interactions and feedbacks involving radiative transfer, cloud/radiation, and dynamic processes. The objective is to gain physical understanding of and insight into the role of radiation and clouds in general circulation models (GCMs). Since the reliable prediction of weather and climate change requires a physically based GCM, it is critically important to understand the processes involved in such models. In order to place this subject in a proper perspective with respect to radiation and clouds, we shall first present a set of equations that govern the behavior of fluids in the atmosphere.

7.4.1   Introduction to general circulation models

The equation set that constitutes a GCM contains the mathematical definitions of horizontal velocity fields, the first law of thermodynamics, the equation of continuity, the hydrostatic equation, and the equation of state. These equations have been introduced in Section 4.4. This system of equations consists of six unknowns: u, v, w, ρ, p, and T (see Subsections 4.4.1 – 4.4.6). The system is not closed because the heating and frictional terms must be determined from knowledge of other variables. An additional equation may be introduced to include moisture variation based on the conservation of water vapor in terms of specific humidity, q (Subsection 4.4.4). Thus, a GCM is represented by seven equations that are used to predict seven variables. Specific prognostic equations for the prediction of cloud and precipitation fields have not been incorporated in GCMs. Rather, cloud cover and precipitation are computed from the known variables in time and space.

To describe a GCM, the representation in the vertical direction must be introduced. There are several alternative ways of treating the vertical coordinate in numerical models. Because of the hydrostatic nature of large-scale motion, the height coordinate can be converted to a pressure coordinate to simplify the continuity equation and the density variation that occurs in other equations. However, the pressure coordinate encounters difficulties over a mountain, since a particular constant-pressure surface may intersect this mountain at certain times but not other times. For this reason, the σ-coordinate system, introduced by Phillips (1957), has been used to remove the difficulty in the representation of lower-boundary conditions. The term $\sigma = p/p_*$, where p_* is the surface pressure; $\sigma = 0$ when $p = 0$;

and $\sigma = 1$ when $p = p_*$. The surface pressure in the σ coordinate follows the terrain height at a given point in time, and hence it becomes a variable. While the σ coordinate removes the problem in the representation of the lower-boundary conditions, it generates another problem in the computation of the pressure gradient force term, which is a small quantity but splits into two terms with large values. Numerical methods have been developed to circumvent this problem by interpolating between σ and p coordinates. Since the observed data are normally available in the p coordinate, numerical interpolations are also required for the data corresponding to the initial field. For verification purposes, it is required that the predicted variables in the σ coordinate from the model be transformed to the p coordinate.

The position of the earth and the atmosphere can be defined by the spherical coordinates. In the spherical coordinate system $(\theta, \lambda, \sigma)$, where θ denotes the co-latitude ($= \pi/2$ - latitude) and λ is the longitude, we have

$$\mathbf{v} \cdot \nabla \chi = \frac{u}{a_e \sin\theta} \frac{\partial \chi}{\partial \lambda} + \frac{v}{a_e} \frac{\partial \chi}{\partial \theta} + \dot{\sigma} \frac{\partial \chi}{\partial \sigma}, \tag{7.4.1}$$

where χ can be u, v, T, q, etc., $\dot{\sigma} = d\sigma/dt$, the vertical velocity in the σ coordinate, and a_e is the radius of the earth. Thus, the equations of momentum, as defined in Eqs. (4.4.12) and (4.4.13), may be expressed by

$$\frac{\partial u}{\partial t} + \frac{u}{a_e \sin\theta} \frac{\partial u}{\partial \lambda} + \frac{v}{a_e} \frac{\partial u}{\partial \theta} + \dot{\sigma} \frac{\partial u}{\partial \sigma} - fv + \frac{uv}{a_e} \cot\theta$$
$$= -\frac{1}{a_e \sin\theta} \frac{\partial \Phi}{\partial \lambda} - \frac{RT}{a_e \sin\theta} \frac{\partial \ln p_*}{\partial \lambda} + F^u + \frac{g}{p_*} \frac{\partial \tau^u}{\partial \sigma}, \tag{7.4.2}$$

$$\frac{\partial v}{\partial t} + \frac{u}{a_e \sin\theta} \frac{\partial v}{\partial \lambda} + \frac{v}{a_e} \frac{\partial v}{\partial \theta} + \dot{\sigma} \frac{\partial v}{\partial \sigma} + fu - \frac{u^2}{a_e} \cot\theta$$
$$= -\frac{1}{a_e} \frac{\partial \Phi}{\partial \theta} - \frac{RT}{a_e} \frac{\partial \ln p_*}{\partial \theta} + F^v + \frac{g}{p_*} \frac{\partial \tau^v}{\partial \sigma}, \tag{7.4.3}$$

where $\Phi = gz$ is the geopotential height, and $F^{u,v}$ and $\tau^{u,v}$ are the horizontal and vertical eddy viscosities, respectively, in the u and v directions. The equation of continuity may be written

$$\frac{1}{a \sin\theta} \left(\frac{\partial u}{\partial \lambda} + \frac{\partial (v \sin\theta)}{\partial \theta} \right) + \frac{\partial \dot{\sigma}}{\partial \sigma} + \frac{d \ln p_*}{dt} = 0. \tag{7.4.4}$$

Using the boundary conditions that $\dot{\sigma} = 0$ at $\sigma = 0$ and 1, an equation for the solution of the surface pressure may be derived and is given by

$$\frac{\partial \ln p_*}{\partial t} = -\int_0^1 \left[\frac{1}{a_e \sin\theta} \left(\frac{\partial u}{\partial \lambda} + \frac{\partial (v \sin\theta)}{\partial \theta} \right) \right.$$
$$\left. + \frac{u}{a_e \sin\theta} \frac{\partial \ln p_*}{\partial \lambda} + \frac{v}{a_e} \frac{\partial \ln p_*}{\partial \theta} \right] d\sigma. \tag{7.4.5}$$

Combining the equation of state and the hydrostatic equation defined in Eqs. (4.4.3) and (4.4.15), we have

$$T = -\frac{\sigma}{R}\frac{\partial \Phi}{\partial \sigma}, \tag{7.4.6}$$

where R is the gas constant for air. This equation relates the geopotential height that is involved in the momentum equations to temperature. The vertical velocity in the pressure coordinate is given by

$$\omega = \sigma \dot{p}_* + p_* \dot{\sigma}. \tag{7.4.7}$$

The first law of thermodynamics in terms of temperature, defined in Eq. (4.4.47), may be expressed by

$$\frac{\partial T}{\partial t} + \frac{u}{a_e \sin\theta}\frac{\partial T}{\partial \lambda} + \frac{v}{a_e}\frac{\partial T}{\partial \theta} - \left(\frac{RT}{C_p \sigma} - \frac{\partial T}{\partial \sigma}\right)\dot{\sigma} - \frac{RT}{C_p}\frac{d\ln p_*}{dt}$$
$$= \frac{Q}{C_p} + \frac{Q_R}{C_p} + F^T + \frac{g}{p_*}\frac{\partial \tau^T}{\partial \sigma}, \tag{7.4.8}$$

where Q and Q_R represent the condensational and radiative heating rates, respectively, and F^T and τ^T denote the horizontal and vertical eddy thermal diffusion terms, respectively. Radiative heating affects wind fields through Eqs. (7.4.8) and (7.4.6). Finally, using Eq. (4.4.4), the humidity equation may be written

$$\frac{\partial q}{\partial t} + \frac{u}{a_e \sin\theta}\frac{\partial q}{\partial \lambda} + \frac{v}{a_e}\frac{\partial q}{\partial \theta} + \dot{\sigma}\frac{\partial q}{\partial \sigma} = \frac{Q}{L} + F^q + \frac{g}{P_*}\frac{\partial \tau^q}{\partial \sigma}, \tag{7.4.9}$$

where F^q and τ^q represent the horizontal and vertical water vapor diffusion terms, respectively. The condensational heating rate Q occurs in both the temperature and humidity equations. The horizontal diffusion terms must be parameterized in terms of the large-scale variables (see, e.g., Smagorinsky, 1963). Computations of Q, Q_R and $(\tau^{(u,v)}, \tau^T, \tau^q)$ in terms of resolvable parameters in atmospheric models require a comprehensive understanding and knowledge of the processes governing cloud physics, radiative transfer and planetary boundary layers.

Equations (7.4.2)–(7.4.9) represent the set of equations that constitute a GCM. Solutions for Φ, T, u, v, ω, and q as functions of time (t) and space (λ, θ, σ), based on these nonlinear partial differential equations, require appropriate numerical methods and initial conditions. The evolution of the atmospheric flow patterns may be determined by integrating the prediction equations beginning at some initial time. Because of the nature of the nonlinearity in these partial and ordinary differential equations, analytic solutions are not practical, and numerical methods for the solutions must be used. In the design of numerical methods, appropriate time and spatial differencing schemes are required to ensure computational stability. In the finite-difference approach, any dependent variable is approximated in terms of its values at discrete points in space and time. An alternate approach is to represent the flow field in space as a finite series of smooth and orthogonal functions. The prediction equations are then expressed in terms of the expansion coefficients, which are dependent on time. This method is referred to as the *spectral method*. Spectral models have emerged as an attractive means of simulating the general circulation of the atmosphere. It has been shown that the spectral method has several advantages with respect to the finite-difference method: the elimination of aliasing, the accuracy of horizontal advection computations, and the ease of global modeling in which a semi-implicit time integration scheme can be incorporated. The spectral method encounters difficulties in the computations of nonlinear terms. However, these difficulties appear to have been largely resolved through the advent of transform methods (Orszag, 1970).

An initial condition, consisting of three-dimensional distributions of dependent variables, is required to perform numerical integration of a prediction model. Initial conditions for weather prediction are based on global observations at a specific time from which interpolations of u, v, p, T, and q are performed to model grid points by means of the objective analysis. Because of the initial imbalance of pressure and wind fields, due to the presence of gravity waves, the initial data must be adjusted by numerical methods in order to eliminate meteorological noise (see a review by Kasahara, 1977). The prediction of large-scale geopotential height fields for a time period of about 5 days is largely influenced by initial conditions and the data initialization procedure. Because of the random turbulent nature of atmospheric motion, observational uncertainty in the initial state grows during model prediction. There is an inherent time limit of about a few weeks for predictability of the atmospheric state by means of a numerical integration of atmospheric models (Lorenz, 1969). This predictability limit is associated with the ability of atmospheric models to predict the day-to-day evolution of the atmosphere in a deterministic sense. It is feasible, however, for models to predict the statistical properties of the atmosphere. Models may be "tuned" to the mean or equilibrium state of the atmosphere, that is, the present climate, in terms of general circulation, temperature, radiative balance, cloud, and other pertinent patterns. Through the process of tuning, all the relevant coefficients and/or computational schemes are adjusted to present climate conditions. When climate is perturbed by external forcings, such as the increase of greenhouse gases, model coefficients (e.g., drag coefficient) and computational schemes (e.g., cloud formation) may no longer be representative of the perturbed climate. Thus it is critically important to understand the physical processes involved with the model and to incorporate physical feedbacks in the model.

Having introduced the general structure of a GCM, we shall now focus our discussion on aspects of radiation and cloud physics in the context of the model. Cloud cover and precipitation are generally computed from specific humidity, based on empirical threshold methods. Empirical equations for the evaluation of

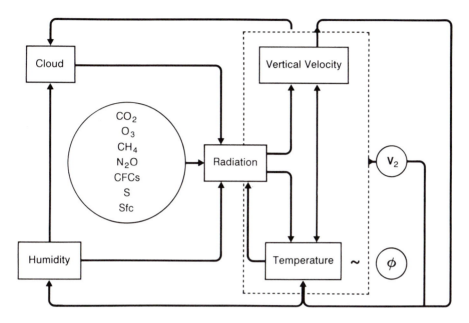

FIG. 7.14   Interactions and feedbacks involving radiative processes, clouds, temperature, and dynamic processes.

cloud cover are tuned to available climatological data. The global cloud liquid water content (LWC) data base has not been available and, therefore, is not a diagnostic variable in GCMs. Interactive radiation calculations require information concerning cloud position, cloud cover, and cloud LWC. Some measure of cloud particle size distribution is also important. Cloud position and cover are given from the model, but cloud LWC and cloud type (e.g., ice, water, or mixed clouds) must be assumed.

Figure 7.14 is a flow chart illustrating the interactions among radiative processes, vertical velocity, temperature, humidity, external perturbation parameters (CO_2, O_3, solar constant, aerosols, surfaces, etc.), clouds, and large-scale motion. These interactions can be understood from the physical equations that define dynamic and thermodynamic processes. First, radiative processes are driven by the energy emitted from the sun in terms of the solar constant and are governed by the compositions of the earth and the atmosphere, including the surface albedo, absorbing gaseous profiles, and clouds (aerosols). Second, temperature affects the radiative processes through thermal emission and the temperature dependence of absorption coefficients. Radiative flux exchanges, in turn, produce atmospheric heating/cooling that directly affects the local time rate of change of temperature and indirectly affects the vertical velocity. Finally, temperature fields and vertical velocity are linked to the large-scale horizontal winds that, in turn, affect temperature, humidity, and cloud distributions. Basically, clouds and radiation interact with dynamic processes through the temperature equation. The preceding discussion suffices to show the complexity of the feedbacks involving radiative transfer, cloud fields, and temperature distributions in atmospheric models. It is important to quantify these feedbacks in terms of time and spatial scales. The formation of clouds leads to the release of latent heat, which also has a significant effect on dynamic processes.

7.4.2   *The effect of radiative heating and cloud radiative forcing*

The possibility of improving mean zonal simulations by virtue of refinements and improvements in radiative processes within the context of a spectral GCM has been explored by Ramanathan et al. (1983). The radiative processes that are responsible for improved GCM simulations include a careful treatment of O_3 solar heating calculations in the upper atmosphere, the incorporation of temperature dependence of ir cooling by the CO_2 15 μm band, and the incorporation of the dependence of cirrus emissivity on IWC. The spectral GCM with the improved cloud/radiation model has reproduced a clear separation between the wintertime tropospheric and polar night jets, temperatures of $\sim$200 K in the winter polar stratosphere, and the asymmetry in the zonal winds in terms of interhemispheric and seasonal variations. In particular, the radiative effects of high-level model clouds are shown to be significant in the simulation of zonal wind patterns. In the numerical experiments, the ir emissivity of cirrus clouds is either assumed black or given by Eq. (5.3.24b), with a wavelength integrated absorption coefficient of $0.1 \, m^2 g^{-1}$. These are referred to as the variable nonblack and black cirrus experiments. Figure 7.15 shows the zonal mean winds produced from these two experiments. In the latter experiment both the winter and summer jets increase in strength unrealistically when compared with the results from the control run. In the former experiment, the simulated summer hemisphere jet agrees much better with the observed value. Although the control run was not performed with more realistic cloud formations in upper levels and ir emissivity coupling, the aforementioned two experiments illustrate the importance of the radiative properties of cirrus clouds in general circulation models designed for climate sensitivity experiments.

In an attempt to understand the intricate interactions of radiation, clouds, and dynamic processes, Liou and Zheng (1984) have carried out numerical experiments using a GCM that is appropriate for the performance of short- and medium-range weather prediction. They have illustrated the quantitative effects of radiative transfer and cloud–radiation interactions on temperature and cloud prediction in a 10-day prediction experiment for the Northern Hemisphere. Features of the general circulation of the model atmosphere are related to whether or not radiative transfer programs are included in the model. Figure 7.16 shows the predicted vertical velocity zonal average profiles at day five with (lower diagram) and without (upper diagram) the incorporation of radiative heating rates. In refer-

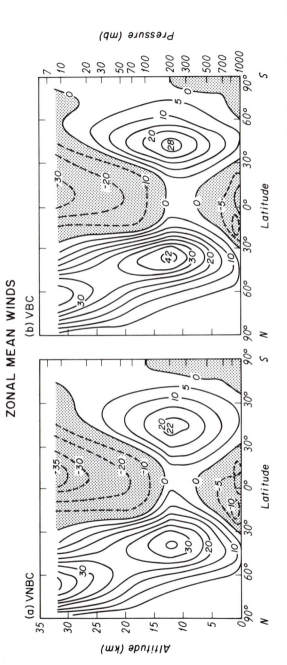

FIG. 7.15 Zonal mean winds generated from a GCM for (a) variable nonblack cirrus (VNBC) and (b) variable black cirrus (VBC) (after Ramanathan et al., 1983).

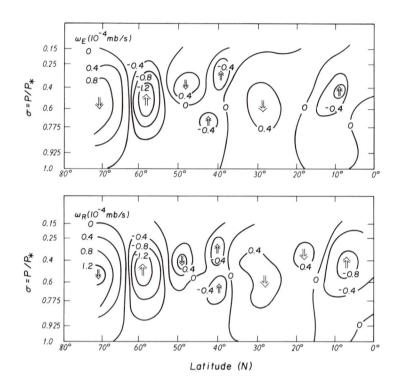

FIG. 7.16   The predicted, zonally averaged vertical velocity profiles with (lower diagram) and without (upper diagram) the inclusion of atmospheric radiative heating–cooling calculations in a GCM after 5 days (after Liou and Zheng, 1984).

ence to the lower diagram, the maximum upward motion is observed at 10° N near about 600 mb, while downward motion is seen at 30° N. This pattern constitutes the Hadley circulation in the tropical region. There is also a strong maximum upward motion at ∼60° N. On both sides of this maximum, downward motions are shown with maxima located at ∼50 and 70° N. These patterns constitute a basic meridional circulation in middle latitudes and another Hadley cell in high latitudes. When atmospheric radiative heating processes are removed in the model, there are a number of noticeable modifications to the vertical velocity patterns. The Hadley circulation in the tropical region is significantly weakened in both magnitude and extent. Also, the upward and downward motions located at ∼60 and 50° N, respectively, decrease slightly in magnitude, indicating the possibility of a reduction of intensity in the meridional circulation. The decreases in the zonal vertical velocities are due to the absence of radiative cooling rates in the model calculations, which produce significant vertical and horizontal differentials under cloudy conditions. In turn, these horizontal and vertical nonhomogeneities in radiative cooling lead to the intensification of the vertical velocity. This dynamic intensification appears to be closely related to ir radiative flux exchanges involving high clouds in the model.

As illustrated in Subsection 6.4.2, cloud radiative forcing information has been extracted from ERB data. It should be noted that such forcing does not represent an external influence that induces a response in a weather or climate system. However, clouds are highly interactive with the system because they are produced by upward motion and the hydrological cycle. Atmospheric and surface temperatures and humidity are in turn affected by the presence of clouds. The degree of cloud radiative forcing is influenced by internal interactions and feedbacks, as illustrated in Fig. 7.14. This forcing represents a measure of the instantaneous impact of clouds on radiative fluxes and heating rates. Thus, cloud radiative forcing is a useful diagnostic parameter from the standpoint of a GCM (Charlock and Ramanathan, 1985).

The foregoing discussion of cloud radiative forcing is in reference to radiative fluxes at TOA. However, it is important to recognize that in principle there are an infinite number of cloud profiles in the atmosphere that can give a consistent ERB, and that the general circulation of the atmosphere is not directly related to ERB but to the radiative heating within the atmosphere. A realistic definition and understanding of cloud radiative forcing must begin with a consideration of radiative heating fields. Analogous to the definitions of ERB cloud radiative forcing in Eqs. (6.4.3b)–(6.4.4b), cloud radiative forcing with respect to internal radiative heating may be defined as

$$\dot{C}_{ir} = \dot{T}_{ir} - \dot{T}_{ir}^{cl}, \tag{7.4.10a}$$

$$\dot{C}_s = \dot{T}_s - \dot{T}_s^{cl}, \tag{7.4.10b}$$

where $\dot{T} = \partial T/\partial t$, the radiative heating rate. In view of the discussion presented in Subsection 6.5.4, it is clear that the presence of clouds produces strong ir cooling at cloud tops and significant heating at cloud bases, and that the intensity of cooling/heating configurations is dependent on the position and optical depth of clouds.

Atmospheric cloud radiative forcing has been investigated using GCMs (Slingo and Slingo, 1988; Randall et al., 1989). Figure 7.17 shows the latitudinal distribution of zonal-mean cloud radiative forcing computed from the difference between the atmospheric (including clouds) and clear-sky heating rates for the perpetual January simulation. Zonally averaged solar radiative forcing is rather small; a $0.4 \, \mathrm{K} \, \mathrm{d}^{-1}$ warming in the upper tropical troposphere and weak negative values at low levels in the tropics and the Northern Hemisphere are shown. The latter feature is due to decreased gaseous absorption of solar flux in the shadows of upper-level clouds. On the other hand, ir radiative forcing shows pronounced features. These include (1) strong cooling associated with low-level cloud tops at low levels in the midlatitudes; (2) maximum warming of about $2 \, \mathrm{K} \, \mathrm{d}^{-1}$ at the 7 km level in the tropics; and (3) maximum cooling aloft of about $1 \, \mathrm{K} \, \mathrm{d}^{-1}$ at 16 km. The latter two features constitute a radiative dipole pattern that signifies the existence of optically thick convective anvil clouds, as noted by Ackerman et al. (1988). Anvils tend to destabilize the upper troposphere because of cloud-top cooling. Warming in the middle troposphere would tend to suppress shallow convection. The cloud-induced radiative destabilization of deep convection suggests that cloud ir radiative forcing has a positive feedback on the processes that generate high clouds. Because of the domination of ir radiative cooling, the pattern of cloud net radiative forcing resembles that of ir radiative forcing, as indicated in Fig. 7.17.

7.4.3   Sensitivity of cloud cover and radiation to climate

Cloud cover can exert a large influence on climate, as shown by the results of one-dimensional radiative–convective models. This is due to the competing effects of (1) the decrease in solar fluxes available to the earth and the atmosphere for absorption (solar albedo effect) and (2) the increased trapping of outgoing ir fluxes (greenhouse effect). Cloud cover sensitivity experiments using GCMs have been conducted by a number of researchers. Smagorinsky (1978) has suggested that the increase in downward ir fluxes due to increased CO_2 will enhance evaporation from the earth's surface; in turn, this enhanced evaporation will increase the amount of low clouds and, therefore, will exert a cooling effect on the climate. Thus, the possible warming effects of the CO_2 increase may be compensated for by this negative feedback process. Schneider et al. (1978), on the other hand, have speculated that cloud variations may have a positive feedback effect on the sensitivity of the global mean climate. Higher sea surface temperatures that result in a larger variance of the vertical velocity and greater efficiency of moisture removal through precipitation lead to lower relative humidity and less cloudiness. Wetherald and Manabe (1980) have indicated that the influence of the cloud feedback mechanism on the sensitivity of the global mean climate, due to a radiative forcing such as a change in the solar constant or CO_2 concentrations, is small because of compensation by the solar albedo and ir greenhouse effects. Hansen et al. (1984) have suggested that the interactions between cloud cover and radiative transfer markedly amplify the surface warming induced by the doubling of atmospheric CO_2. This amplification has also been noted in the experiments carried out by Washington and Meehl (1984). Despite the fact that the preceding GCMs produced similar features of cloud cover change, there are significant variations in the degree and extent of cloud cover feedback to global warming. This variability appears to suggest that the cloud feedback produced by radiative forcing in atmospheric models could well be model dependent. It is critically important to investigate the feedback processes that are allowed in a model setting.

Wetherald and Manabe (1988) have performed comprehensive experiments to study cloud cover feedback processes in a GCM. In order to facilitate the interpretation and analysis of the results from numerical experiments, a simple scheme for the computation of cloud cover was used. At each grid point, cloud was placed in the layer where the relative humidity exceeded 99%; otherwise,

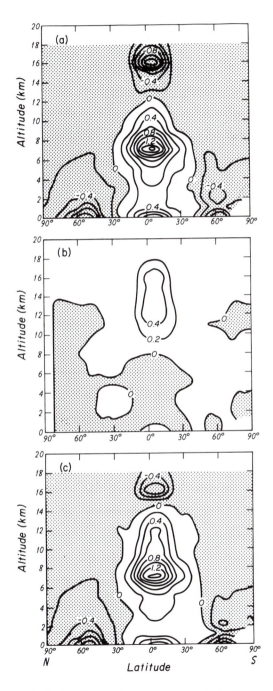

Fig. 7.17 Latitudinal distribution of the zonal mean cross section for (a) cloud longwave (ir) radiative forcing, (b) cloud shortwave (solar) radiative forcing, and (c) cloud net radiative forcing. The results are obtained from the perpetual January control simulation, and the contour interval is 0.5 K d^{-1} (after Randall et al. 1989).

no cloud was forecast. This relatively primitive scheme gives a global integral of total cloud amount approximately equal to 50%. Cloud may occur in a single (thin) layer or in multiple contiguous vertical layers (thick). The radiative properties of these clouds were prescribed. Two experiments were carried out: one prescribed the cloud cover, while the other predicted the distribution of cloud cover. The latter experiment incorporated the interaction between cloud cover and radiative transfer in the atmosphere in a simple way. This experiment showed that there is a positive feedback process in the model, enhancing the CO_2-induced warming of the model troposphere.

In response to the increase of CO_2 concentrations, cloud cover increases around the tropopause but decreases in the upper troposphere. Cloud height increases, and the outgoing ir flux from the top of the model atmosphere decreases. At most latitudes, the effect of reduced cloud cover in the upper troposphere overshadows that of increased cloud cover around the tropopause in terms of the reduction of solar albedo effects. The change in the distribution of high clouds also increases the heat gain due to the absorption of solar fluxes. The latitudinal profiles of the cloud-induced changes of outgoing ir fluxes show a positive feedback to the increase of CO_2 concentrations. The increase of the net incoming solar flux coupled with the reduction of the outgoing ir flux, due to the change in cloud cover, leads to a significant positive feedback to greenhouse warming. Wetherald and Manabe point out that the preceding results based on a seasonal model differ from their 1980 results partly because the negative feedback effect involving stratus clouds is significantly smaller than in the earlier model, which used annually averaged solar insolation.

On the basis of the preceding discussion, it is evident that the cloud feedback associated with radiative forcing depends on the performance of the model with respect to the formation of clouds, which, in turn, is a function of such factors as the vertical layer, numerical method, parameterization of moist convection and cloud formation, and the radiative properties of clouds. A cloud cover feedback study should be regarded as a study of the possible mechanisms involved in a model setting rather than a quantitative assessment of the influence of cloud cover on the sensitivity of climate. To reduce the large uncertainty in the estimate of climate sensitivity, it appears that the treatment of the cloud feedback processes in climate models should be improved. The model should be capable of reproducing the horizontal distributions of cloud cover and radiative fluxes that are available from satellite observations as discussed in Sections 4.2 and 6.4. As demonstrated by the one-dimensional models, cloud LWC and mean effective particle size, two parameters that are dependent on temperature and other relevant variables, have significant feedbacks to greenhouse perturbations. These parameters must eventually be included in GCM simulations to test their sensitivity in climate. The solution for the problem of clouds in climate depends on the success of constructing a physically based cloud formation model that can effectively couple with GCMs.

7.4.4 *Cloud formation schemes associated with general circulation models*

7.4.4.1 Cumulus convection

A number of parameterization schemes for cumulus convection have been discussed in some detail in Section 4.8. These schemes have been incorporated into GCMs by many researchers in connection with the physical improvement of GCMs. Donner et al. (1982) has studied the effect of cumulus convection on the general circulation of the atmosphere using Kuo's scheme in a GCM. Through condensation and cumulus vertical flux convergences of heat and moisture, cumulus convection affects the large-scale flow. Numerical experiments show that cumulus convection warms the upper troposphere, cools and dries the lower tropical troposphere, and weakens the Hadley circulation.

Tiedtke (1984) has undertaken an investigation on the response of large-scale atmospheric flow to cumulus convective heating, as well as cumulus cloud-radiation interactions. The convective adjustment method, Kuo's scheme, and Arakawa and Schubert's theory have been used in the large-scale simulations. A strong link is shown between tropical diabatic heat sources and planetary-scale circulations.

Kuo's and Arakawa and Schubert's penetrating convective schemes appear to be similar in their overall performance. These two schemes produce more convective heating in the higher troposphere and less heating below, compared with the results obtained from the convective adjustment scheme. The major role of cumulus convection in GCMs is to modify the diabatic heat sources. Cumulus cloud-radiation interactions produce a large sensitivity on the mean circulation directly through the solar albedo and ir greenhouse effects and indirectly via feedbacks with the large-scale flow, as illustrated in Fig. 7.14. To account for these interactions and feedbacks, it is important to relate the total convective cloud cover properly to the activity generated by the convective scheme. In the following subsections, we introduce a number of cloud formation schemes associated with GCMs.

7.4.4.2 Relative humidity scheme

Cloudiness and precipitation are generally found to occur at space-averaged relative humidities considerably less than 100%. Clouds are usually subgrid processes in large-scale models, and therefore it is necessary to consider partial cloudiness, or cloud cover, in a model grid box. Let the specific humidity in cloudy and clear regions be denoted q_c and q_0, respectively. Then the averaged specific humidity for the grid box having cloud cover η is the weighted average, given by

$$q = \eta q_c + (1 - \eta)q_0. \tag{7.4.11}$$

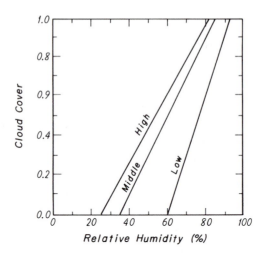

FIG. 7.18   Empirical relationships between cloud cover and relative humidity for high (300–550 mb), middle (550–800 mb), and low (800–1000 mb) clouds based on synoptic data obtained in midlatitudes (after Smagorinsky, 1960).

After rearranging terms, we have

$$\eta = (q - q_0)/(q_c - q_0) = (q/q_s - q_0/q_s)/(q_c/q_s - q_0/q_s), \qquad (7.4.12)$$

where we have introduced the saturation specific humidity q_s, which is a function of the large-scale temperature T. Since the temperatures within the cloud and in the surrounding air do not differ significantly, the specific humidity within the cloud is approximately equal to the saturation specific humidity. Moreover, if we define a threshold relative humidity for clear regions such that $h_0 = q_0/q_s(T)$, and note that the large-scale relative humidity $h = q/q_s(T)$, Eq. (7.4.12) may be written

$$\eta \cong \frac{1}{(1 - h_0)}(h - h_0). \qquad (7.4.13)$$

Having introduced the threshold relative humidity, clouds can be formed in the grid box when the large-scale relative humidity is less than 100%. The threshold relative humidity is a closure parameter for the computation of cloud cover and must be parameterized. From the observed cloud cover and relative humidity in midlatitudes, Smagorinsky (1960) has developed empirical relations for high (300–550 mb), middle (550–800 mb) and low (800–1000 mb) clouds for use in GCMs. Figure 7.18 shows the empirical relationships between cloud cover and relative humidity for the three cloud types. The relationships are apparently linear. Low clouds require the highest relative humidity, whereas a much lower relative humidity is required to form high-level clouds. These relationships imply that h_0 is a constant, according to Eq. (7.4.13).

Cloud LWC within the grid box is related to the saturation water vapor specific humidity. A simple approach to relate the two may be written as follows:

$$q_m = \gamma q_s(T). \tag{7.4.14}$$

Again, γ denotes an empirical coefficient that must be obtained by comparing the computed planetary albedo and radiation budget with observational data. Precipitation is computed based on the assumption that the condensed matter falls as soon as it has condensed and is related to condensational heating rates in a manner denoted in Eq. (4.8.22).

7.4.4.3 Parcel method

Consider a thin geometric slab in the atmosphere at level z above the surface, as shown in Fig. 7.19. This slab consists of a number of air parcels that originate independently from different depths. Assume that a parcel moves upward to a distance ξ', which equals zero at $z - \xi$. At that distance the parcel, with an infinitesimal geometric slab $d\xi'$, may undergo changes in temperature and pressure. Because the specific humidity remains constant during parcel motion, the change in relative humidity $h = q/q_s$ at position ξ' is given by

$$dh = -q \, dq_s/q_s^2. \tag{7.4.15}$$

The saturation specific humidity can be expressed in terms of the saturation vapor pressure as follows:

$$q_s = \epsilon e_s/p, \tag{7.4.16}$$

where the ratio of the molecular weight of water vapor to that of dry air, $\epsilon = 0.622$. The saturation water vapor density can be derived from the Clausius-Clapeyron equation, given in Eq. (4.5.42), in the form

$$\frac{1}{e_s} \frac{de_s}{dt} = \frac{L}{R_v T^2} \frac{dT}{dt}. \tag{7.4.17a}$$

A proper integration and empirical fitting yields

$$e_s = 6.11 \exp\left(\frac{a(T - T_0)}{T - b}\right), \tag{7.4.17b}$$

where e_s is in units of mb, $T_0 = 273.16\,\mathrm{K}$, $a = 21.874$ and $b = 7.66$ if $T \leq 273.16\,\mathrm{K}$, and $a = 17.269$ and $b = 35.860$ if $T > 273.16\,\mathrm{K}$.

Using the definition of relative humidity and Eq. (7.4.16), we find

$$dh = -h\left(\frac{d\ell n e_s}{dT} dT - \frac{dp}{p}\right). \tag{7.4.18}$$

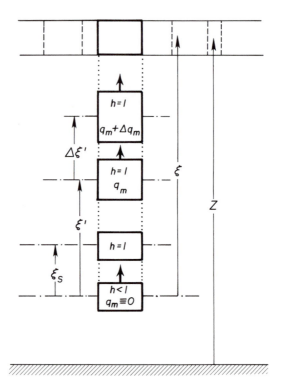

FIG. 7.19 Schematic geometry of the vertical coordinates in the parcel method for cloud formation. h is the relative humidity, q_m the LWC, ξ' the vertical displacement, and ξ_s is the position at which saturation occurs (after Sasamori, 1975).

We may assume that the motion of the parcel is governed by adiabatic and hydrostatic conditions so that

$$dT = -\gamma_d \, d\xi', \tag{7.4.19a}$$

$$dp = -\rho g \, d\xi', \tag{7.4.19b}$$

where $\gamma_d(= g/C_p)$ is the dry adiabatic lapse rate. Inserting Eqs. (7.4.19a) and (7.4.19b) into Eq. (7.4.18) yields

$$\frac{dh}{d\xi'} = \left(\frac{d\,\ell ne_s}{dT} \gamma_d - \frac{g}{RT} \right) h = fh. \tag{7.4.20}$$

Equation (7.4.20) defines the parameter f. The change in relative humidity is related to the change in temperature and the expansion of the air parcel as it moves upward.

If the parcel, which is not completely dry, becomes saturated through upward displacement on the order of about 2 km, the dependence of f on ξ' may be negligible. Thus the solution of Eq. (7.4.20) may be written

$$h(\xi') = h(0)e^{f\xi'}. \tag{7.4.21}$$

By defining the position at which saturation occurs [i.e., $h(\xi_s) = 1$], we have

$$\xi_s = -\ell n h(0)/f. \tag{7.4.22}$$

At the position where $\xi' > \xi_s$, condensation may take place within the air parcel. Subsequently, a fraction of the cloud LWC may begin to precipitate out of the parcel in the form of raindrops or snowflakes. The heat exchange between the parcel and the surrounding air is relatively small. Thus, increase in cloud LWC per unit air mass based on the parcel method may be expressed by

$$dq_m^{(1)} = C_p(\gamma_d - \gamma_s)\,d\xi'/L. \tag{7.4.23}$$

From the first law of thermodynamics discussed in Subsection 4.4.3, the saturation lapse rate is given by

$$\gamma_s = \left(\frac{1 + Lq_s}{RT}\right)\gamma_d \Big/ \left(1 + \frac{Lq_s}{C_p}\frac{d\ell n e_s}{dT}\right). \tag{7.4.24}$$

A portion of the cloud LWC may precipitate out of the parcel, representing a sink for LWC. Based on the growth of raindrops and snowflakes in clouds, the growth rate in terms of the rate of precipitation must be proportional to the cloud LWC (see Subsection 4.5.2). Moreover, if a characteristic time scale τ_p, can be defined for the conversion of liquid water to precipitation, then the decrease in cloud LWC due to precipitation is directly proportional to LWC as follows:

$$dq_m^{(2)} = -q_m\,dt/\tau_p, \tag{7.4.25}$$

where τ_p must be estimated from observational data. Combining Eqs. (7.4.23) and (7.4.24), the change in the cloud LWC with respect to height for an ascending motion may be written

$$\frac{dq_m}{d\xi'} = \frac{C_p}{L}(\gamma_d - \gamma_s) - \frac{q_m}{\tau_p w}, \tag{7.4.26}$$

where $w = d\xi'/dt$ is the vertical velocity. The vertical velocity and the saturation lapse rate do not change significantly in the course of upward displacement in the case of stratiform clouds. Using the boundary condition that $q_m = 0$ at $\xi' = \xi_s$, the solution for Eq. (7.4.26) is given by

$$q_m = \frac{C_p}{L}(\gamma_d - \gamma_s)w\tau_p\left[1 - \exp\left(\frac{(\xi_s - \xi)}{w\tau_p}\right)\right]. \tag{7.4.27}$$

We now consider a grid box that consists of many air parcels originating from different levels. Let the probability of air parcels that reach the saturation level ξ_s be $p(\xi)$. The mean cloud LWC and cloud cover may then be expressed in the forms

$$\bar{q}_m = \int_{-z_\infty}^{z} q_m(\xi)p(\xi)\,d\xi, \tag{7.4.28}$$

$$\bar{\eta} = \int_{-z_\infty}^{z} \eta(\xi)p(\xi)\,d\xi, \qquad \begin{cases} \eta = 1, & \text{if } \xi > \xi_s \\ \eta = 0, & \text{if } \xi < \xi_s. \end{cases} \tag{7.4.29}$$

At this point, a probability function must be defined. By inspecting the general circulation statistics, Sasamori (1975) has suggested the following normal distribution:

$$p(\xi) = \frac{1}{A} \exp[-(\bar{\xi} - \xi)^2/2\sigma^2], \tag{7.4.30}$$

where A is a normalization factor such that

$$\int_{-\infty}^{z} p(\xi)\,d\xi = A.$$

The mean displacement $\bar{\xi}$ and its standard deviation σ are related to the mean vertical velocity $\bar{w}$ and its standard deviation σ_w through a time constant, τ_w, in the forms

$$\bar{\xi} = \bar{w}\tau_w, \qquad \sigma^2 = \sigma_w^2 \tau_w^2, \tag{7.4.31}$$

and

$$\sigma_w^2 = \int_S (w - \bar{w})^2\,dS, \tag{7.4.32}$$

where S denotes a certain horizontal area in which an averaging procedure is performed. Given the temperature, vertical velocity, and relative humidity at the surface, $\bar{q}_m$ and $\bar{\eta}$ can be computed from Eqs. (7.4.28) and (7.4.29), where γ_s and ξ_s are given in Eqs. (7.4.24) and (7.4.22), respectively. The computations of $\bar{q}_m$ and $\bar{\eta}$ are subject to two tuning parameters: τ_w and τ_p. Finally the rate of precipitation from Eq. (7.4.25) is

$$\bar{P} = \frac{d\bar{q}_m}{dt} = \frac{\bar{q}_m}{\tau_p}. \tag{7.4.33}$$

Hense and Heise (1984) have used the parcel method to investigate the global cloud cover and radiation budget derived from a general circulation model for the Northern Hemisphere in January. The parcel scheme for cloud formation is much better than the relative humidity scheme in terms of the simulated cloud cover in middle and high latitudes because the vertical velocity field is explicitly used. More importantly, because the cloud LWC is computed in the model, radiation fields are interactive with the diagnosed cloud fields. The crucial parameter in the parcel method is the standard deviation of mean vertical velocity σ_w, which should be deduced from the meso-(subgrid)scale variation of vertical velocity.

7.4.4.4   Physical method

In this method, the governing equations for moisture fields are explicitly used in connection with a large-scale model. The specific humidity equation in the σ coordinate has been introduced in Eq. (7.4.9). Similar forms can be written for the cloud LWC and IWC equations denoted in Eqs. (4.4.61)–(4.4.64). Taking into account the bulk terminal velocity for cloud particles, the general equations for the LWC (IWC) mixing ratio may be written in the form

$$\frac{\partial q_\chi}{\partial t} = A(q_\chi) - (\dot{\sigma} + \dot{\sigma}_\chi) \frac{\partial q_\chi}{\partial \sigma} + (F^\chi + \frac{g}{p_*} \frac{\partial \tau^\chi}{\partial \sigma}) + S_\chi, \qquad (7.4.34)$$

where F^χ and τ^χ denote the horizontal and vertical eddy diffusion terms, respectively, χ can be m (LWC), or mi (IWC), and the mean horizontal advection term

$$-A(q_\chi) = \frac{u}{a_e \sin\theta} \frac{\partial q_\chi}{\partial \lambda} + \frac{v}{a_e} \frac{\partial q_\chi}{\partial \theta}. \qquad (7.4.35)$$

To solve Eq. (7.4.34), the source and sink term S_χ must be expressed in terms of the basic variables, as are the vertical and horizontal diffusion terms, and the bulk terminal velocity $\dot{\sigma}_\chi$.

The terminal velocity for a cloud droplet less than about 50 μm is on the order of $\sim 10^{-2}$ m s^{-1} [Eq. (4.5.5)]. The large-scale upward vertical velocity is also on the order of $\sim 10^{-2}$ m s^{-1}. Thus, the second term on the right-hand side of Eq. (7.4.34) may be ignored. The eddy diffusion terms involving cloud particles have not been discussed in the past and are generally disregarded in the cloud LWC calculation. In Section 4.6, eddy terms for cloud fields have been expressed in terms of the mixing length theory. Alternately, it may be appropriate to assume, as a first approximation, that the diffusion terms for cloud droplets are proportional to those for water vapor. The proportionality coefficients may be determined by the ratio of mean values or by the ratio of their gradients. The transports of large precipitation particles by eddies and the horizontal transports of these particles by mean motions should both be small, compared with the transport by terminal velocity. Moreover, because the large terminal velocity of precipitation particles transports these particles from high layers to low layers, local time rate of accumulation of precipitation may be ignored. It follows that the vertical divergence of the precipitation rate must be largely balanced by the source and sink terms.

In view of the preceding discussion, a set of equations that govern the large-scale moisture field in a grid box may be written as follows: Water vapor

$$\frac{\partial q}{\partial t} = A(q) - \dot{\sigma} \frac{\partial q}{\partial \sigma} + \text{eddy diffusion} + S_v. \qquad (7.4.36)$$

Cloud LWC

$$\frac{\partial q_m}{\partial t} \cong A(q_m) + \text{eddy diffusion} + S_m. \tag{7.4.37}$$

Cloud IWC

$$\frac{\partial q_{mi}}{\partial t} = A(q_{mi}) - (\dot{\sigma} + \dot{\sigma}_{mi})\frac{\partial q_{mi}}{\partial \sigma} + \text{eddy diffusion} + S_{mi}. \tag{7.4.38}$$

Precipitation

$$-\frac{\partial \tilde{P}}{\partial \sigma} \cong S_{mr}, \tag{7.4.39}$$

where the precipitation flux, $\tilde{P} = p_* \dot{\sigma}_{mr} q_{mr}/g$. The sources and sinks and their parameterizations have been discussed in some detail in Section 4.6.

To introduce cloud cover associated with the source and sink terms, we shall consider only the water vapor and liquid water transitions. Let Q_c denote the condensational heating rate (energy per mass per time), E_r and E_c the cooling rates due to evaporation of raindrops and cloud drops, respectively, and P the rate of rainwater production (per time). Then, from Eqs. (4.4.65) and (4.4.66), written in the height coordinate, we have $S_1/\rho = \eta Q_c/L$, $S_6/\rho = (1-\eta)E_c/L$, $S_7/\rho = (1-\eta)E_r/L$, and $S_2/\rho = \eta P$ (see Fig. 4.14). The net heating rate due to condensation and evaporation in a grid box is given by

$$Q = \eta Q_c - (1 - \eta)(E_r + E_c), \tag{7.4.40}$$

where Q_c, E_r and E_c are in-cloud values. The source and sink terms in Eqs. (7.4.36) and (7.4.37) for water vapor and cloud LWC can then be written

$$S_v = -\eta Q_c/L + (1 - \eta)(E_r + E_c)/L = -Q/L, \tag{7.4.41}$$

$$S_m = \eta Q_c/L - (1 - \eta)E_c/L - \eta P$$
$$= Q/L + (1 - \eta)E_r - \eta P. \tag{7.4.42}$$

From the standpoint of microphysical cloud processes, Q_c and E_c may be evaluated from the parameterization schemes described in Subsections 4.5.1 and 4.5.3. However, for application to large-scale processes, clouds must be allowed to form in a grid box when the model relative humidity is much less than 1, as pointed out in Subsection 7.4.4.2. It follows that alternate means are needed to obtain the net condensational heating rate. Two approaches may be undertaken to determine this heating rate.

(1). The most straightforward method is based on the thermodynamic principle that the condensational heating rate is related to the time rate of change of saturation specific humidity; that is, $Q_c = dq_s/dt$. Based on the definition of q_s

in Eq. (7.4.16), the Clausius–Clapeyron equation [Eq. (7.4.17a)], and the first law of thermodynamics [Eq. (4.4.39)] without radiative heating, we find

$$\eta Q_c \cong -\eta q_s T \frac{LR - C_p R_v T}{C_p R_v T^2 + q_s L^2} \frac{\omega}{p}, \tag{7.4.43}$$

where R and R_v are the gas constants for air and water vapor, respectively. A factor η (cloud cover or amount) is added because the condensational heating rate over a grid box is proportional to cloud cover. Strictly speaking, condensational heating is directly related to the in-cloud vertical velocity. A precise determination of this velocity would be extremely involved. As a first approximation, however, we may use the large-scale vertical velocity ω to evaluate the condensational heating rate. Clouds form in a grid box if $\omega < 0$. A fraction of this box is covered by clouds having a cloud amount η, and latent heat is released and distributed to the entire box. A criterion is required to determine cloud amount from large-scale relative humidity. An empirical determination has been illustrated in Fig. 7.19. On the other hand, if $\omega > 0$, the downward motion of cloud particles would lead to evaporation. The physical shortcoming of this approach is that it does not account for the condensational heating produced by the horizontal advection of water vapor by mean motion and eddy diffusion.

(2). Consider the thermodynamic and specific humidity equations for a grid box and rewrite Eqs. (7.4.8) and (7.4.9) in the forms

$$\frac{\partial T}{\partial t} = A_*(T) + \frac{Q}{C_p}, \tag{7.4.44}$$

$$\frac{\partial q}{\partial t} = A_*(q) - \frac{Q}{L}, \tag{7.4.45}$$

where A_* represents all terms other than the condensational heating terms in Eqs. (7.4.8) and (7.4.9). In principle, $A_*(T)$ may include the radiative heating rate. In the stratiform region, the occurrence of condensation should be associated with relative humidity. Using the definition of saturation specific humidity in Eq. (7.4.16), the Clausius–Clapeyron equation in Eq. (7.4.17a) and the first law of thermodynamics, we can obtain an independent equation for the local time rate of change of specific humidity as follows:

$$\frac{\partial q}{\partial t} = \frac{hq_s \epsilon L}{RT^2} \frac{\partial T}{\partial t} + q_s \frac{\partial h}{\partial t}. \tag{7.4.46}$$

Equations (7.4.44)–(7.4.46) can be used to eliminate the $\partial T/\partial t$ and $\partial q/\partial t$ terms resulting from the following net condensational heating expression:

$$Q = \left(M - q_s \frac{\partial h}{\partial t} \right) \bigg/ \left(1 + \frac{Lh}{C_p} \right), \tag{7.4.47}$$

where the equation

$$M + A_*(q) - \frac{hq_s \epsilon L}{RT^2} A_*(T) \tag{7.4.48}$$

represents the amount of water vapor available for condensation and moistening (first term) and the possible expansion or compression of air (second term). Note that in the pressure coordinate the term involving $\partial p / \partial t$ vanishes.

The solution for the net condensational heating in a grid box is related to the local time rate of change of relative humidity, which, from Eq. (7.4.13) is given by

$$\frac{\partial h}{\partial t} = (1 - \eta) \frac{\partial h_0}{\partial t} + (1 - h_o) \frac{\partial \eta}{\partial t}. \tag{7.4.49}$$

In this equation, two additional tendency terms involving threshold relative humidity and cloud cover are introduced. We must develop two independent equations to close the system of equations. First, let us consider the available moisture source, which includes the source M defined in Eq. (7.4.48) and the evaporation terms $E_r + E_c$. Let $M^* = M + E_r + E_c$. A portion of this source, ηM^*, would be used to enhance condensation in the cloudy area. The remaining portion, $(1 - \eta)M^*$, would enter the clear area. A part of this moisture would cause changes in the specific humidity, $(1 - \eta)\partial q_0 / \partial t$; the rest would generate changes in cloud cover, $\partial \eta / \partial t$. The increase in humidity associated with the cloud cover changes would be $(q_s + q_m / \eta - q_0)$, where q_m / η denotes the in-cloud mixing ratio for LWC. Moisture conservation requires that

$$(1 - \eta)M^* = (1 - \eta) \frac{\partial q_0}{\partial t} + (q_s - q_0 + q_m / \eta) \frac{\partial \eta}{\partial t}. \tag{7.4.50}$$

The clear-sky specific humidity is related to threshold relative humidity by $q_0 = q_s h_0$. The saturation relative humidity is a function of temperature only, and changes in it are relatively insignificant in comparison with changes in clear-sky relative humidity. It follows that

$$\frac{\partial q_0}{\partial t} \cong q_s \frac{\partial h_0}{\partial t}. \tag{7.4.51}$$

Combining Eqs. (7.4.49)–(7.4.51) yields

$$\frac{(1 - \eta)}{q_s} M^* = \frac{\partial h}{\partial t} - \frac{q_m}{\eta q_s} \frac{\partial \eta}{\partial t}. \tag{7.4.52}$$

This equation links the moisture source in the clear region to the averaged relative humidity for the grid box.

However, we must now find a specific relation between $\partial \eta / \partial t$ and $\partial h / \partial t$ in order to close the system. A simple approach is to assume that the threshold relative humidity is a constant value (say 0.8). In this case, cloud cover can be directly computed from Eq. (7.4.13) and, as a result, the relative humidity tendency can be expressed in terms of the available moisture source M^*. This idea was originally proposed by Sundqvist (1978). If condensation occurs in the grid box, it is conceivable that the clear-sky specific humidity must be related to cloud cover. We may assume that

$$h_0 = h_{00} + f(\eta), \tag{7.4.53}$$

where the closure parameter h_{00} is an absolute constant and f is a functional form involving cloud cover. Sundqvist (1988) has proposed a linear relation in the form: $f(\eta) = \eta(1 - h_{00})$. Using Eqs. (7.4.53) and (7.4.52) we find,

$$\frac{\partial h}{\partial t} = \frac{f^*}{f^* + q_m/\eta q_s} \frac{1 - \eta}{q_s} M^*, \tag{7.4.54}$$

where $f^* = (1 - \eta)f' + (1 - h_{00} - f)$, and $f' = \partial f/\partial \eta$. From Eqs. (7.4.13) and (7.4.53), cloud cover is given by

$$\eta = 1 - [(1 - h)/(1 - h_{00})]^{1/2}. \tag{7.4.55}$$

Once the large-scale relative humidity tendency and cloud cover are given, the net condensational heating given in Eq. (7.4.47) can be evaluated. Evaporation due to cloud droplets is very small, and, for all practical purposes, cooling rates produced by this evaporation may be disregarded. Parameterizations of evaporation due to raindrops and precipitation have been detailed in Subsections 4.5.2 and 4.5.3. Incorporation of cloud LWC as a prognostic variable may follow Eq. (7.4.37).

The preceding two physical approaches describe the manner in which large-scale condensation and cloud cover may be evaluated from resolvable variables. The advantage of the first is its simplicity, but it does not account for condensation due to horizontal transports of moisture. The second method is more complicated in terms of numerical computations, but it offers a systematic development in which all the relevant large-scale processes are encompassed in cloud formation. Both methods, however, require the definition of a closure parameter, namely, the threshold relative humidity.

Parameterization of stratiform cloud condensation based on the second approach with a constant threshold relative humidity has been incorporated in a GCM by Sundqvist (1981), in which cloud LWC is a prognostic variable. The ice-related transitions and radiative heating rates were not accounted for in the cloud model. Some encouraging results concerning the 5-day prediction of cloud cover and LWC have been illustrated. Zheng and Liou (1986) have used a prognostic cloud LWC equation and the first method to compute stratiform cloud condensation in connection with the investigation of the effects of mountain and cloud–radiation interactions on the prediction of cloud LWC and precipitation patterns over the Tibetan Plateau. The incorporation of cloud LWC and cloud cover in the model allows the radiative heating rates to be interactively computed. This interaction noticeably affects temperature and precipitation predictions in the vicinity of the plateau. Sundqvist et al. (1989) have implemented a stratiform cloud condensation by combining the second approach and a cloud LWC prediction equation into a fine mesh numerical weather prediction model. The results from a 36-hour integration reveal that the inclusion of cloud LWC prediction improves forecasts and that cloud cover prediction compares reasonably well with the satellite ir cloud picture.

Two critical issues remain in the physically based cloud formation scheme in GCMs: verification and coupling with radiation calculations. With respect to verification, the threshold relative humidity that is used as the closure parameter must be validated based on observations. In Fig. 7.18, cloud cover appears linearly dependent on the relative humidity, implying that the threshold relative humidity is constant. However, if this humidity is permitted to vary as a function of cloud cover, in a form denoted in Eq. (7.4.53), the relationship between cloud cover and relative humidity would be nonlinear. Determination of the closure parameter in the cloud formation scheme based on physical principles and global observations is a subject requiring extensive research efforts. Furthermore, the model-generated LWCs/IWCs must be verified against observed values. Data obtained from satellite microwave channels have been used to infer LWCs over the oceans, but the uncertainties in the retrieved LWCs are still very large. Improvements are needed. Application of satellite microwave techniques to infer LWC over land requires an innovative approach to circumvent the variation of the surface emissivity with moisture content. Since ice crystals are opaque to microwaves, the inference of IWC from space, especially for cirrus clouds, requires the development of new instrumentation and retrieval methodologies, possibly using millimeter and infrared wavelengths.

With respect to the coupling of the cloud formation and radiation calculations, not only LWC/IWC but also cloud particle size distribution must be properly taken into account in the model. The significance of particle size can be understood from the following example: For a given LWC, small cloud droplets reflect more sunlight than large cloud droplets. In the case of ice crystals, fundamental data on the scattering and absorption of nonspherical ice particles are needed in order to implement a reliable radiation parameterization scheme involving cirrus clouds. In the context of radiative parameterization, the asymmetry factor that determines the amount of forward scattering for irregular ice crystals is generally smaller than that for spherical droplets and regular ice hexagonal crystals. Modeling of the cloud particle size distribution in terms of mean effective size based on the principle of light scattering by particles has been discussed in Section 5.3. It appears to be impractical to include the activities of individual cloud particles in large-scale weather prediction and climate models. The cloud particle size distribution in terms of mean effective size would be best inferred from prognostic variables, including LWC/IWC, humidity and temperature, based on the thermodynamic principles as well as on analyses of microphysical cloud data determined from reliable aircraft observations. The incorporation of a physically based cloud prediction scheme in terms of cloud cover and LWC/IWC, and of an interactive radiation method, based on the principles of radiative transfer, in connection with GCMs for weather and climate studies will be a challenging research subject for years to come.

REFERENCES

Ackerman, T. P., K. N. Liou, F. P. J. Valero, and L. Pfister, 1988: Heating rates in tropical anvils. *J. Atmos. Sci.*, **45**, 1606–1623.

Albrecht, B. A., 1989: Aerosols, cloud microphysics, and fractional cloudiness. *Science*, **245**, 1227–1230.

Bacastow, R., and C. D. Keeling, 1973: Atmospheric carbon dioxide and radiocarbon in the natural carbon cycle. II. Changes from A.D. 1700 to 2070 as deduced from a geochemical model. In *Carbon and the Biosphere*, G. M. Woodwell and E. V. Decan, Eds., U.S. Atomic Energy Commission, CONF-720510, pp. 86–135.

Braham, R., 1974: Cloud physics of urban weather modification: A preliminary report. *Bull. Amer. Meteor. Soc.*, **55**, 100–106.

Budyko, M. I., 1969: The effect of solar radiation variations on the climate of the earth. *Tellus*, **21**, 611–619.

Cess, R. D., 1976: Climate change: An appraisal of atmospheric feedback mechanisms employing zonal climatology. *J. Atmos. Sci.*, **33**, 1831–1843.

Changnon, S. A., Jr., 1981: Midwestern cloud, sunshine and temperature trends since 1901: Possible evidence of jet contrail effects. *J. Appl. Meteor.*, **20**, 496–508.

Charlock, T. P., 1982: Cloud optical feedback and climatic stability in a radiative-convective model. *Tellus*, **34**, 245–254.

Charlock, T. P., and V. Ramanathan, 1985: The albedo field and cloud radiative forcing produced by a general circulation model with internally generated cloud optics. *J. Atmos. Sci.*, **42**, 1408–1429.

Charlock, T. P., and W. D. Sellers, 1980: Aerosol effects on climate: Calculations with time-dependent and steady-state radiative convective models. *J. Atmos. Sci.*, **37**, 1327–1341.

Charlson, R. J., J. E. Lovelock, M.O., Andreae, and S.G. Warren, 1987: Oceanic phytoplankton, atmospheric sulphur, cloud albedo and climate. *Nature*, **326**, 655–661.

Charney, J. G., 1975: Dynamics of deserts and drought in the Sahel. *Quart. J. Roy. Meteor. Soc.*, **101**, 193–202.

Charney, J. G., W. J. Quirk, S. H. Chow, and J. Kornfield, 1977: A comparative study of the effects of albedo change on drought in semi-arid regions. *J. Atmos. Sci.*, **34**, 1366–1385.

Chýlek, P., and J. A. Coakley, 1975: Analytical analysis of a Budyko-type climate model. *J. Atmos. Sci.*, **32**, 675–679.

CLIMAP Project Members, 1976: The surface of the ice-age earth. *Science*, **191**, 1131–1137.

Cox, S. K., 1971: Cirrus clouds and climate. *J. Atmos. Sci.*, **28**, 1513–1515.

Crutzen, P. J., 1983: Atmospheric interactions: Homogeneous gas reactions of C, N, and S containg compounds. In *The Major Biogeophysical Cycles and Their Interactions*, B. Bolin and R.B. Cook, Eds., John Wiley & Sons, New York, pp. 67–114.

Dickinson, R. E., 1984: Modeling evapotranspiration for three-dimensional global climate models. In *Climate Processes and Climate Sensitivity*, J. E. Hansen and T. Takahashi, Eds., *Geophys. Monogr. Ser.,* **29**, Amer. Geophys. Union, Washington, D.C., pp. 58–72.

Donner, L. J., H. L. Kuo, and E. J. Pitcher, 1982: The significance of thermodynamic forcing by cumulus convection in a general circulation model. *J. Atmos. Sci.*, **391**, 2159–2181.

Elsaesser, H. W., M. C. MacCracken, G. L. Potter, and F. M. Luther, 1976: An additional model test of positive feedback from high desert albedo. *Quart. J. Roy. Meteor. Soc.*, **102**, 655–666.

Feigelson, E. M., 1981: *Radiation in Cloudy Atmosphere.* Section of Meteorology and Atmospheric Physics of Soviet Geophysical Committees, Hydrometeorology Press (IZD), 375 pp.

Freeman, K. P., and K. N. Liou, 1979: Climate effects of cirrus clouds. *Adv. Geophys.*, **21**, 231–287.

Hansen, J., D. Johnson, A. Lacis, S. Lebedeff, P. Lee, D. Rind, and G. Russell, 1981: Climate impact of increasing atmospheric carbon dioxide. *Science*, **213**, 957–966.

Hansen, J. E., A. Lacis, D. Rind, G. Russell, P. Stone, I. Fund, R. Ruedy, and J. Lerner, 1984: Climate sensitivity: Analysis of feedback mechanisms. In *Climate Processes and Climate Sensitivities*, J.E. Hansen, and T. Takahashi, Eds., *Geophys. Monogr. Ser.,* **29**, Amer. Geophys. Union, Washingon, D.C., pp. 130–163.

Hansen, J., W. C. Wang, and A. Lacis, 1978: Mount Agung eruption provides test of a global climate perturbation. *Science*, **199**, 1065–1068.

Hays, J. D., J. Imbrie, and N. J. Shackleton, 1976: Variations in the earth's orbit: Pacemaker of the ice ages. *Science*, **194**, 1121–1132.

Held, I. M, and M. J. Suarez, 1974: Simple albedo feedback models of the icecaps. *Tellus*, **26**, 613–62.

Hense, A., and E. Heise, 1984: A sensitivity study of cloud parameterizations in general circulation models. *Contrib. Atmos. Phys.*, **57**, 240–258.

Hobbs, P. V., H. Harrison, and E. Robinson, 1974: Atmospheric effects of pollutants. *Science*, **183**, 909–915.

Hoskins, B. J., 1983: Modelling of the transient eddies and their feedback on the mean flow. In *Large-Scale Dynamic Processes in the Atmosphere*, B.J. Hoskins and R. P. Pearce, Eds., Academic Press, New York, pp. 169–199.

Imbrie, J., J. D. Hays, D. G. Martinson, A. McIntyre, A. G. Mix, J.J. Morley, N. G. Pisias, W. L. Prell, and N. J. Shackleton, 1984: The orbital theory of Pleistocene climate: Support from a revised chronology of marine $\delta^{18}O$ record. In *Milankovitch and Climate*, Part 1, A. L. Berger et al., Eds., Reidel, Boston, pp. 269–305.

Kasahara, A., 1977: Computational aspects of numerical models for weather prediction and climate simulation. In *Methods in Computational Physics*, Vol. 17, J. Chang, Ed., Academic Press, New York, pp. 1–66.

Lamb, P.J., 1982: Persistence of Subsaharan drought. *Nature,* **299**, 46–48.

Lenoble, J., 1984: A general survey of the problem of aerosol climatic impact. In *Aerosols and Their Climatic Effects*, H. E. Gerber and A. Deepak, Eds., A. Deepak Publishing, Hampton, Virginia, pp. 279–294.

Li, P., M. D. Chou, and A. Arking, 1982: Climate studies with a multi-layer energy balance model. I. Model description and sensitivity to the solar constant. *J. Atmos. Sci.*, **39**, 2639–2656.

Li, P., M. D. Chou, and A. Arking, 1987: Climate warming due to increasing atmospheric CO_2: Simulations with a multilayer coupled atmosphere-ocean seasonal energy balance model. *J. Geophys. Res.*, **92**, 5505–5521.

Lindzen, R. S., and B. Farrell, 1977: Some realistic modifications of simple climate models. *J. Atmos. Sci.*, **34**, 1487–1501.

Liou, K. N., 1986: Influence of cirrus clouds on weather and climate processes: A global perspective. *Mon. Wea. Rev.*, **114**, 1167–1199.

Liou, K. N., and K. L. Gebhart, 1982: Numerical experiments on the thermal equilibrium temperature in cirrus cloudy atmospheres. *J. Meteor. Soc. Japan*, **60**, 570–582.

Liou, K. N., and S. C. Ou, 1989: The role of cloud microphysical processes in climate: An assessment from a one-dimensional perspective. *J. Geophys. Res*, **94**, 8599–8607

Liou, K. N., and Q. Zheng, 1984: A numerical experiment on the interactions of radiation, clouds and dynamic processes in a general circulation model. *J. Atmos. Sci.*, **41**, 1513–1535.

Liou, K. N., S. C. Ou, and G. Koenig, 1990: An investigation on the climatic effect of contrail cirrus. In *Air Traffic and the Environment-Background, Tendencies and Potential Global Atmospheric Effects*, U. Schumann, Ed., Springer-Verlag, Berlin, pp. 154–169.

Liou, K. N., S. C. S. Ou, and P. J. Lu, 1985: Interactive cloud formation and climatic temperature perturbations. *J. Atmos. Sci.*, **42**, 1969–1981.

Lorenz, E. N., 1969: The predictability of a flow which possesses many scales of motion. *Tellus*, **21**, 289–307.

Lorenz, E. N., 1979: Forced and free variations of weather and climate. *J. Atmos. Sci.*, **36**, 1367–1376.

Machta, L., and T. Carpenter, 1971: Trends in high cloudiness at Denver and Salt Lake City. In *Man's Impact on the Climate*, W. H. Mathews, W. W. Kellogg, and G. D. Robinson, Eds., MIT Press, Boston, pp. 410–415.

Manabe, S., 1975: Cloudiness and the radiative convective equilibrium. In *The Changing Global Environment*, S. F. Singer, Ed., Reidel, Dordrecht, Holland, pp. 175–176.

Manabe, S., and R. T. Wetherald, 1967: Thermal equilibrium of the atmosphere with a given distribution of the relative humidity. *J. Atmos. Sci.*, **24**, 241–259.

Matveev, L. T., 1984: *Cloud Dynamics*. Kluwer Group, Dordrecht, Holland, 340 pp.

Milankovitch, M., 1941: *Canon of Insolation and the Ice-Age Problem*. Königlich Serbische Akademie, Belgrade, 484 pp. (English translation by the Israel Program for Scientific Translation, published by the U.S. Department of Commerce and the National Science Foundation).

Nicholson, S. E., 1983: Sub-Saharan rainfall in the years 1976–80: Evidence of continued drought. *Mon. Wea. Rev.*, **111**, 1646–1654.

North, G. R., 1975: Theory of energy-balance climatic models. *J. Atmos. Sci.*, **32**, 2033–2043.

North, G. R., R. F. Calahan, and J. A. Coakley, 1981: Energy balance climate models. *Rev. Geophys. Space Phys.*, **19**, 91–121.

Orszag, S. A., 1970: Transfer method for the calculation of vector-coupled sums: Applications to the spectral form of the vorticity equation. *J. Atmos. Sci.*, **27**, 890–895.

Ou, S. C., and K. N. Liou, 1984: A two-dimensional radiative-turbulent climate model. I. Sensitivity to cirrus radiative properties. *J. Atmos. Sci.*, **41**, 2289–2309.

Phillips, N. A., 1957: A coordinate system having some special advantages for numerical forecasting. *J. Meteor.*, **14**, 184–185.

Ramanathan, V., 1981: The role of ocean-atmosphere interactions in the CO_2 climate problem. *J. Atmos. Sci.*, **38**, 918–930.

Ramanathan, V., L. Callis, R. Cess, J. Hansen, I. Isaksen, W. Kuhn, A. Lacis, F. Luther, J. Mahlman, R. Reck, and M. Schlesinger, 1987: Climate-chemical interactions and effects of changing atmospheric trace gases. *Rev. Geophys.*, **25**, 1441–1482

Ramanathan, V., and R. E. Dickinson, 1979: The role of stratospheric ozone in the zonal and seasonal radiative energy balance of the earth-atmosphere system. *J. Atmos. Sci.*, **36**, 1084–1104.

Ramanathan, V., E. Pitcher, R. Malone, M. Blackmon, 1983: The response of a spectral general circulation model to refinements in radiative processes. *J. Atmos. Sci.*, **40**, 605–630.

Randall, D. A., Harshvardhan, D. A. Dazlich, and T. G. Corsetti, 1989: Interactions among radiation, convection, and large-scale dynamics in a general
circulation model. *J. Atmos. Sci.*, **46**, 1943–1970.

Saltzman, B., 1983: A survey of statistical-dynamical models of the terrestrial
climate. *Adv. Geophys.*, **20**, 183–304.

Sasamori, T., 1975: A statistical model for stationary atmospheric cloudiness,
liquid water content, and rate of precipitation. *Mon. Wea. Rev.*, **103**, 1037–
1049.

Schneider, S. H., W. M. Washington, and R. M. Chervin, 1978: Cloudiness as a
climatic feedback mechanism: Effects on cloud amounts of prescribed global
and regional surface temperature changes in the NCAR GCM. *J. Atmos. Sci.*,
35, 2207–2221.

Sellers, W. D., 1969: A global climatic model based on the energy balance of the
earth-atmosphere system. *J. Appl. Meteor.*, **8**, 392–400.

Slingo, A., and J. M. Slingo, 1988: The response of a general circulation model
to cloud longwave radiative forcing. I: Introduction and initial experiments.
Quart. J. Roy. Meteor. Soc., **114**, 1027–1062.

Smagorinsky, J., 1960: On the dynamic prediction of large-scale condensation by
numerical methods. In *Physics of Precipitation*, H. Weickman, Ed., Geophys.
Monogr. No.5, Amer. Geophys. Union, Washington, D.C., pp. 71–78.

Smagorinsky, J., 1963: General circulation experiments with the primitive equations. I. The basic experiment. *Mon. Wea. Rev.*, **91**, 99–165.

Smagorinsky, J., 1978: Modeling and predictability. In *Studies in Geophysics:
Energy and Climate*, National Academy of Science, Washington, D.C. pp.
133–139.

Smith, E. A., and B. J. Sohn, 1990: Surface forcing of interannual variations in
the radiation balance over North Africa. Part I. Partitioning the surface and
cloud forcing. *Climate Change*, **17**, 147–192.

Somerville, R., and L. Remer, 1984: Cloud optical thickness feedbacks in the CO_2
climate problem. *J. Geophys. Res.*, **89**, 9668–9672.

Stolarski, R., 1988: The Antarctic ozone hole. *Sci. Amer.*, **258**, 30–36.

Stone, P. H., 1974: The meridional variation of the eddy heat fluxes by baroclinic
waves and their parameterization. *J. Atmos. Sci.*, **31**, 444–456.

Study of Man's Impact on Climate, 1971: *Inadvertent Climate Modification*. MIT
Press, Boston, 308 pp.

Suarez, M. J., and I. M. Held, 1979: The sensitivity of an energy balance climate
model to variations in the orbital parameters. *J. Geophys. Res.*, **84**, 4825–
4836.

Sundqvist, H., 1978: A parameterization scheme for non-convective condensation
including prediction of cloud water content. *Quart. J. Roy. Meteor. Soc.*, **104**,
677–690.

Sundqvist, H., 1981: Prediction of stratoform clouds: Results from a 5-day forecast with a global model. *Tellus*, **33**, 242–253.

Sundqvist, H., 1988: Parameterization of condensation and associated clouds in models for weather prediction and general circulation simulation. In *Physically-Based Modelling and Simulation of Climate and Climate Change*, M. E. Schlesinger, Ed., Reidel, Dordrecht, Holland, pp. 433–461.

Sundqvist, H., E. Berge, and J. E. Kristjansson, 1989: Condensation and cloud parameterization studies with a mesoscale numerical weather prediction model. *Mon. Wea. Rev.*, **117**, 1641–1657.

Tanaka, M., B. C. Weare, A. R. Navato, and R. E. Newell, 1975: Recent African rainfall patterns. *Nature*, **255**, 201–203.

Tiedtke, M., 1984: The effect of penetrative cumulus convection on the large-scale flow in a general circulation model. *Beitr. Phys. Atmos.*, **57**, 216–239.

Toon, O. B., and J. B. Pollack, 1976: A global average model of atmospheric aerosols for radiative transfer calculations. *J. Appl. Meteor.*, **15**, 225–246.

Turco, R. P., O. B. Toon, T. P. Ackerman, J. B. Pollack, and C. Sagan, 1983: Nuclear winter: Global consequences of multiple nuclear explosions. *Science*, **222**, 1283–1292.

Twomey, S. A., M. Piepgrass, and T. L. Wolfe, 1984: An assessment of the impact of pollution on global cloud albedo. *Tellus*, **36B**, 356–366.

Wang, W. C., D. J. Wuebbles, W. M. Washington, R. G. Issacs, and G. Molnar, 1986: Trace gases and other potential perturbations to global climate. *Rev. Geophys.*, **24**, 110–140.

Washington, W. M., and G. A. Meehl, 1984: Seasonal cycle experiment on the climate sensitivity due to a doubling of CO_2 with an atmospheric general circulation model coupled to a simple mixed-layer ocean model. *J. Geophys. Res.*, **89**, 9475–9503.

Wetherald, R. T., and S. Manabe, 1980: Cloud cover and climate sensitivity. *J. Atmos. Sci.*, **37**, 1485–1510.

Wetherald, R. T., and S. Manabe, 1988: Cloud feedback processes in a general circulation model. *J. Atmos. Sci.*, **45**, 1397–1415.

Xue, Y., K. N. Liou, and A. Kasahara, 1990: Investigation of biogeophysical feedback on the African climate using a two-dimensional model. *J. Climate*, **3**, 337–352.

Zheng, Q., and K. N. Liou, 1986: Dynamic and thermodynamic influences of the Tibetan Plateau on the atmosphere in a general circulation model. *J. Atmos. Sci.*, **43**, 1340–1354.

APPENDIX A
PHYSICAL CONSTANTS

Acceleration of gravity (at sea level and 45° latitude)	$g = 9.80616 \times 10^2 \text{ cm s}^{-2}$
Angular velocity of rotation of the earth	$\omega = 7.29221 \times 10^{-5} \text{ rad s}^{-1}$
Avogadro's number	$N_0 = 6.02297 \times 10^{23} \text{ mole}^{-1}$
Boltzmann's constant	$K = 1.38062 \times 10^{-16} \text{ erg K}^{-1}$
Density of air at standard pressure and temperature	$\rho = 1.273 \times 10^{-3} \text{ g cm}^{-3}$
Density of ice (0° C)	$\rho_i = 0.917 \text{ g cm}^{-3}$
Density of liquid water (4° C)	$\rho_\ell = 1 \text{ g cm}^{-3}$
Latent heat of sublimation	$L_s(T) = 2.839 \times 10^{10} - 3.6 \times 10^4 (T + 35)^2 \text{ erg g}^{-1}, T(°C)$
Latent heat of vaporization	$L(T) = 2.5008 \times 10^{10} - 2.3 \times 10^7 T \text{ erg g}^{-1}, T(°C)$
Mean distance between the earth and the sun	$r_0 = 1.49598 \times 10^{13} \text{ cm}$
Mean radius of the earth	$a_e = 6.37120 \times 10^8 \text{ cm}$
Mean radius of the sun (visible disk)	$a_s = 6.96000 \times 10^{10} \text{ cm}$
Molecular weight of dry air	$M = 28.97 \text{ g mole}^{-1}$
Planck's constant	$h = 6.62620 \times 10^{-27} \text{ erg s}$
Saturation vapor pressure (0° C)	$e_0 = 6.1078 \text{ mb}$
Solar constant	$S \cong 1.365 \times 10^6 \text{ erg cm}^{-2} \text{ s}^{-1}$
Specific heat of air at constant pressure	$C_p = 10.04 \times 10^6 \text{ cm}^2 \text{ s}^{-2} \text{ K}^{-1}$
Specific heat of air at constant volume	$C_v = 7.17 \times 10^6 \text{ cm}^2 \text{ s}^{-2} \text{ K}^{-1}$
Standard pressure	$p_0 = 1.01325 \times 10^6 \text{ dyn cm}^{-2}$ $= 1013.25 \text{ mb}$
Standard temperature	$T_0 = 273.16 \text{ K}$
Stefan–Boltzmann constant	$\sigma = 5.66961 \times 10^{-5} \text{ erg cm}^{-2} \text{ s}^{-1} \text{ K}^{-4}$
Universal gas constant	$R^* = 8.31432 \times 10^7 \text{ erg mole}^{-1} \text{ K}^{-1}$
Velocity of light	$c = 2.99792458 \times 10^{10} \text{ cm s}^{-1}$
Wien's displacement constant	$\alpha = 0.2897 \text{ cm K}$

APPENDIX B
STANDARD ATMOSPHERIC PROFILES

Altitude (km)	Pressure (mb)	Temperature (K)	Density (g m^{-3})	Water Vapor (g m^{-3})	Ozone (g m^{-3})
0	1.013×10^3	288.1	1.225×10^3	5.9×10^0	5.4×10^{-5}
1	8.986×10^2	281.6	1.111×10^3	4.2×10^0	5.4×10^{-5}
2	7.950×10^2	275.1	1.007×10^3	2.9×10^0	5.4×10^{-5}
3	7.012×10^2	268.7	9.093×10^2	1.8×10^0	5.0×10^{-5}
4	6.166×10^2	262.2	8.193×10^2	1.1×10^0	4.6×10^{-5}
5	5.405×10^2	255.7	7.364×10^2	6.4×10^{-1}	4.5×10^{-5}
6	4.722×10^2	249.2	6.601×10^2	3.8×10^{-1}	4.5×10^{-5}
7	4.111×10^2	242.7	5.900×10^2	2.1×10^{-1}	4.8×10^{-5}
8	3.565×10^2	236.2	5.258×10^2	1.2×10^{-1}	5.2×10^{-5}
9	3.080×10^2	229.7	4.671×10^2	4.6×10^{-2}	7.1×10^{-5}
10	2.650×10^2	223.2	4.135×10^2	1.8×10^{-2}	9.0×10^{-5}
11	2.270×10^2	216.8	3.648×10^2	8.2×10^{-3}	1.3×10^{-4}
12	1.940×10^2	216.6	3.119×10^2	3.7×10^{-3}	1.6×10^{-4}
13	1.658×10^2	216.6	2.666×10^2	1.8×10^{-3}	1.7×10^{-4}
14	1.417×10^2	216.6	2.279×10^2	8.4×10^{-4}	1.9×10^{-4}
15	1.211×10^2	216.6	1.948×10^2	7.2×10^{-4}	2.1×10^{-4}
16	1.035×10^2	216.6	1.665×10^2	6.1×10^{-4}	2.3×10^{-4}
17	8.850×10^1	216.6	1.423×10^2	5.2×10^{-4}	2.8×10^{-4}
18	7.565×10^1	216.6	1.216×10^2	4.4×10^{-4}	3.2×10^{-4}
19	6.467×10^1	216.6	1.040×10^2	4.4×10^{-4}	3.5×10^{-4}
20	5.529×10^1	216.6	8.891×10^1	4.4×10^{-4}	3.8×10^{-4}
21	4.729×10^1	217.6	7.572×10^1	4.8×10^{-4}	3.8×10^{-4}
22	4.047×10^1	218.6	6.451×10^1	5.2×10^{-4}	3.9×10^{-4}
23	3.467×10^1	219.6	5.500×10^1	5.7×10^{-4}	3.8×10^{-4}
24	2.972×10^1	220.6	4.694×10^1	6.1×10^{-4}	3.6×10^{-4}
25	2.549×10^1	221.6	4.008×10^1	6.6×10^{-4}	3.4×10^{-4}
30	1.197×10^1	226.5	1.841×10^1	3.8×10^{-4}	2.0×10^{-4}
35	5.746×10^0	236.5	8.463×10^0	1.6×10^{-4}	1.1×10^{-4}
40	2.871×10^0	250.4	3.996×10^0	6.7×10^{-5}	4.9×10^{-5}
45	1.491×10^0	264.2	1.966×10^0	3.2×10^{-5}	1.7×10^{-5}
50	7.978×10^{-1}	270.6	1.027×10^0	1.2×10^{-5}	4.0×10^{-6}
75	5.520×10^{-2}	219.7	8.754×10^{-2}	1.5×10^{-7}	8.6×10^{-8}
100	3.008×10^{-4}	210.0	4.989×10^{-4}	1.0×10^{-9}	4.3×10^{-11}

Source: U.S. Standard Atmosphere, 1976.

APPENDIX C
COMPLEX REFRACTIVE INDEX
OF WATER AND ICE

a.   *Water*

$\lambda(\mu m)$	m_r	m_i	$\lambda(\mu m)$	m_r	m_i	$\lambda(\mu m)$	m_r	m_i
0.200	1.396	1.10×10^{-7}	1.80	1.312	1.15×10^{-4}	4.9	1.328	1.37×10^{-2}
0.225	1.373	4.90×10^{-8}	2.00	1.306	1.10×10^{-3}	5.0	1.325	1.24×10^{-2}
0.250	1.362	3.35×10^{-8}	2.20	1.296	2.89×10^{-4}	5.1	1.322	1.11×10^{-2}
0.275	1.354	2.35×10^{-8}	2.40	1.279	9.56×10^{-4}	5.2	1.317	1.01×10^{-2}
0.300	1.349	1.60×10^{-8}	2.60	1.242	3.17×10^{-3}	5.3	1.312	9.80×10^{-3}
0.325	1.346	1.08×10^{-8}	2.65	1.219	6.70×10^{-3}	5.4	1.305	1.03×10^{-2}
0.350	1.343	6.50×10^{-9}	2.70	1.188	1.90×10^{-2}	5.5	1.298	1.16×10^{-2}
0.375	1.341	3.50×10^{-9}	2.75	1.157	5.90×10^{-2}	5.6	1.289	1.42×10^{-2}
0.400	1.339	1.86×10^{-9}	2.80	1.142	1.15×10^{-1}	5.7	1.277	2.03×10^{-2}
0.425	1.338	1.30×10^{-9}	2.85	1.149	1.85×10^{-1}	5.8	1.262	3.30×10^{-2}
0.450	1.337	1.02×10^{-9}	2.90	1.201	2.68×10^{-1}	5.9	1.248	6.22×10^{-2}
0.475	1.336	9.35×10^{-10}	2.95	1.292	2.98×10^{-1}	6.0	1.265	1.07×10^{-1}
0.500	1.335	1.00×10^{-9}	3.00	1.371	2.72×10^{-1}	6.1	1.319	1.31×10^{-1}
0.525	1.334	1.32×10^{-9}	3.05	1.426	2.40×10^{-1}	6.2	1.363	8.80×10^{-2}
0.550	1.333	1.96×10^{-9}	3.10	1.467	1.92×10^{-1}	6.3	1.357	5.70×10^{-2}
0.575	1.333	3.60×10^{-9}	3.15	1.483	1.35×10^{-1}	6.4	1.347	4.49×10^{-2}
0.600	1.332	1.09×10^{-8}	3.20	1.478	9.24×10^{-2}	6.5	1.339	3.92×10^{-2}
0.625	1.332	1.39×10^{-8}	3.25	1.467	6.10×10^{-2}	6.6	1.334	3.56×10^{-2}
0.650	1.331	1.64×10^{-8}	3.30	1.450	3.68×10^{-2}	6.7	1.329	3.37×10^{-2}
0.675	1.331	2.23×10^{-8}	3.35	1.432	2.61×10^{-2}	6.8	1.324	3.27×10^{-2}
0.700	1.331	3.35×10^{-8}	3.40	1.420	1.95×10^{-2}	6.9	1.321	3.22×10^{-2}
0.725	1.330	9.15×10^{-8}	3.45	1.410	1.32×10^{-2}	7.0	1.317	3.20×10^{-2}
0.750	1.330	1.56×10^{-7}	3.50	1.400	9.40×10^{-3}	7.1	1.314	3.20×10^{-2}
0.775	1.330	1.48×10^{-7}	3.60	1.385	5.15×10^{-3}	7.2	1.312	3.21×10^{-2}
0.800	1.329	1.25×10^{-7}	3.70	1.374	3.60×10^{-3}	7.3	1.309	3.22×10^{-2}
0.825	1.329	1.82×10^{-7}	3.80	1.364	3.40×10^{-3}	7.4	1.307	3.24×10^{-2}
0.850	1.329	2.93×10^{-7}	3.90	1.357	3.80×10^{-3}	7.5	1.304	3.26×10^{-2}
0.875	1.328	3.91×10^{-7}	4.00	1.351	4.60×10^{-3}	7.6	1.302	3.28×10^{-2}
0.900	1.328	4.86×10^{-7}	4.10	1.346	5.62×10^{-3}	7.7	1.299	3.31×10^{-2}
0.925	1.328	1.06×10^{-6}	4.20	1.342	6.88×10^{-3}	7.8	1.297	3.35×10^{-2}
0.950	1.327	2.93×10^{-6}	4.30	1.338	8.45×10^{-3}	7.9	1.294	3.39×10^{-2}
0.975	1.327	3.48×10^{-6}	4.40	1.334	1.03×10^{-2}	8.0	1.291	3.43×10^{-2}
1.000	1.327	2.89×10^{-6}	4.50	1.332	1.34×10^{-2}	8.2	1.286	3.51×10^{-2}
1.200	1.324	9.89×10^{-6}	4.60	1.330	1.47×10^{-2}	8.4	1.281	3.61×10^{-2}
1.400	1.321	1.38×10^{-4}	4.70	1.330	1.57×10^{-2}	8.6	1.275	3.72×10^{-2}
1.600	1.317	8.55×10^{-5}	4.80	1.330	1.50×10^{-2}	8.8	1.269	3.85×10^{-2}

9.0	1.262	3.99×10^{-2}	16.0	1.325	4.22×10^{-1}	29.0	1.551	3.33×10^{-1}
9.2	1.255	4.15×10^{-2}	16.5	1.351	4.28×10^{-1}	30.0	1.551	3.28×10^{-1}
9.4	1.247	4.33×10^{-2}	17.0	1.376	4.29×10^{-1}	32.0	1.546	3.24×10^{-1}
9.6	1.239	4.54×10^{-2}	17.5	1.401	4.29×10^{-1}	34.0	1.536	3.29×10^{-1}
9.8	1.229	4.79×10^{-2}	18.0	1.423	4.26×10^{-1}	36.0	1.527	3.43×10^{-1}
10.0	1.218	5.08×10^{-2}	18.5	1.443	4.21×10^{-1}	38.0	1.522	3.61×10^{-1}
10.5	1.185	6.62×10^{-2}	19.0	1.461	4.14×10^{-1}	40.0	1.519	3.85×10^{-1}
11.0	1.153	9.68×10^{-2}	19.5	1.476	4.04×10^{-1}	42.0	1.522	4.09×10^{-1}
11.5	1.126	1.42×10^{-1}	20.0	1.480	3.93×10^{-1}	44.0	1.530	4.36×10^{-1}
12.0	1.111	1.99×10^{-1}	21.0	1.487	3.82×10^{-1}	46.0	1.541	4.62×10^{-1}
12.5	1.123	2.59×10^{-1}	22.0	1.500	3.73×10^{-1}	48.0	1.555	4.88×10^{-1}
13.0	1.146	3.05×10^{-1}	23.0	1.511	3.67×10^{-1}	50.0	1.587	5.14×10^{-1}
13.5	1.177	3.43×10^{-1}	24.0	1.521	3.61×10^{-1}	60.0	1.703	5.87×10^{-1}
14.0	1.210	3.70×10^{-1}	25.0	1.531	3.56×10^{-1}	70.0	1.821	5.76×10^{-1}
14.5	1.241	3.88×10^{-1}	26.0	1.539	3.50×10^{-1}	80.0	1.886	5.47×10^{-1}
15.0	1.270	4.02×10^{-1}	27.0	1.545	3.44×10^{-1}	90.0	1.924	5.36×10^{-1}
15.5	1.297	4.14×10^{-1}	28.0	1.549	3.38×10^{-1}	100.0	1.957	5.32×10^{-1}

Source: Hale and Querry (1973), reference in Chapter 5.

b. Ice

$\lambda(\mu m)$	m_r	m_i	$\lambda(\mu m)$	m_r	m_i	$\lambda(\mu m)$	m_r	m_i
0.210	1.3800	1.325×10^{-8}	0.940	1.3025	5.530×10^{-7}	1.280	1.2965	1.330×10^{-5}
0.250	1.3509	8.623×10^{-9}	0.960	1.3022	7.550×10^{-7}	1.290	1.2963	1.320×10^{-5}
0.300	1.3339	5.504×10^{-9}	0.980	1.3018	1.120×10^{-6}	1.300	1.2961	1.320×10^{-5}
0.350	1.3249	3.765×10^{-9}	1.000	1.3015	1.620×10^{-6}	1.310	1.2958	1.310×10^{-5}
0.400	1.3194	2.710×10^{-9}	1.010	1.3013	2.000×10^{-6}	1.320	1.2956	1.320×10^{-5}
0.420	1.3177	2.260×10^{-9}	1.020	1.3012	2.250×10^{-6}	1.330	1.2954	1.320×10^{-5}
0.440	1.3163	1.910×10^{-9}	1.030	1.3010	2.330×10^{-6}	1.340	1.2952	1.340×10^{-5}
0.460	1.3151	1.530×10^{-9}	1.040	1.3008	2.330×10^{-6}	1.350	1.2950	1.390×10^{-5}
0.480	1.3140	1.640×10^{-9}	1.050	1.3006	2.170×10^{-6}	1.360	1.2948	1.420×10^{-5}
0.500	1.3130	1.910×10^{-9}	1.060	1.3005	1.960×10^{-6}	1.370	1.2945	1.480×10^{-5}
0.520	1.3122	2.260×10^{-9}	1.070	1.3003	1.810×10^{-6}	1.380	1.2943	1.580×10^{-5}
0.540	1.3114	2.930×10^{-9}	1.080	1.3001	1.740×10^{-6}	1.390	1.2941	1.740×10^{-5}
0.560	1.3106	3.290×10^{-9}	1.090	1.3000	1.730×10^{-6}	1.400	1.2938	1.980×10^{-5}
0.580	1.3100	4.040×10^{-9}	1.100	1.2998	1.700×10^{-6}	1.410	1.2936	2.500×10^{-5}
0.600	1.3094	5.730×10^{-9}	1.110	1.2996	1.760×10^{-6}	1.420	1.2933	5.400×10^{-5}
0.620	1.3088	8.580×10^{-9}	1.120	1.2995	1.820×10^{-6}	1.430	1.2930	1.040×10^{-4}
0.640	1.3083	1.220×10^{-8}	1.130	1.2993	2.040×10^{-6}	1.440	1.2927	2.030×10^{-4}
0.660	1.3078	1.660×10^{-8}	1.140	1.2991	2.250×10^{-6}	1.450	1.2925	2.708×10^{-4}
0.680	1.3073	2.090×10^{-8}	1.150	1.2989	2.290×10^{-6}	1.460	1.2923	3.511×10^{-4}
0.700	1.3069	2.900×10^{-8}	1.160	1.2987	3.040×10^{-6}	1.471	1.2921	4.299×10^{-4}
0.720	1.3065	4.030×10^{-8}	1.170	1.2985	3.840×10^{-6}	1.481	1.2919	5.181×10^{-4}
0.740	1.3060	4.920×10^{-8}	1.180	1.2984	4.770×10^{-6}	1.493	1.2917	5.855×10^{-4}
0.760	1.3057	7.080×10^{-8}	1.190	1.2982	5.760×10^{-6}	1.504	1.2915	5.899×10^{-4}
0.780	1.3053	1.020×10^{-7}	1.200	1.2980	6.710×10^{-6}	1.515	1.2913	5.635×10^{-4}
0.800	1.3049	1.340×10^{-7}	1.210	1.2978	8.660×10^{-6}	1.527	1.2911	5.480×10^{-4}
0.820	1.3045	1.430×10^{-7}	1.220	1.2976	1.020×10^{-5}	1.538	1.2908	5.266×10^{-4}
0.840	1.3042	1.510×10^{-7}	1.230	1.2974	1.130×10^{-5}	1.563	1.2903	4.394×10^{-4}
0.860	1.3038	2.150×10^{-7}	1.240	1.2972	1.220×10^{-5}	1.587	1.2896	3.701×10^{-4}
0.880	1.3035	3.350×10^{-7}	1.250	1.2970	1.290×10^{-5}	1.613	1.2889	3.372×10^{-4}
0.900	1.3032	4.200×10^{-7}	1.260	1.2969	1.320×10^{-5}	1.650	1.2878	2.410×10^{-4}
0.920	1.3028	4.740×10^{-7}	1.270	1.2967	1.350×10^{-5}	1.680	1.2869	1.890×10^{-4}

$\lambda(\mu m)$	m_r	m_i	$\lambda(\mu m)$	m_r	m_i	$\lambda(\mu m)$	m_r	m_i
1.700	1.2862	1.660×10^{-4}	2.778	1.1077	2.000×10^{-2}	4.560	1.3442	3.000×10^{-2}
1.730	1.2852	1.450×10^{-4}	2.817	1.0674	3.800×10^{-2}	4.580	1.3463	2.850×10^{-2}
1.760	1.2841	1.280×10^{-4}	2.833	1.0476	5.200×10^{-2}	4.719	1.3442	1.730×10^{-2}
1.800	1.2826	1.030×10^{-4}	2.849	1.0265	6.800×10^{-2}	4.904	1.3345	1.290×10^{-2}
1.830	1.2814	8.600×10^{-5}	2.865	1.0036	9.230×10^{-2}	5.000	1.3290	1.200×10^{-2}
1.840	1.2809	8.220×10^{-5}	2.882	0.9820	1.270×10^{-1}	5.100	1.3233	1.250×10^{-2}
1.850	1.2805	8.030×10^{-5}	2.899	0.9650	1.690×10^{-1}	5.200	1.3180	1.340×10^{-2}
1.855	1.2802	8.500×10^{-5}	2.915	0.9596	2.210×10^{-1}	5.263	1.3143	1.400×10^{-2}
1.860	1.2800	9.900×10^{-5}	2.933	0.9727	2.760×10^{-1}	5.400	1.3062	1.750×10^{-2}
1.870	1.2795	1.500×10^{-4}	2.950	0.9917	3.120×10^{-1}	5.556	1.2972	2.400×10^{-2}
1.890	1.2785	2.950×10^{-4}	2.967	1.0067	3.470×10^{-1}	5.714	1.2890	3.500×10^{-2}
1.905	1.2777	4.687×10^{-4}	2.985	1.0219	3.880×10^{-1}	5.747	1.2873	3.800×10^{-2}
1.923	1.2769	7.615×10^{-4}	3.003	1.0427	4.380×10^{-1}	5.780	1.2860	4.200×10^{-2}
1.942	1.2761	1.010×10^{-3}	3.021	1.0760	4.930×10^{-1}	5.814	1.2851	4.600×10^{-2}
1.961	1.2754	1.313×10^{-3}	3.040	1.1295	5.540×10^{-1}	5.848	1.2854	5.200×10^{-2}
1.980	1.2747	1.539×10^{-3}	3.058	1.2127	6.120×10^{-1}	5.882	1.2881	5.700×10^{-2}
2.000	1.2740	1.588×10^{-3}	3.077	1.3251	6.250×10^{-1}	6.061	1.3016	6.900×10^{-2}
2.020	1.2733	1.540×10^{-3}	3.096	1.4260	5.930×10^{-1}	6.135	1.3090	7.000×10^{-2}
2.041	1.2724	1.412×10^{-3}	3.115	1.4966	5.390×10^{-1}	6.250	1.3172	6.700×10^{-2}
2.062	1.2714	1.244×10^{-3}	3.135	1.5510	4.910×10^{-1}	6.289	1.3189	6.500×10^{-2}
2.083	1.2703	1.068×10^{-3}	3.155	1.5999	4.380×10^{-1}	6.329	1.3204	6.400×10^{-2}
2.105	1.2690	8.414×10^{-4}	3.175	1.6363	3.720×10^{-1}	6.369	1.3220	6.200×10^{-2}
2.130	1.2674	5.650×10^{-4}	3.195	1.6502	3.000×10^{-1}	6.410	1.3224	5.900×10^{-2}
2.150	1.2659	4.320×10^{-4}	3.215	1.6428	2.380×10^{-1}	6.452	1.3215	5.700×10^{-2}
2.170	1.2644	3.500×10^{-4}	3.236	1.6269	1.930×10^{-1}	6.494	1.3204	5.600×10^{-2}
2.190	1.2628	2.870×10^{-4}	3.257	1.6128	1.580×10^{-1}	6.579	1.3181	5.500×10^{-2}
2.220	1.2604	2.210×10^{-4}	3.279	1.5924	1.210×10^{-1}	6.667	1.3171	5.700×10^{-2}
2.240	1.2586	2.030×10^{-4}	3.300	1.5733	1.030×10^{-1}	6.757	1.3181	5.800×10^{-2}
2.245	1.2582	2.010×10^{-4}	3.322	1.5577	8.360×10^{-2}	6.897	1.3195	5.700×10^{-2}
2.250	1.2577	2.030×10^{-4}	3.345	1.5413	6.680×10^{-2}	7.042	1.3193	5.500×10^{-2}
2.260	1.2567	2.140×10^{-4}	3.367	1.5265	5.400×10^{-2}	7.143	1.3190	5.500×10^{-2}
2.270	1.2558	2.320×10^{-4}	3.390	1.5114	4.220×10^{-2}	7.246	1.3191	5.400×10^{-2}
2.290	1.2538	2.890×10^{-4}	3.413	1.4973	3.420×10^{-2}	7.353	1.3180	5.200×10^{-2}
2.310	1.2518	3.810×10^{-4}	3.436	1.4845	2.740×10^{-2}	7.463	1.3163	5.200×10^{-2}
2.330	1.2497	4.620×10^{-4}	3.460	1.4721	2.200×10^{-2}	7.576	1.3154	5.200×10^{-2}
2.350	1.2475	5.480×10^{-4}	3.484	1.4612	1.860×10^{-2}	7.692	1.3154	5.200×10^{-2}
2.370	1.2451	6.180×10^{-4}	3.509	1.4513	1.520×10^{-2}	7.812	1.3155	5.000×10^{-2}
2.390	1.2427	6.800×10^{-4}	3.534	1.4421	1.260×10^{-2}	7.937	1.3145	4.700×10^{-2}
2.410	1.2400	7.300×10^{-4}	3.559	1.4337	1.060×10^{-2}	8.065	1.3119	4.300×10^{-2}
2.430	1.2373	7.820×10^{-4}	3.624	1.4155	8.020×10^{-3}	8.197	1.3068	3.900×10^{-2}
2.460	1.2327	8.480×10^{-4}	3.732	1.3942	6.850×10^{-3}	8.333	1.2993	3.700×10^{-2}
2.500	1.2258	9.250×10^{-4}	3.775	1.3873	6.600×10^{-3}	8.475	1.2925	3.900×10^{-2}
2.520	1.2220	9.200×10^{-4}	3.847	1.3773	6.960×10^{-3}	8.696	1.2839	4.000×10^{-2}
2.550	1.2155	8.920×10^{-4}	3.969	1.3645	9.160×10^{-3}	8.929	1.2740	4.200×10^{-2}
2.565	1.2118	8.700×10^{-4}	4.099	1.3541	1.110×10^{-2}	9.091	1.2672	4.400×10^{-2}
2.580	1.2079	8.900×10^{-4}	4.239	1.3446	1.450×10^{-2}	9.259	1.2599	4.500×10^{-2}
2.590	1.2051	9.300×10^{-4}	4.348	1.3388	2.000×10^{-2}	9.524	1.2451	4.600×10^{-2}
2.600	1.2021	1.010×10^{-3}	4.387	1.3381	2.300×10^{-2}	9.804	1.2224	4.700×10^{-2}
2.620	1.1957	1.350×10^{-3}	4.444	1.3385	2.600×10^{-2}	10.000	1.1991	5.100×10^{-2}
2.675	1.1741	3.420×10^{-3}	4.505	1.3405	2.900×10^{-2}	10.200	1.1715	6.500×10^{-2}
2.725	1.1473	7.920×10^{-3}	4.547	1.3429	2.930×10^{-2}	10.310	1.1553	7.500×10^{-2}

$\lambda(\mu m)$	m_r	m_i	$\lambda(\mu m)$	m_r	m_i	$\lambda(\mu m)$	m_r	m_i
10.420	1.1370	8.800×10^{-2}	19.230	1.4968	7.600×10^{-2}	52.750	1.6981	5.070×10^{-1}
10.530	1.1181	1.080×10^{-1}	19.610	1.4993	7.500×10^{-2}	53.500	1.7206	4.883×10^{-1}
10.640	1.1013	1.340×10^{-1}	20.000	1.5015	6.700×10^{-2}	54.240	1.7486	4.707×10^{-1}
10.750	1.0908	1.680×10^{-1}	20.410	1.4986	5.500×10^{-2}	55.000	1.7674	4.203×10^{-1}
10.870	1.0873	2.040×10^{-1}	20.830	1.4905	4.500×10^{-2}	55.740	1.7648	3.771×10^{-1}
11.000	1.0925	2.480×10^{-1}	22.220	1.4607	2.900×10^{-2}	56.400	1.7501	3.376×10^{-1}
11.110	1.1065	2.800×10^{-1}	22.600	1.4518	2.750×10^{-2}	57.000	1.7233	3.056×10^{-1}
11.360	1.1478	3.410×10^{-1}	23.050	1.4422	2.700×10^{-2}	57.460	1.6849	2.835×10^{-1}
11.630	1.2020	3.790×10^{-1}	23.600	1.4316	2.730×10^{-2}	58.400	1.6240	3.170×10^{-1}
11.900	1.2582	4.090×10^{-1}	24.600	1.4138	2.890×10^{-2}	59.290	1.5960	3.517×10^{-1}
12.200	1.3231	4.220×10^{-1}	25.000	1.4068	3.000×10^{-2}	60.000	1.5851	3.902×10^{-1}
12.500	1.3857	4.220×10^{-1}	26.000	1.3895	3.400×10^{-2}	61.000	1.5992	4.509×10^{-1}
12.820	1.4448	4.030×10^{-1}	28.570	1.3489	5.300×10^{-2}	61.250	1.6140	4.671×10^{-1}
12.990	1.4717	3.890×10^{-1}	31.000	1.3104	7.550×10^{-2}	62.500	1.6662	4.779×10^{-1}
13.160	1.4962	3.740×10^{-1}	33.330	1.2642	1.060×10^{-1}	63.780	1.7066	4.890×10^{-1}
13.330	1.5165	3.540×10^{-1}	34.480	1.2366	1.350×10^{-1}	64.670	1.7371	4.899×10^{-1}
13.510	1.5333	3.350×10^{-1}	35.640	1.2166	1.761×10^{-1}	65.580	1.7686	4.873×10^{-1}
13.700	1.5490	3.150×10^{-1}	37.000	1.2023	2.229×10^{-1}	66.550	1.8034	4.766×10^{-1}
13.890	1.5628	2.940×10^{-1}	38.240	1.1964	2.746×10^{-1}	67.600	1.8330	4.508×10^{-1}
14.080	1.5732	2.710×10^{-1}	39.600	1.1997	3.280×10^{-1}	69.000	1.8568	4.193×10^{-1}
14.290	1.5803	2.460×10^{-1}	41.140	1.2086	3.906×10^{-1}	70.530	1.8741	3.880×10^{-1}
14.710	1.5792	1.980×10^{-1}	42.760	1.2217	4.642×10^{-1}	73.000	1.8911	3.433×10^{-1}
15.150	1.5667	1.640×10^{-1}	43.580	1.2417	5.247×10^{-1}	75.000	1.8992	3.118×10^{-1}
15.380	1.5587	1.520×10^{-1}	44.580	1.2818	5.731×10^{-1}	76.290	1.9043	2.935×10^{-1}
15.630	1.5508	1.420×10^{-1}	45.500	1.3278	6.362×10^{-1}	80.000	1.9033	2.350×10^{-1}
16.130	1.5381	1.280×10^{-1}	46.150	1.3866	6.839×10^{-1}	82.970	1.8874	1.981×10^{-1}
16.390	1.5330	1.250×10^{-1}	46.710	1.4649	7.091×10^{-1}	85.000	1.8750	1.865×10^{-1}
16.670	1.5322	1.230×10^{-1}	47.360	1.5532	6.790×10^{-1}	86.800	1.8670	1.771×10^{-1}
16.950	1.5334	1.160×10^{-1}	48.000	1.6038	6.250×10^{-1}	90.800	1.8536	1.620×10^{-1}
17.240	1.5329	1.070×10^{-1}	48.780	1.6188	5.654×10^{-1}	95.170	1.8425	1.490×10^{-1}
18.180	1.5170	7.900×10^{-2}	50.030	1.6296	5.433×10^{-1}	100.000	1.8323	1.390×10^{-1}
18.870	1.5010	7.200×10^{-2}	51.280	1.6571	5.292×10^{-1}			

Source: Warren (1984), reference in Chapter 5.

INDEX